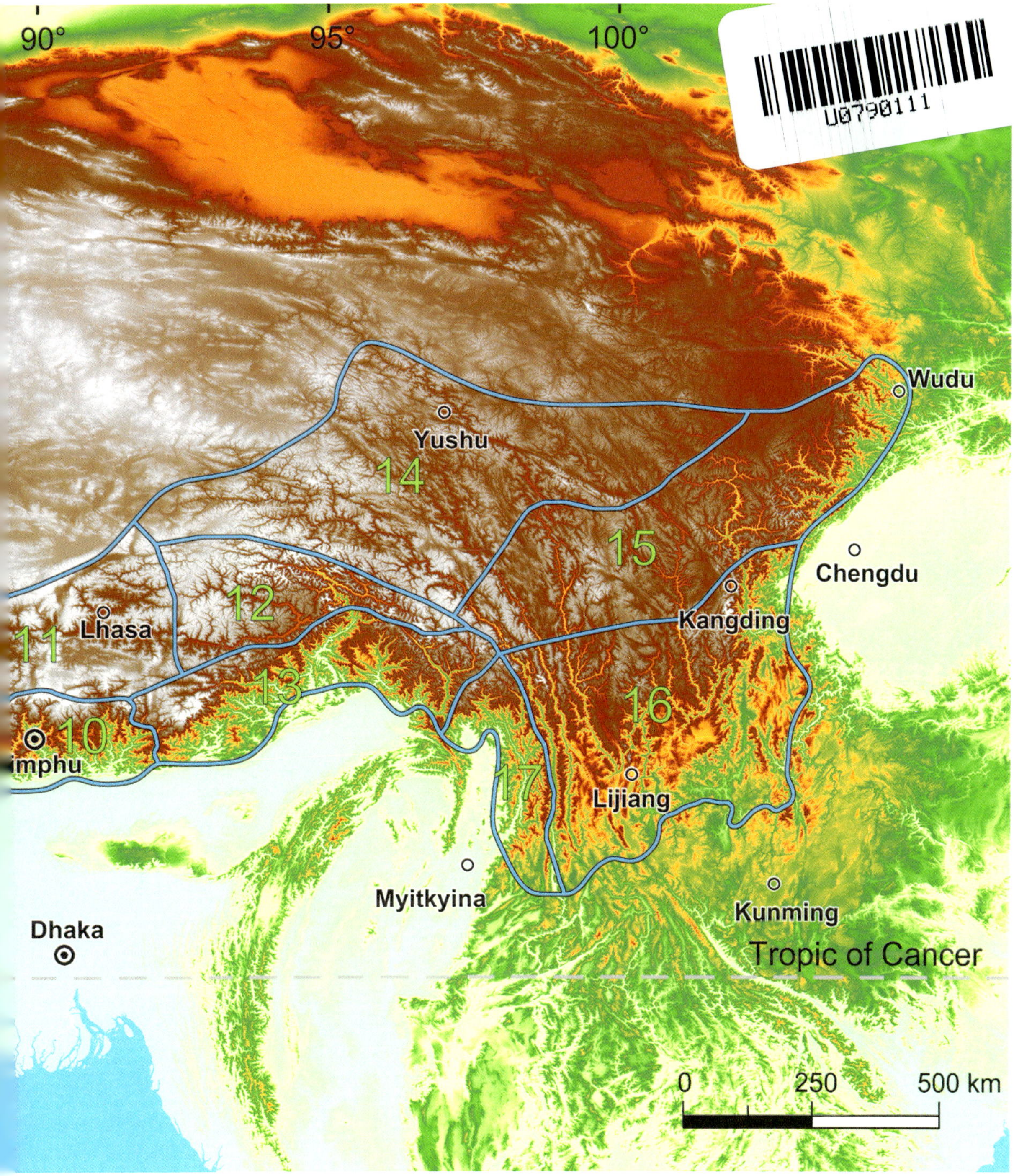

Pan-Himalaya

ng Zangbo; 6. W Nepal; 7. C Nepal; 8. E Nepal; 9. Sikkim & Darjiling; ahmaputra; 14. Tangut; 15. N Hengduan; 16. S Hengduan; 17. U Irrawaddy.

国家出版基金项目
NATIONAL PUBLICATION FOUNDATION

Flora *of* Pan-Himalaya

A Preliminary Catalogue of Vascular Plants in the Pan-Himalaya

You-Sheng CHEN

South China Botanical Garden, Chinese Academy of Sciences
Email: yschen@scbg.ac.cn

SCIENCE PRESS
Beijing

CAMBRIDGE
UNIVERSITY PRESS

Flora *of* Pan-Himalaya

A Preliminary Catalogue of Vascular Plants in the Pan-Himalaya

Author:
You-Sheng CHEN

Editor-in-Chief:
De-Yuan HONG

Map Approval No. : GS (2019) 1135

Published by Science Press (Beijing)

ISBN 978-7-03-061953-2

Flora *of* Pan-Himalaya

Office: Dan PENG
Tel: 86-10-62836084
Fax: 86-10-62590843

Email: flph@ibcas.ac.cn
Website: www.flph.org

It is with great pleasure that we acknowledge the generous financial support from the following:

Ministry of Science and Technology of the People's Republic of China
National Natural Science Foundation of China
Chinese Academy of Sciences

Foreword

We are delighted to welcome the appearance of this highly significant regional flora, and congratulate De-Yuan HONG and all of the others that worked together to make it possible. The Himalaya is one of the most beautiful, extensive, and revered ranges on earth, containing nine of our ten highest peaks, and its plants deserve the kind of inclusive and collaborative treatment that we have been presented in these fine volumes.

The mighty Himalaya was formed as the basaltic Indian plate plowed into the underbelly of Asia. Carrying the lighter rocks that would become the mountain ranges we see now, it broke out of the gigantic southern continent of Gondwanaland approximately 90 million years ago and started its journey northward to its current position. In the south, any part of India that rose above the ocean would have been covered with archaic southern vegetation similar to that which survives today in southern Australia and South America, New Zealand, and New Caledonia. As the fragment that includes India journeyed northward, it passed from the south across the equator and started crowding into the Asian plate perhaps 55-50 million years ago, continuing to thrust up the Himalaya from that time onward. The same movement folded what is now China into a series of accordion-like pleats that became the current mountain ranges of southern and southwestern China and its neighboring countries.

Crossing so many latitudes seems to have led to the extinction of virtually all of the plants and animals that occupied India when its position lay south of the equator. However, when it came within reach of Asia and under the influence of a tropical and subtropical climate from the early Eocene Period onward, the plants and animals of southern Asia colonized the new lands, and more migrated from the north as the mountains of the region were thrust ever upward. Importantly, there was a great opportunity for the evolution of endemic plants and animals in the new mountains, and the mixture of all of these rich biological strands have combined to form the incredible biological richness of the area as we see it today.

Both because of the political subdivisions of the region and its complex floristic composition, it took very courageous and forward-looking leadership for De-Yuan HONG and his colleagues to undertake this monumental flora. Except for China, there are relatively few endemics in most of the countries of the Himalayan ranges and their bordering lowlands, but many in the region. This lack of national endemics, however, stems mainly from the artificial presence of countries in the area, which taken as a whole is an extremely rich one for endemics in virtually all groups of plants, animals, and fungi that occur there. The plants of some of the countries are reasonably well known, those of others poorly known, and our historical efforts to bring the available information together and improve its quality by transnational comparisons have been inadequate.

That situation is all the more unfortunate because of the severe threats that the ecosystems of the regions facing from rapid human population growth and the even more rapidly rising expectations for increased consumption and especially from the climate change that threatens all of the ecosystems of the area, including the possible destruction of its glaciers and the disruptions of the monsoons on which so many depend for the food and livelihood. A sound knowledge of the plants and ecosystems of the region is the only way to win sustainability, even if population, consumption levels, and the runaway use of malign technologies like burning coal are brought under control.

For all of these reasons, we declare once more our delight in the appearance of this flora and congratulate those who have worked together to make it a reality. It is a critically important first step in improving our knowledge of the plants of the area on a regional basis, and one of great and general importance.

Peter H. Raven,
President Emeritus,
Missouri Botanical Garden

Introduction

—General Guideline of the *Flora of Pan-Himalaya*

The Himalaya and adjacent regions (Pan-Himalaya) form a natural phytogeographical unit, from the Vakhan Corridor eastwards to the Hengduan Mountains via the Karakorum and the Himalaya. This region covers the northeastern corner of Afghanistan, northern Pakistan, northern India, Nepal, Bhutan, northern Myanmar, and southwestern China (S Tibet, SE Qinghai, SE Gansu, W Sichuan and NW Yunnan). This region is divided into 17 subregions: Vakhan, N Pakistan, Jammu & Kashmir, U Ganga & Indus, U Yarlung Zangbo, W Nepal, C Nepal, E Nepal, Sikkim & Darjiling, Bhutan, M Yarlung Zangbo, L Yarlung Zangbo, Yarlung Zangbo-Brahmaputra, Tangut, N Hengduan, S Hengduan, and U Irrawaddy (see the maps).

The *Flora of Pan-Himalaya* (FLPH) will be published in English in two editions: printed and online. It treats all the native and naturalized vascular plants of this region.

The Flora will be published in 50 volumes with ca. 80 books, with the larger families divided into two to seven parts. The classification systems used in this Flora will reflect our current understanding of phylogenetic or phylogenomic relationships of the relevant groups. The third edition of the Angiosperm Phylogeny Group system (APG III) will be adopted for the angiosperms, and up-to-date phylogenetic or phylogenomic systems for the gymnosperms, ferns, and lycophytes will be followed for the treatments of those groups.

The natural and rational delimitation of species is the critical task of taxonomy, and is the most important criterion for judging the scientific value and standard of a Flora. Population concepts should be employed in determining the taxonomy; character analysis should be considered as the basis for a rational taxonomic treatment. Therefore, field observations at the population level, and statistic analysis based on broad population sampling and collections are greatly encouraged to investigate the variation of characters and evaluate their taxonomic significance for the delimitation of taxa. Herbarium specimens kept at the following herbaria must be examined and identified: PE, KUN, K, BM, E, CDBI, and SZ. The authors are also encouraged to visit the herbaria of A, B, CAL, CAS, G, KATH, LE, P, TI, W, and HNWP.

Delimitation of closely related species should be based on correlated discontinuous (including statistically discontinuous) variation of at least two characters. As to apomicts (such as *Taraxacum* spp., *Sorbus* spp.) and vegetatively reproducing plants, "macrospecies" should be adopted instead of "microspecies".

The use of subspecies is encouraged in infraspecific classifications, but varieties only in special cases. Subspecies are geographical races and horizontally or vertically vicarious, whilst varieties are prominent ecotypes. Genotype variants within populations should not be recognized as different taxa. The description of a species should cover all its elements. Variants which do not merit formal taxonomic recognition may be noted below the description of the species.

Larger families and genera may be concisely subdivided. These systems should reflect the current understanding of phylogenetic or phylogenomic relationships, especially those generated from solid molecular and genomic data.

The concept of the genus should reflect the current understanding of phylogenies. Generic concepts strongly supported by both molecular/genomic and morphological evidence should be adopted, but those with weak support in molecular/genomic analysis and have no sound morphological evidence should not be accepted. The treatment of more poorly known groups should be conservative and any perceived problems indicated.

Identification keys are one of the most important products of classification, and the best reflection of its usefulness. Authors should construct a key based on the examination of all relevant specimens.Where possible, keys should not be based on a single character, and must avoid vague words, such as "relatively large" vs. "relatively small", "longer" vs. "shorter". It is important to use easily visible characters in keys. Dichotomous indented keys such as those in the *Flora of China* are to be used in this Flora.

Description of species should be concise with emphasis on diagnostic characters, and should include habitat information and the geographical distribution, plus chromosome numbers, phenology, and other biological features, as well as reliable economic uses and conservation information where available.

Literature citation for accepted names and their basionyms must be provided. Important references, such as monographs and Floras, should be cited, but not more than three titles (at most five if absolutely necessary). All synonyms based on types from within the Flora area should be listed along with other synonyms that have been widely used in relevant publications.

Detailed type information should be indicated after the nomenclatural citation of taxa. Designation of types is highly encouraged in this Flora.

A distribution map is given for each species. This will be a dot map based on the specimens seen for the Flora. The vouchers are to be cited to record geographical distribution for each species, subspecies (or variety). At least one representative specimen is needed for each distribution unit (county in China, district in the other countries). For the more widely distributed taxa (across nine or more subdivisions), a larger distribution unit can be used, no citation of vouchers is needed, and the distribution outlines are also acceptable.

The nomenclatural novelties (new taxa, new combinations, and *nomina nova*) may be published in the Flora or elsewhere.

Sterile hybrids, and those hybrids not forming populations, are to be briefly discussed after the descriptions of their parents. Plants only known from cultivation may be indicated at the end of the treatment of the genus, but should not be included in the identification keys.

All genera, and at least one third of the species, should be illustrated with good quality line drawings with scale bars to indicate magnifications. The illustrations can be drawn from herbarium materials, living materials or images available, selected by the authors, or taken from publications with permission. Wherever possible the material used should be indicated. Diagnostic characters of the species should be clearly illustrated. One plate may hold illustrations of one, two, or three species. For larger genera composite plates of diagnostic characters are encouraged.

The web edition of the Flora will follow the corresponding printed edition. Color photographs, detailed citation of vouchers, and other sources of information are to be included only in the web edition.

The physical geography, history of studies of vascular plants from this region, biogeography, biodiversity conservation, general references, and key to families are to be included in the first (introductory) volume. A comprehensive index to taxa and other subjects are to be published in the last volume.

More detailed guidelines and a model treatment for preparation of Flora accounts will be provided by the Editorial Committee.

The Editorial Committee of *Flora of Pan-Himalaya*

Preface

The Pan-Himalaya, including the Himalaya, the Hengduan Mountains, the Karakoran, and NE Hendu Kosh, constitutes a unique geographic unit of the earth. Mittermeier et al. (2004) identified 34 biodiversity hotspots for the world. Of these, "Himalaya" and "Mountains of Southwestern China"(nearly corresponding to the Hengduan Mountains) fall almost entirely within the Pan-Himalaya, and "Indo-Burma" falls partly within this area.

The classification systems used in *Flora of Pan-Himalaya* are hoped to reflect our current understanding of phylogenetic relationships of the relevant groups. The third edition of the Angiosperm Phylogeny Group system (APG III) is basically adopted for the angiosperms, as it has been used in several major herbaria of the world, and up-to-date phylogenetic systems for the gymnosperms, ferns, and lycophytes are followed for the treatments of those groups. The more recent APG IV system (The Angiosperm Phylogeny Group, 2016) made a few changes after APG III, but it is too recent for the major herbaria to adopt. In fact I had already arranged the catalogue according to APG IV system, but it would involve changes of some families in planned volumes. Therefore the editor-in-chief asked me for still using APG III system. This Flora is probably the first one using the achievements of phylogenetic / phylogenomic analyses. There are many differences between APG III system and the Engler or the Hutchinson classification system adopted in most Floras, so that it is a great challenge for FLPH authors to grasp the scopes and placements of the families recognized. To avoid confusion and omissions, it is urgent to produce a preliminary catalogue for this Flora. The purpose of the compilation of this catalogue is to provide a framework for the compilation of *Flora of Pan-Himalaya*, rather than to provide a complete list of accepted scientific names and synonyms of plants in this region.

The work on this catalogue started in 2010. I consulted major floristic works, checklists, important website resources, and recent papers or monographs related to the Pan-Himalayan region. In the present catalogue, 25 families of angiosperms are added, mainly because of the changes made by the APG III system. However, for gymnosperms, two families are reduced. The Taxodiaceae and Cephalotaxaceae are merged to the Cupressaceae and Taxaceae respectively. More changes have occurred for many genera in the systematic position in their families.

Earlier this year, I sent a draft of the catalogue to the authors of *Flora of Pan-Himalaya*. With kind help of many authors, treatments of many families have been revised. As for those already published volumes, the present catalogue accepts their treatments. Added to the catalogue are many new species published in recent years, such as *Amomum nimkeyense* M. Sabu, Hareesh, Tatum & A. K. Das (2018), *Amomum riwatchii* M. Sabu & Hareesh (2018), *Amomum sabuanum* V. P. Thomas, Nissar & U. Gupta (2014), *Hedychium putaoense* Y. H. Tan & H. B. Ding (2018), *Monolophus arunachalensis* Bhaumik, S. Dey & Langhu (2017), *Zingiber arunachalensis* A. Joe, T. Jayakr., Hareesh & M. Sabu (2017) in the Zingiberaceae. However, if I found a new scientific name already treated as synonym, the "new species" is not listed, e.g. *Salweenia bouffordiana* H. Sun, Zhi M. Li & J. P. Yue (2011), which is treated as a synonym of *S. wardii* Baker f. by Zhu (2015). Some newly published generic names are not recognized because of recent researches or not recognized by new references. For instances, *Nujiangia* X. H. Jin & D. Z. Li (2012) is included in *Gennaria* Parl. by Jin et al. (2015); *Kuepferia* Adr. Favre and *Sinogentiana* Adr. Favre & Y. M. Yuan published by Favre et al. (2014) are included in *Gentiana* L. in the Catalogue of

Life website. Some old generic names were reduced by new treatments, e.g. *Kobresia* Willd., which is included in *Carex* L. following Global Carex Group (2015), and *Nomocharis* Franch. is included in *Lilium* L. following Gao Y D & Gao X F (2016).

Generally speaking, the eastern Pan-Himalaya is much richer than the western Pan-Himalaya in species diversity. There are only 617 species of flowering plants recorded from the Wakhan corridor in NE Afghanistan, with 103 species of the Asteraceae, 87 species of the Poaceae, 67 species of the Fabaceae, and only two species of the Orchidaceae (Podlech, 2012). On the other side, the eastern Pan-Himalaya is a biodiversity hotspot, with new species being continuously described even in recent years. For example, more than 50 new species of *Saussurea* DC. have been described from the eastern Pan-Himalaya within the last ten years.

This catalogue includes 282 families, 2535 genera and 19,462 species of vascular plants in the Pan-Himalaya. The Pan-Himalayan region is renowned and characterized by its distinct mountain flora, with plants ranging from tropical, subtropical, temperate, to arctic on the same mountain at different altitudes. This region holds the richest temperate and alpine flora in the world, with tremendously differentiated genera, such as *Pedicularis* (414/600) *, *Rhododendron* (342/1000), *Corydalis* (301/465), *Primula* (285/500), *Saxifraga* (256/450), *Carex* (240/2000), *Saussurea* (236/460), *Gentiana* (218/360), *Impatiens* (191/900), *Berberis* (175/500), *Delphinium* (144/350), *Aconitum* (127/400), *Salix* (118/520), *Artemisia* (111/380), *Rubus* (106/700), *Arenaria* (99/300), *Ligularia* (94/140), *Polystichum* (94/500), *Dryopteris* (91/400), *Athyrium* (89/220), *Androsace* (86/100), *Strobilanthes* (85/400), *Oxytropis* (80/310), *Clematis* (77/300), *Cremanthodium* (74/76), *Meconopsis* (73/74), *Anaphalis* (59/110), *Rhodiola* (54/90).

I try to list only accepted scientific names in this catalogue. However, some uncertain species or unresolved names are listed because they have not been revised by any expert. For those species without the type variety or subspecies within our region, one of the infraspecific taxa is regarded as representative of the species in statistics. The list of the subspecies and varieties in this catalogue is not necessarily complete. It is difficult to edit a catalogue accepted by all people, especially for pteridophytes, where things become more complicated. For example, a specific name has many synonyms from different genera, or even from different families, or an expert accepted a species that is distributed in India, but another expert treated such material as a different species. In this case, I usually adopt the treatment of *Flora of China*, or new taxonomic revisions, or the Catalogue of Life website.

---

* The numerator is the number of species in the Pan-Himalaya, while the denominator is the total species number of the genus.

Acknowledgements

Partial funding of the *Flora of Pan-Himalaya* is provided by Ministry of Science and Technology of the People's Republic of China (grant 2013FY112100), National Natural Science Foundation of China (grant 31110103911; 31620103902), the International Partnership Program of Chinese Academy of Sciences (grant 151111KYSB20170021), Chinese Academy of Sciences (grant KSCX2-EW-Z-1), and the State Key Laboratory of Systematic and Evolutionary Botany, Institute of Botany, Chinese Academy of Sciences.

A special thank is due to Mr. Michael GILBERT (K) and Prof. De-Yuan HONG (PE) for their assistance in improving English and providing helpful suggestions, to Prof. Xian-Chun ZHANG (PE) for answering my questions about pteridophytes, to Prof. Cheng-Xin FU (HU), Xiao-Hua JIN (PE) and Fu-Sheng YANG (PE) for providing literature.

I am much indebted to the experts who have participated in the revision of the catalogue. They are listed as follows:

Acanthaceae: Yun-Fei DENG (IBSC)

Annonaceae: Xue-Liang HOU (AU)

Apiaceae: Xing-Jin HE (SZ)

Araceae (partly): Fu-Hua BING (Yantai University)

Aspleniaceae: Shi-Yong DONG (IBSC)

Asteraceae (*Synotis*): Ming TANG (JXAU)

Balsaminaceae: Sheng-Xiang YU (PE)

Betulaceae: Bao-Qing REN (Taiyuan Botanical Garden)

Blechnaceae: Fa-Guo WANG (IBSC)

Boraginaceae: Quan-Ru LIU (BNU), Jia-Chen HAO (BNU)

Burmanniaceae: Yue-Yu HANG (NAS)

Clethraceae: Shi-Xiong YANG (KUN)

Crassulaceae: Shi-Yong MENG (PEY)

Cucurbitaceae: Jian-Qiang LI (HIB)

Cyperaceae (*Carex*): Xiao-Feng JIN (HTC), Shu-Ren ZHANG (PE)

Davalliaceae: Fa-Guo WANG (IBSC)

Diapensiaceae: Shi-Xiong YANG (KUN)

Dioscoreaceae: Yue-Yu HANG (NAS)

Dryopteridaceae: Fa-Guo WANG (IBSC)

Equisetaceae: Fa-Guo WANG (IBSC)

Fabaceae (partly): SAREN (PE), Huang SUN (KUN), Chao-Yang CHANG (WUK), Yun KANG (FUS)

Fagaceae: Min DENG (CSH)

Hydrangeaceae: Sheng-Xiang YU (PE)
Isoëtaceae: Fa-Guo WANG (IBSC)
Lamiaceae: Chun-Lei XIANG (KUN), Qiang WANG (PE), Guo-Xiong HU (GZAC)
Lauraceae: Chen REN (IBSC)
Leeaceae: Li-Min LU (PE)
Lentibulariaceae: Zhen-Yu LI (PE)
Nephrolepidaceae: Fa-Guo WANG (IBSC)
Onocleaceae: Fa-Guo WANG (IBSC)
Ophioglossaceae: Fa-Guo WANG (IBSC)
Orchidaceae: Rong LI (KUN), Xiao-Hua JIN (PE), Lin LI (IBSC)
Orobanchaceae (*Pedicularis*): Fu-Sheng YANG (PE)
Pandanaceae: Yue-Yu HANG (NAS)
Pentaphylacaceae: Shi-Xiong YANG (KUN)
Poaceae (*Poa*): Marina OLONOVA (TK)
Primulaceae (*Lysimachia*): Jian-Wen SHAO (ANUB)
Ranunculaceae (partly): Lei XIE (BJFC), Yan LUO (HITBC)
Salicaceae (*Salix*): Li HE (FJFC)
Styracaceae: Shi-Xiong YANG (KUN)
Symplocaceae: Shi-Xiong YANG (KUN)
Theaceae: Shi-Xiong YANG (KUN)
Verbenaceae: Qiang WANG (PE)
Vitaceae: Li-Min LU (PE)
Zygophyllaceae: Shi-Yong MENG (PEY)

Contents

Chapter 1. Families and genera

Families of ferns and its allies (Pteridophyta) are arranged according to Christenhusz *et al.* (2011a). There are 38 families, 132 genera and 1221 species of pteridophytes (see Table 1).

Families of gymnosperms (Gymnosperma) are arranged according to Christenhusz *et al.* (2011b). There are 8 families with 24 genera and 109 species in gymnosperms (see Table 1).

Families of angiosperms (Angiosperma) are arranged according to APG III system (Haston *et al.*, 2009). The present catalogue accepts four additional families (Lobeliaceae, Leeaceae, Linderniaceae, Stixaceae); the former two follow the opinions of De-Yuan HONG (2015) and Li-Min LU (2019, in the present volume), while the latter two are adopted according to *A Dictionary of the Families and Genera of Chinese Vascular Plants* edited by De-Zhu LI (2018). There are 236 families, 2405 genera and 18,132 species of angiosperms (see Table 1).

In the catalogue, there are 45 families in angiosperms and pteridophytes with more than 100 species (see Table 2). These large families include 15,451 species, which account for 79.4% of the total number of vascular plants in the Pan-Himalaya. Obviously, these large families are the backbone and characteristic groups in this region. Among these large families, five families (Athyriaceae, Dryopteridaceae, Polypodiaceae, Pteridaceae, Thelypteridaceae) belong to pteridophytes.

There are 46 families of vascular plants with only one species in the Pan-Himalaya. They are Altingiaceae, Basellaceae, Bromeliaceae, Cannaceae, Cardiopteridaceae, Caricaceae, Centroplacaceae, Ceratophyllaceae, Cercidiphyllaceae, Cibotiaceae, Crypteroniaceae, Datiscaceae, Dichapetalaceae, Dipentodontaceae, Erythroxylaceae, Eucommiaceae, Eupteleaceae, Garryaceae, Ginkgoaceae, Gisekiaceae, Hydroleaceae, Lomariopsidaceae, Martyniaceae, Nelumbonaceae, Pedaliaceae, Penthoraceae, Petrosaviaceae, Platanaceae, Polemoniaceae, Psilotaceae, Rafflesiaceae, Rhachidosoraceae, Rhizophoraceae, Saururaceae, Scheuchzeriaceae, Schizaeaceae, Sladeniaceae, Sphenocleaceae, Stemonuraceae, Stixaceae, Talinaceae, Tetramelaceae, Triuridaceae, Trochodendraceae, Tropaeolaceae, and Velloziaceae.

Twenty-nine families each has only two species in the Pan-Himalaya. They are Achariaceae, Acoraceae, Aizoaceae, Aponogetonaceae, Biebersteiniaceae, Blechnaceae, Butomaceae, Cactaceae, Cannaceae, Circaeasteraceae, Clethraceae, Coriariaceae, Cycadaceae, Daphniphyllaceae, Diplaziopsidaceae, Dipteridaceae, Droseraceae, Gnetaceae, Isoëtaceae, Ixioliriaceae, Juncaginaceae, Moringaceae, Myricaceae, Opiliaceae, Podocarpaceae, Podostemaceae, Resedaceae, Stemonaceae, and Torricelliaceae.

For the convenience of the authors of *Flora of Pan-Himalaya*, we here provide a linear sequence of families of vascular plants in the Pan-Himalaya, with numbers of genera and species of each family, and with planned volume arrangement. These are shown in Table 3.

Table 1. The rich diversity of vascular plants in the Pan-Himalaya and China is shown in the three major plant groups. The data for numbers of families and genera in China come from *A Dictionary of the Families and Genera of Chinese Vascular Plants* (Li, 2018), and the data for the number of species in China come from *Plants of China* (Hong & Blackmore, 2013).

Groups	**Number of families**		**Number of genera**		**Number of species**	
	Pan-Himalaya	China	Pan-Himalaya	China	Pan-Himalaya	China
Pteridophyta	38	39	132	162	1,221	2,278
Gymnosperma	8	10	24	44	109	207
Angiosperma	236	265	2,405	3,040	18,132	29,611
Total	282	314	2,561	3,246	19,462	32,096

Table 2. Large families with more than 100 species in the Pan-Himalaya, compared with the data for China. The data for number of species in China come from *A Dictionary of the Families and Genera of Chinese Vascular Plants* (Li, 2018).

Families	Number of species in Pan-Himalaya	Number of species in China	Families	Number of species in Pan-Himalaya	Number of species in China
Asteraceae	1,686	2,350	Urticaceae	197	430
Poaceae	1,096	1,797	Balsaminaceae	191	271
Orchidaceae	1,051	1,350	Pteridaceae	185	233
Fabaceae	952	1,673	Acanthaceae	184	310
Ranunculaceae	632	923	Salicaceae	171	385
Lamiaceae	584	970	Crassulaceae	164	232
Rosaceae	575	942	Gesneriaceae	160	421
Ericaceae	529	837	Polypodiaceae	159	260
Primulaceae	522	652	Polygonaceae	157	236
Orobanchaceae	504	471	Athyriaceae	157	278
Apiaceae	437	616	Celastraceae	142	257
Cyperaceae	428	865	Euphorbiaceae	128	253
Papaveraceae	415	443	Araceae	127	190
Gentianaceae	398	420	Campanulaceae	127	159
Brassicaceae	373	400	Plantaginaceae	125	165
Caryophyllaceae	349	396	Liliaceae	125	148
Rubiaceae	331	743	Fagaceae	121	295
Saxifragaceae	301	620	Zingiberaceae	117	216
Boraginaceae	262	300	Malvaceae	116	246
Dryopteridaceae	225	500	Thelypteridaceae	115	197
Apocynaceae	222	423	Caprifoliaceae	107	144
Lauraceae	200	445	Asparagaceae	106	258
Berberidaceae	198	303	Total	15,451	

Table 3. Linear sequence of families of vascular plants in the Pan-Himalaya with numbers of genera and species, and with the planned volume arrangement of *Flora of Pan-Himalaya* (FLPH) indicated.

Number in sequence	Families	Number of genera	Number of species	Volume in *Flora of Pan-Himalaya*
1	Lycopodiaceae	3	40	2
2	Isoëtaceae	2	2	
3	Selaginellaceae	1	48	
4	Equisetaceae	1	6	
5	Ophioglossaceae	3	16	
6	Psilotaceae	1	1	
7	Marattiaceae	2	6	
8	Osmundaceae	2	5	
9	Hymenophyllaceae	3	27	
10	Gleicheniaceae	3	13	
11	Dipteridaceae	1	2	
12	Lygodiaceae	1	5	
13	Schizaeaceae	1	1	
14	Marsileaceae	1	3	
15	Salviniaceae	2	4	
16	Plagiogyriaceae	1	5	
17	Cibotiaceae	1	1	
18	Cyatheaceae	1	7	
19	Lindsaeaceae	3	8	
20	Dennstaedtiaceae	6	22	
21	Pteridaceae	18	185	
22	Cystopteridaceae	4	17	3
23	Aspleniaceae	2	77	
24	Diplaziopsidaceae	1	2	
25	Thelypteridaceae	18	115	
26	Woodsiaceae	1	11	
27	Rhachidosoraceae	1	1	
28	Onocleaceae	1	3	
29	Blechnaceae	2	2	
30	Athyriaceae	5	157	
31	Hypodematiaceae	2	4	
32	Dryopteridaceae	8	225	4
33	Lomariopsidaceae	1	1	
34	Nephrolepidaceae	1	6	

Continued

Number in sequence	Families	Number of genera	Number of species	Volume in *Flora of Pan-Himalaya*
35	Tectariaceae	2	19	4
36	Oleandraceae	1	3	
37	Davalliaceae	1	12	
38	Polypodiaceae	24	159	
39	Cycadaceae	1	2	5
40	Ginkgoaceae	1	1	
41	Gnetaceae	1	2	
42	Ephedraceae	1	14	
43	Pinaceae	8	44	
44	Podocarpaceae	1	2	
45	Cupressaceae	7	34	
46	Taxaceae	4	10	
47	Nymphaeaceae	2	6	6
48	Schisandraceae	3	19	
49	Chloranthaceae	2	6	
50	Saururaceae	1	1	
51	Piperaceae	2	44	
52	Aristolochiaceae	2	22	
53	Myristicaceae	2	6	
54	Magnoliaceae	7	32	
55	Annonaceae	15	32	
56	Hernandiaceae	1	5	
57	Lauraceae	19	200	
58	Acoraceae	1	2	7
59	Araceae	22	127	
60	Tofieldiaceae	1	3	
61	Alismataceae	4	12	
62	Butomaceae	2	2	
63	Hydrocharitaceae	7	14	
64	Scheuchzeriaceae	1	1	
65	Aponogetonaceae	1	2	
66	Juncaginaceae	1	2	
67	Potamogetonaceae	2	18	
68	Petrosaviaceae	1	1	
69	Nartheciaceae	1	11	
70	Burmanniaceae	1	4	
71	Dioscoreaceae	2	40	
72	Triuridaceae	1	1	
73	Velloziaceae	1	1	

Continued

Number in sequence	Families	Number of genera	Number of species	Volume in *Flora of Pan-Himalaya*
74	Stemonaceae	1	2	7
75	Pandanaceae	1	5	
76	Melanthiaceae	4	27	8
77	Colchicaceae	4	10	
78	Smilacaceae	1	56	
79	Liliaceae	10	125	
80	Orchidaceae	153	1,051	9
81	Hypoxidaceae	2	7	10
82	Ixioliriaceae	1	2	
83	Iridaceae	4	40	
84	Asphodelaceae	5	13	
85	Amaryllidaceae	8	78	
86	Asparagaceae	21	106	
87	Palmae	13	39	
88	Commelinaceae	13	45	11
89	Pontederiaceae	2	3	
90	Musaceae	3	18	
91	Cannaceae	1	1	
92	Marantaceae	2	3	
93	Costaceae	1	3	
94	Zingiberaceae	19	117	
95	Typhaceae	2	12	
96	Bromeliaceae	1	1	
97	Xyridaceae	1	3	
98	Eriocaulaceae	1	23	
99	Juncaceae	2	93	
100	Cyperaceae	30	428	12
101	Poaceae	188	1,096	13
102	Ceratophyllaceae	1	1	14
103	Eupteleaceae	1	1	
104	Papaveraceae	13	415	
105	Circaeasteraceae	2	2	15
106	Lardizabalaceae	4	11	
107	Menispermaceae	17	35	
108	Berberidaceae	7	198	
109	Ranunculaceae	35	632	16
110	Sabiaceae	2	20	17
111	Nelumbonaceae	1	1	
112	Platanaceae	1	1	

Continued

Number in sequence	Families	Number of genera	Number of species	Volume in *Flora of Pan-Himalaya*
113	Proteaceae	3	8	17
114	Trochodendraceae	1	1	
115	Buxaceae	3	13	
116	Dilleniaceae	2	6	
117	Paeoniaceae	1	10	
118	Altingiaceae	1	1	
119	Hamamelidaceae	5	13	
120	Cercidiphyllaceae	1	1	
121	Daphniphyllaceae	1	2	
122	Iteaceae	1	5	
123	Grossulariaceae	1	24	
124	Saxifragaceae	7	301	
125	Crassulaceae	10	164	18
126	Penthoraceae	1	1	
127	Haloragaceae	2	3	
128	Vitaceae	11	72	
129	Leeaceae	1	8	
130	Zygophyllaceae	4	13	
131	Fabaceae	130	952	19
132	Polygalaceae	5	32	20
133	Rosaceae	39	575	
134	Elaeagnaceae	2	26	21
135	Rhamnaceae	10	85	
136	Ulmaceae	3	12	
137	Cannabaceae	5	19	
138	Moraceae	7	81	
139	Urticaceae	21	197	22
140	Fagaceae	6	121	23
141	Myricaceae	1	2	
142	Juglandaceae	5	9	
143	Betulaceae	6	45	
144	Coriariaceae	1	2	
145	Cucurbitaceae	23	69	
146	Tetramelaceae	1	1	
147	Datiscaceae	1	1	
148	Begoniaceae	1	63	24
149	Celastraceae	14	142	
150	Connaraceae	2	3	
151	Oxalidaceae	3	15	

Continued

Number in sequence	Families	Number of genera	Number of species	Volume in *Flora of Pan-Himalaya*
152	Elaeocarpaceae	2	31	24
153	Rhizophoraceae	1	1	25
154	Erythroxylaceae	1	1	
155	Rafflesiaceae	1	1	
156	Euphorbiaceae	28	128	
157	Centroplacaceae	1	1	
158	Ochnaceae	1	3	
159	Phyllanthaceae	15	88	
160	Elatinaceae	2	4	
161	Malpighiaceae	2	9	
162	Dichapetalaceae	1	1	
163	Putranjivaceae	2	5	
164	Passifloraceae	2	9	
165	Salicaceae	11	171	26
166	Violaceae	2	56	
167	Achariaceae	2	2	
168	Linaceae	3	11	27
169	Calophyllaceae	2	4	
170	Clusiaceae	2	13	
171	Podostemaceae	2	2	
172	Hypericaceae	1	44	
173	Geraniaceae	2	50	
174	Combretaceae	3	23	
175	Lythraceae	10	24	
176	Onagraceae	5	54	
177	Myrtaceae	7	38	
178	Melastomataceae	10	36	
179	Crypteroniaceae	1	1	28
180	Staphyleaceae	3	9	
181	Stachyuraceae	1	4	
182	Biebersteiniaceae	1	2	
183	Nitrariaceae	2	3	
184	Burseraceae	3	6	
185	Anacardiaceae	16	45	
186	Sapindaceae	16	76	
187	Rutaceae	22	82	
188	Simaroubaceae	3	8	29
189	Meliaceae	12	32	
190	Dipentodontaceae	1	1	

Continued

Number in sequence	Families	Number of genera	Number of species	Volume in *Flora of Pan-Himalaya*
191	Malvaceae	33	116	29
192	Thymelaeaceae	8	58	
193	Dipterocarpaceae	4	6	
194	Tropaeolaceae	1	1	
195	Moringaceae	2	2	
196	Caricaceae	1	1	
197	Resedaceae	2	2	
198	Stixaceae	1	1	
199	Capparaceae	3	22	
200	Cleomaceae	6	9	
201	Brassicaceae	79	373	30
202	Balanophoraceae	2	8	31
203	Olacaceae	2	5	
204	Opiliaceae	2	2	
205	Santalaceae	9	34	
206	Loranthaceae	8	43	
207	Schoepfiaceae	1	3	
208	Tamaricaceae	2	15	
209	Plumbaginaceae	5	18	
210	Polygonaceae	11	157	32
211	Droseraceae	1	2	
212	Caryophyllaceae	28	349	33
213	Amaranthaceae	37	85	34
214	Gisekiaceae	1	1	
215	Aizoaceae	2	2	
216	Phytolaccaceae	1	4	
217	Nyctaginaceae	5	8	
218	Molluginaceae	2	5	
219	Basellaceae	1	1	
220	Talinaceae	1	1	
221	Portulacaceae	2	4	
222	Cactaceae	2	2	
223	Cornaceae	6	24	
224	Hydrangeaceae	5	58	
225	Balsaminaceae	1	191	
226	Polemoniaceae	1	1	
227	Lecythidaceae	2	3	
228	Sladeniaceae	1	1	
229	Pentaphylacaceae	5	34	

Continued

Number in sequence	Families	Number of genera	Number of species	Volume in *Flora of Pan-Himalaya*
230	Sapotaceae	8	11	34
231	Ebenaceae	1	12	
232	Primulaceae	16	522	35
233	Theaceae	6	35	
234	Symplocaceae	1	29	
235	Diapensiaceae	3	7	
236	Styracaceae	8	19	
237	Actinidiaceae	3	29	
238	Clethraceae	1	2	
239	Ericaceae	18	529	36
240	Icacinaceae	7	11	37
241	Eucommiaceae	1	1	
242	Garryaceae	1	1	
243	Rubiaceae	72	331	
244	Gentianaceae	20	398	38
245	Loganiaceae	4	8	39
246	Apocynaceae	67	222	
247	Boraginaceae	41	262	40
248	Convolvulaceae	14	81	41
249	Solanaceae	19	83	
250	Sphenocleaceae	1	1	
251	Hydroleaceae	1	1	
252	Carlemanniaceae	2	4	
253	Oleaceae	10	86	
254	Gesneriaceae	29	160	42
255	Plantaginaceae	23	125	43
256	Scrophulariaceae	5	57	
257	Linderniaceae	3	24	
258	Pedaliaceae	1	1	
259	Lamiaceae	93	584	44
260	Mazaceae	3	10	45
261	Phrymaceae	3	8	
262	Paulowniaceae	2	3	
263	Orobanchaceae	24	504	
264	Lentibulariaceae	2	20	46
265	Acanthaceae	31	184	
266	Bignoniaceae	8	25	
267	Verbenaceae	8	11	
268	Martyniaceae	1	1	

Continued

Number in sequence	Families	Number of genera	Number of species	Volume in *Flora of Pan-Himalaya*
269	Stemonuraceae	1	1	46
270	Cardiopteridaceae	1	1	
271	Helwingiaceae	1	3	47
272	Aquifoliaceae	1	41	
273	Campanulaceae	13	127	
274	Lobeliaceae	1	19	
275	Menyanthaceae	2	4	
276	Asteraceae	195	1,686	48
277	Adoxaceae	4	44	49
278	Caprifoliaceae	16	107	
279	Torricelliaceae	1	2	
280	Pittosporaceae	1	12	
281	Araliaceae	15	80	
282	Apiaceae	79	437	50

Chapter 2. Catalogue

PTERIDOPHYTA

1. LYCOPODIACEAE

Huperzia appressa (Desv.) Á. Löve & D. Löve

Huperzia arunachalensis (D. D. Pant & P. S. Pandey) Fraser-Jenk.

Huperzia bucahwangensis Ching

Huperzia cancellata (Spring) Trevis.

Huperzia cavei Fraser-Junk. & B. S. Kholia

Huperzia chinensis (Herter ex Nessel) Ching

Huperzia crispata (Ching) Ching

Huperzia delavayi (Christ & Herter) Ching

Huperzia emeiensis (Ching & H. S. Kung) Ching & H. S. Kung

Huperzia fordii (Baker) R. D. Dixit

Huperzia hamiltonii (Spreng.) Trevis.

Huperzia herteriana (Kumm.) T. Sen & U. Sen

Huperzia indica S. R. Ghosh

Huperzia javanica (Sw.) Fraser-Jenk.

Huperzia kangdingensis (Ching) Ching

Huperzia laipoensis Ching

Huperzia liangshanica (H. S. Kung) Ching & H. S. Kung

Huperzia medogensis Ching & Y. X. Lin

Huperzia muscicola Ching ex W. M. Chu

Huperzia nyalamensis Ching & S. K. Wu

Huperzia petiolata (C. B. Clarke) R. D. Dixit

Huperzia phlegmaria (L.) Rothm.

Huperzia pulcherrima (Wall. ex Hook. & Grev.) Pic. Serm.

Huperzia rubicaulis S. K. Wu & X. Cheng

Huperzia selago (L.) Bernh. ex Schrank & Mart.

Huperzia squarrosa (G. Forst.) Trevis.

Huperzia subulifolia (Wall. ex Hook. & Grev.) Trevis.

Huperzia sutchueniana (Herter) Ching

Huperzia tibetica (Ching) Ching

Huperzia yunnanensis (Ching) Holub

Lycopodiella cernua (L.) Pic. Serm.

Lycopodium alpinum L.

Lycopodium annotinum L.

Lycopodium casuarinoides Spring

Lycopodium complanatum L.

Lycopodium japonicum Thunb.

Lycopodium multispicatum J. H. Wilce

Lycopodium obscurum L.

Lycopodium veitchii Christ

Lycopodium zonatum Ching

2. ISOËTACEAE

Calamaria coromandelina (L. f.) Kuntze

Isoëtes hypsophila Hand.-Mazz.

3. SELAGINELLACEAE

Selaginella adunca A. Braun ex Hieron.

Selaginella aitchisonii Hieron.

Selaginella albocincta Ching

Selaginella amblyphylla Alston

Selaginella bisulcata Spring

Selaginella braunii Baker

Selaginella bryopteris (L.) Baker

Selaginella chrysocaulos (Hook. & Grev.) Spring

Selaginella chrysorrhizos Spring

Selaginella ciliaris (Retz.) Spring

Selaginella davidii Franch.

Selaginella decipiens Warb.

Selaginella delicatula (Desv. ex Poir.) Alston

Selaginella effusa Alston

Selaginella emodi Fraser-Jenk.

Selaginella fulcrata (Buch.-Ham. ex D. Don) Spring

Selaginella helferi Warb.

Selaginella helvetica (L.) Link

Selaginella heterostachys Baker

Selaginella indica (Milde) R. M. Tryon

Selaginella involvens (Sw.) Spring

Selaginella jacquemontii Spring

Selaginella jugorum Hand.-Mazz.

Selaginella kurzii Baker

Selaginella labordei Hieron. ex Christ

Selaginella laxistrobila K. H. Shing

Selaginella leptophylla Baker

Selaginella mairei H. Lév.

Selaginella megaphylla Baker

Selaginella monospora Spring

Selaginella nipponica Franch. & Sav.

Selaginella nummularifolia Ching

Selaginella pallida (Hook. & Grev.) Spring

Selaginella pallidissima Spring

Selaginella pennata (D. Don) Spring

Selaginella pentagona Spring

Selaginella picta A. Braun ex Baker

Selaginella pulvinata (Hook. & Grev.) Maxim.

Selaginella remotifolia Spring

Selaginella repanda (Desv. ex Poir.) Spring

Selaginella reticulata (Hook. & Grev.) Spring

Selaginella semicordata (Wall. ex Hook. & Grev.) Spring

Selaginella subdiaphana Spring

Selaginella tenuifolia Spring

Selaginella tibetica Ching & S. K. Wu

Selaginella trichophylla K. H. Shing

Selaginella vaginata Spring

Selaginella vardei H. Lév.

4. EQUISETACEAE

Equisetum arvense L.

Equisetum debile Roxb. ex Vaucher

Equisetum diffusum D. Don

Equisetum hyemale L.

Equisetum palustre L.

Equisetum ramosissimum Desf.

Equisetum ramosissimum subsp. **debile** (Roxb. ex Vaucher) Hauke

5. OPHIOGLOSSACEAE

Botrychium daucifolium Wall. ex Hook. & Grev.

Botrychium lanuginosum Wall. ex Hook. & Grev.

Botrychium lunaria (L.) Sw.

Botrychium multifidum (S. G. Gmel.) Rupr.

Botrychium ternatum (Thunb.) Sw.

Botrychium virginianum (L.) Sw.

Helminthostachys zeylanica (L.) Hook.

Ophioglossum costatum R. Br.

Ophioglossum gramineum Willd.

Ophioglossum nudicaule L. f.

Ophioglossum pendulum L.

Ophioglossum petiolatum Hook.

Ophioglossum polyphyllum A. Braun ex Schub.

Ophioglossum reticulatum L.

Ophioglossum thermale Kom.

Ophioglossum vulgatum L.

6. PSILOTACEAE

Psilotum nudum (L.) P. Beauv.

7. MARATTIACEAE

Angiopteris esculenta Ching

Angiopteris evecta (G. Forst.) Hoffm.

Angiopteris helferiana C. Presl

Angiopteris indica Desv.

Angiopteris wallichiana C. Presl

Christensenia aesculifolia (Blume) Maxon

8. OSMUNDACEAE

Osmunda claytoniana L.

Osmunda japonica Thunb.

Osmunda regalis L.

Osmunda vachellii Hook.

Osmundastrum cinnamomeum (L.) C. Presl

9. HYMENOPHYLLACEAE

Crepidomanes auriculatum (Blume) K. Iwats.

Crepidomanes bipunctatum (Poir.) Copel.

Crepidomanes birmanicum (Bedd.) K. Iwats.

Crepidomanes campanulatum (Roxb.) Panigrahi & Sarn. Singh

Crepidomanes latealatum (Bosch) Copel.

Crepidomanes maximum (Blume) K. Iwats.

Crepidomanes minutum (Blume) K. Iwats.

Crepidomanes parvifolium (Baker) K. Iwats.

Crepidomanes schmidianum (Zenker ex Taschner) K. Iwats.

Crepidomanes schmidianum var. **latifrons** (Bosch) K. Iwats.

Crepidomanes striatum (D. Don) Thapa

Crepidomanes zayuense Ching & S. K. Wu

Hymenophyllum badium Hook. & Grev.

Hymenophyllum barbatum (Bosch) Baker

Hymenophyllum corrugatum Christ

Hymenophyllum denticulatum Sw.

Hymenophyllum exsertum Wall. ex Hook.

Hymenophyllum javanicum Spreng.

Hymenophyllum levingei C. B. Clarke

Hymenophyllum longissimum (Ching & P. S. Chiu) K. Iwats.

Hymenophyllum polyanthos Sw.

Hymenophyllum simonsianum Hook.

Hymenophyllum stenocladum (Ching & P. S. Chiu) K. Iwats.

Hymenophyllum tenellum (Jacq.) Kuhn

Vandenboschia auriculata (Blume) Copel.

Vandenboschia gigantea (Bory ex Willd.) Pic. Serm.

Vandenboschia radicans (Sw.) Copel.

Vandenboschia striata (D. Don) Ebihara

10. GLEICHENIACEAE

Dicranopteris ampla Ching & P. S. Chiu
Dicranopteris lanigera (D. Don) Fraser-Jenk.
Dicranopteris linearis (Burm. f.) Underw.
Dicranopteris montana (Holttum) S. R. Ghosh
Dicranopteris pedata (Houttuyn) Nakaike
Dicranopteris splendida (Hand.-Mazz.) Tagawa
Dicranopteris taiwanensis Ching & P. S. Chiu
Diplopterygium blotianum (C. Chr.) Nakai
Diplopterygium chinensis (Rosenst.) De Vol
Diplopterygium giganteum (Wall. ex Hook.) Nakai
Diplopterygium glaucum (Thunb. ex Houtt.) Nakai
Diplopterygium volubile (Jungh.) Nakai
Gleichenia dichotoma (Thunb.) Hook.

11. DIPTERIDACEAE

Dipteris chinensis Christ
Dipteris wallichii (R. Br.) T. Moore

12. LYGODIACEAE

Lygodium circinatum (Burm. f.) Sw.
Lygodium flexuosum (L.) Sw.
Lygodium japonicum (Thunb.) Sw.
Lygodium microphyllum (Cav.) R. Br.
Lygodium salicifolium C. Presl

13. SCHIZAEACEAE

Schizaea digitata (L.) Sw.

14. MARSILEACEAE

Marsilea aegyptiaca Willd.
Marsilea minuta L.
Marsilea quadrifolia L.

15. SALVINIACEAE

Azolla filiculoides Lam.
Azolla pinnata R. Br.
Salvinia molesta D. S. Mitch.
Salvinia natans (L.) All.

16. PLAGIOGYRIACEAE

Plagiogyria adnata (Blume) Bedd.
Plagiogyria euphlebia (Kunze) Mett.
Plagiogyria glauca (Blume) Mett.
Plagiogyria japonica Nakai
Plagiogyria pycnophylla (Kunze) Mett.

17. CIBOTIACEAE

Cibotium barometz (L.) J. Sm.

18. CYATHEACEAE

Alsophila andersonii J. Scott ex Bedd.

Alsophila brunoniana Wall. ex Hook.

Alsophila contaminans Wall. ex Hook.

Alsophila gigantea Wall. ex Hook.

Alsophila henryi Baker

Alsophila khasyana Moore

Alsophila spinulosa (Wall. ex Hook.) R. M. Tryon

19. LINDSAEACEAE

Lindsaea commixta Tagawa

Lindsaea ensifolia Sw.

Lindsaea himalaica K. U. Kramer

Lindsaea lucida Blume

Lindsaea odorata Roxb.

Lindsaea repens (Bory) Thwaites

Odontosoria chinensis (L.) J. Sm.

Osmolindsaea odorata (Roxb.) Lehtonen & Christenh.

20. DENNSTAEDTIACEAE

Dennstaedtia appendiculata (Wall. ex Hook.) J. Sm.

Dennstaedtia wilfordii (T. Moore) H. Christ

Dennstaedtia zeylanica (Sw.) Zink ex Fraser-Jenk. & Kandel

Histiopteris incisa (Thunb.) J. Sm.

Hypolepis alpina (Blume) Hook.

Hypolepis polypodioides (Blume) Hook.

Hypolepis punctata (Thunb.) Mett. ex Kuhn

Hypolepis resistens Hook.

Microlepia calvescens (Wall. ex Hook.) C. Presl

Microlepia firma Mett. ex Kuhn

Microlepia hallbergii (d'Almeida) C. Chr.

Microlepia hookeriana (Wall. ex Hook.) C. Presl

Microlepia marginata (Panzer) C. Chr.

Microlepia marginata var. **bipinnata** Makino

Microlepia marginata var. **villosa** (C. Presl) Y. C. Wu

Microlepia nepalensis (Spreng.) Fraser-Jenk., Kandel & Pariyar

Microlepia platyphylla (D. Don) J. Sm.

Microlepia rhomboidea (Wall. ex Kunze) Prantl

Microlepia setosa (Sm.) Alston

Microlepia speluncae (L.) T. Moore

Microlepia strigosa (Thunb.) C. Presl

Microlepia trichocarpa Hayata

Monachosorum henryi Christ

Monachosorum subdigitatum (Blume) Kuhn

Pteridium revolutum (Blume) Nakai

Pteridium revolutum var. **muricatulum** Ching & S. H. Wu

21. PTERIDACEAE

Actiniopteris radiata (Sw.) Link

Actiniopteris semiflabellata Pic. Serm.

Adiantum breviserratum (Ching) Ching & Y. X. Lin

Adiantum capillus-veneris L.

Adiantum caudatum L.

Adiantum davidii Franch.

Adiantum edgeworthii Hook.

Adiantum erythrochlamys Diels

Adiantum fengianum Ching

Adiantum flabellulatum L.

Adiantum incisum Forssk.

Adiantum lunulatum Burm. f.

Adiantum myriosorum Baker

Adiantum pedatum L.

Adiantum philippense L.

Adiantum proliferum Roxb.

Adiantum raddianum C. Presl

Adiantum roborowskii Maxim.

Adiantum sinicum Ching

Adiantum soboliferum Wall. ex Hook.

Adiantum tibeticum Ching

Adiantum venustum D. Don

Adiantum wattii Baker

Aleuritopteris albofusca (Baker) Pic. Serm.

Aleuritopteris albomarginata (C. B. Clarke) Ching

Aleuritopteris anceps (Blanford) Panigrahi

Aleuritopteris argentea (S. G. Gmel.) Fée

Aleuritopteris argentea var. **obscura** (Christ) Ching

Aleuritopteris bicolor (Roxb.) Fraser-Jenk.

Aleuritopteris chrysophylla (Hook.) Ching

Aleuritopteris dealbata (C. Presl) Fée

Aleuritopteris dubia (C. Hope) Ching

Aleuritopteris duclouxii (Christ) Ching

Aleuritopteris duthiei (Baker) Ching

Aleuritopteris farinosa (Forssk.) Fée

Aleuritopteris formosana (Hayata) Tagawa

Aleuritopteris gongshanensis G. M. Zhang

Aleuritopteris grevilleoides (Christ) G. M. Zhang ex X. C. Zhang

Aleuritopteris grisea (Blanford) Panigrahi

Aleuritopteris kuhnii (Milde) Ching

Aleuritopteris leptolepis (Fraser-Jenk.) Fraser-Jenk.

Aleuritopteris likiangensis Ching

Aleuritopteris niphobola (C. Chr.) Ching

Aleuritopteris pygmaea Ching

Aleuritopteris rosulata (C. Chr.) Ching

Aleuritopteris rufa (D. Don) Ching

Aleuritopteris speciosa Ching & S. K. Wu

Aleuritopteris stenochlamys Ching

Aleuritopteris subdimorpha (C. B. Clarke & Baker) Fraser-Jenk.

Aleuritopteris subvillosa (Hook.) Ching

Aleuritopteris subvillosa var. **tibetica** (Ching & S. K. Wu) H. S. Kung

Aleuritopteris tamburii (Hook.) Ching

Aleuritopteris tamburii var. **viridis** H. S. Kung

Aleuritopteris veitchii (Christ) Ching

Aleuritopteris yalungensis H. S. Kung

Anogramma leptophylla (L.) Link

Anogramma microphylla (Hook.) Diels

Anogramma reichsteinii Fraser-Jenk.

Antrophyum coriaceum (D. Don) Wall. ex T. Moore

Antrophyum obovatum Baker

Antrophyum parvulum Blume

Antrophyum plantagineum (Cav.) Kaulf.

Antrophyum reticulatum (G. Forst.) Kaulf.

Argyrochosma connectens (C. Chr.) G. M. Zhang

Calciphilopteris ludens (Wall. ex Hook.) Yesilyurt & H. Schneid.

Ceratopteris thalictroides (L.) Brongn.

Cerosora microphylla (Hook.) R. M. Tryon

Cheilanthes acrostica (Balb.) Tod.

Cheilanthes albomarginata C. B. Clarke

Cheilanthes anceps Blanf.

Cheilanthes argentea (S. G. Gmel.) Kunze

Cheilanthes belangeri (Bory) C. Chr.

Cheilanthes bhutanica Fraser-Jenk. & Wangdi

Cheilanthes brausei Fraser-Jenk.

Cheilanthes christii Fraser-Jenk. & Yatsk.

Cheilanthes hancockii Baker

Cheilanthes insignis Ching

Cheilanthes nitidula Wall. ex Hook.

Cheilanthes persica (Bory) Mett. ex Kuhn

Cheilanthes pteridioides (Reichard) C. Chr.

Cheilanthes smithii (C. Chr.) R. M. Tryon

Cheilanthes subvillosa Hook.

Cheilanthes tenuifolia (Burm. f.) Sw.

Cheilanthes tibetica Fraser-Jenk. & Wangdi

Cheilanthes trichophylla Baker

Coniogramme affinis (C. Presl) Hieron.

Coniogramme denticulato-serrata (Hieron.) R. D. Dixit & A. Das

Coniogramme emeiensis Ching & K. H. Shing

Coniogramme falcata (D. Don) Salomon

Coniogramme fauriei Hieron.

Coniogramme fraxinea (D. Don) Fée ex Diels

Coniogramme indica Fée

Coniogramme intermedia Hieron.

Coniogramme intermedia var. **glabra** Ching

Coniogramme javanica (Blume) Fée

Coniogramme merrillii Ching

Coniogramme ovata S. K. Wu

Coniogramme procera Fée

Coniogramme pubescens Hieron.

Coniogramme purpurea R. D. Dixit & A. Das

Coniogramme rosthornii Hieron.

Coniogramme serrulata (Blume) Fée

Coniogramme subcordata Copel.

Cryptogramma brunoniana Wall. ex Hook. & Grev.

Cryptogramma raddeana Fomin

Cryptogramma stelleri (S. G. Gmel.) Prantl

Haplopteris amboinensis (Fée) X. C. Zhang

Haplopteris doniana (Mett. ex Hieron.) E. H. Crane

Haplopteris elongata (Sw.) E. H. Crane

Haplopteris ensiformis (Sw.) E. H. Crane

Haplopteris flexuosa (Fée) E. H. Crane

Haplopteris himalayensis (Ching) E. H. Crane

Haplopteris linearifolia (Ching) X. C. Zhang

Haplopteris mediosora (Hayata) X. C. Zhang

Haplopteris plurisulcata (Ching) X. C. Zhang

Haplopteris scolopendrina (Bory) C. Presl

Haplopteris sikkimensis (Kuhn) E. H. Crane

Haplopteris zosterifolia (Willd.) E. H. Crane

Onychium cryptogrammoides Christ

Onychium fragile S. Verma & Khullar

Onychium japonicum (Thunb.) Kunze

Onychium japonicum var. **lucidum** (D. Don) Christ

Onychium kholianum Fraser-Jenk.

Onychium moupinense Ching

Onychium plumosum Ching

Onychium siliculosum (Desv.) C. Chr.

Onychium tenuifrons Ching

Onychium tibeticum Ching & S. K. Wu

Onychium vermae Fraser-Jenk.

Paragymnopteris bipinnata (Christ) K. H. Shing

Paragymnopteris bipinnata var. **auriculata** (Franch.) K. H. Shing

Paragymnopteris borealisinensis (Kitag.) M. Hassler

Paragymnopteris delavayi (Baker) K. H. Shing

Paragymnopteris marantae (L.) K. H. Shing

Paragymnopteris sargentii (Christ) K. H. Shing

Paragymnopteris vestita (Wall. ex C. Presl) K. H. Shing

Pellaea calomelanos (Sw.) Link

Pityrogramma calomelanos (L.) Link

Pteris actiniopteroides Christ

Pteris alata Lam.

Pteris amoena Blume

Pteris arisanensis Tagawa

Pteris aspericaulis Wall. ex J. Agardh

Pteris aspericaulis var. **cuspigera** Ching

Pteris aspericaulis var. **subindivisa** (C. B. Clarke) Ching ex S. H. Wu

Pteris assamica Fraser-Jenk. & T. G. Walker

Pteris barbigera Ching

Pteris beddomei Sarn. Singh & Panigrahi

Pteris biaurita L.

Pteris bomiensis Ching & S. K. Wu

Pteris cretica L.

Pteris cretica var. **laeta** (Wall. ex Ettingsh.) C. Chr. & Tardieu

Pteris dactylina Hook.

Pteris digitata Wall.

Pteris dixitii Fraser-Jenk. & Pariyar

Pteris ensiformis Burm.

Pteris fauriei Hieron.

Pteris gallinopes Ching

Pteris grevilleana Wall. ex J. Agardh

Pteris grevilleana var. **ornata** Alderwerelt

Pteris griffithii Hook.

Pteris hirtula (C. Chr.) C. V. Morton

Pteris hirsutissima Ching

Pteris inaequalis Baker

Pteris kathmanduensis Fraser-Jenk. & T. G. Walker

Pteris khasiana (C. B. Clarke) Hieron.

Pteris longifolia L.

Pteris longipes D. Don

Pteris longipinnula Wall. ex J. Agardh

Pteris medogensis Ching & S. K. Wu

Pteris mertensioides Willd.

Pteris multifida Poir.

Pteris panigrahiana Sarn. Singh

Pteris pellucens J. Agardh

Pteris pellucida C. Presl

Pteris pseudoconfusa Sarn. Singh & Panigrahi

Pteris pseudopellucida Ching

Pteris pseudoquadriaurita Khullar

Pteris puberula Ching

Pteris quadriaurita Retz.

Pteris rigida Sw.

Pteris roseolilacina Hieron.

Pteris rufopilosa Ching & Y. X. Lin

Pteris scabririgens Fraser-Jenk., S. C. Verma & T. G. Walker

Pteris semipinnata L.

Pteris spinescens C. Presl

Pteris stenophylla Wall. ex Hook. & Grev.

Pteris submiaoensis Sarn. Singh & Panigrahi

Pteris subquinata Wall. ex J. Agardh

Pteris terminalis Wall. ex J. Agardh

Pteris tirapensis Panigrahi & Sarn. Singh

Pteris tricolor Linden

Pteris venusta Kunze

Pteris vittata L.

Pteris wallichiana J. Agardh

Pteris wallichiana var. **yunnanensis** (Christ) Ching & S. H. Wu

22. CYSTOPTERIDACEAE

Acystopteris tenuisecta (Blume) Tagawa

Cystoathyrium chinense Ching

Cystopteris deqinensis Z. R. Wang

Cystopteris dickieana R. Sim

Cystopteris fragilis (L.) Bernh.

Cystopteris kansuana C. Chr.

Cystopteris modesta Ching

Cystopteris montana (Lam.) Bernh. ex Desv.

Cystopteris moupinensis Franch.

Cystopteris pellucida (Franch.) Ching

Cystopteris sudetica A. Braun & Milde

Cystopteris tibetica Z. R. Wang

Gymnocarpium dryopteris (L.) Newman

Gymnocarpium fedtschenkoanum Pojark.

Gymnocarpium jessoense (Koidz.) Koidz.

Gymnocarpium oyamense (Baker) Ching

Gymnocarpium remotepinnatum (Hayata) Ching

23. ASPLENIACEAE

Asplenium adiantum-nigrum L.

Asplenium aethiopicum (Burm. f.) Becherer

Asplenium affine Sw.

Asplenium aitchisonii Fraser-Jenk. & Reichst.

Asplenium × alternifolium Wulfen

Asplenium anogrammoides H. Christ

Asplenium austrochinense Ching

Asplenium bullatum Wall. ex Mett.

Asplenium capillipes Makino

Asplenium ceterach L.

Asplenium caucasicum (Fraser-Jenk. & Lovis) Viane

Asplenium crinicaule Hance

Asplenium crinicaule var. **sikkimense** Bir

Asplenium daghestanicum Christ

Asplenium dalhousiae Hook.

Asplenium delavayi (Franch.) Copel.

Asplenium dolomiticum (Lovis & Reichst.) Á. Löve & D. Löve

Asplenium ensiforme Wall. ex Hook. & Grev.

Asplenium exiguum Bedd.

Asplenium falcatum Lam.

Asplenium finlaysonianum Wall. ex Hook.

Asplenium fontanum (L.) Bernh.

Asplenium fontanum subsp. **pseudofontanum** (Kossinsky) Reichst. & Schneller

Asplenium formosum Willd.

Asplenium fugax Christ

Asplenium × geni-coalitum Fraser-Jenk.

Asplenium × germanicum Weiss

Asplenium gracilipes Ching & Y. X. Lin

Asplenium griffithianum Hook.

Asplenium gueinzianum Mett. ex Kuhn

Asplenium humistratum Ching ex K. S. Kung

Asplenium indicum Sledge

Asplenium kansuense Ching

Asplenium khasianum Sledge

Asplenium khullarii T. Reichst. & Rasbach ex Fraser-Jenk.

Asplenium kukkonenii Viane & Reichst.

Asplenium lushanense C. Chr.

Asplenium macrophyllum Sw.

Asplenium magnificum (Ching) Bir, Fraser-Jenk. & Lovis

Asplenium × mainlingense Ching & S. K. Wu

Asplenium × mitsutae Viane & Reichst.

Asplenium neovarians Ching
Asplenium nesii Christ
Asplenium nidus L.
Asplenium nitidum Sw.
Asplenium nitidum var. **obtusum** Bedd.
Asplenium normale D. Don
Asplenium paucivenosum (Ching) Bir
Asplenium pekinense Hance
Asplenium pellucidum Lam.
Asplenium phyllitidis D. Don
Asplenium planicaule var. **obtusa** Bir
Asplenium prolongatum Hook.
Asplenium punjabense Bir, Fraser-Jenk. & Lovis
Asplenium qiujiangense (Ching & S. H. Fu) Nakaike
Asplenium rockii C. Chr.
Asplenium ruta-muraria L.
Asplenium sarelii Hook.
Asplenium septentrionale (L.) Hoffm.
Asplenium spathulinum J. Sm.
Asplenium subspathulinum X. C. Zhang
Asplenium subvarians Ching
Asplenium tenuicaule Hayata
Asplenium tenuicaule var. **argutum** Viane
Asplenium tenuicaule var. **subvarians** (Ching) Viane
Asplenium tenuifolium D. Don
Asplenium trapezoideum Ching
Asplenium trichomanes L.
Asplenium tripteropus Nakai
Asplenium varians Wall. ex Hook. & Grev.
Asplenium viride Huds.
Asplenium yoshinagae Makino
Asplenium yoshinagae var. **planicaule** C. V. Morton
Hymenasplenium changputungense (Ching) Viane & S. Y. Dong
Hymenasplenium cheilosorum (Kunze ex Mett.) Tagawa
Hymenasplenium excisum (C. Presl) S. Lindsay
Hymenasplenium furfuraceum (Ching) Viane & S. Y. Dong
Hymenasplenium obliquissimum (Hayata) Sugim.
Hymenasplenium quercicolum (Ching) Viane & S. Y. Dong
Hymenasplenium retusulum (Ching) Viane & S. Y. Dong
Hymenasplenium rivulare (Fraser-Jenk.) Viane & S. Y. Dong
Hymenasplenium szechuanse (Ching) Viane & S. Y. Dong
Hymenasplenium wuliangshanense (Ching) Viane & S. Y. Dong

24. DIPLAZIOPSIDACEAE

Diplaziopsis cavaleriana (Christ) C. Chr.
Diplaziopsis javanica (Blume) C. Chr.

25. THELYPTERIDACEAE

Ampelopteris prolifera (Retz.) Copel.
Cyclogramma auriculata (J. Sm.) Ching
Cyclogramma flexilis (Christ) Tagawa
Cyclogramma squamaestipes (C. B. Clarke) Tagawa
Cyclosorus appendiculatus (C. Presl) K. H. Shing
Cyclosorus aridus (D. Don) Ching
Cyclosorus articulatus (Houlston & T. Moore) Panigrahi

Cyclosorus assamicus (Bedd.) Ching

Cyclosorus clarkei (Bedd.) Ching

Cyclosorus crinipes (Hook.) Ching

Cyclosorus cylindrothrix (Rosenst.) Ching

Cyclosorus dentatus (Forssk.) Ching

Cyclosorus erubescens (Wall. ex Hook.) C. M. Kuo

Cyclosorus evolutus (Bedd.) Ching

Cyclosorus gustavi (Bedd.) Ching

Cyclosorus hispidulus (Decne.) Ching

Cyclosorus holttumii Sarn. Singh & Panigrahi

Cyclosorus jaculosus (Christ) H. Itô

Cyclosorus krameri Sarn. Singh & Panigrahi

Cyclosorus × **kumaunicus** (Holttum) Mazumdar

Cyclosorus mantoniae Sarn. Singh & Panigrahi

Cyclosorus medogensis Ching & S. K. Wu

Cyclosorus meeboldii (Rosenst.) Ching

Cyclosorus molliusculus (Wall. ex Kuhn) Ching

Cyclosorus multilineatus (Wall. ex Hook.) Tardieu & C. Chr.

Cyclosorus namburensis (Bedd.) Ching

Cyclosorus occultus (Hope) Ching

Cyclosorus papilio (C. Hope) Ching

Cyclosorus parasiticus (L.) Farw.

Cyclosorus procerus (D. Don) S. Lindsay & D. J. Middleton

Cyclosorus pseudoacuminatus Panigrahi & Sarn. Singh

Cyclosorus pseudobalansae Panigrahi & Sarn. Singh

Cyclosorus repandus (Fée) B. K. Nayar & S. Kaur

Cyclosorus siamensis (Tagawa & K. Iwats.) Panigrahi

Cyclosorus siamensis var. **khonsaensis** Sarn. Singh

Cyclosorus sledgei Panigrahi & Sarn. Singh

Cyclosorus subpubescens (Blume) Ching

Cyclosorus truncatus (Poiret) Farwell

Cyclosorus wulingshanensis C. M. Zhang

Glaphyropteridopsis erubescens (Wall. ex Hook.) Ching

Goniopteris vivipara (Raddi) Brack.

Leptogramma himalaica Ching

Macrothelypteris ornata (Wall. ex J. Sm.) Ching

Macrothelypteris torresiana (Gaudich.) Ching

Metathelypteris decipiens (C. B. Clarke) Ching

Metathelypteris flaccida (Blume) Ching

Metathelypteris gracilescens (Blume) Ching

Metathelypteris krameri Sarn. Singh & Panigrahi

Metathelypteris laxa (Franch. & Savat.) Ching

Metathelypteris uraiensis (Rosenst.) Ching

Metathelypteris uraiensis var. **tibetica** (Ching & S. K. Wu) K. H. Shing

Oreopteris elwesii (Hooker & Baker) Holttum

Parathelypteris glanduligera (Kunze) Ching

Parathelypteris hirsutipes (C. B. Clarke) Ching

Parathelypteris japonica (Baker) Ching

Parathelypteris nipponica (Franch. & Savat.) Ching

Parathelypteris serrutula (Ching) Ching

Phegopteris connectilis (Michx.) Watt

Phegopteris decursivepinnata (H. C. Hall) Fée

Phegopteris hexagonoptera (Michx.) Fée

Phegopteris tibetica Ching

Pronephrium lakhimpurense (Rosenst.) Holttum

Pronephrium medogensis Y. X. Lin

Pronephrium menisciicarpon (Blume) Holttum

Pronephrium nudatum (Roxb. ex Griff.) Holtt.

Pronephrium parishii (Bedd.) Holttum

Pronephrium penangianum (Hook.) Holttum

Pronephrium triphyllum (Sw.) Holttum

Pseudocyclosorus canus (Baker) Holttum & Jeff W. Grimes

Pseudocyclosorus ciliatus (Wall. ex Benth.) Ching

Pseudocyclosorus duclouxii (Christ) Ching

Pseudocyclosorus dulongjiangensis W. M. Chu

Pseudocyclosorus esquirolii (Christ) Ching

Pseudocyclosorus falcilobus (Hook.) Ching

Pseudocyclosorus fugongensis Y. X. Lin

Pseudocyclosorus gongshanensis Y. X. Lin

Pseudocyclosorus lushuiensis Y. X. Lin

Pseudocyclosorus ochthodes (Kunze) Holttum

Pseudocyclosorus ornatipes Holttum & Jeff W. Grimes

Pseudocyclosorus pseudorepens Ching ex K. H. Shing

Pseudocyclosorus repens (C. Hope) Ching

Pseudocyclosorus stramineus Ching ex Y. X. Lin

Pseudocyclosorus subfalcilobus Ching ex K. H. Shing

Pseudocyclosorus subochthodes (Ching) Ching

Pseudocyclosorus tuberculifer (C. Chr.) Ching

Pseudocyclosorus tylodes (Kunze) Ching

Pseudocyclosorus zayuensis Ching & S. K. Wu

Pseudophegopteris aurita (Hook.) Ching

Pseudophegopteris brevipes Ching & S. K. Wu

Pseudophegopteris hirtirachis (C. Chr.) Holttum

Pseudophegopteris levingei (C. B. Clarke) Ching

Pseudophegopteris microstegia (Hook.) Ching

Pseudophegopteris pyrrhorhachis (Kunze) Ching

Pseudophegopteris pyrrhorhachis var. **glabrata** (C. B. Clarke) Holttum

Pseudophegopteris rectangularis (Zoll.) Holttum

Pseudophegopteris tibetana Ching & S. K. Wu

Pseudophegopteris yigongensis Ching

Pseudophegopteris zayuensis Ching & S. K. Wu

Stegnogramma aspleniodes (C. Chr.) J. Sm. ex Ching

Stegnogramma cyrtomioides (C. Chr.) Ching

Stegnogramma griffithii (T. Moore) K. Iwats.

Stegnogramma leptogrammoides K. Iwats.

Stegnogramma mollissima (Kunze) Fraser-Jenk.

Stegnogramma pozoi (Lag.) K. Iwats.

Thelypteris birii (R. D. Dixit & Balkr.) Fraser-Jenk.

Thelypteris calcarata (Blume) Ching

Thelypteris confluens (Thunb.) C. V. Morton

Thelypteris gracilis (Heward) Proctor

Thelypteris nudata (Roxb.) C. V. Morton

Thelypteris rubra (Ching) K. Iwats.

Thelypteris thelypterioides (Michx.) Holub

Thelypteris urophylla (Mett.) K. Iwats.

Thelypteris xylodes (Kunze) Ching

Trigonospora caudipinna (Ching) Sledge

Trigonospora loyalii Panigrahi & Sarn. Singh

Trigonospora sericea (J. Scott ex Bedd.) Holttum

Trigonospora subcaudipinna Sarn. Singh & Panigrahi

Trigonospora zeylanica (Ching) Sledge

26. WOODSIACEAE

Woodsia alpina (Bolton) Gray

Woodsia andersonii (Bedd.) Christ

Woodsia cinnamomea Christ

Woodsia cycloloba Hand.-Mazz.

Woodsia elongata Hook.

Woodsia glabella R. Br. ex Richardson

Woodsia hancockii Baker

Woodsia indusiosa Christ

Woodsia kangdingensis H. S. Kung, L. B. Zhang & X. S. Guo

Woodsia lanosa Hook.

Woodsia rosthorniana Diels

27. RHACHIDOSORACEAE

Rhachidosorus pulcher (Tagawa) Ching

28. ONOCLEACEAE

Matteuccia intermedia C. Chr.

Matteuccia orientalis (Hook.) Trevis.

Matteuccia struthiopteris (L.) Tod.

29. BLECHNACEAE

Blechnopsis orientalis (L.) Presl

Woodwardia unigemmata (Makino) Nakai

30. ATHYRIACEAE

Anisocampium cuspidatum (Bedd.) Yea C. Liu, W. L. Chiou & M. Kato

Anisocampium niponicum (Mett.) Yea C. Liu, W. L. Chiou & M. Kato

Athyrium adscendens Ching

Athyrium anisopterum Christ

Athyrium araiostegioides Ching

Athyrium arborescens (Bory) Milde

Athyrium atkinsonii Bedd.

Athyrium attenuatum (Wall. ex C. B. Clarke) Tagawa

Athyrium atuntzeense (Ching) Z. R. Wang & Z. R. He

Athyrium austro-orientale Ching

Athyrium baoxingense Ching

Athyrium biserrulatum Christ

Athyrium bomicola Ching

Athyrium brevisorum (Wall. ex Hook.) T. Moore

Athyrium bucahwangense Ching

Athyrium caudatum Ching

Athyrium caudiforme Ching

Athyrium chingianum Z. R. Wang & X. C. Zhang

Athyrium chungtienense Ching

Athyrium clarkei Bedd.

Athyrium contingens Ching & S. K. Wu

Athyrium costulalisorum Ching

Athyrium crassipes Ching

Athyrium criticum Ching

Athyrium davidii (Franch.) Christ

Athyrium daxianglingense Ching & H. S. Kung

Athyrium decorum Ching

Athyrium delavayi Christ

Athyrium delicatulum Ching & S. K. Wu

Athyrium dentigerum (Wall. ex C. B. Clarke) Mehra & Bir

Athyrium dentilobum Ching & S. K. Wu

Athyrium devolii Ching

Athyrium dissitifolium (Baker) C. Chr.

Athyrium dissitifolium var. **kulhaitense** (W. S. Atkinson ex C. B. Clarke) Ching

Athyrium distans (D. Don) T. Moore

Athyrium drepanopterum (Kunze) A. Braun ex Milde

Athyrium dubium Ching

Athyrium falcatum Bedd.

Athyrium fangii Ching

Athyrium fimbriatum Hook. ex T. Moore

Athyrium flabellulatum (C. B. Clarke) Tardieu

Athyrium foliolosum T. Moore ex R. Sim

Athyrium glandulosum Ching

Athyrium himalaicum Ching ex Mehra & Bir

Athyrium hirtirachis Ching & Y. P. Hsu

Athyrium interjectum Ching

Athyrium iseanum Rosenst.

Athyrium iseanum var. **chuanqianense** Z. R. Wang

Athyrium jinshajiangense Ching & K. H. Shing

Athyrium kumaonicum Punetha

Athyrium lineare Ching

Athyrium ludingense Z. R. Wang & Li Bing Zhang

Athyrium mackinnonii (C. Hope) C. Chr.

Athyrium mackinnonii var. **glabratum** Y. T. Hsieh & Z. R. Wang

Athyrium mackinnonii var. **yigongense** Ching & S. K. Wu

Athyrium mackinnoniorum (C. Hope) C. Chr.

Athyrium medogense X. C. Zhang

Athyrium mehrae Bir

Athyrium mengtzeense Hieron.

Athyrium micropterum Fraser-Jenk.

Athyrium nakanoi Makino

Athyrium nephrodioides (Baker) Christ

Athyrium nigripes (Blume) T. Moore

Athyrium nudifrons Ching

Athyrium nyalamense Y. T. Hsieh & Z. R. Wang

Athyrium nyalamense var. **puberulum** Z. R. Wang

Athyrium obtusilimbum Ching

Athyrium omeiense Ching

Athyrium pectinatum (Wall. ex Mett.) T. Moore

Athyrium puncticaule (Blume) T. Moore

Athyrium repens (Ching) Fraser-Jenk.

Athyrium rhachidosorum (Hand.-Mazz.) Ching

Athyrium roseum var. **fugongense** Z. R. Wang

Athyrium rubricaule (Edgew. ex C. B. Clarke) Bir

Athyrium rupicola (Edgew. ex C. Hope) C. Chr.

Athyrium schimperi Moug. ex Fée

Athyrium schizochlamys (Ching) K. Iwats.

Athyrium setiferum C. Chr.

Athyrium silvicola Tagawa

Athyrium spinulosum (Maxim.) Milde

Athyrium strigillosum (E. J. Lowe) T. Moore ex Salomon

Athyrium subtriangulare (Hook.) Bedd.

Athyrium suprapuberulum Ching

Athyrium suprapubescens Ching

Athyrium supraspinescens C. Chr.

Athyrium tarulakaense Ching

Athyrium tenuifrons (Wall. ex C. Hope) Punetha

Athyrium umbrosum C. Presl

Athyrium uniforme Ching

Athyrium vidalii (Franch. & Sav.) Nakai

Athyrium viviparum Christ

Athyrium wallichianum Ching

Athyrium wardii (Hook.) Makino

Athyrium wardii var. **densipinnum** Z. R. Wang & Li Bing Zhang

Athyrium xichouense Y. T. Hsieh & Z. R. Wang

Athyrium yui Ching

Athyrium zayuense Z. R. Wang

Cornopteris badia Ching

Cornopteris banajaoensis (C. Chr.) K. Iwats. & M. G. Price

Cornopteris decurrenti-alata (Hook.) Nakai

Cornopteris latibasis W. M. Chu

Cornopteris latiloba Ching

Cornopteris omeiensis Ching

Cornopteris opaca (D. Don) Tagawa

Deparia acrostichoides (Sw.) M. Kato

Deparia allantodioides (Bedd.) M. Kato

Deparia auriculata (W. M. Chu & Z. R. Wang) Z. R. Wang

Deparia auriculata var. **zhongdianensis** (Z. R. Wang) Z. R. Wang

Deparia boryana (Willd.) M. Kato

Deparia brevipinna (Ching & K. H. Shing ex Z. R. Wang) Z. R. Wang

Deparia dolosa (Christ) M. Kato

Deparia henryi (Baker) M. Kato

Deparia heterophlebia (Mett. ex Baker) R. Sano

Deparia hirtirachis (Ching ex Z. R. Wang) Z. R. Wang

Deparia japonica (Thunb.) M. Kato

Deparia jiulungensis (Ching) Z. R. Wang

Deparia lancea (Thunb. ex Murray) Fraser-Jenkins

Deparia liangshanensis (Ching ex Z. R. Wang) Z. R. Wang

Deparia liangshanensis var. **sericea** (Ching & Z. R. Wang) Z. R. Wang

Deparia longipes (Ching) Shinohara

Deparia ludingensis (Z. R. Wang & Li Bing Zhang) Z. R. Wang

Deparia macdonellii (Bedd.) M. Kato

Deparia medogensis (Ching & S. K. Wu) Z. R. Wang

Deparia medogensis var. **glandulifera** (W. M. Chu) Z. R. Wang

Deparia petersenii (Kunze) M. Kato

Deparia sichuanensis (Z. R. Wang) Z. R. Wang

Deparia sichuanensis var. **gongshanensis** (Z. R. Wang) Z. R. Wang

Deparia subsimilis (Christ) Fraser-Jenk.

Deparia truncata (Ching ex Z. R. Wang) Z. R. Wang

Deparia wilsonii (Christ) X. C. Zhang

Deparia wilsonii var. **habaensis** (Ching & Z. R. Wang) Z. R. Wang

Deparia wilsonii var. **incisoserrata** (Ching & Z. R. Wang) Z. R. Wang

Deparia wilsonii var. **maxima** (Ching & Z. R. Wang) Z. R. Wang

Deparia wilsonii var. **muliensis** (Z. R. Wang) Z. R. Wang

Diplazium australe (R. Br.) N. A. Wakef.

Diplazium axillare Ching

Diplazium bellum (C. B. Clarke) Bir

Diplazium dilatatum Blume

Diplazium doederleinii (Luerss.) Makino

Diplazium donianum (Mett.) Tardieu

Diplazium dulongjiangense (W. M. Chu) Z. R. He

Diplazium esculentum (Retz.) Sw.

Diplazium forrestii (Ching ex Z. R. Wang) Fraser-Jenk.

Diplazium giganteum (Baker) Ching

Diplazium glingense (Ching & Y. X. Lin) Z. R. He

Diplazium heterophlebium (Mett. ex Baker) Diels

Diplazium himalayense (Ching) Panigrahi

Diplazium hirsutipes (Bedd.) B. K. Nayar & S. Kaur

Diplazium kawakamii Hayata

Diplazium latifolium (Don) T. Moore

Diplazium latipinnulum (Ching & W. M. Chu) Z. R. He

Diplazium laxifrons Rosenst.

Diplazium leptophyllum (Baker) Christ

Diplazium lobulosum (Wall. ex Mett.) C. Presl

Diplazium maximum (D. Don) C. Chr.

Diplazium medogense (Ching & S. K. Wu) Fraser-Jenk.

Diplazium multicaudatum (Wall. ex C. B. Clarke) Z. R. He

Diplazium muricatum (Mett.) Alderw.

Diplazium procerum (Sod) C. Chr.

Diplazium proliferum (Lam.) Thouars

Diplazium sibiricum (Turcz. ex Kunze) Sa. Kurata

Diplazium sibiricum var. **glabrum** (Tagawa) Sa. Kurata

Diplazium sikkimense (C. B. Clarke) C. Chr.

Diplazium spectabile (Wall. ex Mett.) Ching

Diplazium squamigerum (Mett.) Christ

Diplazium stenochlamys C. Chr.

Diplazium stoliczkae Bedd.

Diplazium subsinuatum (Wall. ex Hook. & Grev.) Tagawa

Diplazium subspectabile (Ching & W. M. Chu) Z. R. He

Diplazium succulentum (C. B. Clarke) C. Chr.

Diplazium tibeticum (Ching & S. K. Wu) Z. R. He

Diplazium torrentium (C. B. Clarke) Tardieu

31. HYPODEMATIACEAE

Hypodematium crenatum (Forssk.) Kuhn & Decken

Hypodematium daochengense K. H. Shing

Hypodematium hirsutum (D. Don) Ching

Leucostegia truncata (D. Don) Fraser-Jenk.

32. DRYOPTERIDACEAE

Arachniodes amabilis (Blume) Tind.

Arachniodes aristata (G. Forst.) Tind.

Arachniodes assamica (Kuhn) Ohwi

Arachniodes nipponica (Rosenst.) Ohwi

Arachniodes simplicior (Makino) Ohwi

Arachniodes simulans (Ching) Ching

Arachniodes speciosa (D. Don) Ching

Arachniodes superba Fraser-Jenk.

Bolbitis appendiculata (Willd.) K. Iwats.

Bolbitis contaminans (Wall.) Ching

Bolbitis costata (C. Presl) Ching

Bolbitis crispatula (C. B. Clarke) Ching

Bolbitis deltigera (Bedd.) C. Chr.

Bolbitis heteroclita (C. Presl) Ching

Bolbitis major (Bedd.) Hennipman

Bolbitis medogensis (Ching & S. K. Wu) S. Y. Dong

Bolbitis sinensis (Baker) K. Iwats.

Bolbitis tibetica Ching & S. K. Wu

Bolbitis virens (Wall. ex Hook. & Grev.) Schott

Ctenitis holttumii Sarn. Singh & Panigrahi

Ctenitis subglandulosa (Hance) Ching

Cyrtomium anomophyllum (Zenker) Fraser-Jenk.

Cyrtomium falcatum (L. f.) C. Presl

Cyrtomium fortunei J. Sm.

Cyrtomium hookerianum C. Chr.

Cyrtomium lonchitoides (Christ) Christ

Cyrtomium macrophyllum (Makino) Tagawa

Cyrtomium omeiense Ching & K. H. Shing

Cyrtomium tsinglingense Ching & K. H. Shing

Cyrtomium tukusicola Tagawa

Cyrtomium urophyllum Ching

Cyrtomium yamamotoi Tagawa

Dryopteris acutodentata Ching

Dryopteris alpestris Tagawa

Dryopteris alpicola Ching & Z. R. Wang

Dryopteris angustifrons (Hook.) Kuntze

Dryopteris annamensis (Tagawa) Li Bing Zhang

Dryopteris apiciflora (Wall. ex Mett.) Kuntze

Dryopteris assamensis (C. Hope) C. Chr. & Ching

Dryopteris atrata (Wall. ex Kunze) Ching

Dryopteris barbigera (T. Moore ex Hook.) Kuntze

Dryopteris basisora Christ

Dryopteris blanfordii (C. Hope) C. Chr.

Dryopteris blanfordii subsp. **nigrosquamosa** (Ching) Fraser-Jenk.

Dryopteris caroli-hopei Fraser-Jenk.

Dryopteris chrysocoma (Christ) C. Chr.

Dryopteris chrysocoma var. **squamosa** (C. Chr.) Ching

Dryopteris clarkei (Baker) Kuntze

Dryopteris cochleata (Buch.-Ham. ex D. Don) C. Chr.

Dryopteris conjugata Ching

Dryopteris costalisora Tagawa

Dryopteris cycadina (Franch. & Sav.) C. Chr.

Dryopteris diacalpe Li Bing Zhang

Dryopteris dickinsii (Franch. & Sav.) C. Chr.

Dryopteris diffracta (Baker) C. Chr.

Dryopteris dulongensis (S. K. Wu & X. Cheng) Li Bing Zhang

Dryopteris edwardsii Fraser-Jenk.

Dryopteris exstipellata (Ching & S. H. Wu) Li Bing Zhang

Dryopteris fibrillosissima Ching

Dryopteris filix-mas (L.) Schott

Dryopteris fructuosa (Christ) C. Chr.

Dryopteris gonggaensis H. S. Kung, Li Bing Zhang & X. S. Guo

Dryopteris gracillima Ching

Dryopteris grandifrons Li Bing Zhang

Dryopteris habaensis Ching

Dryopteris hasseltii (Blume) C. Chr.

Dryopteris hendersonii (Bedd.) C. Chr.

Dryopteris heterolaena C. Chr.

Dryopteris himachalensis Fraser-Jenk.

Dryopteris hirtipes (Blume) Kuntze

Dryopteris hookeriana (T. Moore) Li Bing Zhang

Dryopteris immixta Ching

Dryopteris juxtaposita Christ

Dryopteris kawakamii Hayata

Dryopteris komarovii Kossinsky

Dryopteris lachoongensis (Bedd.) B. K. Nayar & S. Kaur

Dryopteris late-repens (Trotter & C. Hope) C. Chr.

Dryopteris latibasis Ching

Dryopteris lepidopoda Hayata

Dryopteris lunanensis (Christ) C. Chr.

Dryopteris marginata (C. B. Clarke) Christ

Dryopteris medogensis (Ching & S. K. Wu) Li Bing Zhang

Dryopteris montigena Ching

Dryopteris neorosthornii Ching

Dryopteris nigropaleacea (Fraser-Jenk.) Fraser-Jenk.

Dryopteris nobilis Ching

Dryopteris nobilis var. **fengiana** Ching

Dryopteris nyingchiensis Ching

Dryopteris paleolata (Pichi Sermolli) Li Bing Zhang

Dryopteris panda (C. B. Clarke) C. Chr.

Dryopteris papilio (C. Hope) C. Chr.

Dryopteris pauliae Fraser-Jenk., Widén & Gibby

Dryopteris peranema Li Bing Zhang

Dryopteris porosa Ching

Dryopteris pseudocaenopteris (Kunze) Li Bing Zhang

Dryopteris pteridoformis Christ

Dryopteris pulcherrima Ching

Dryopteris pulvinulifera (Bedd.) Kuntze

Dryopteris pycnopteroides (Christ) C. Chr.

Dryopteris ramosa (C. Hope) C. Chr.

Dryopteris redactopinnata Soumen K. Basu & Panigrahi

Dryopteris rosthornii (Diels) C. Chr.

Dryopteris rubrobrunnea W. M. Chu

Dryopteris scottii (Bedd.) Ching ex C. Chr.

Dryopteris serratodentata (Bedd.) Hayata

Dryopteris setosa (Thunb.) Akas.

Dryopteris sikkimensis (Bedd.) Kuntze

Dryopteris sinofibrillosa Ching

Dryopteris sparsa (D. Don) Kuntze

Dryopteris sphaeropteroides (Baker) C. Chr.

Dryopteris splendens (Hook.) Kuntze

Dryopteris squamifera Ching & S. K. Wu

Dryopteris squamiseta (Hook.) Kuntze

Dryopteris stenolepis (Baker) C. Chr.

Dryopteris stewartii Fraser-Jenk.

Dryopteris sublacera C. Chr.

Dryopteris subpycnopteroides Ching ex Fraser-Jenk.

Dryopteris subtriangularis (C. Hope) C. Chr.

Dryopteris thibetica (Franch.) C. Chr.

Dryopteris tingiensis Ching & S. K. Wu ex Fraser-Jenk.

Dryopteris transmorrisonense (Hayata) Hayata

Dryopteris wallichiana (Spreng.) Hyl.

Dryopteris woodsiisora Hayata

Dryopteris xanthomelas (Christ) C. Chr.

Dryopteris yigongensis Ching

Dryopteris yungtzeensis Ching

Elaphoglossum ceylanicum Krajina ex Sledge

Elaphoglossum marginatum T. Moore

Elaphoglossum stelligerum Elaphoglossum stelligerum (Wall. ex Baker) T. Moore ex Alston & Bonner

Elaphoglossum thomsonii S. R. Ghosh & A. Biswas

Elaphoglossum yunnanense (Baker) C. Chr.

Lomagramma medogensis Ching & Y. X. Ling

Polystichum acanthophyllum (Franch.) Christ

Polystichum acutidens Christ

Polystichum adungense Ching & Fraser-Jenk. ex H. S. Kung & Li Bing Zhang

Polystichum annapurnicola Fraser-Jenk. ex Thapa

Polystichum atkinsonii Bedd.

Polystichum bakerianum (Atkinson ex C. B. Clarke) Diels

Polystichum baoxingense Ching & H. S. Kung

Polystichum bigemmatum Ching ex L. L. Xiang

Polystichum bomiense Ching & S. K. Wu

Polystichum braunii (Spenn.) Fée

Polystichum capillipes (Baker) Diels

Polystichum caruifolium (Baker) Diels

Polystichum castaneum (C. B. Clarke) B. K. Nayar & S. Kaur

Polystichum centronepalense Fraser-Jenk.

Polystichum costularisorum Ching ex W. M. Chu & Z. R. He

Polystichum cyclolobum C. Chr.

Polystichum deflexum Ching ex W. M. Chu

Polystichum delavayi (Christ) Ching ex Li Bing Zhang & H. S. Kung

Polystichum deltodon (Baker) Diels

Polystichum diffundens H. S. Kung & Li Bing Zhang

Polystichum discretum (D. Don) J. Sm.

Polystichum duthiei (C. Hope) C. Chr.

Polystichum elevatovenusum Ching ex W. M. Chu & Z. R. He

Polystichum exauriforme H. S. Kung & Li Bing Zhang

Polystichum frigidicola H. S. Kung & Li Bing Zhang

Polystichum fugongense Ching & W. M. Chu ex H. S. Kung & Li Bing Zhang

Polystichum garhwalicum N. C. Nair & Nag

Polystichum glaciale Christ

Polystichum gongboense Ching & S. K. Wu

Polystichum habaense Ching & H. S. Kung

Polystichum harpophyllum (Zenker ex Kunze) Sledge

Polystichum hecatopterum Diels

Polystichum hookerianum (C. Presl) C. Chr.

Polystichum huae H. S. Kung & Li Bing Zhang

Polystichum incisopinnulum H. S. Kung & Li Bing Zhang

Polystichum integrilimbum Ching & H. S. Kung

Polystichum kangdingense H. S. Kung & Li Bing Zhang ex Li Bing Zhang

Polystichum lachenense (Hook.) Bedd.

Polystichum langchungense Ching ex H. S. Kung

Polystichum lentum (D. Don) T. Moore

Polystichum lonchitis (L.) Roth

Polystichum longipaleatum Christ

Polystichum longipinnulum N. C. Nair

Polystichum luctuosum (Kunze) T. Moore

Polystichum makinoi (Tagawa) Tagawa

Polystichum manmeiense (Christ) Nakaike

Polystichum marginatum Schott

Polystichum mayebarae Tagawa

Polystichum mehrae Fraser-Jenk. & Khullar

Polystichum meiguense Ching & H. S. Kung

Polystichum melanostipes Ching & H. S. Kung

Polystichum mollissimum Ching

Polystichum moupinense (Franch.) Bedd.

Polystichum mucronifolium (Blume) C. Presl

Polystichum muscicola Ching ex W. M. Chu & Z. R. He

Polystichum neolobatum Nakai

Polystichum nepalense (Spreng.) C. Chr.

Polystichum nigrum Ching & H. S. Kung

Polystichum nudisorum Ching

Polystichum obliquum (D. Don) T. Moore

Polystichum oreodoxa Ching ex H. S. Kung & Li Bing Zhang

Polystichum orientalitibeticum Ching

Polystichum otophorum (Franch.) Bedd.

Polystichum paramoupinense Ching

Polystichum pianmaense W. M. Chu

Polystichum piceopaleaceum Tagawa

Polystichum prescottianum (Wall. ex Mett.) T. Moore

Polystichum pseudoacutidens Ching ex W. M. Chu & Z. R. He

Polystichum pseudocastaneum Ching & S. K. Wu

Polystichum punctiferum C. Chr.

Polystichum qamdoense Ching & S. K. Wu

Polystichum rhombiforme Ching & S. K. Wu

Polystichum rufopaleaceum Ching ex Li Bing Zhang & H. S. Kung

Polystichum rupicola Ching ex W. M. Chu

Polystichum salwinense Ching & H. S. Kung

Polystichum semifertile (C. B. Clarke) Ching

Polystichum setillosum Ching

Polystichum shensiense Christ

Polystichum sinense Christ

Polystichum squarrosum (D. Don) Fée

Polystichum stenophyllum (Franch.) Christ

Polystichum stimulans (Kunze ex Mett.) Bedd.

Polystichum submarginale (Baker) Ching ex P. S. Wang

Polystichum submite (Christ) Diels

Polystichum tangmaiense H. S. Kung & Tateishi

Polystichum tenuius (Ching) Li Bing Zhang

Polystichum thomsonii (Hook. f.) Bedd.

Polystichum tibeticum Ching

Polystichum tsus-simense (Hook.) J. Sm.

Polystichum wattii (Bedd.) C. Chr.

Polystichum yadongense Ching & S. K. Wu

Polystichum yigongense Ching & S. K. Wu

Polystichum yuanum Ching

Polystichum yunnanense Christ

Polystichum zayuense W. M. Chu & Z. R. He

33. LOMARIOPSIDACEAE

Lomariopsis lineata (C. Presl) Holttum

34. NEPHROLEPIDACEAE

Nephrolepis biserrata (Sw.) Schott

Nephrolepis cordifolia (L.) C. Presl

Nephrolepis delicatula (Decne.) Pic. Serm.

Nephrolepis duffii T. Moore

Nephrolepis falcata (Cav.) C. Chr.

Nephrolepis hirsutula (G. Forst.) C. Presl

35. TECTARIACEAE

Pleocnemia submembranacea (Hayata) Tagawa & K. Iwats.

Tectaria cicutaria (L.) Copel.

Tectaria coadunata (J. Sm.) C. Chr.

Tectaria decurrens (C. Presl) Copel.

Tectaria dubia (C. B. Clarke & Baker) Ching

Tectaria fuscipes (Wall. ex Bedd.) C. Chr.

Tectaria gemmifera (Fée) Alston

Tectaria griffithii (Baker) C. Chr.

Tectaria heterocarpa (Bedd.) C. V. Morton

Tectaria impressa (Fée) Holttum

Tectaria ingens (Atkinson ex C. B. Clarke) Holttum

Tectaria khonsaensis Sarn. Singh & Panigrahi

Tectaria mehrae Panigrahi & Sarn. Singh

Tectaria pandurifolia (C. Chr.) C. Chr.

Tectaria polymorpha (Wall. ex Hook.) Copel.

Tectaria rockii Christ

Tectaria setulosa (Baker) Holttum

Tectaria simonsii (Baker) Ching

Tectaria vasta (Blume) Copel.

36. OLEANDRACEAE

Oleandra neriiformis Cav.

Oleandra pistillaris (Sw.) C. Chr.

Oleandra wallichii (Hook.) C. Presl

37. DAVALLIACEAE

Davallia assamica (Bedd.) Baker

Davallia clarkei Baker

Davallia divaricata Blume

Davallia griffithiana Hook.

Davallia mariesii T. Moore ex Baker

Davallia membranulosa Wall. ex Hook.

Davallia multidentata Hook.

Davallia perdurans Christ

Davallia pulchra D. Don

Davallia repens (L. f.) Kuhn

Davallia solida (G. Forst.) Sw.

Davallia trichomanoides Blume

38. POLYPODIACEAE

Aglaomorpha coronans (Wall. ex Mett.) Copeland

Arthromeris caudata Ching & Y. X. Lin

Arthromeris elegans Ching

Arthromeris himalayensis (Hook.) Ching

Arthromeris himalayensis var. **niphoboloides** (C. B. Clarke) S. G. Lu

Arthromeris intermedia Ching

Arthromeris lehmannii (Mett.) Ching

Arthromeris lungtauensis Ching

Arthromeris mairei (Brause) Ching

Arthromeris medogensis Ching & Y. X. Lin

Arthromeris nigropaleacea S. G. Lu

Arthromeris notholaenoides V. K. Rawat & Fraser-Jenk.

Arthromeris salicifolia Ching & Y. X. Lin

Arthromeris tatsienensis (Franch. & Bur.) Ching

Arthromeris tenuicauda (Hook.) Ching

Arthromeris tomentosa W. M. Chu

Arthromeris wallichiana (Spreng.) Ching

Arthromeris wardii (C. B. Clarke) Ching

Christopteris tricuspis (Hook.) Christ

Drynaria baronii Diels

Drynaria delavayi Christ

Drynaria laurentii (Christ) Hieron.

Drynaria mollis Bedd.

Drynaria propinqua (Wall. ex Mett.) J. Sm.

Drynaria quercifolia (L.) J. Sm.

Gymnogrammitis dareiformis (Hook.) Ching ex Tardieu & C. Chr.

Himalayopteris erythrocarpa (Mett. ex Kuhn) W. Shao & S. G. Lu

Lemmaphyllum carnosum (Wall. ex J. Sm.) C. Presl

Lemmaphyllum microphyllum C. Presl

Lemmaphyllum rostratum (Bedd.) Tagawa

Lepisorus albertii (Regel) Ching

Lepisorus amaurolepidus (Sledge) Bir & Trikha

Lepisorus angustus Ching

Lepisorus asterolepis (Baker) Ching ex S. X. Xu

Lepisorus bicolor (Takeda) Ching

Lepisorus cespitosus Y. X. Lin

Lepisorus clathratus (C. B. Clarke) Ching

Lepisorus confluens W. M. Chu

Lepisorus contortus (Christ) Ching

Lepisorus eilophyllus (Diels) Ching

Lepisorus elegans Ching & W. M. Chu

Lepisorus henryi (Hieron. ex C. Chr.) Li Wang

Lepisorus heterolepis (Rosenst.) Ching

Lepisorus kuchenensis (Y. C. Wu) Ching

Lepisorus likiangensis Ching & S. K. Wu

Lepisorus lineariformis Ching & S. K. Wu

Lepisorus loriformis (Wall. ex Mett.) Ching

Lepisorus macrosphaerus (Baker) Ching

Lepisorus marginatus Ching

Lepisorus medogensis Ching & Y. X. Lin

Lepisorus mehrae Fraser-Jenk.

Lepisorus miyoshianus (Makino) Fraser-Jenk. & Subh. Chandra

Lepisorus morrisonensis (Hayata) H. Itô

Lepisorus mucronatus (Fée) Li Wang

Lepisorus nudus (Hook.) Ching

Lepisorus oligolepidus (Baker) Ching

Lepisorus pseudonudus Ching

Lepisorus scolopendrium (Buch.-Ham. ex D. Don) Mehra & Bir

Lepisorus sinensis (Christ) Ching

Lepisorus sordidus (C. Chr.) Ching

Lepisorus stenistis (C. B. Clarke) Y. X. Lin

Lepisorus subconfluens Ching

Lepisorus sublinearis (Baker ex Takeda) Ching

Lepisorus suboligolepidus Ching

Lepisorus thunbergianus (Kaulfuss) Ching

Lepisorus tibeticus Ching & S. K. Wu

Lepisorus tosaensis (Makino) H. Itô

Lepisorus tricholepis K. H. Shing & Y. X. Lin

Lepisorus variabilis Ching & S. K. Wu

Lepisorus waltonii (Ching) S. L. Yu

Leptochilus axillaris (Cav.) Kaulf.

Leptochilus decurrens Blume

Leptochilus ellipticus (Thunb.) Nooteb.

Leptochilus ellipticus var. **flexilobus** (Christ) X. C. Zhang

Leptochilus ellipticus var. **pentaphyllus** (Baker) X. C. Zhang & Nooteboom

Leptochilus ellipticus var. **pothifolius** (D. Don) X. C. Zhang

Leptochilus hemionitideus (C. Presl) Nooteb.

Leptochilus insignis (Blume) Fraser-Jenk.

Leptochilus pedunculatus (Hook. & Grev.) Fraser-Jenk.

Leptochilus pteropus (Blume) Fraser-Jenk.

Loxogramme avenia (Blume) C. Presl

Loxogramme carinata M. G. Price

Loxogramme chinensis Ching

Loxogramme cuspidata (Zenker) M. G. Price

Loxogramme duclouxii Christ

Loxogramme grammitoides (Baker) C. Chr.

Loxogramme involuta (D. Don) C. Presl

Loxogramme lankokiensis (Rosenst.) C. Chr.

Loxogramme porcata M. G. Price

Micropolypodium sikkimense (Hieron.) X. C. Zhang

Microsorum membranaceum (D. Don) Ching

Microsorum punctatum (L.) Copel.

Microsorum zippelii (Bl.) Ching

Neocheiropteris ovata (Wall. ex Hook. & Grev.) Fraser-Jenk.

Phanerophlebia juglandifolia (Humb. & Bonpl. ex Willd.) J. Sm.

Phymatosorus cuspidatus (D. Don) Pic. Serm.

Pleopeltis macrocarpa (Bory ex Willd.) Kaulf.

Pleurosoriopsis makinoi (Maxim. ex Makino) Fomin

Polypodiastrum argutum (Wall. ex Hook.) Ching

Polypodiastrum dielseanum (C. Chr.) Ching

Polypodiastrum mengtzeense (Christ) Ching

Polypodiodes amoena (Wall. ex Mett.) Ching

Polypodiodes amoena var. **duclouxii** (Christ) Ching ex S. G. Lu

Polypodiodes amoena var. **pilosa** (C. B. Clarke & Baker) S. R. Ghosh

Polypodiodes hendersonii (Bedd.) Fraser-Jenk.

Polypodiodes lachnopus (Wall. ex Hook.) Ching

Polypodiodes microrhizoma (C. B. Clarke ex Baker) Ching

Polypodiodes niponica (Mett.) Ching

Polypodiodes paramoena Ching & T. X. Lin

Polypodiodes pseudolachnopus S. G. Lu

Polypodiodes subamoena (C. B. Clarke) Ching

Polypodiodes wattii (Bedd.) Ching

Pyrrosia adnascens (Sw.) Ching

Pyrrosia boothii (Hook.) Ching

Pyrrosia costata (Wall. ex C. Presl) Tagawa & K. Iwats.

Pyrrosia davidii (Giesenh. ex Diels) Ching

Pyrrosia drakeana (Franch.) Ching

Pyrrosia ensata Ching ex K. H. Shing

Pyrrosia flocculosa (D. Don) Ching

Pyrrosia heteractis (Mett. ex Kuhn) Ching

Pyrrosia lanceolata (Wall.) Farw.

Pyrrosia lingua (Thunb.) Farw.

Pyrrosia longifolia (Burm. f.) C. V. Morton

Pyrrosia mannii (Giesenh.) Ching

Pyrrosia nuda (Giesenh.) Ching

Pyrrosia nummulariifolia (Sw.) Ching

Pyrrosia obovata (Blume) Ching

Pyrrosia pannosa (Mett. ex Kuhn) Ching

Pyrrosia petiolosa (Christ) Ching

Pyrrosia piloselloides (L.) M. G. Price

Pyrrosia porosa (C. Presl) Hovenkamp

Pyrrosia stenophylla (Bedd.) Ching

Pyrrosia subfurfuracea (Hook.) Ching

Selliguea albidoglauca (C. Chr.) S. G. Lu, Hovenkamp & M. G. Gilbert

Selliguea chenopus (Christ) S. G. Lu, Hovenkamp & M. G. Gilbert

Selliguea chrysotricha (C. Chr.) Fraser-Jenk.

Selliguea conjuncta (Ching) S. G. Lu, Hovenkamp & M. G. Gilbert

Selliguea conmixta (Ching) S. G. Lu, Hovenkamp & M. G. Gilbert

Selliguea connexa (Ching) S. G. Lu, Hovenkamp & M. G. Gilbert

Selliguea crenatopinnata (C. B. Clarke) S. G. Lu, Hovenkamp & M. G. Gilbert

Selliguea ebenipes (Hook.) S. Linds.

Selliguea glaucopsis (Franch.) S. G. Lu, Hovenkamp & M. G. Gilbert

Selliguea griffithiana (Hook.) Fraser-Jenk.

Selliguea hastata (Thunb.) Fraser-Jenk.

Selliguea incisocrenata (Ching ex W. M. Chu & S. G. Lu) S. G. Lu, Hovenkamp & M. G. Gilbert

Selliguea likiangensis (Ching) S. G. Lu, Hovenkamp & M. G. Gilbert

Selliguea malacodon (Hooker) S. G. Lu, Hovenkamp & M. G. Gilbert

Selliguea nigropaleacea (Ching) S. G. Lu, Hovenkamp & M. G. Gilbert

Selliguea obtusa (Ching) S. G. Lu, Hovenkamp & M. G. Gilbert

Selliguea oxyloba (Wall. ex Kunze) Fraser-Jenk.

Selliguea pianmaensis (W. M. Chu) S. G. Lu, Hovenkamp & M. G. Gilbert

Selliguea quasidivaricata (Hayata) H. Ohashi & K. Ohashi

Selliguea rhynchophylla (Hook.) Fraser-Jenk.

Selliguea roseomarginata (Ching) S. G. Lu, Hovenkamp & M. G. Gilbert

Selliguea senanensis (Maxim.) S. G. Lu, Hovenkamp & M. G. Gilbert

Selliguea stewartii (Bedd.) S. G. Lu, Hovenkamp & M. G. Gilbert

Selliguea stracheyi (Ching) S. G. Lu, Hovenkamp & M. G. Gilbert

Selliguea tibetana (Ching & S. K. Wu) S. G. Lu, Hovenkamp & M. G. Gilbert

Selliguea trisecta (Baker) Fraser-Jenk.

Thylacopteris papillosa (Blume) J. Sm.

Tomophyllum donianum (Spreng.) Fraser-Jenk. & Parris

Tomophyllum subfalcatum (Blume) Parris

Tricholepidium normale (D. Don) Ching

Tricholepidium subhemionitideum (Christ) Fraser-Jenk.

Tricholepidium superficiale (Blume) Fraser-Jenk.

GYMNOSPERMA

39. CYCADACEAE

Cycas panzhihuaensis L. Zhou & S. Y. Yang

Cycas pectinata Buch.-Ham.

40. GINKGOACEAE

Ginkgo biloba L.

41. GNETACEAE

Gnetum montanum Markgraf

Gnetum pendulum C. Y. Cheng

42. EPHEDRACEAE

Ephedra dawuensis Y. Yang

Ephedra gerardiana Wall. ex C. A. Mey.

Ephedra gerardiana var. **sikkimensis** Wall. ex Stapf

Ephedra intermedia Schrenk ex C. A. Mey.

Ephedra intermedia var. **tibetica** Schrenk & Meyer

Ephedra likiangensis Florin

Ephedra minuta Florin

Ephedra monosperma Gmelin ex C. A. Mey.

Ephedra oxyphylla Riedl

Ephedra pachyclada Boiss.

Ephedra pangiensis Rita Singh & P. Sharma

Ephedra procera C. A. Mey.

Ephedra regeliana Florin

Ephedra rituensis Y. Yang, D. Z. Fu & G. H. Zhu

Ephedra sarcocarpa Aitch. & Hemsl.

Ephedra saxatilis (Stapf) Royle ex Florin

43. PINACEAE

Abies chayuensis W. C. Cheng & L. K. Fu

Abies chensiensis Tiegh.

Abies delavayi Franch.

Abies delavayi var. **motuoensis** W. C. Cheng & L. K. Fu

Abies densa Griff.

Abies ernestii Rehder

Abies ernestii var. **salouenensis** (Borderes-Rey & Gaussen) W. C. Cheng & L. K. Fu

Abies fabri (Mast.) Craib

Abies fargesii Franch.

Abies fargesii var. **faxoniana** (Rehder & E. H. Wilson) Tang S. Liu

Abies ferreana Bordères & Gaussen

Abies ferreana var. **longibracteata** L. K. Fu & Nan Li

Abies forrestii Coltm.-Rog.

Abies georgei Orr

Abies georgei var. **smithii** (Viguie & Gaussen) W. C. Cheng & L. K. Fu

Abies nukiangensis W. C. Cheng & L. K. Fu

Abies pindrow (D. Don) Royle

Abies recurvata Masters

Abies spectabilis (D. Don) Spach

Abies squamata Masters

Cedrus deodara (Roxb.) G. Don

Keteleeria evelyniana Masters

Larix griffithii Hook. f.

Larix himalaica W. C. Cheng & L. K. Fu

Larix kongboensis R. R. Mill

Larix mastersiana Rehder & E. H. Wilson

Larix potaninii Batalin

Larix potaninii var. **australis** A. Henry ex Hand.-Mazz.

Larix speciosa W. C. Cheng & Y. W. Law

Picea asperata Masters

Picea asperata var. **aurantiaca** (Masters) Boom

Picea aurantiaca Mast.

Picea brachytyla (Franch.) E. Pritz.

Picea brachytyla var. **complanata** (Masters) W. C. Cheng ex Rehder

Picea farreri C. N. Page & Rushforth

Picea likiangensis (Franch.) E. Pritz.

Picea likiangensis var. **hirtella** (Rehder & E. H. Wilson) W. C. Cheng

Picea likiangensis var. **linzhiensis** W. C. Cheng & L. K. Fu

Picea likiangensis var. **montigena** (Masters) W. C. Cheng

Picea likiangensis var. **rubescens** Rehder & E. H. Wilson

Picea purpurea Masters

Picea smithiana (Wall.) Boiss.

Picea smithiana var. **nepalensis** Franco

Picea spinulosa (Griff.) A. Henry

Picea wilsonii Masters

Pinus armandii Franch.

Pinus bhutanica Grierson, D. G. Long & C. N. Page

Pinus densata Masters

Pinus densata subsp. **tibetica** Businský

Pinus gerardiana Wall. ex D. Don

Pinus kesiya Royle ex Gordon

Pinus merkusii Jungh. & de Vriese

Pinus roxburghii Sarg.

Pinus wallichiana A. B. Jacks.

Pinus yunnanensis Franch.

Pinus yunnanensis var. **pygmaea** (Hsüeh) Hsüeh

Pseudotsuga forrestii Craib

Tsuga chinensis (Franch.) E. Pritzel

Tsuga chinensis var. **forrestii** (Downie) Silba

Tsuga chinensis var. **robusta** W. C. Cheng & L. K. Fu

Tsuga dumosa (D. Don) Eichl.

44. PODOCARPACEAE

Podocarpus forrestii Craib & W. W. Sm.

Podocarpus neriifolius D. Don

45. CUPRESSACEAE

Calocedrus macrolepis Kurz

Cryptomeria japonica (Thunb. ex L. f.) D. Don

Cunninghamia lanceolata (Lamb.) Hook.

Cupressus cashmeriana Royle ex Carrière

Cupressus chengiana S. Y. Hu

Cupressus corneyana Knight & Perry ex Carrière

Cupressus duclouxiana Hickel

Cupressus funebris Endl.

Cupressus gigantea W. C. Cheng & L. K. Fu

Cupressus sempervirens L.

Cupressus torulosa D. Don

Juniperus baimashanensis Y. F. Yu & L. K. Fu

Juniperus centrasiatica Kom.

Juniperus chengii L. K. Fu & Y. F. Yu

Juniperus chinensis L.

Juniperus communis var. **saxatilis** Pallas

Juniperus convallium Rehder & E. H. Wilson

Juniperus convallium var. **microsperma** (W. C. Cheng & L. K. Fu) Silba

Juniperus excelsa subsp. **polycarpos** (K. Koch) Takhtajan

Juniperus formosana Hayata

Juniperus indica Bertoloni

Juniperus komarovii Florin

Juniperus pingii W. C. Cheng ex Ferre

Juniperus pingii var. **carinata** Y. F. Yu & L. K. Fu

Juniperus pingii var. **wilsonii** (Rehder) Silba

Juniperus przewalskii Kom.

Juniperus pseudosabina Fisch. & C. A. Mey.

Juniperus pseudosabina var. **turkestanica** (Kom.) Silba

Juniperus recurva Buch.-Ham. ex D. Don

Juniperus recurva var. **coxii** (A. B. Jackson) Melville

Juniperus sabina var. **erectopatens** (W. C. Cheng & L. K. Fu) Y. F. Yu & L. K. Fu

Juniperus saltuaria Rehder & E. H. Wilson

Juniperus semiglobosa Regel

Juniperus sibirica Burgsd.

Juniperus squamata Buch.-Ham. ex D. Don

Juniperus squamata var. **fargesii** Rehder & E. H. Wilson

Juniperus squamata var. **hongxiensis** Y. F. Yu & L. K. Fu

Juniperus squamata var. **parviflora** Y. F. Yu & L. K. Fu

Juniperus tibetica Kom.

Juniperus turkestanica Kom.

Platycladus orientalis (L.) Franco

Taiwania cryptomerioides Hayata

46. TAXACEAE

Amentotaxus argotaenia (Hance) Pilg.

Amentotaxus assamica D. K. Ferguson

Cephalotaxus fortunei Hook.

Cephalotaxus fortunei var. **alpina** H. L. Li

Cephalotaxus lanceolata K. M. Feng ex W. C. Cheng, L. K. Fu & C. Y. Cheng

Cephalotaxus mannii Hook. f.

Cephalotaxus sinensis (Rehder & E. H. Wilson) H. L. Li

Taxus fuana Nan Li & R. R. Mill

Taxus wallichiana Zucc.

Taxus wallichiana var. **mairei** (Lemée & H. Lév.) L. K. Fu & Nan Li

Torreya fargesii var. **yunnanensis** (W. C. Cheng & L. K. Fu) N. Kang

Torreya parvifolia T. P. Yi, Lin Yang & T. L. Long

ANGIOSPERMA

47. NYMPHAEACEAE

Euryale ferox Salisb.

Nymphaea alba L.

Nymphaea candida C. Presl

Nymphaea lotus L.

Nymphaea nouchali Burm. f.

Nymphaea nouchali var. **pubescens** (Willd.) Hook. f. & Thoms.

Nymphaea nouchali var. **versicolor** (Sims) R. Ansari & Jeeja

Nymphaea tetragona Georgi

48. SCHISANDRACEAE

Illicium burmanicum E. H. Wilson

Illicium cambodianum Hance

Illicium griffithii Hook. f. & Thomson

Illicium henryi Diels

Illicium macranthum A. C. Sm.

Illicium manipurense Watt. ex King

Illicium merrillianum A. C. Sm.

Illicium micranthum Dunn

Illicium simonsii Maxim.

Illicium wardii A. C. Sm.

Kadsura heteroclita (Roxb.) Craib

Schisandra grandiflora (Wall.) Hook. f. & Thomson

Schisandra lancifolia (Rehder & E. H. Wilson) A. C. Sm.

Schisandra micrantha A. C. Sm.

Schisandra neglecta A. C. Sm.

Schisandra plena A. C. Sm.

Schisandra propinqua (Wall.) Baillon

Schisandra propinqua subsp. **intermedia** (A. C. Sm.) R. M. K. Saunders

Schisandra propinqua subsp. **sinensis** (Oliv.) R. M. K. Saunders

Schisandra rubriflora Rehder & E. H. Wilson

Schisandra sphaerandra Stapf

49. CHLORANTHACEAE

Chloranthus elatior Link

Chloranthus erectus (Buch.-Ham.) Verdc.

Chloranthus henryi Hemsl.

Chloranthus holostegius (Hand.-Mazz.) C. Pei & Shan

Chloranthus multistachys C. Pei

Sarcandra glabra (Thunb.) Nakai

50. SAURURACEAE

Houttuynia cordata Thunb.

51. PIPERACEAE

Peperomia heyneana Miq.

Peperomia pellucida (L.) Kunth

Peperomia tetraphylla (G. Forst.) Hook. & Arn.

Piper acutistigmum C. DC.

Piper anisotis Hook. f.

Piper attenuatum Buch.-Ham. ex Miq.

Piper betel L.

Piper betleoides C. DC.

Piper boehmeriifolium (Miq.) C. DC.

Piper calvilimbum C. DC.

Piper chaba Hunter

Piper chuvya Hunter ex C. DC.

Piper curtistipes C. DC.

Piper diffusum Vahl

Piper filipedunculum C. DC.

Piper griffithii C. DC.

Piper hamiltonii C. DC.

Piper khasianum C. DC.

Piper laxivenum C. DC.

Piper leptostachyum Wall. ex Miq.
Piper longum L.
Piper macropodum C. DC.
Piper makruense C. DC.
Piper mullesua Buch.-Ham. ex D. Don
Piper nigrum L.
Piper ovatistigmum C. DC.
Piper pedicellatum C. DC.
Piper peepuloides Roxb.
Piper phalangense C. DC.
Piper pothoides (Wall. ex Miq.) C. DC.
Piper rhytidocarpum Hook. f.
Piper sarmentosum Roxb.
Piper sonadense C. DC.
Piper suipigua Buch.-Ham. ex D. Don
Piper sumatranum (Miq.) C. DC.
Piper sylvaticum Roxb.
Piper tenuibracteum C. DC.
Piper thomsonii (C. DC.) Hook. f.
Piper trioicum Roxb.
Piper tumidonodosum P. K. Mukh.
Piper voigtii C. DC.
Piper wallichii (Miq.) Hand.-Mazz.
Piper yui M. G. Gilbert & N. H. Xia
Piper yunnanense Y. C. Tseng

52. ARISTOLOCHIACEAE

Aristolochia cucurbitoides C. F. Liang
Aristolochia delavayi Franch.
Aristolochia forrestiana J. S. Ma
Aristolochia gentilis Franch.
Aristolochia griffithii Hook. f. & Thomson ex Duch.
Aristolochia indica L.
Aristolochia kaempferi Willd.
Aristolochia moupinensis Franch.
Aristolochia nakaoi Maek.
Aristolochia obliqua S. M. Hwang
Aristolochia ovatifolia S. M. Hwang
Aristolochia platanifolia (Klotzsch) Duch.
Aristolochia punjabensis Lace
Aristolochia saccata Wall.
Aristolochia sinoburmanica Y. H. Tan & B. Yang
Aristolochia tagala Chamisso
Aristolochia thibetica Franch.
Aristolochia transsecta (Chatterjee) C. Y. Wu ex S. M. Hwang
Aristolochia wardiana J. S. Ma
Aristolochia wuana Zhen W. Liu & Y. F. Deng
Aristolochia zhongdianensis J. S. Ma
Asarum himalaicum Hook. f. & Thomson ex Klotzsch

53. MYRISTICACEAE

Horsfieldia amygdalina (Wall.) Warb.
Horsfieldia kingii (Hook. f.) Warburg
Knema cinerea Warb.
Knema globularia (Lam.) Warb.
Knema linifolia (Roxb.) Warb.
Knema tenuinervia W. J. de Wilde

54. MAGNOLIACEAE

Alcimandra cathcartii (Hook. f. & Thomson) Dandy

Magnolia campbellii Hook. f. & Thomson

Magnolia globosa Hook. f. & Thomson

Magnolia gustavii King

Magnolia hodgsonii (Hook. f. & Thomson) H. Keng

Magnolia pealiana King

Magnolia pterocarpa Roxb.

Magnolia rabaniana (Hook. f. & Thomson) D. C. S. Raju & M. P. Nayar

Magnolia rostrata W. W. Sm.

Manglietia caveana Hook. f. & Thomson

Manglietia forrestii W. W. Sm. ex Dandy

Manglietia hookeri Cubitt & W. W. Sm.

Manglietia insignis (Wall.) Blume

Michelia baillonii (Pierre) Finet & Gagnep.

Michelia cathcartii Hook. f. & Thomson

Michelia champaca L.

Michelia doltsopa Buch.-Ham. ex DC.

Michelia floribunda Finet & Gagnep.

Michelia glabra P. Parm.

Michelia griffithii (Hook. f. & Thomson) Finet & Gagnep.

Michelia kingii Dandy

Michelia kisopa Buch.-Ham. ex DC.

Michelia mannii King

Michelia oblonga Wall. ex Hook. f. & Thomson

Michelia punduana Hook. f. & Thomson

Michelia velutina DC.

Michelia wardii Dandy

Michelia wilsonii Finet & Gagnep.

Michelia yunnanensis Franch. ex Finet & Gagnep.

Pachylarnax pleiocarpa Dandy

Parakmeria nitida (W. W. Sm.) Y. W. Law

Parakmeria yunnanensis Hu

Talauma hodgsonii Hook. f. & Thomson

55. ANNONACEAE

Annona reticulata L.

Annona squamosa L.

Artabotrys caudatus Wall. ex Hook. f. & Thomson

Artabotrys hexapetalus (L. f.) Bhandari

Dasymaschalon longiflorum (Roxb.) Finet & Gagnep.

Dasymaschalon tibetense X. L. Hou

Desmos chinensis Lour.

Desmos dumosus (Roxb.) Saff.

Fissistigma bicolor (Roxb.) Merr.

Fissistigma polyanthum Hook. f. & Thomson Merr.

Fissistigma retusum (H. Lév.) Rehder

Friesodielsia fornicata (Roxb.) D. Das

Goniothalamus sesquipedalis (Wall.) Hook. f. & Thomson

Hubera cerasoides (Roxb.) Chaowasku

Miliusa dioeca (Roxb.) Chaowasku

Miliusa dolichantha Craib

Miliusa macrocarpa Hook. f. & Thomson

Miliusa sclerocarpa (A. DC.) Kurz

Miliusa tenuistipitata W. T. Wang

Miliusa tomentosa (Roxb.) J. Sinclair

Miliusa velutina (A. DC.) Hook. f. & Thomson

Mitrephora harae H. Ohashi

Monoon fragrans (Dalzell) B. Xue & R. M. K. Saunders

Monoon longifolium (Sonn.) B. Xue & R. M. K. Saunders

Monoon simiarum (Wall. ex Hook. f. & Thomson) B. Xue & R. M. K. Saunders

Polyalthia chinensis S. K. Wu & P. T. Li

Trivalvaria argentea (Hook. f. & Thomson) J. Sinclair

Trivalvaria casseabriae Y. H. Tan, S. S. Zhou & B. Yang

Trivalvaria rubra Y. H. Tan, S. S. Zhou & B. Yang

Uvaria concava Teijsm. & Binn.

Uvaria hamiltonii Hook. f. & Thomson

Wuodendron praecox (Hook. f. & Thomson) B. Xue, Y. H. Tan & X. L. Hou

56. HERNANDIACEAE

Illigera cordata var. **mollissima** (W. W. Sm.) Kubitzki

Illigera grandiflora W. W. Sm. & Jeffrey

Illigera khasiana C. B. Clarke

Illigera nervosa Merr.

Illigera villosa C. B. Clarke

57. LAURACEAE

Actinodaphne angusifolia Nees

Actinodaphne gullavara (Buch.-Ham. ex Nees) M. R. Almeida

Actinodaphne longipes Kosterm.

Actinodaphne obovata (Nees) Blume

Actinodaphne reticulata Meisn.

Actinodaphne sikkimensis Meisn.

Actinodaphne trichocarpa C. K. Allen

Alseodaphne andersonii (King ex Hook. f.) Kosterm.

Alseodaphne owdenii R. Parker

Beilschmiedia assamica Meisn.

Beilschmiedia brandisii Hook. f.

Beilschmiedia clarkei Hook. f.

Beilschmiedia dalzellii (Meisn.) Kosterm.

Beilschmiedia gammieana King ex Hook. f.

Beilschmiedia kweichowensis W. C. Cheng

Beilschmiedia robusta C. K. Allen

Beilschmiedia roxburghiana Nees

Beilschmiedia sikkimensis King ex Hook. f.

Beilschmiedia xizangensis H. P. Tsui

Beilschmiedia yaanica (N. Chao) N. Chao

Cinnadenia paniculata (Hook. f.) Kosterm.

Cinnamomum albiflorum Nees

Cinnamomum bazania (Buch.-Ham.) Nees

Cinnamomum bejolghota (Buch.-Ham.) Sweet

Cinnamomum camphora (L.) J. Presl

Cinnamomum cecicodaphne Meisn.

Cinnamomum chenii Nakai

Cinnamomum contractum H. W. Li

Cinnamomum glanduliferum (Wall.) Meisn.

Cinnamomum glaucescens (Nees) Hand.-Mazz.

Cinnamomum impressinervium Meisn.

Cinnamomum iners Reinw. ex Blume

Cinnamomum longepaniculatum (Gamble) N. Chao ex H. W. Li

Cinnamomum mairei H. Lév.

Cinnamomum obtusifolium Roxb. ex Nees

Cinnamomum parthenoxylon (Jack) Meisn.

Cinnamomum pauciflorum Nees

Cinnamomum pittosporoides Hand.-Mazz.

Cinnamomum porrectum (Roxb.) Kosterm.

Cinnamomum septentrionale Hand.-Mazz.

Cinnamomum soncaurium (Buch.-Ham.) Nees & Eberm.

Cinnamomum subavenium Miq.

Cinnamomum szechuanense Y. C. Yang

Cinnamomum tamala (Buch.-Ham.) T. Nees & C. H. Eberm.

Cinnamomum tenuipile Kosterm.

Cinnamomum wilsonii Gamble

Cryptocarya amygdalina Nees

Cryptocarya bhutanica D. G. Long

Cryptocarya floribunda Nees

Dodecadenia grandiflora Nees

Dodecadenia paniculata Hook. f.

Endiandra deomalica (Bennet & Chandra) Kosterm.

Iteadaphne caudatum (Nees) H. W. Li

Lindera assamica (Meisn.) Kurz

Lindera bifarium (Nees) Hook. f.

Lindera botanica Meisn.

Lindera cercidifolia Hemsl.

Lindera communis Hemsl.

Lindera doniana C. K. Allen

Lindera floribunda (C. K. Allen) H. P. Tsui

Lindera fragrans Oliv.

Lindera fruticosa Hemsl.

Lindera gracilipes H. W. Li

Lindera gracilis H. P. Tsui

Lindera griffithii Meisn.

Lindera hamiltonii Kosterm.

Lindera hemsleyana (Diels) C. K. Allen

Lindera heterophylla Meisn.

Lindera kariensis W. W. Sm.

Lindera latifolia Hook. f.

Lindera limprichtii H. Winkl.

Lindera longipedunculata C. K. Allen

Lindera malaccensis Hook. f.

Lindera megaphylla Hemsl.

Lindera melastomacea (Nees) Villar

Lindera motuoensis H. P. Tsui

Lindera nacusua (D. Don) Merr.

Lindera neesiana (Wall. ex Nees) Kurz

Lindera obtusiloba Blume

Lindera paxiana H. Winkl.

Lindera praetermissa Grierson & D. C. Long

Lindera prattii Gamble

Lindera pulcherrima (Nees) Hook. f.

Lindera sanjappae Bhaumik, M. K. Pathak & Chakrab.

Lindera setchuanensis Gamble

Lindera strychnifolia (Siebold & Zucc.) Fern.-Vill.

Lindera supracostata Lecomte

Lindera thomsonii C. K. Allen

Lindera tienchuanensis W. P. Fang & H. S. Kung ex Yen C. Yang et al.

Lindera tonkinensis var. **subsessilis** H. W. Li

Lindera triloba Hook. f. & Thomson ex Meisn.

Lindera varmae M. K. Pathak, Bhaumik & Chakrab.

Lindera venosa (Meisn.) Benth. ex Hook. f.

Lindera villipes H. P. Tsui

Litsea albescens (Hook. f.) D. G. Long

Litsea albicans Hook. f.

Litsea biflora H. P. Tsui

Litsea chartacea (Wall. ex Nees) Hook. f.

Litsea chengshuzhii H. P. Tsui

Litsea chinpingensis Yen C. Yang & P. H. Huang

Litsea chunii W. C. Cheng

Litsea citrata Blume

Litsea coelestis H. P. Tsui

Litsea cubeba (Lour.) Pers.

Litsea depressa H. P. Tsui

Litsea doshia (Buch.-Ham. ex D. Don) Kosterm.

Litsea elongata (Nees) Hook. f.

Litsea faberi Hemsl.

Litsea glabrata (Wall. ex Nees) Hook. f.

Litsea glutinosa (Lour.) C. B. Robinson

Litsea gongshanensis H. W. Li

Litsea hookeri (Meisn.) D. G. Long

Litsea khasyana Meisn.

Litsea kingii Hook. f.

Litsea laeta (Nees) Trimen

Litsea lancifolia (Roxb. ex Nees) Fern.-Vill.

Litsea liboshengii H. P. Tsui

Litsea longistaminata (H. Liu) Kosterm.

Litsea mishmiensis Hook. f.

Litsea mollis Hemsl.

Litsea monantha Y. C. Yang & P. H. Huang

Litsea monopetala (Roxb.) Pers.

Litsea moupinensis Lecomte

Litsea nervosa (Kurz) Grierson & D. G. Long

Litsea nitida (Roxb.) Hook. f.

Litsea oblonga (Nees) Hook. f.

Litsea panamanja (Buch.-Ham. ex Nees) Hook. f.

Litsea populifolia (Hemsl.) Gamble

Litsea pungens Hemsl.

Litsea rubescens Lecomte

Litsea salicifolia (Roxb. ex Nees) Hook. f.

Litsea sebifera (Willd.) Pers.

Litsea sericea (Wall. ex Nees) Hook. f.

Litsea taronensis H. W. Li

Litsea tibetana Yen C. Yang & P. H. Huang

Litsea tsinlingensis Yen C. Yang & P. H. Huang

Litsea veitchiana Gamble

Litsea wallichii Hook. f.

Litsea wilsonii Gamble

Machilus chayuensis S. K. Lee

Machilus chrysotricha H. W. Li

Machilus clarkeana King ex Hook. f

Machilus duthiei King ex Hook. f.

Machilus edulis King ex Hook. f.

Machilus glaucescens (Nees) Wight

Machilus ichangensis Rehder & E. H. Wilson

Machilus longipedicellata var. **synechothrix** (Hand.-Mazz.) Hand.-Mazz.

Machilus mairei H. Lév.

Machilus obscurinervis S. K. Lee

Machilus odoratissima Nees

Machilus ovatiloba S. K. Lee

Machilus parviflora Meisn.

Machilus pubescens Blume

Machilus rufipes H. W. Li

Machilus sericea Blume

Machilus tenuipilis H. W. Li

Machilus viridis Hand.-Mazz.

Machilus yunnanensis Lecomte

Neocinnamomum caudatum (Nees) Merr.

Neocinnamomum cavaleriei H. Lév.

Neocinnamomum delavayi (Lecomte) H. Liu

Neocinnamomum mekongense (Hand.-Mazz.) Kosterm.

Neolitsea chrysotricha H. W. Li

Neolitsea confertifolia (Hemsl.) Merr.

Neolitsea cuipala (Buch.-Ham. ex D. Don) Kosterm.

Neolitsea foliosa (Nees) Gamble

Neolitsea homilantha C. K. Allen

Neolitsea lanuginosa (Nees) Gamble

Neolitsea levinei Merr.

Neolitsea pallens (D. Don) Momiy. & Hara

Neolitsea sanjappae M. K. Pathak, Bhaumik & Chakrab.

Neolitsea sutchuanensis Yen C. Yang

Neolitsea umbrosa (Nees) Gamble

Parasassafras confertiflorum (Meisn.) D. G. Long

Persea bombycina (King ex Hook. f.) Kosterm.

Persea bootanica (Meisn.) Kosterm.

Persea fructifera Kosterm.

Persea gamblei (King ex Hook. f.) Kosterm.

Persea kurzii (Hook. f.) Kosterm.

Persea minutiflora Kosterm.

Persea robusta (W. W. Sm.) Kosterm.

Phoebe attenuata (Nees) Nees

Phoebe cathia (D. Don) Kosterm.

Phoebe chinensis Chun

Phoebe forrestii W. W. Sm.

Phoebe glaucifolia S. K. Lee & F. N. Wei

Phoebe hainesiana Brandis

Phoebe hui W. C. Cheng ex Y. C. Yang

Phoebe lanceolata (Nees) Nees

Phoebe legendrei Lecomte

Phoebe motuonan S. K. Lee & F. N. Wei

Phoebe nanmu (Oliv.) Gamble

Phoebe neurantha (Hemsl.) Gamble

Phoebe neuranthoides S. K. Lee & F. N. Wei

Phoebe pallida (Nees) Nees

Phoebe paniculata (Nees) Nees

Phoebe zhennan S. K. Lee & F. N. Wei

Potameia paradoxa (Hook. f.) Kosterm.

Sinosassafras flavinervium (C. K. Allen) H. W. Li

58. ACORACEAE

Acorus calamus L.

Acorus gramineus Aiton

59. ARACEAE

Aglaonema hookerianum Schott

Alocasia acuminata Schott

Alocasia cucullata (Lour.) G. Don

Alocasia fallax Schott

Alocasia fornicata (Roxb.) Schott

Alocasia macrorrhizos (L.) G. Don

Alocasia navicularis (K. Koch & C. D. Bouché) K. Koch & C. D. Bouché

Alocasia odora (Roxb.) K. Koch

Amorphophallus bulbifer (Schott) Blume

Amorphophallus campanulatus Decne.

Amorphophallus commutatus (Schott) Engl.

Amorphophallus kachinensis Engl. & Gehrm.

Amorphophallus konjac K. Koch

Amorphophallus napalensis (Wall.) Bogner & Mayo

Amorphophallus paeoniifolius (Dennst.) Nicolson

Amorphophallus yunnanensis Engl.

Ariopsis peltata Nimmo

Arisaema aridum H. Li

Arisaema bonatianum Engl.

Arisaema brucei H. Li, R. Li & J. Murata

Arisaema candidissimum W. W. Sm.

Arisaema ciliatum H. Li

Arisaema concinnum Schott

Arisaema consanguineum Schott

Arisaema costatum (Wall.) Martius ex Schott & Endl.

Arisaema decipiens Schott

Arisaema echinatum (Wall.) Schott

Arisaema echinoides H. Li

Arisaema elephas Buchet

Arisaema erubescens (Wall.) Schott

Arisaema exappendiculatum Hara

Arisaema fargesii Buchet

Arisaema flavum subsp. **tibeticum** J. Murata

Arisaema franchetianum Engl.

Arisaema galeatum N. E. Br.

Arisaema gracilentum Brugg.

Arisaema griffithii Schott

Arisaema griffithii var. **verrucosum** (Schott) Hara

Arisaema handelii Stapf ex Hand.-Mazz.

Arisaema heterophyllum Blume
Arisaema intermedium Blume
Arisaema intermedium var. **biflagellatum** (Hara) Hara
Arisaema jacquemontii Blume
Arisaema jingdongense H. Peng & H. Li
Arisaema lichiangense W. W. Sm.
Arisaema linearifolium Gusman & J. T. Yin
Arisaema lobatum Engl.
Arisaema longitubum Z. X. Ma
Arisaema lushuiense G. W. Hu & H. Li
Arisaema meleagris Buchet
Arisaema nepenthoides (Wall.) Mart.
Arisaema parvum N. E. Br. ex Hemsl.
Arisaema petiolulatum Hook. f.
Arisaema pianmaense H. Li
Arisaema propinquum Schott
Arisaema quinquelobatum H. Li & J. Murata
Arisaema roxburghii Kunth
Arisaema saxatile Buchet
Arisaema setosum A. S. Rao & Verma
Arisaema siangense Gusman
Arisaema sinii K. Krause
Arisaema smitinandii S. Y. Hu
Arisaema souliei Buchet
Arisaema speciosum (Wall.) Martius ex Schott & Endl.
Arisaema speciosum var. **mirabile** (Schott) Hara
Arisaema tengtsungense H. Li
Arisaema tortuosum (Wall.) Schott
Arisaema tortuosum var. **curvatum** (Roxb.) Engler
Arisaema tsangpoense J. T. Yin & Gusman
Arisaema utile Schott
Arisaema vexillatum Hara & H. Ohashi
Arisaema wardii C. Marquand & Airy Shaw
Arisaema wattii Hook. f.
Arisaema wilsonii Engl.
Arisaema yunnanense Buchet
Arisaema zhui H. Li
Arum jacquemontii Blume
Colocasia affinis Schott
Colocasia antiquorum Schott
Colocasia esculenta (L.) Schott
Colocasia fallax Schott
Colocasia mannii Hook. f.
Cryptocoryne cruddasiana Prain
Cryptocoryne retrospiralis (Roxb.) Kunth
Homalomena aromatica (Spreng.) Schott
Homalomena pendula (Blume) Bakh. f.
Homalomena rubescens (Roxb.) Kunth.
Lagenandra undulata Sastry
Lasia spinosa (L.) Thwaites
Lemna minor L.
Lemna perpusilla Torr.
Pistia stratiotes L.
Pothos chinensis (Raf.) Merr.
Pothos macrocephalus Scort. ex Hook. f.
Pothos repens (Lour.) Druce
Pothos scandens L.
Pothos vrieseanus Schott
Remusatia hookeriana Schott
Remusatia pumila (D. Don) H. Li & A. Hay
Remusatia vivipara (Roxb.) Schott
Rhaphidophora beccarii (Engl.) Engl.
Rhaphidophora calophylla Schott
Rhaphidophora crassicaulis Engl. & K. Krause
Rhaphidophora decursiva (Roxb.) Schott
Rhaphidophora dulongensis H. Li
Rhaphidophora glauca (Wall.) Schott
Rhaphidophora hongkongensis Schott
Rhaphidophora hookeri Schott
Rhaphidophora korthalsii Schott
Rhaphidophora lancifolia Schott

Rhaphidophora luchunensis H. Li

Rhaphidophora peepla (Roxb.) Schott

Sauromatum brevipes (Hook. f.) N. E. Br.

Sauromatum diversifolium (Wall. ex Schott) Cusimano & Hett.

Sauromatum gaoligongense Z. L. Wang & H. Li

Sauromatum horsfieldii Miq.

Sauromatum venosum (Aiton) Kunth

Scindapsus officinalis (Roxb.) Schott

Spirodela polyrrhiza (L.) Schleid.

Steudnera assamica Hook. f.

Steudnera capitellata Hook. f.

Steudnera discolor W. Bull

Steudnera griffithii (Schott) Hook. f.

Typhonium baoshanense Z. L. Dao & H. Li

Typhonium blumei Nicolson & Sivad.

Typhonium flagelliforme (Lodd.) Blume

Typhonium roxburghii Schott

Typhonium trilobatum (L.) Schott

Wolffia arrhiza (L.) Horkel ex Wimm.

Wolffia brasiliensis Wedd.

Wolffia columbiana H. Karst.

Wolffia globosa (Roxb.) Hartog & Plas

60. TOFIELDIACEAE

Tofieldia divergens Bureau & Franch.

Tofieldia himalaica Baker

Tofieldia thibetica Franch.

61. ALISMATACEAE

Alisma gramineum Lej.

Alisma lanceolatum With.

Alisma orientale (Sam.) Juz.

Alisma plantago-aquatica L.

Butomopsis latifolia (D. Don) Kunth

Caldesia parnassifolia (L.) Parl.

Sagittaria guayanensis subsp. **lappula** (D. Don) Bogin

Sagittaria latifolia Willd.

Sagittaria longiloba Engelm. ex J. G. Sm.

Sagittaria sagittifolia L.

Sagittaria tengtsungensis H. Li

Sagittaria trifolia L.

62. BUTOMACEAE

Butomopsis latifolia (D. Don) Kunth

Butomus umbellatus L.

63. HYDROCHARITACEAE

Blyxa aubertii Rich.

Blyxa echinosperma (C. B. Clarke) Hook. f.

Blyxa japonica (Miq.) Maxim. ex Ascherson & Gürke

Hydrilla verticillata (L. f.) Royle

Hydrilla verticillata var. **roxburghii** Caspary

Hydrocharis dubia (Blume) Backer

Najas graminea Delile

Najas marina L.

Najas minor All.

Najas oguraensis Miki

Nechamandra alternifolia (Roxb. ex R. Wight) Thwaites

Ottelia acuminata (Gagnep.) Dandy

Ottelia acuminata var. **crispa** (Hand.-Mazz.) H. Li
Ottelia alismoides (L.) Pers.
Vallisneria natans (Lour.) Hara
Vallisneria spiralis L.

64. SCHEUCHZERIACEAE

Scheuchzeria palustris L.

65. APONOGETONACEAE

Aponogeton natans (L.) Engl. & K. Krause
Aponogeton undulatus Roxb.

66. JUNCAGINACEAE

Triglochin maritima L.
Triglochin palustris L.

67. POTAMOGETONACEAE

Potamogeton alpinus Balbis
Potamogeton amblyphyllus C. A. Mey.
Potamogeton berchtoldii Fieber
Potamogeton crispus L.
Potamogeton distinctus A. Benn.
Potamogeton filiformis Pers.
Potamogeton friesii Rupr.
Potamogeton gramineus L.
Potamogeton lucens L.
Potamogeton maackianus A. Benn.
Potamogeton natans L.
Potamogeton nodosus Poir.
Potamogeton octandrus Poir.
Potamogeton oxyphyllus Miq.
Potamogeton pectinatus L.
Potamogeton perfoliatus L.
Potamogeton pusillus L.
Zannichellia palustris L.

68. PETROSAVIACEAE

Petrosavia sakuraii (Makino) J. J. Sm. ex Steenis

69. NARTHECIACEAE

Aletris alpestris Diels
Aletris alpestris var. **occidentalis** Hara
Aletris capitata F. T. Wang & Tang
Aletris glabra Burcau & Franch.
Aletris glandulifera Bureau & Franch.
Aletris gracilis Rendle
Aletris laxiflora Bureau & Franch.
Aletris nana S. C. Chen
Aletris pauciflora (Klotzsch) Hand.-Mazz.
Aletris pauciflora var. **khasiana** (Hook. f.) F. T. Wang & Tang
Aletris simpliciflora R. Li & S. D. Zhang
Aletris spicata (Thunb.) Franch.
Aletris stenoloba Franch.

70. BURMANNIACEAE

Burmannia chinensis Gand.

Burmannia coelestis D. Don

Burmannia disticha L.

Burmannia nepalensis (Miers) Hook. f.

71. DIOSCOREACEAE

Dioscorea alata L.

Dioscorea althaeoides R. Knuth

Dioscorea belophylla (Prain) Voigt ex Haines

Dioscorea bicolor Prain & Burkill

Dioscorea biformifolia C. P'ei & C. T. Ting

Dioscorea birmanica Prain & Burkill

Dioscorea bulbifera L.

Dioscorea cirrhosa Lour.

Dioscorea collettii Hook. f.

Dioscorea collettii var. **hypoglauca** (Palib.) S. J. Pei & C. T. Ting

Dioscorea delavayi Franch.

Dioscorea deltoidea Wall. ex Griseb.

Dioscorea glabra Roxb.

Dioscorea hamiltonii Hook. f.

Dioscorea hemsleyi Prain & Burkil

Dioscorea hispida Dennst.

Dioscorea japonica Thunb.

Dioscorea kamoonensis Kunth

Dioscorea laurifolia Wall. ex Hook. f.

Dioscorea lepcharum Prain & Burkill

Dioscorea listeri Prain & Burkill

Dioscorea martini Prain & Burkill

Dioscorea melanophyma Prain & Burkill

Dioscorea nipponica Makino

Dioscorea nipponica subsp. **rosthornii** (Diels) C. T. Ting

Dioscorea nitens Prain & Burkill

Dioscorea orbiculata Hook. f.

Dioscorea panthaica Prain & Burkil

Dioscorea pentaphylla L.

Dioscorea prazeri Prain & Burkill

Dioscorea pubera Blume

Dioscorea scortechinii Prain & Burkill

Dioscorea sinoparviflora C. T. Ting, M. G. Gilbert & Turland

Dioscorea subcalva Prain & Burkill

Dioscorea trinervia Roxb. ex Prain & Burkill

Dioscorea velutipes Prain & Burkill

Dioscorea wattii Prain & Burkill

Dioscorea xizangensis C. T. Ting

Dioscorea yunnanensis Prain & Burkill

Dioscorea zingiberensis C. H. Wright

Tacca chantrieri André

Tacca integrifolia Ker Gawl.

72. TRIURIDACEAE

Sciaphila nana Blume

73. VELLOZIACEAE

Acanthochlamys bracteata P. C. Kao

74. STEMONACEAE

Stemona mairei (H. Lév.) K. Krause

Stemona tuberosa Lour.

75. PANDANACEAE

Pandanus fascicularis Lam.

Pandanus furcatus Roxb.

Pandanus tectorius Parkinson ex Du Roi

Pandanus unguifer Hook. f.

Pandanus urophyllus Hance

76. MELANTHIACEAE

Paris axialis H. Li

Paris bashanensis F. T. Wang & Tang

Paris daliensis H. Li & V. G. Soukup

Paris delavayi Franch.

Paris dulongensis H. Li & S. Kurita

Paris fargesii Franch.

Paris forrestii (Takht.) H. Li

Paris mairei H. Lév.

Paris marmorata Stearn

Paris polyandra S. F. Wang

Paris polyphylla Sm.

Paris polyphylla var. **chinensis** (Franch.) Hara

Paris polyphylla var. **stenophylla** Franch.

Paris polyphylla var. **wallichii** Hara

Paris polyphylla var. **yunnanensis** (Franch.) Hand.-Mazz.

Paris rugosa H. Li & S. Kurita

Paris thibetica Franch.

Paris thibetica var. **apetala** Hand.-Mazz.

Paris verticillata M. Bieb.

Paris wenxianensis Z. X. Peng & R. N. Zhao

Trillium govanianum Wall. ex D. Don

Trillium smallii Maxim.

Trillium tschonoskii Maxim.

Veratrum grandiflorum (Maxim. ex Baker) O. Loes.

Veratrum mengtzeanum O. Loes.

Veratrum micranthum F. T. Wang & Tang

Veratrum nigrum L.

Veratrum stenophyllum Diels

Veratrum stenophyllum var. **taronense** F. T. Wang & Z. H. Tsi

Veratrum taliense O. Loes.

Ypsilandra alpina F. T. Wang & Tang

Ypsilandra thibetica Franch.

Ypsilandra yunnanensis W. W. Sm. & Jeffrey

77. COLCHICACEAE

Colchicum kesselringii Regel

Colchicum luteum Baker

Disporum bodinieri (H. Lév. & Vaniot) F. T. Wang & Tang

Disporum calcaratum D. Don

Disporum cantoniense (Lour.) Merr.

Disporum leucanthum Hara

Disporum longistylum (H. Lév. & Vaniot) Hara

Disporum megalanthum F. T. Wang & Tang

Gloriosa superba L.

Iphigenia indica (L.) A. Gray ex Kunth

78. SMILACACEAE

Smilax arisanensis Hayata

Smilax aspera L.

Smilax aspericaulis Wall.

Smilax bapouensis H. Li

Smilax binchuanensis P. Li & C. X. Fu

Smilax biumbellata Koyama

Smilax bracteata C. Presl

Smilax castaneiflora H. Lév. & Vaniot

Smilax chapaensis Gagnep.

Smilax china L.

Smilax chinensis (F. T. Wang) P. Li & C. X. Fu

Smilax chingii F. T. Wang & Tang

Smilax cocculoides Warb.

Smilax darrisii H. Lév.

Smilax discotis Warb.

Smilax elegans Wall.

Smilax ferox Wall.

Smilax glabra Roxb.

Smilax griffithii A. DC.

Smilax hemsleyana Craib

Smilax japonica (Kunth) P. Li & C. X. Fu

Smilax kingii Hook. f.

Smilax lanceifolia Roxb.

Smilax lebrunii H. Lév.

Smilax longebracteolata Hook. f.

Smilax lunglingensis F. T. Wang & Tang

Smilax lushuiensis S. C. Chen

Smilax luzonensis C. Presl

Smilax macrocarpa Blume

Smilax mairei H. Lév.

Smilax megacarpa A. DC.

Smilax megalantha C. H. Wright

Smilax menispermoidea A. DC.

Smilax microphylla C. H. Wright

Smilax minutiflora A. DC.

Smilax munita S. C. Chen

Smilax myrtillus A. DC.

Smilax nana F. T. Wang

Smilax ocreata A. DC.

Smilax orthoptera A. DC.

Smilax osmastonii F. T. Wang & Tang

Smilax outanscianensis Pamp.

Smilax ovalifolia Roxb.

Smilax perfoliata Lour.

Smilax polyacantha Wall.

Smilax polyandra (Gagnep.) P. Li & C. X. Fu

Smilax polycolea Warb.

Smilax pottingeri Prain

Smilax prolifera Roxb.

Smilax quadrata A. DC.

Smilax riparia A. DC.

Smilax scobinicaulis C. H. Wright

Smilax stemonifolia H. Lév. & Vaniot

Smilax vaginata Decne.

Smilax wightii A. DC.

Smilax zeylanica L.

79. LILIACEAE

Cardiocrinum giganteum (Wall.) Makino

Cardiocrinum giganteum var. **yunnanense** (Leichtlin ex Elwes) Stearn

Clintonia udensis Trautv. & C. A. Mey.

Fritillaria chitralensis (hort.) B. Mathew

Fritillaria cirrhosa D. Don

Fritillaria cirrhosa subsp. **roylei** (Hook.) Ali

Fritillaria crassicaulis S. C. Chen
Fritillaria dajinensis S. C. Chen
Fritillaria davidii Franch.
Fritillaria delavayi Franch.
Fritillaria fusca Turrill
Fritillaria imperialis L.
Fritillaria przewalskii Maxim.
Fritillaria raddeana Regel
Fritillaria sichuanica S. C. Chen
Fritillaria sinica S. C. Chen
Fritillaria taipaiensis P. Y. Li
Fritillaria unibracteata P. K. Hsiao & K. C. Hsia
Gagea absurda Levichev
Gagea alexii Ali & Levichev
Gagea anisanthos C. Koch
Gagea bashoensis Ali
Gagea bowes-lyonii Levichev
Gagea bulbifera (Pall.) Salisb.
Gagea chitralensis S. Dasgupta & Deb
Gagea chloroneura Rech. f.
Gagea davlianidzeae Levichev
Gagea drummondii Levichev & Ali
Gagea dschungarica Regel
Gagea elegans Wall. ex D. Don
Gagea exilis Vved.
Gagea gageoides (Zucc.) Vved.
Gagea jaeschkei Pasch.
Gagea jispensis Ali & Levichev
Gagea kunawurensis (Royle) Greuter
Gagea leosii Ali & Levichev
Gagea liotardii (Sternb.) Schult. & Schult. f.
Gagea lowariensis Pascher
Gagea lutea (L.) Ker Gawl.
Gagea mergalahensis Ali & Levichev
Gagea nakaiana Kitag.
Gagea noltiei Peruzzi, J.-M. Tison, A. Peterson & J. Peterson
Gagea olgae Regel
Gagea ova Stapf
Gagea pakistanica Levichev & Ali
Gagea pauciflora (Turcz. ex Trautv.) Ledeb.
Gagea persica Boiss.
Gagea persica var. **kashmiriensis** (Turrill) S. Dasgupta & Deb
Gagea punjabica Levichev & Ali
Gagea rawalpindica Levichev & Ali
Gagea reticulata (Pall.) Schult. & Schult. f.
Gagea serotina (L.) Ker Gawl.
Gagea setifolia Baker
Gagea siphonantha Rech. f.
Gagea spumosa Levichev
Gagea stipitata Bunge
Gagea tenera Pasch.
Gagea vvedenskyi Grossh.
Gagea wallichii Levichev & Ali
Lilium amoenum E. H. Wilson ex Sealy
Lilium apertum Franch.
Lilium bakerianum Collett & Hemsl.
Lilium bakerianum var. **aureum** Grove & Cotton
Lilium bakerianum var. **delavayi** (Franch.) E. H. Wilson
Lilium bakerianum var. **rubrum** Stearn
Lilium bakerianum var. **yunnanense** (Franch.) Sealy ex Woodcock & Stearn
Lilium basilissum (W. E. Evans) Y. D. Gao
Lilium brevistylum (S. Yun Liang) S. Yun Liang
Lilium brownii F. E. Brown ex Miellez
Lilium davidii Duchartre ex Elwes
Lilium duchartrei Franch.
Lilium farreri (Harrow ex W. E. Evans) Y. D. Gao

Lilium gongshanense (Y. D. Gao & X. J. He) Y. D. Gao

Lilium habaense F. T. Wang & Tang

Lilium henrici Franch.

Lilium henrici var. **maculatum** (W. E. Evans) Woodcock & Stearn

Lilium lankongense Franch.

Lilium lijiangense L. J. Peng

Lilium lophophorum (Bureau & Franch.) Franch.

Lilium lophophorum var. **linearifolium** (Sealy) S. Yun Liang

Lilium matangense J. M. Xu

Lilium medogense S. Yun Liang

Lilium meleagrinum (Franch.) Y. D. Gao

Lilium nanum Klotzsch

Lilium nanum var. **flavidum** (Rendle) Sealy

Lilium nepalense D. Don

Lilium oxypetalum (D. Don) Baker

Lilium papilliferum Franch.

Lilium paradoxum Stearn

Lilium pardanthinum (Franch.) Y. D. Gao

Lilium pinifolium L. J. Peng

Lilium polyphyllum D. Don

Lilium primulinum Baker

Lilium primulinum var. **burmanicum** (W. W. Sm.) Stearn

Lilium pyi H. Lév.

Lilium regale E. H. Wilson

Lilium saccatum S. Yun Liang

Lilium saluenense (Balf. f.) S. Yun Liang

Lilium sargentiae E. H. Wilson

Lilium sempervivoideum H. Lév.

Lilium sherriffiae Stearn

Lilium souliei (Franch.) Sealy

Lilium stewartianum Balf. f. & W. W. Sm.

Lilium sulphureum Baker ex Hook. f.

Lilium synapticum (Sealy) Y. D. Gao

Lilium taliense Franch.

Lilium wallichianum Schult. & Schult. f.

Lilium wardii Stapf ex F. C. Stern

Lilium xanthellum F. T. Wang & Tang

Lilium xanthellum var. **luteum** S. Yun Liang

Lilium yapingense Y. D. Gao & X. J. He

Lloydia delavayi Franch.

Lloydia flavonutans Hara

Lloydia himalensis Royle

Lloydia ixiolirioides Baker ex Oliv.

Lloydia longiscapa Hook.

Lloydia nana R. Li & H. Li

Lloydia oxycarpa Franch.

Lloydia serotina (L.) Rchb.

Lloydia serotina var. **parva** (C. Marquand & Airy Shaw) Hara

Lloydia tibetica Baker ex Oliv.

Lloydia yunnanensis Franch.

Notholirion bulbuliferum (Lingelsh. ex H. Limpr.) Stearn

Notholirion campanulatum Cotton & Stearn

Notholirion macrophyllum (D. Don) Boiss.

Notholirion thomsonianum (Royle) Stapf

Streptopus chatterjeeanus S. Dasgupta

Streptopus obtusatus Fassett

Streptopus parasimplex Hara & H. Ohashi

Streptopus parviflorus Franch.

Streptopus simplex D. Don

Tricyrtis maculata (D. Don) J. F. Macbr.

Tulipa clusiana DC.

Tulipa heteropetala Ledeb.

Tulipa heterophylla (Regel) Baker

Tulipa lanata Regel

Tulipa stellata Hook.

80. ORCHIDACEAE

Acampe papillosa (Lindl.) Lindl.

Acampe praemorsa (Roxb.) Blatt. & McCann

Acampe rigida (Buch.-Ham. ex Sm.) P. F. Hunt

Acanthephippium striatum Lindl.

Acanthephippium sylhetense Lindl.

Acriopsis indica Wight

Acriopsis liliifolia (J. Koenig) Seidenf.

Acrochaene punctata Lindl.

Aerides falcata Lindl. & Paxton

Aerides multiflora Roxb.

Aerides odorata Lour.

Aerides rosea Lodd. ex Lindl. & Paxton

Agrostophyllum brevipes King & Pantl.

Agrostophyllum callosum Rchb. f.

Agrostophyllum myrianthum King & Pantl.

Agrostophyllum planicaule (Wall. ex Lindl.) Rchb. f.

Ania penangiana (Hook. f.) Summerh.

Ania viridifusca (Hook.) Tang & F. T. Wang ex Summerh.

Anoectochilus brevilabris Lindl.

Anoectochilus grandiflorus Lindl.

Anoectochilus roxburghii (Wall.) Lindl.

Anthogonium gracile Wall. ex Lindl.

Aphyllorchis alpina King & Pantl.

Aphyllorchis gollanii Duthie

Aphyllorchis montana Rchb. f.

Apostasia odorata Blume

Apostasia wallichii R. Br.

Appendicula cornuta Blume

Arachnis labrosa (Lindl. & Paxton) Rchb. f.

Arundina graminifolia (D. Don) Hochr.

Ascocentrum ampullaceum (Roxb.) Schltr.

Ascocentrum curvifolium (Lindl.) Schltr.

Ascocentrum himalaicum (Deb, Sengupta & Malick) Christenson

Ascocentrum semiteretifolium Seidenf.

Biermannia arunachalensis A. N. Rao

Biermannia bimaculata (King & Pantl.) King & Pantl.

Biermannia jainiana S. N. Hegde & A. N. Rao

Bletilla formosana (Hayata) Schltr.

Bletilla ochracea Schltr.

Bletilla striata (Thunb. ex A. Murray) Rchb. f.

Brachycorythis galeandra (Rchb. f.) Summerh.

Brachycorythis iantha (Wight) Summerh.

Brachycorythis obcordata (Lindl.) Summerh.

Bryobium pudicum (Ridl.) Y. P. Ng & P. J. Cribb

Bulbophyllum acutiflorum A. Rich.

Bulbophyllum affine Lindl.

Bulbophyllum ambrosia (Hance) Schltr.

Bulbophyllum ambrosia subsp. **nepalensis** J. J. Wood

Bulbophyllum amplifolium (Rolfe) N. P. Balakr. & Sud. Chowdhury

Bulbophyllum andersonii (Hook. f.) J. J. Sm.

Bulbophyllum apodum Hook. f.

Bulbophyllum appendiculatum (Rolfe) J. J. Sm.

Bulbophyllum bisetum Lindl.

Bulbophyllum bomiense Z. H. Tsi

Bulbophyllum brienianum (Rolfe) Merr.

Bulbophyllum capillipes C. S. P. Parish & Rchb. f.

Bulbophyllum careyanum (Hook.) Spreng.

Bulbophyllum cariniflorum Rchb. f.

Bulbophyllum cauliflorum Hook. f.

Bulbophyllum cornu-cervi King

Bulbophyllum crassipes Hook. f.

Bulbophyllum cupreum Lindl.

Bulbophyllum cylindraceum Lindl.

Bulbophyllum delitescens Hance

Bulbophyllum depressum King & Pantl.

Bulbophyllum disciflorum Rolfe

Bulbophyllum dulongjiangense X. H. Jin

Bulbophyllum elatum (Hook. f.) J. J. Sm.

Bulbophyllum emarginatum (Finet) J. J. Sm.

Bulbophyllum eublepharum Rchb. f.

Bulbophyllum forrestii Seidenf.

Bulbophyllum gamblei (Hook. f.) Hook. f.

Bulbophyllum gongshanense Z. H. Tsi

Bulbophyllum gracilipes King & Pantl.

Bulbophyllum griffithii (Lindl.) Rchb. f.

Bulbophyllum guttulatum (Hook. f.) N. P. Balakr.

Bulbophyllum gymnopus Hook. f.

Bulbophyllum helenae (Kuntze) J. J. Sm.

Bulbophyllum hirtum (Sm.) Lindl.

Bulbophyllum hymenanthum Hook. f.

Bulbophyllum interpositum J. J. Verm., Schuit. & de Vogel

Bulbophyllum jainii (Hynn. & Malhotra) J. J. Verm., Schuit. & de Vogel

Bulbophyllum khasyanum Griff.

Bulbophyllum leopardinum (Wall.) Lindl. ex Wall.

Bulbophyllum leptanthum Hook. f.

Bulbophyllum lobbii Lindl.

Bulbophyllum longiflorum Thouars

Bulbophyllum macraei (Lindl.) Rchb. f.

Bulbophyllum moniliforme E. C. Parish & Rchb. f.

Bulbophyllum muscicola Rchb. f.

Bulbophyllum nasutum Rchb. f.

Bulbophyllum nodosum (Rolfe) J. J. Sm.

Bulbophyllum nujiangense X. H. Jin & W. T. Jin

Bulbophyllum odoratissimum (Sm.) Lindl.

Bulbophyllum ornatissimum (Rchb. f.) J. J. Sm.

Bulbophyllum paramjitii Agrawala, M. U. Sharief & B. K. Singh

Bulbophyllum parviflorum C. S. P. Parish & Rchb. f.

Bulbophyllum penicillium C. S. P. Parish & Rchb. f.

Bulbophyllum piluliferum King & Pantl.

Bulbophyllum polyrrhizum Lindl.

Bulbophyllum protractum Hook. f.

Bulbophyllum pteroglossum Schltr.

Bulbophyllum putaoensis Q. Liu

Bulbophyllum raskotii J. J. Verm., Schuit. & de Vogel

Bulbophyllum repens Griff.

Bulbophyllum reptans (Lindl.) Lindl.

Bulbophyllum retusiusculum Rchb. f.

Bulbophyllum rigidum King & Pantl.

Bulbophyllum rolfei (Kuntze) Seidenf.

Bulbophyllum roseopictum J. J. Verm., Schuit. & de Vogel

Bulbophyllum rothschildianum (O'Brien) J. J. Sm.

Bulbophyllum roxburghii (Lindl.) Rchb. f.

Bulbophyllum salweenensis X. H. Jin

Bulbophyllum sarcophylloides Garay, Hamer & Siegerist

Bulbophyllum sarcophyllum J. J. Sm.

Bulbophyllum scabratum Rchb. f.

Bulbophyllum secundum Hook. f.

Bulbophyllum shweliense W. W. Sm.

Bulbophyllum spathulatum (Rolfe ex E. W. Cooper) Seidenf.

Bulbophyllum stenobulbon C. S. P. Parish & Rchb. f.

Bulbophyllum sterile (Lam.) Suresh

Bulbophyllum stocksii (Hook. f.) J. J. Verm., Schuit. & de Vogel

Bulbophyllum striatum (Griff.) Rchb. f.

Bulbophyllum sunipia J. J. Verm., Schuit. & de Vogel

Bulbophyllum tengchongense Z. H. Tsi

Bulbophyllum tortuosum (Blume) Lindl.

Bulbophyllum triste Rchb. f.

Bulbophyllum umbellatum Lindl.

Bulbophyllum viridiflorum (Hook. f.) Schltr.

Bulbophyllum wallichii (Lindl.) Rchb. f.

Bulbophyllum xylophyllum C. S. P. Parish & Rchb. f.

Bulbophyllum yoksunense J. J. Sm.

Bulbophyllum yunnanense Rolfe

Bulleyia yunnanensis Schltr.

Calanthe alpina Hook. f. ex Lindl.

Calanthe arcuata Rolfe

Calanthe biloba Lindl.

Calanthe brevicornu Lindl.

Calanthe chloroleuca Lindl.

Calanthe clavata Lindl.

Calanthe davidii Franch.

Calanthe densiflora Lindl.

Calanthe dulongensis H. Li, R. Li & Z. L. Dao

Calanthe ecarinata Rolfe

Calanthe fugongensis X. H. Jin & S. C. Chen

Calanthe fulgens Lindl.

Calanthe furcata var. **alismatifolia** (Lindl.) M. Hiroe

Calanthe graciliflora Hayata

Calanthe griffithii Lindl.

Calanthe hancockii Rolfe

Calanthe herbacea Lindl.

Calanthe keshabii Lucksom

Calanthe mannii Hook. f.

Calanthe metoensis Z. H. Tsi & K. Y. Lang

Calanthe nipponica Makino

Calanthe odora Griff.

Calanthe plantaginea Lindl.

Calanthe plantaginea var. **lushuiensis** K. Y. Lang & Z. H. Tsi

Calanthe puberula Lindl.

Calanthe reflexa Maxim.

Calanthe sylvatica (Thouars) Lindl.

Calanthe tricarinata Lindl.

Calanthe triplicata (Willemet) Ames

Calanthe trulliformis King & Pantl.

Calanthe whiteana King & Pantl.

Calanthe yuksomnensis Lucksom

Callostylis bambusifolia (Lindl.) S. C. Chen & J. J. Wood

Callostylis rigida Blume

Calypso bulbosa var. **speciosa** (Schltr.) Makino

Cephalanthera calcarata S. C. Chen & K. Y. Lang

Cephalanthera damasonium (Mill.) Druce

Cephalanthera falcata (Thunb.) Blume

Cephalanthera gracilis S. C. Chen & G. H. Zhu

Cephalanthera humilis X. H. Jin

Cephalanthera longifolia (L.) Fritsch

Cephalantheropsis longipes (Hook. f.) Ormerod

Cephalantheropsis obcordata (Lindl.) Ormerod

Ceratostylis himalaica Hook. f.

Ceratostylis subulata Blume

Chamaegastrodia inverta (W. W. Sm.) Seidenf.

Chamaegastrodia shikokiana Makino & F. Maek.

Cheirostylis griffithii Lindl.

Cheirostylis gunnarii A. N. Rao

Cheirostylis moniliformis (Griff.) Seidenf.

Cheirostylis orobanchoides (F. Muell.) D. L. Jones & M. A. Clem.

Cheirostylis pusilla Lindl.

Cheirostylis sessanica A. N. Rao

Cheirostylis sherriffii N. Pearce & P. J. Cribb

Cheirostylis tippica A. N. Rao
Cheirostylis yunnanensis Rolfe
Chiloschista lunifera (Rchb. f.) J. J. Sm.
Chiloschista parishii Seidenf.
Chiloschista usneoides (D. Don) Lindl.
Chiloschista yunnanensis Schltr.
Chrysoglossum assamicum Hook. f.
Chrysoglossum ornatum Blume
Cleisocentron pallens (Cathcart ex Lindl.) N. Pearce & P. J. Cribb
Cleisostoma appendiculatum (Lindl.) Benth. & Hook. f. ex B. D. Jacks.
Cleisostoma aspersum (Rchb. f.) Garay
Cleisostoma discolor Lindl.
Cleisostoma filiforme (Lindl.) Garay
Cleisostoma linearilobatum (Seidenf. & Smitinand) Garay
Cleisostoma medogense Z. H. Tsi
Cleisostoma paniculatum (Ker Gawl.) Garay
Cleisostoma racemiferum (Lindl.) Garay
Cleisostoma simondii (Gagnep.) Seidenf.
Cleisostoma striatum (Rchb. f.) N. E. Br.
Cleisostoma subulatum Blume
Cleisostoma tricallosum S. N. Hegde & A. N. Rao
Cleisostoma williamsonii (Rchb. f.) Garay
Coelogyne assamica Linden & Rchb. f.
Coelogyne barbata Lindl. ex Griff.
Coelogyne calcicola Kerr
Coelogyne corymbosa Lindl.
Coelogyne cristata Lindl.
Coelogyne ecarinata C. Schweinf.
Coelogyne fimbriata Lindl.
Coelogyne flaccida Lindl.
Coelogyne fuscescens Lindl.
Coelogyne gongshanensis H. Li ex S. C. Chen
Coelogyne griffithii Hook. f.
Coelogyne hajrae Phukan
Coelogyne holochila P. F. Hunt & Summerh.
Coelogyne leucantha W. W. Sm.
Coelogyne longipes Lindl.
Coelogyne magnifica Y. H. Tan, S. S. Zhou & B. Yang
Coelogyne micrantha Lindl.
Coelogyne nitida (Wall. ex D. Don) Lindl.
Coelogyne occultata Hook. f.
Coelogyne ovalis Lindl.
Coelogyne pianmaensis R. Li & Z. L. Dao
Coelogyne prolifera Lindl.
Coelogyne punctulata Lindl.
Coelogyne putaoensis X. H. Jin, L. A. Ye & Schuit.
Coelogyne radicosa Ridl.
Coelogyne raizadae S. K. Jain & S. Das
Coelogyne rigida C. S. P. Parish & Rchb. f.
Coelogyne schultesii S. K. Jain & S. Das
Coelogyne stricta (D. Don) Schltr.
Coelogyne suaveolens (Lindl.) Hook. f.
Coelogyne taronensis Hand.-Mazz.
Coelogyne viscosa Rchb. f.
Coelogyne weixiensis X. H. Jin
Collabium chinense (Rolfe) Tang & F. T. Wang
Collabium yunnanense Ormerod
Conchidium muscicola (Lindl.) Rauschert
Conchidium pusillum Griff.
Corallorhiza trifida Chatelain
Corybas himalaicus (King & Pantl.) Schltr.
Corybas taliensis Tang & F. T. Wang
Corymborkis veratrifolia (Reinw.) Blume
Cremastra appendiculata (D. Don) Makino
Crepidium acuminatum (D. Don) Szlach.
Crepidium bahanense (Hand.-Mazz.) S. C. Chen & J. J. Wood
Crepidium biauritum (Lindl.) Szlach.
Crepidium calophyllum (Rchb. f.) Szlach.
Crepidium josephianum (Rchb. f.) Marg.
Crepidium khasianum (Hook. f.) Szlach.

Crepidium orbiculare (W. W. Sm. & Jeffrey) Seidenf.

Crepidium purpureum (Lindl.) Szlach.

Cryptochilus luteus Lindl.

Cryptochilus sanguineus Wall.

Cymbidium aloifolium (L.) Sw.

Cymbidium cochleare Lindl.

Cymbidium concinnum Z. J. Liu & S. C. Chen

Cymbidium cyperifolium Wall. ex Lindl.

Cymbidium dayanum Rchb. f.

Cymbidium devonianum Paxton

Cymbidium eburneum Lindl.

Cymbidium elegans Lindl.

Cymbidium elegans var. **lushuiense** (Z. J. Liu, S. C. Chen & X. C. Shi) Z. J. Liu & S. C. Chen

Cymbidium ensifolium (L.) Sw.

Cymbidium ensifolium var. **munronianum** (King & Pantl.) T. Tang & F. T. Wang

Cymbidium erythraeum Lindl.

Cymbidium erythraeum var. **flavum** (Z. J. Liu & J. Yong Zhang) Z. J. Liu, S. C. Chen & P. J. Cribb

Cymbidium faberi Rolfe

Cymbidium floribundum Lindl.

Cymbidium gammieanum King & Pantl.

Cymbidium goeringii (Rchb. f.) Rchb. f.

Cymbidium gongshanense H. Li & G. H. Feng

Cymbidium hookerianum Rchb. f.

Cymbidium iridioides D. Don

Cymbidium kanran Makino

Cymbidium lancifolium Hook.

Cymbidium lowianum (Rchb. f.) Rchb. f.

Cymbidium macrorhizon Lindl.

Cymbidium mannii Rchb. f.

Cymbidium mastersii Griff. ex Lindl.

Cymbidium parishii Rchb. f.

Cymbidium sichuanicum Z. J. Liu & S. C. Chen

Cymbidium sinense (Jacks.) Willd.

Cymbidium tigrinum Parish ex Hook.

Cymbidium tracyanum L. Castle

Cymbidium whiteae King & Pantl.

Cypripedium bardolphianum W. W. Sm. & Farrer

Cypripedium calcicola Schltr.

Cypripedium cordigerum D. Don

Cypripedium debile Rchb. f.

Cypripedium elegans Rchb. f.

Cypripedium fargesii Franch.

Cypripedium farreri W. W. Sm.

Cypripedium fasciolatum Franch.

Cypripedium flavum P. F. Hunt & Summerh.

Cypripedium forrestii P. J. Cribb

Cypripedium franchetii E. H. Wilson

Cypripedium guttatum Sw.

Cypripedium henryi Rolfe

Cypripedium himalaicum Rolfe

Cypripedium lichiangense S. C. Chen & P. J. Cribb

Cypripedium ludlowii P. J. Cribb

Cypripedium margaritaceum Franch.

Cypripedium micranthum Franch.

Cypripedium palangshanense Tang & F. T. Wang

Cypripedium plectrochilum Franch.

Cypripedium sichuanense Perner

Cypripedium subtropicum S. C. Chen & K. Y. Lang

Cypripedium tibeticum King ex Rolfe

Cypripedium wardii Rolfe

Cypripedium yunnanense Franch.

Cyrtosia javanica Blume

Dactylorhiza hatagirea (D. Don) Soó

Dactylorhiza kafiriana Renz

Dactylorhiza umbrosa (Kar. & Kir.) Nevski

Dactylorhiza viridis (L.) R. M. Bateman, Pridgeon & M. W. Chase

Dendrobium acinaciforme Roxb.
Dendrobium aduncum Wall. ex Lindl.
Dendrobium albopurpureum (Seidenf.) Schuit. & Peter B. Adams
Dendrobium amoenum Wall. ex Lindl.
Dendrobium amplum Lindl.
Dendrobium anceps Sw.
Dendrobium aphyllum (Roxb.) C. E. C. Fisch.
Dendrobium arunachalense C. Deori, S. K. Sarma, Phukan & A. A. Mao
Dendrobium bicameratum Lindl.
Dendrobium brunneum Schuit. & Peter B. Adams
Dendrobium capillipes Rchb. f.
Dendrobium chrysanthum Wall. ex Lindl.
Dendrobium chryseum Rolfe
Dendrobium crepidatum Lindl. & Paxton
Dendrobium cumulatum Lindl.
Dendrobium darjeelingense Pradhan
Dendrobium denneanum Kerr
Dendrobium densiflorum Wall.
Dendrobium denudans D. Don
Dendrobium devonianum Paxton
Dendrobium eriiflorum Griff.
Dendrobium falconeri Hook.
Dendrobium fargesii Finet
Dendrobium farmeri Paxton
Dendrobium fimbriatum Hook.
Dendrobium formosum Roxb. ex Lindl.
Dendrobium forrestii (Ormerod) Schuit. & Peter B. Adams
Dendrobium fugax Rchb. f.
Dendrobium fuscescens Griff.
Dendrobium gaoligongense (Hong Yu & S. G. Zhang) Schuit. & Peter B. Adams
Dendrobium gibsonii Lindl.
Dendrobium heterocarpum Wall. ex Lindl.
Dendrobium hookerianum Lindl.
Dendrobium jenkinsii Wall. ex Lindl.
Dendrobium khasianum Deori
Dendrobium lindleyi Steudel
Dendrobium lituiflorum Lindl.
Dendrobium longicornu Lindl.
Dendrobium macraei Lindl.
Dendrobium macrostachyum Lindl.
Dendrobium moniliforme (L.) Sw.
Dendrobium monticola P. F. Hunt & Summerh.
Dendrobium moschatum (Buch.-Ham.) Sw.
Dendrobium nareshbahadurii H. B. Naithani
Dendrobium naungmungense Q. Liu & X. H. Jin
Dendrobium nobile Lindl.
Dendrobium palpebrae Lindl.
Dendrobium parciflorum Lindl.
Dendrobium parishii Rchb. f.
Dendrobium peguanum Lindl.
Dendrobium pendulum Roxb.
Dendrobium polyanthum Wall. ex Lindl.
Dendrobium porphyrochilum Lindl.
Dendrobium praecinctum Rchb. f.
Dendrobium pulchellum Roxb. ex Lindl.
Dendrobium ritaeanum King & Pantl.
Dendrobium rotundatum (Lindl.) Hook. f.
Dendrobium ruckeri Lindl.
Dendrobium salaccense (Blume) Lindl.
Dendrobium spatella Rchb. f.
Dendrobium stuposum Lindl.
Dendrobium sulcatum Lindl.
Dendrobium terminale C. S. P. Parish & Rchb. f.
Dendrobium transparens Wall. ex Lindl.
Dendrobium treutleri (Hook. f.) Schuit. & Peter B. Adams
Dendrobium vexabile Rchb. f.
Dendrobium wardianum R. Warner
Dickasonia vernicosa L. O. Williams

Didymoplexis pallens Griff.

Dienia cylindrostachya Lindl.

Dienia ophrydis (J. Koenig) Seidenf.

Diglyphosa latifolia Blume

Diplomeris hirsuta (Lindl.) Lindl.

Diplomeris josephi A. N. Rao & Swamin.

Diplomeris pulchella D. Don

Diploprora truncata Rolfe ex Downie

Epipactis helleborine (L.) Crantz

Epipactis humilior (Tang & F. T. Wang) S. C. Chen & G. H. Zhu

Epipactis mairei Schltr.

Epipactis persica (Soó) Hausskn. ex Nannf.

Epipactis royleana Lindl.

Epipactis veratrifolia Boiss. & Hohen.

Epipogium aphyllum Sw.

Epipogium japonicum Makino

Epipogium roseum (D. Don) Lindl.

Eria apertiflora Summerh.

Eria arunachalensis A. N. Rao

Eria baniae Bajrach., L. R. Shakya & Chettri

Eria biflora Griff.

Eria carinata Gibson ex Lindl.

Eria clausa King & Pantl.

Eria clavicaulis Wall. ex Lindl.

Eria coronaria (Lindl.) Rchb. f.

Eria cristata Rolfe

Eria ferruginea Lindl.

Eria gagnepainii A. D. Hawkes & A. H. Heller

Eria globulifera Seidenf.

Eria gloensis Ormerod & Agrawala

Eria javanica (Sw.) Blume

Eria jengingensis S. N. Hegde

Eria kamlangensis A. N. Rao

Eria lasiopetala (Willd.) Ormerod

Eria lohitensis A. N. Rao, Harid. & S. N. Hegde

Eria nepalensis Bajrach. & K. K. Shrestha

Eria sharmae H. J. Chowdhery, G. S. Giri & G. D. Pal

Eria tenuicaulis S. C. Chen & Z. H. Tsi

Eria tomentosa (J. Koenig) Hook. f.

Eria vittata Lindl.

Eriodes barbata (Lindl.) Rolfe

Erythrodes blumei (Lindl.) Schltr.

Erythrodes hirsuta (Griff.) Ormerod

Erythrorchis altissima (Blume) Blume

Esmeralda bella Rchb. f.

Esmeralda cathcartii (Lindl.) Rchb. f.

Esmeralda clarkei Rchb. f.

Eulophia bicallosa (D. Don) P. F. Hunt & Summerh.

Eulophia bracteosa Lindl.

Eulophia burmanica Hook. f.

Eulophia dabia (D. Don) Hochr.

Eulophia densiflora Lindl.

Eulophia explanata Lindl.

Eulophia flava (Lindl.) Hook. f.

Eulophia graminea Lindl.

Eulophia herbacea Lindl.

Eulophia mackinnonii Duthie

Eulophia macrobulbon (Parish & Rchb. f.) Hook. f.

Eulophia mannii (Rchb. f.) Hook. f.

Eulophia monantha W. W. Sm.

Eulophia obtusa (Lindl.) Hook. f.

Eulophia promensis Lindl.

Eulophia spectabilis (Dennst.) Suresh

Eulophia stenopetala Lindl.

Eulophia zollingeri (Rchb. f.) J. J. Sm.

Galearis roborowskyi (Maxim.) S. C. Chen, P. J. Cribb & S. W. Gale

Galearis spathulata (Lindl.) P. F. Hunt

Galearis tschiliensis (Schltr.) S. C. Chen, P. J. Cribb & S. W. Gale

Galearis wardii (W. W. Sm.) P. F. Hunt

Galeola cathcartii Hook. f.

Galeola faberi Rolfe

Galeola falconeri Hook. f.

Galeola lindleyana Hook. f. & Thomson Rchb. f.

Galeola nudifolia Lour.

Gastrochilus acutifolius (Lindl.) Kuntze

Gastrochilus affinis (King & Pantl.) Schltr.

Gastrochilus alatus X. H. Jin & S. C. Chen

Gastrochilus arunachalensis A. N. Rao

Gastrochilus calceolaris (Buch.-Ham. ex Sm.) D. Don

Gastrochilus corymbosus A. P. Das & S. Chanda

Gastrochilus dasypogon (Sm.) Kuntze

Gastrochilus distichus (Lindl.) Kuntze

Gastrochilus dulongjiangensis Qiang Liu & J. Y. Gao

Gastrochilus inconspicuus (Hook. f.) Kuntze

Gastrochilus intermedius (Griff. ex Lindl.) Kuntze

Gastrochilus linearifolius Z. H. Tsi & Garay

Gastrochilus obliquus (Lindl.) Kuntze

Gastrochilus obliquus var. **suavis** (Seidenf.) Z. H. Tsi

Gastrochilus pseudodistichus (King & Pantl.) Schltr.

Gastrochilus rutilans Seidenf.

Gastrochilus sessanicus A. N. Rao

Gastrochilus sinensis Z. H. Tsi

Gastrodia angusta S. Chow & S. C. Chen

Gastrodia arunachalensis S. N. Hegde & A. N. Rao

Gastrodia dyeriana King & Pantl.

Gastrodia elata Blume

Gastrodia kachinensis X. H. Jin & L. A. Ye

Gastrodia mishmensis A. N. Rao, Harid. & S. N. Hegde

Gastrodia putaoensis X. H. Jin

Gennaria griffithii (Hook. f.) X. H. Jin & D. Z. Li

Geodorum densiflorum (Lam.) Schltr.

Geodorum pallidum D. Don

Goodyera alveolata Pradhan

Goodyera biflora (Lindl.) Hook. f.

Goodyera bomiensis K. Y. Lang

Goodyera clavata N. Pearce & P. J. Cribb

Goodyera dongchenii Lucksom

Goodyera dongchenii var. **gongligongensis** X. H. Jin & S. C. Chen

Goodyera fumata Thwaites

Goodyera fusca (Lindl.) Hook. f.

Goodyera hemsleyana King & Pantl.

Goodyera henryi Rolfe

Goodyera hispida Lindl.

Goodyera makuensis Ormerod

Goodyera nankoensis Fukuy.

Goodyera procera (Ker Gawler) Hook.

Goodyera recurva Lindl.

Goodyera repens (L.) R. Br.

Goodyera robusta Hook. f.

Goodyera rubicunda (Blume) Lindl.

Goodyera schlechtendaliana Rchb. f.

Goodyera velutina Maxim. ex Regel

Goodyera viridiflora (Blume) Blume

Goodyera vittata (Lindl.) Benth. ex Hook. f.

Goodyera wolongensis K. Y. Lang

Goodyera wuana Tang & F. T. Wang

Goodyera yunnanensis Schltr.

Gymnadenia bicornis Tang & K. Y. Lang

Gymnadenia conopsea (L.) R. Br.

Gymnadenia crassinervis Finet

Gymnadenia orchidis Lindl.

Habenaria acianthoides Schltr.

Habenaria acuifera Wall. ex Lindl.

Habenaria aitchisonii Rchb. f.

Habenaria arietina Hook. f.

Habenaria balfouriana Schltr.

Habenaria commelinifolia (Roxb.) Wall. ex Lindl.

Habenaria davidii Franch.

Habenaria delavayi Finet

Habenaria dentata (Sw.) Schltr.

Habenaria digitata Lindl.

Habenaria diphylla (Nimmo) Dalzell

Habenaria diplonema Schltr.

Habenaria fargesii Finet

Habenaria finetiana Schltr.

Habenaria furcifera Lindl.

Habenaria intermedia D. Don

Habenaria limprichtii Schltr.

Habenaria linguella Lindl.

Habenaria longifolia Buch.-Ham. ex Lindl.

Habenaria mairei Schltr.

Habenaria malintana (Blanco) Merr.

Habenaria marginata Colebr.

Habenaria nepalensis Kolan.

Habenaria pantlingiana Kraenzl.

Habenaria pectinata D. Don

Habenaria pubescens Lindl.

Habenaria reniformis (D. Don) Hook. f.

Habenaria shweliensis W. W. Sm. & S. C. Banerji

Habenaria stenopetala Lindl.

Habenaria szechuanica Schltr.

Habenaria tibetica Schltr.

Habenaria trichosantha Lindl.

Habenaria wolongensis K. Y. Lang

Habenaria yuana Tang & F. T. Wang

Hemipilia cordifolia Lindl.

Hemipilia crassicalcarata S. S. Chien

Hemipilia flabellata Bureau & Franch.

Hemipilia forrestii Rolfe

Hemipilia henryi Rolfe

Hemipilia limprichtii Schltr.

Hemipilia purpureopunctata (K. Y. Lang) X. H. Jin, Schuit. & W. T. Jin

Herminium alaschanicum Maxim.

Herminium albomarginatum (King & Pantl.) X. H. Jin, Schuit., Raskoti & L. Q. Huang

Herminium albosanguineum (Renz) X. H. Jin, Schuit., Raskoti & L. Q. Huang

Herminium albovirens (Renz) X. H. Jin, Schuit., Raskoti & L. Q. Huang

Herminium biporosum Maxim.

Herminium bulleyi (Rolfe) Tang & F. T. Wang

Herminium chloranthum Tang & F. T. Wang

Herminium clavigerum (Lindl.) X. H. Jin, Schuit., Raskoti & L. Q. Huang

Herminium coiloglossum Schltr.

Herminium ecalcaratum (Finet) Schltr.

Herminium edgeworthii (Hook. f. ex Collett) X. H. Jin, Schuit., Raskoti & L. Q. Huang

Herminium elisabethae (Duthie) Tang & F. T. Wang

Herminium fallax (Lindl.) Hook. f.

Herminium fimbriatum (Raskoti) X. H. Jin, Schuit., Raskoti & L. Q. Huang

Herminium forceps (Finet) Schltr.

Herminium glossophyllum Tang & F. T. Wang

Herminium gongganum Ormerod

Herminium gracile King & Pantl.

Herminium handelii X. H. Jin, Schuit., Raskoti & L. Q. Huang

Herminium himalayanum (Renz) X. H. Jin, Schuit., Raskoti & L. Q. Huang

Herminium hongdeyuanii Raskoti

Herminium jaffreyanum King & Pantl.

Herminium josephi Rchb. f.

Herminium kalimpongense Pradhan

Herminium kamengense A. N. Rao

Herminium kumaunensis Deva & H. B. Naithani

Herminium lanceum (Thunb. ex Sw.) Vuijk

Herminium latilabre (Lindl.) X. H. Jin, Schuit., Raskoti & L. Q. Huang

Herminium longilobatum S. N. Hegde & A. N. Rao

Herminium mackinonii Duthie

Herminium macrophyllum (D. Don) Dandy

Herminium mannii (Rchb. f.) Tang & F. T. Wang

Herminium monophyllum (D. Don) P. F. Hunt & Summerh.

Herminium monorchis (L.) R. Br.

Herminium motuoensis X. H. Jin

Herminium neotineoides Ames & Schltr.

Herminium ophioglossoides Schltr.

Herminium oxysepalum (K. Y. Lang) X. H. Jin, Schuit., Raskoti & L. Q. Huang

Herminium pugioniforme Lindl. ex Hook. f.

Herminium pygmaeum Renz

Herminium quinquelobum King & Pantl.

Herminium singulum Tang & F. T. Wang

Herminium souliei (Finet) Rolfe

Herminium suave Tang & F. T. Wang

Herminium tangianum (S. Y. Hu) K. Y. Lang

Herminium tibeticum X. H. Jin, Schuit. & Raskoti

Herminium wangianum X. H. Jin, Schuit., Raskoti & L. Q. Huang

Herminium yunnanense Rolfe

Herpysma longicaulis Lindl.

Hetaeria affinis (Griff.) Seidenf. & Ormerod

Hetaeria pelota N. Pearce & P. J. Cribb

Holcoglossum flavescens (Schltr.) Z. H. Tsi

Holcoglossum gaoligongense (G. Q. Zhang, K. Wei Liu & Z. J. Liu) Kumar & S. W. Gale

Holcoglossum kimballianum (Rchb. f.) Garay

Holcoglossum nagalandensis (Phukan & Odyuo) X. H. Jin

Holcoglossum nujiangense X. H. Jin & S. C. Chen

Holcoglossum rupestre (Hand.-Mazz.) Garay

Holcoglossum weixiense X. H. Jin & S. C. Chen

Hsenhsua chrysea (W. W. Sm) X. H. Jin, Schuit., W. T. Jin & L. Q. Huang

Ischnogyne mandarinorum (Kraenzl.) Schltr.

Kingidium taenialis P. F. Hunt

Lecanorchis sikkimensis N. Pearce & P. J. Cribb

Liparis assamica King & Pantl.

Liparis balansae Gagnep.

Liparis barbata Lindl.

Liparis bistriata C. S. P. Parish & Rchb. f.

Liparis bootanensis Griff.

Liparis campylostalix Rchb. f.

Liparis cathcartii Hook. f.

Liparis cespitosa (Lam.) Lindl.

Liparis chapaensis Gagnep.

Liparis cheniana X. H. Jin

Liparis cordifolia Hook. f.

Liparis deflexa Hook. f.

Liparis delicatula Hook. f.

Liparis distans C. B. Clarke

Liparis elliptica Wight

Liparis fargesii Finet

Liparis gamblei Hook. f.

Liparis gigantea C. L. Tso

Liparis glossula Rchb. f.

Liparis grossa Rchb. f.

Liparis luteola Lindl.

Liparis lydiae Lucksom

Liparis mannii Rchb. f.

Liparis nervosa (Thunb.) Lindl.

Liparis odorata (Willd.) Lindl.

Liparis perpusilla Hook. f.

Liparis petiolata (D. Don) P. F. Hunt & Summerh.

Liparis plantaginea Lindl.

Liparis platyrachis Hook. f.

Liparis pygmaea King & Pantl.

Liparis resupinata Ridl.

Liparis rockii Ormerod

Liparis rostrata Rchb. f.

Liparis somae Hayata

Liparis stricklandiana Rchb. f.

Liparis tigerhillensis A. P. Das & S. Chanda

Liparis tschangii Schltr.

Liparis viridiflora (Blume) Lindl.

Listera fugongensis X. H. Jin

Luisia brachystachys (Lindl.) Blume

Luisia filiformis Hook. f.

Luisia psyche Rchb. f.

Luisia teres (Thunb.) Blume

Luisia trichorrhiza (Hook.) Blume

Luisia tristis (G. Forst.) Hook. f.

Malaxis monophyllos (L.) Sw.

Malaxis muscifera (Lindl.) Kuntze

Malleola tibetica W. C. Huang & X. H. Jin

Micropera mannii (Hook. f.) Tang & F. T. Wang

Micropera obtusa (Lindl.) Tang & F. T. Wang

Micropera rostrata (Roxb.) N. P. Balakr.

Micropera tibetica X. H. Jin & Y. J. Lai

Monomeria barbata Lindl.

Mycaranthes floribunda (D. Don) S. C. Chen & J. J. Wood

Mycaranthes pannea (Lindl.) S. C. Chen & J. J. Wood

Myrmechis chinensis Rolfe

Myrmechis gracilis (Blume) Blume

Myrmechis japonica (Rchb. f.) Rolfe

Myrmechis pumila (Hook. f.) Tang & F. T. Wang

Neofinetia falcata (Thunb.) Hu

Neofinetia xichangensis Z. J. Liu & S. C. Chen

Neogyna gardneriana (Lindl.) Rchb. f.

Neottia acuminata Schltr.

Neottia alternifolia (King & Pantl.) Szlach.

Neottia bambusetorum (Hand.-Mazz.) Szlach.

Neottia biflora (Schltr.) Szlach.

Neottia brevicaulis (King & Pantl.) Szlach.

Neottia chenii S. W. Gale & P. J. Cribb

Neottia confusa M. Bhaumik

Neottia dentata (King & Pantl.) Szlach.

Neottia dihangensis M. Bhaumik

Neottia divaricata (Panigrahi & P. Taylor) Szlach.

Neottia fangii (Tang & F. T. Wang ex S. C. Chen & G. H. Zhu) S. C. Chen, S. W. Gale & P. J. Cribb

Neottia gaudissartii Hand.-Mazz.

Neottia karoana Szlach.

Neottia listeroides Lindl.

Neottia longicaulis (King & Pantl.) Szlach.

Neottia megalochila S. C. Chen

Neottia microphylla (S. C. Chen & Y. B. Luo) S. C. Chen, S. W. Gale & P. J. Cribb

Neottia mucronata (Panigrahi & J. J. Wood) Szlach.

Neottia nepalensis (N. P. Balakr.) Szlach.

Neottia ovata (L.) Bluff & Fingerh.

Neottia pantlingii (W. W. Sm.) Tang & F. T. Wang

Neottia pinetorum (Lindl.) Szlach.

Neottia puberula (Maxim.) Szlach.

Neottia smithiana Schltr.

Neottia smithii (Schltr.) Szlach.

Neottia tenuis (Lindl.) Szlach.

Neottia wardii (Rolfe) Szlach.

Nervilia concolor (Blume) Schltr.

Nervilia crociformis (Zoll. & Mor.) Seidenf.

Nervilia falcata (King & Pantl.) Schltr.

Nervilia fordii (Hance) Schltr.

Nervilia gammieana (Hook. f.) Schltr.

Nervilia hookeriana (King & Pantl.) Schltr.

Nervilia infundibulifolia Blatt. & McCann

Nervilia juliana (Roxb.) Schltr.

Nervilia mackinnonii (Duthie) Schltr.

Nervilia macroglossa (Hook. f.) Schltr.

Nervilia petraea (Afzel. ex Sw.) Summerh.

Nervilia plicata (Andrews) Schltr.

Oberonia acaulis Griff.

Oberonia angustifolia Lindl.

Oberonia anthropophora Lindl.

Oberonia arunachalensis A. N. Rao

Oberonia bopannae Chowlu & Kumar

Oberonia caulescens Lindl.

Oberonia cavaleriei Finet

Oberonia emarginata King & Pantl.

Oberonia ensiformis (Sm.) Lindl.

Oberonia falcata King & Pantl.

Oberonia falconeri Hook. f.

Oberonia gongshanensis Ormerod

Oberonia helferi Hook. f.

Oberonia jenkinsiana Griff. ex Lindl.

Oberonia jhae Chowlu & Rab

Oberonia kamlangensis A. N. Rao

Oberonia katakiana A. N. Rao

Oberonia kwangsiensis Seidenf.

Oberonia latipetala L. O. Williams

Oberonia lobulata King & Pantl.

Oberonia mannii Hook. f.

Oberonia maxima C. S. P. Parish ex Hook. f.

Oberonia mucronata (D. Don) Ormerod & Seidenf.

Oberonia obcordata Lindl.

Oberonia pachyphylla King & Pantl.

Oberonia pachyrachis Hook. f.

Oberonia prainiana King & Pantl.

Oberonia pumilio Rchb. f.

Oberonia pyrulifera Lindl.

Oberonia recurva Lindl.

Oberonia rufilabris Lindl.

Oberonia sulcata Joseph & Sud. Chowdhury

Oberonia teres Kerr

Odontochilus asraoa (J. Joseph & Abbar.) Ormerod

Odontochilus brevistylus Hook. f.

Odontochilus clarkei Hook. f.

Odontochilus crispus (Lindl.) Hook. f.

Odontochilus elwesii C. B. Clarke ex Hook. f.

Odontochilus lanceolatus (Lindl.) Blume

Odontochilus nandae Raskoti & Kurzweil

Odontochilus poilanei (Gagnep.) Ormerod

Odontochilus putaoensis X. H. Jin, L. A. Ye & A. T. Mu

Odontochilus tortus King & Pantl.

Oreorchis angustata L. O. Williams ex N. Pearce & P. J. Cribb

Oreorchis aurantiaca P. J. Cribb & N. Pearce

Oreorchis discigera W. W. Sm.

Oreorchis erythrochrysea Hand.-Mazz.

Oreorchis fargesii Finet

Oreorchis foliosa (Lindl.) Lindl.

Oreorchis indica (Lindl.) Hook. f.

Oreorchis micrantha Lindl.

Oreorchis nana Schltr.

Oreorchis nepalensis N. Pearce & P. J. Cribb

Oreorchis oligantha Schltr.

Oreorchis parvula Schltr.

Oreorchis patens (Lindl.) Lindl.

Oreorchis porphyranthes Tuyama

Oreorchis sanguinea (N. Pearce & P. J. Cribb) N. Pearce & P. J. Cribb

Ornithochilus difformis (Wall. ex Lindl.) Schltr.

Otochilus albus Lindl.

Otochilus fuscus Lindl.

Otochilus lancilabius Seidenf.

Otochilus porrectus Lindl.

Pachystoma pubescens Blume

Panisea demissa (D. Don) Pfitzer

Panisea tricallosa Rolfe

Panisea uniflora (Lindl.) Lindl.

Paphiopedilum areeanum O. Gruss

Paphiopedilum armeniacum S. C. Chen & F. Y. Liu

Paphiopedilum charlesworthii (Rolfe) Pfitzer

Paphiopedilum fairrieanum (Lindl.) Stein

Paphiopedilum insigne (Wall. ex Lindl.) Pfitzer

Paphiopedilum spicerianum (Rchb. f.) Pfitzer

Paphiopedilum tigrinum Koop. & N. Haseg.

Paphiopedilum venustum (Wall. ex Sims) Pfitzer

Paphiopedilum wardii Summerh.

Papilionanthe biswasiana (Ghose & Mukerjee) Garay

Papilionanthe greenii (W. W. Sm.) Garay

Papilionanthe subulata (Willd.) Garay

Papilionanthe teres (Roxb.) Schltr.

Papilionanthe uniflora (Lindl.) Garay

Papilionanthe vandarum (Rchb. f.) Garay

Pecteilis susannae (L.) Rafinesque

Pecteilis triflora (D. Don) Tang & F. T. Wang

Pennilabium proboscidcum A. S. Rao & J. Joseph

Pennilabium struthio Carr

Peristylus affinis (D. Don) Seidenf.

Peristylus aristatus Lindl.

Peristylus calcaratus (Rolfe) S. Y. Hu

Peristylus coeloceras Finet

Peristylus constrictus (Lindl.) Lindl.

Peristylus densus (Lindl.) Santapau & Kapadia

Peristylus goodyeroides (D. Don) Lindl.

Peristylus hamiltonianus (Lindl.) Lindl.

Peristylus jinchuanicus K. Y. Lang

Peristylus lacertifer (Lindl.) J. J. Sm.

Peristylus lawii Wight

Peristylus nematocaulon (Hook. f.) Banerji & P. Pradhan

Peristylus orbicularis (Hook. f.) Agrawala, H. J. Chowdhery & S. Choudhury

Peristylus parishii Rchb. f.

Peristylus prainii Kranzlin

Peristylus pseudophrys (King & Pantl.) Kraenzl.

Peristylus richardianus Wight

Peristylus superanthus J. J. Wood

Peristylus tentaculatus (Lindl.) J. J. Sm.

Peristylus tipuliferus (C. S. P. Parish & Rchb. f.) Mukerjee

Phaius delavayi (Finet) P. J. Cribb & Perner

Phaius flavus (Blume) Lindl.

Phaius mishmensis (Lindl. & Paxton) Rchb. f.

Phaius nanus Hook. f.

Phaius tankervilleae (Banks) Blume

Phaius wallichii Lindl.

Phalaenopsis decumbens (Griff.) Holttum

Phalaenopsis deliciosa Rchb. f.

Phalaenopsis lobbii (Rchb. f.) H. R. Sweet

Phalaenopsis mannii Rchb. f.

Phalaenopsis marriottiana var. **parishii** (Rchb. f.) Kocyan & Schuit.

Phalaenopsis stobartiana Rchb. f.

Phalaenopsis taenialis (Lindl.) Christenson & Pradhan

Phalaenopsis wilsonii Rolfe

Pholidota articulata Lindl.

Pholidota chinensis Lindl.

Pholidota convallariae (E. C. Parish & Rchb. f.) Hook. f.

Pholidota imbricata Hook.

Pholidota missionariorum Gagnep.

Pholidota pallida Lindl.

Pholidota protracta Hook. f.

Pholidota pygmaea H. J. Chowdhery & G. D. Pal

Pholidota recurva Lindl.

Pholidota rubra Lindl.

Pholidota wattii King & Pantl.

Phreatia elegans Lindl.

Pinalia acervata (Lindl.) Kuntze

Pinalia amica (Rchb. f.) Kuntze

Pinalia bipunctata (Lindl.) Kuntze

Pinalia bractescens (Lindl.) Kuntze

Pinalia concolor (E. C. Parish & Rchb. f.) Kuntze

Pinalia conferta (S. C. Chen & Z. H. Tsi) S. C. Chen & J. J. Wood

Pinalia connata (J. Joseph, S. N. Hegde & Abbar.) Ormerod & E. W. Wood

Pinalia excavata (Lindl.) Kuntze

Pinalia graminifolia (Lindl.) Kuntze

Pinalia leucantha Kuntze

Pinalia pumila (Lindl.) Kuntze

Pinalia salwinensis (Hand.-Mazz.) Ormerod

Pinalia spicata (D. Don) S. C. Chen & J. J. Wood

Pinalia stricta (Lindl.) Kuntze

Pinalia szetschuanica (Schltr.) S. C. Chen & J. J. Wood

Platanthera anatina Ormerod

Platanthera angustilabris Seidenf.

Platanthera arcuata Lindl.

Platanthera bakeriana (King & Pantl.) Kraenzl.

Platanthera bhutanica K. Inoue

Platanthera biermanniana (King & Pantl.) Kranzl.

Platanthera calceoliforme (W. W. Sm.) X. H. Jin, Schuit. & W. T. Jin

Platanthera chiloglossa (Tang & F. T. Wang) K. Y. Lang

Platanthera chlorantha (Custer) Rchb.

Platanthera contigua Tang & F. T. Wang

Platanthera curvata K. Y. Lang

Platanthera danghatuensis Ormerod

Platanthera dulongensis X. H. Jin & Efimov

Platanthera dyeriana (King & Pantl.) Kraenzl.

Platanthera exelliana Soó

Platanthera finetiana Schltr.

Platanthera fugongensis Ormerod

Platanthera fuscescens (L.) Kraenzl.

Platanthera handel-mazzettii K. Inoue

Platanthera herminioides Tang & F. T. Wang

Platanthera hologlottis Maxim.

Platanthera japonica (Thunb.) Lindl.

Platanthera latilabris Lindl.

Platanthera leptocaulon (Hook. f.) Soó

Platanthera likiangensis Tang & F. T. Wang

Platanthera minutiflora Schltr.

Platanthera ophiocephala (W. W. Sm.) Tang & F. T. Wang

Platanthera orbicularis (Hook. f.) X. H. Jin, Schuit. & Raskoti

Platanthera oreophila Schltr.

Platanthera ovatilabris X. H. Jin & Efimov

Platanthera roseotincta (W. W. Sm.) Tang & Wang

Platanthera sikkimensis (Hook. f.) Kraenzl.

Platanthera sinica Tang & F. T. Wang

Platanthera stenantha (Hook. f.) Soó

Platanthera stenochila X. H. Jin , Schuit., Raskoti & Lu Q. Huang

Platanthera stenophylla Tang & F. T. Wang

Platanthera uniformis Tang & F. T. Wang

Platanthera urceolata (Hook. f.) R. M. Bateman

Platanthera yadongensis X. H. Jin & W. T. Jin

Pleione × confusa P. J. Cribb & C. Z. Tang

Pleione × taliensis P. J. Cribb & Butterfield

Pleione albiflora P. J. Cribb & C. Z. Tang

Pleione bulbocodioides (Franch.) Rolfe

Pleione coronaria P. J. Cribb & C. Z. Tang

Pleione forrestii Schltr.

Pleione grandiflora (Rolfe) Rolfe

Pleione hookeriana (Lindl.) Rollisson

Pleione humilis (Sm.) D. Don

Pleione kaatiae P. H. Peeters

Pleione limprichtii Schltr.

Pleione maculata (Lindl.) Lindl. & Paxton

Pleione praecox (Sm.) D. Don

Pleione saxicola Tang & F. T. Wang ex S. C. Chen

Pleione scopulorum W. W. Sm.

Pleione yunnanensis (Rolfe) Rolfe

Podochilus cultratus Lindl.

Podochilus khasianus Hook. f.

Pogonia yunnanensis Finet

Polystachya concreta (Jacq.) Garay & H. R. Sweet

Pomatocalpa armigerum (King & Pantl.) Tang & F. T. Wang

Pomatocalpa bambusarum (King & Pantl.) Garay

Pomatocalpa bhutanicum N. P. Balakr.

Pomatocalpa spicatum Breda

Pomatocalpa undulatum (Lindl.) J. J. Sm.

Ponerorchis amplexifolia (Tang & F. T. Wang) X. H. Jin, Schuit. & W. T. Jin

Ponerorchis basifoliata (Finet) X. H. Jin, Schuit. & W. T. Jin

Ponerorchis bifoliata (Tang & F. T. Wang) X. H. Jin, Schuit. & W. T. Jin

Ponerorchis brevicalcarata (Finet) Soó

Ponerorchis camptoceras (Rolfe) X. H. Jin, Schuit. & W. T. Jin

Ponerorchis capitata (Tang & F. T. Wang) X. H. Jin, Schuit. & W. T. Jin

Ponerorchis chusua (D. Don) Soó

Ponerorchis compacta (Schltr.) X. H. Jin, Schuit. & W. T. Jin

Ponerorchis crenulata Soó

Ponerorchis cucullata (L.) X. H. Jin, Schuit. & W. T. Jin

Ponerorchis cucullata var. **calcicola** (W. W. Sm.) X. H. Jin, Schuit. & W. T. Jin

Ponerorchis dolichocentra (Tang, F. T. Wang & K. Y. Lang) X. H. Jin, Schuit. & W. T. Jin

Ponerorchis faberi (Rolfe) X. H. Jin, Schuit. & W. T. Jin

Ponerorchis farreri (Schltr.) X. H. Jin, Schuit. & W. T. Jin

Ponerorchis gonggashanica (K. Y. Lang) X. H. Jin, Schuit. & W. T. Jin

Ponerorchis gracilis (Blume) X. H. Jin, Schuit. & W. T. Jin

Ponerorchis hemipiloides (Finet) Soó

Ponerorchis limprichtii (Schltr.) Soó

Ponerorchis luteola (K. Y. Lang & S. C. Chen) X. H. Jin, Schuit. & W. T. Jin

Ponerorchis monantha (Finet) X. H. Jin, Schuit. & W. T. Jin

Ponerorchis oblonga (K. Y. Lang) X. H. Jin, Schuit. & W. T. Jin

Ponerorchis ovata (K. Y. Lang) X. H. Jin, Schuit. & W. T. Jin

Ponerorchis papilionacea (Tang, F. T. Wang & K. Y. Lang) X. H. Jin, Schuit. & W. T. Jin

Ponerorchis physoceras (Schltr.) X. H. Jin, Schuit. & W. T. Jin

Ponerorchis puberula (King & Pantl.) Verm.

Ponerorchis pugeensis (K. Y. Lang) S. C. Chen, P. J. Cribb & S. W. Gale

Ponerorchis pulchella (Hand.-Mazz.) Soó

Ponerorchis secundiflora (Hook. f.) X. H. Jin

Ponerorchis simplex (Tang & F. T. Wang) X. H. Jin, Schuit. & W. T. Jin

Ponerorchis tetraloba (Finet) X. H. Jin, Schuit. & W. T. Jin

Ponerorchis tibetica (Schltr.) X. H. Jin, Schuit. & W. T. Jin

Ponerorchis trifurcata (Tang, F. T. Wang & K. Y. Lang) X. H. Jin, Schuit. & W. T. Jin

Porpax elwesii Rolfe

Porpax fibuliformis (King & Pantl.) King & Pantl.

Porpax meirax (Parish & Rchb. f.) King & Pantl.

Porpax seidenfadenii A. N. Rao

Pteroceras teres (Blume) Holttum

Renanthera imschootiana Rolfe

Rhomboda abbreviata (Lindl.) Ormerod

Rhomboda arunachalensis A. N. Rao

Rhomboda lanceolata (Lindl.) Ormerod

Rhomboda longifolia Lindl.

Rhomboda moulmeinensis (C. S. P. Parish & Rchb. f.) Ormerod

Rhynchostylis retusa (L.) Blume

Risleya atropurpurea King & Pantl.

Robiquetia arunachalensis (A. N. Rao) Kocyan & Schuit.

Robiquetia spatulata (Blume) J. J. Sm.

Robiquetia succisa (Lindl.) Seidenf. & Garay

Saccolabiopsis pusilla (Lindl.) Seidenf. & Garay

Sarcoglyphis arunachalensis A. N. Rao

Satyrium nepalense D. Don

Satyrium nepalense var. **ciliatum** (Lindl.) Hook. f.

Satyrium yunnanense Rolfe

Schoenorchis gemmata (Lindl.) J. J. Sm.

Seidenfadeniella rosea (Wight) C. S. Kumar

Senghasiella glaucifolia (Bureau & Franch.) Szlach.

Smitinandia micrantha (Lindl.) Holttum

Spathoglottis ixioides (D. Don) Lindl.

Spathoglottis plicata Blume

Spathoglottis pubescens Lindl.

Spiranthes sinensis (Pers.) Ames

Spiranthes sunii Boufford & Wen H. Zhang

Staurochilus ramosus (Lindl.) Seidenf.

Stereochilus hirtus Lindl.

Stereosandra javanica Blume

Stigmatodactylus paradoxus (Prain) Schltr.

Taeniophyllum arunachalense A. N. Rao & J. Lal

Taeniophyllum complanatum Fukuy.

Taeniophyllum crepidiforme (King & Pantl.) King & Pantl.

Taeniophyllum glandulosum Blume

Taeniophyllum retrospiculatum (King & Pantl.) King & Pantl.

Taeniophyllum stella Carr

Tainia latifolia (Lindl.) Rchb. f.

Tainia minor Hook. f.

Tainia pulchra (Blume) Gagnep.

Thelasis khasiana Hook. f.

Thelasis longifolia Hook. f.

Thelasis pygmaea (Griff.) Lindl.

Thrixspermum centipeda Lour.

Thrixspermum crassilabre (King & Pantl.) Ormerod

Thrixspermum japonicum (Miq.) Rchb. f.

Thrixspermum musciflorum A. S. Rao & Joseph

Thrixspermum pygmaeum (King & Pantl.) Holttum

Thunia alba (Lindl.) Rchb. f.

Tipularia cunninghamii (King & Prain) S. C. Chen, S. W. Gale & P. J. Cribb

Tipularia josephii Lindl.

Tipularia szechuanica Schltr.

Trichotosia dasyphylla (E. C. Parish & Rchb. f.) Kraenzl.

Trichotosia pulvinata (Lindl.) Kraenzl.

Trichotosia velutina (Lodd. ex Lindl.) Kraenzl.

Tropidia angulosa (Lindl.) Blume

Tropidia curculigoides Lindl.

Tropidia pedunculata Blume

Uncifera acuminata Lindl.

Uncifera lancifolia (King & Prantl.) Schltr.

Uncifera obtusifolia Lindl.

Vanda alpina (Lindl.) Lindl.
Vanda bicolor Griff.
Vanda chlorosantha (Garay) Christenson
Vanda coerulea Griff. ex Lindl.
Vanda coerulescens Griff.
Vanda cristata Lindl.
Vanda griffithii Lindl.
Vanda pumila Hook. f.
Vanda roxburghii R. Br.
Vanda stangeana Rchb. f.
Vanda teres Lindl.
Vanda testacea (Lindl.) Rchb. f.
Vandopsis undulata (Lindl.) J. J. Sm.
Yoania prainii King & Pantl.
Zeuxine affinis (Lindl.) Benth. ex Hook. f.
Zeuxine clandestina Blume
Zeuxine flava (Wall. ex Lindl.) Trimen
Zeuxine glandulosa King & Pantl.
Zeuxine goodyeroides Lindl.
Zeuxine lindleyana A. N. Rao
Zeuxine longilabris (Lindl.) Trimen
Zeuxine membranacea Lindl.
Zeuxine nervosa (Wall. ex Lindl.) Trimen
Zeuxine pulchra King & Pantl.
Zeuxine reflexa King & Pantl.
Zeuxine strateumatica (L.) Schltr.

81. HYPOXIDACEAE

Curculigo capitulata (Lour.) Kuntze
Curculigo crassifolia (Baker) Hook. f.
Curculigo gracilis (Kurz) Hook. f.
Curculigo oligantha (Fischer) Bennet & Raizada
Curculigo orchioides Gaertn.
Curculigo prainiana (Deb) Bennet & Raizada
Hypoxis aurea Lour.

82. IXIOLIRIACEAE

Ixiolirion karateginum Lipsky
Ixiolirion tataricum (Pall.) Schult. & Schult. f.

83. IRIDACEAE

Crocus korolkowii Regel ex Maw
Crocus sativus L.
Iris aitchisonii (Baker) Boiss.
Iris barbatula Noltie & K. Y. Guan
Iris bulleyana Dykes
Iris chrysographes Dykes
Iris clarkei Baker
Iris collettii Hook. f.
Iris collettii var. **acaulis** Noltie
Iris crocea Jacquem. ex R. C. Foster
Iris cuniculiformis Noltie & K. Y. Guan
Iris decora Wall.
Iris delavayi Micheli
Iris dolichosiphon Noltie
Iris dolichosiphon subsp. **orientalis** Noltie
Iris domestica (L.) Goldblatt & Mabb.
Iris farreri Dykes
Iris forrestii Dykes
Iris germanica L.
Iris goniocarpa Baker

Iris habaensis X. D. Dong

Iris hookeriana Foster

Iris kashmiriana Baker

Iris kemaonensis Wall. ex Royle

Iris lactea Pall.

Iris lactea var. **chrysantha** Y. T. Zhao

Iris latistyla Y. T. Zhao

Iris leptophylla Lingelsh. ex H. Limpr.

Iris loczyi Kanitz

Iris milesii Foster

Iris narcissiflora Diels

Iris potaninii Maxim.

Iris potaninii var. **ionantha** Y. T. Zhao

Iris reticulata M. Bieb.

Iris sikkimensis Dykes

Iris songarica Schrenk

Iris speculatrix Hance

Iris spuria subsp. **notha** (M. Bieb.) Aschers. & Graebn.

Iris spuria subsp. **sogdiana** (Bunge) B. Mathew

Iris staintonii Hara

Iris subdichotoma Y. T. Zhao

Iris tectorum Maxim.

Iris tenuifolia Pall.

Iris wattii Baker ex Hook. f.

Iris wilsonii C. H. Wright

84. ASPHODELACEAE

Aloe vera (L.) Burm. f.

Asphodelus tenuifolius Cav.

Dianella ensifolia (L.) Redouté

Eremurus chinensis O. Fedtsch.

Eremurus dolichomischus Vved. & Wendelbo

Eremurus fuscus (O. Fedtsch.) Vved. ex V. V. Nikitin

Eremurus himalaicus Baker

Eremurus persicus (Jaub. & Spach) Boiss.

Eremurus persicus subsp. **sikesarus** (O. Fedtsch.) Ali

Eremurus stenophyllus (Boiss. & Buhse) Baker

Eremurus stenophyllus subsp. **aurantiacus** (Baker) Wendelbo

Hemerocallis forrestii Diels

Hemerocallis fulva (L.) L.

Hemerocallis nana W. W. Sm. & Forrest

Hemerocallis plicata Stapf

85. AMARYLLIDACEAE

Allium aciphyllum J. M. Xu

Allium ascalonicum L.

Allium atrosanguineum Schrenk

Allium atrosanguineum var. **fedschenkoanum** (Regel) G. H. Zhu & Turland

Allium atrosanguineum var. **tibeticum** (Regel) G. H. Zhu & Turland

Allium auriculatum Kunth

Allium barszczewskii Lipsky

Allium beesianum W. W. Sm.

Allium caesioides Wendelbo

Allium caesium Schrenk

Allium carolinianum Redouté

Allium cepa L.

Allium changduense J. M. Xu

Allium chienchuanense J. M. Xu

Allium chitralicum F. T. Wang & Tang

Allium chrysanthum Regel

Allium consanguineum Kunth

Allium cyaneum Regel

Allium cyathophorum Bureau & Franch.

Allium cyathophorum var. **farreri** (Stearn) Stearn

Allium eusperma Airy Shaw

Allium farctum Wendelbo

Allium fasciculatum Rendle

Allium fedtschenkoanum E. Regel

Allium filifolium Regel

Allium fistulosum L.

Allium forrestii Diels

Allium gilgiticum F. T. Wang & Tang

Allium gillii Wendelbo

Allium griffithianum Boiss.

Allium hookeri Thwaites

Allium hookeri var. **muliensis** Airy Shaw

Allium humile Kunth

Allium hymenorrhizum Ledeb.

Allium hypsistum Stearn

Allium jacquemontii Kunth

Allium kingdonii Stearn

Allium loratum Baker

Allium macleanii Baker

Allium macranthum Baker

Allium macrostemon Bunge

Allium mairei H. Lév.

Allium maowenense J. M. Xu

Allium nanodes Airy Shaw

Allium oreoprasum Schrenk

Allium ovalifolium Hand.-Mazz.

Allium ovalifolium var. **cordifolium** (J. M. Xu) J. M. Xu

Allium ovalifolium var. **leuconeurum** J. M. Xu

Allium pamiricum Wendelbo

Allium phariense Rendle

Allium plurifoliatum var. **zhegushanense** J. M. Xu

Allium prattii C. H. Wright ex Hemsl.

Allium przewalskianum Regel

Allium rhabdotum Stearn

Allium rhynchogynum Diels

Allium roylei Stearn

Allium rude J. M. Xu

Allium schoenoprasum L.

Allium sikkimense Baker

Allium siphonanthum J. M. Xu

Allium songpanicum J. M. Xu

Allium spicatum (Prain) N. Friesen

Allium stoliczki Regel

Allium stracheyi Baker

Allium tanguticum Regel

Allium trifurcatum (F. T. Wang & Tang) J. M. Xu

Allium tripterum Nasir

Allium tuberosum Rottl. ex Spreng.

Allium victorialis L.

Allium wallichii Kunth

Allium wallichii var. **platyphyllum** (Diels) J. M. Xu

Allium xiangchengense J. M. Xu

Allium xichuanense J. M. Xu

Allium yuanum F. T. Wang & Tang

Crinum amoenum Ker Gawl. ex Roxb.

Crinum lorifolium Roxb.

Crinum wattii Baker

Hippeastrum puniceum (Lam.) Voss

Hymenocallis littoralis (Jacq.) Salisb.

Lycoris aurea (L'Hér.) Herb.

Lycoris radiata (L'Hér.) Herb.

Nothoscordum borbonicum Kunth

Pancratium verecundum Aiton

Zephyranthes candida (Lindl.) Herb.

Zephyranthes carinata Herb.

86. ASPARAGACEAE

Agave americana L.

Agave cantula Roxb.

Agave sisalana Perrine ex Engelm.

Agave vera-cruz Mill.

Agave vivipara L.

Agave wightii J. R. Drumm. & Prain

Asparagus adscendens Roxb.

Asparagus capitatus Baker

Asparagus capitatus subsp. **gracilis** Browicz

Asparagus cochinchinensis (Lour.) Merr.

Asparagus curillus Buch.-Ham. ex Roxb.

Asparagus filicinus Buch.-Ham. ex D. Don

Asparagus filicinus var. **brevipes** Baker

Asparagus filicinus var. **giraldii** C. H. Wright

Asparagus griffithii Baker

Asparagus lycopodineus (Baker) F. T. Wang & T. Tang

Asparagus meioclados H. Lév.

Asparagus munitus F. T. Wang & S. C. Chen

Asparagus myriacanthus F. T. Wang & S. C. Chen

Asparagus officinalis L.

Asparagus penicillatus Hara

Asparagus persicus Baker

Asparagus racemosus Willd.

Asparagus racemosus var. **subacerosus** Baker

Asparagus sichuanicus S. C. Chen & D. Q. Liu

Asparagus taliensis F. T. Wang & Tang ex S. C. Chen

Asparagus tibeticus F. T. Wang & S. C. Chen

Asparagus yanbianensis S. C. Chen

Asparagus yanyuanensis S. C. Chen

Campylandra aurantiaca Baker

Campylandra delavayi (Franch.) M. N. Tamura, S. Yun Liang & Turland

Campylandra ensifolia (F. T. Wang & Tang) M. N. Tamura, S. Yun Liang & Turland

Campylandra fimbriata (Hand.-Mazz.) M. N. Tamura, S. Yun Liang & Turland

Campylandra liangshanensis (Z. Y. Zhu) M. N. Tamura, S. Yun Liang & Turland

Campylandra tui (F. T. Wang & Tang) M. N. Tamura, S. Yun Liang & Turland

Campylandra urotepala (Hand.-Mazz.) M. N. Tamura, S. Yun Liang & Turland

Campylandra wattii C. B. Clarke

Chlorophytum arundinaceum Baker

Chlorophytum chinense Bureau & Franch.

Chlorophytum khasianum Hook. f.

Chlorophytum majus (Hemsl.) Marais & Reilly

Chlorophytum minus (C. H. Wright) Marais & Reilly

Chlorophytum nepalense (Lindl.) Baker

Dipcadi reidii Deb & S. Dasgupta

Dipcadi serotinum (L.) Medik.

Disporopsis aspersa (Hua) Engl.

Dracaena angustifolia (Medik.) Roxb.

Dracaena petiolata Hook. f.

Drimia indica (Roxb.) Jessop

Heteropolygonatum pendulum (Z. G. Liu & X. H. Hu) M. N. Tamura & Ogisu

Liriope kansuensis (Batalin) C. H. Wright

Maianthemum atropurpureum (Franch.) LaFrankie

Maianthemum bifolium (L.) F. W. Schmidt

Maianthemum forrestii (W. W. Sm.) LaFrankie

Maianthemum fusciduliflorum (Kawano) S. C. Chen & Kawano

Maianthemum fuscum (Wall.) LaFrankie

Maianthemum fuscum var. **cordatum** R. Li & H. Li

Maianthemum gongshanense (S. Yun Liang) H. Li

Maianthemum henryi (Baker) LaFrankie

Maianthemum oleraceum (Baker) LaFrankie

Maianthemum purpureum (Wall.) LaFrankie

Maianthemum szechuanicum (F. T. Wang & Tang) H. Li

Maianthemum tatsienense (Franch.) LaFrankie

Maianthemum tubiferum (Batalin) LaFrankie

Ophiopogon amblyphyllus F. T. Wang & L. K. Dai

Ophiopogon angustifoliatus (F. T. Wang & Tang) S. C. Chen

Ophiopogon bodinieri H. Lév.

Ophiopogon clarkei Hook. f.

Ophiopogon dracaenoides (Baker) Hook. f.

Ophiopogon grandis W. W. Sm.

Ophiopogon intermedius D. Don

Ophiopogon lushuiensis S. C. Chen

Ophiopogon motouensis S. C. Chen

Ophiopogon yunnanensis S. C. Chen

Peliosanthes griffithii Baker

Peliosanthes macrophylla Wall. ex Baker

Peliosanthes teta Andrews

Polygonatum alternicirrhosum Hand.-Mazz.

Polygonatum brevistylum Baker

Polygonatum cathcartii Baker

Polygonatum cirrhifolium (Wallich) Royle

Polygonatum curvistylum Hua

Polygonatum dolichocarpum M. N. Tamura, Fuse & Y. P. Yang

Polygonatum gongshanense L. H. Zhao & X. J. He

Polygonatum graminifolium Hook.

Polygonatum griffithii Baker

Polygonatum hirtellum Hand.-Mazz.

Polygonatum hookeri Baker

Polygonatum kingianum Collett & Hemsl.

Polygonatum nervulosum Baker

Polygonatum nodosum Hua

Polygonatum oppositifolium (Wall.) Royle

Polygonatum prattii Baker

Polygonatum punctatum Royle ex Kunth

Polygonatum sibiricum Redouté

Polygonatum singalilense Hara

Polygonatum stewartianum Diels

Polygonatum verticillatum (L.) All.

Polygonatum wardii F. T. Wang & Tang

Reineckea carnea (Andrews) Kunth

Rohdea eucomoides (Baker) N. Tanaka

Rohdea nepalensis (Raf.) N. Tanaka

Rohdea wattii (C. B. Clarke) Yamashita & M. N. Tamura

Sansevieria trifasciata Prain

Scilla griffithii Hochr.

Theropogon pallidus Maxim.

Tupistra clarkei Hook. f.

Tupistra nutans Wall. ex Lindl.

Tupistra stoliczana Kurz

Yucca aloifolia L.

87. PALMAE

Areca catechu (L. f.) Willd.

Arenga micrantha C. F. Wei

Arenga pinnata (Wurmb) Merr.

Arenga westerhoutii Griff.

Calamus acanthospathus Griff.

Calamus erectus Roxb.

Calamus flagellum Griff. ex Mart.

Calamus floribundus Griff.

Calamus gracilis Roxb.

Calamus guruba Buch.-Ham. ex Mart.

Calamus latifolius Roxb.

Calamus leptospadix Griff.

Calamus nambariensis Becc.

Calamus tenuis Roxb.

Caryota maxima Blume

Caryota obtusa Griff.

Caryota urens L.

Cocos nucifera L.

Daemonorops jenkinsiana (Griff.) Martius

Licuala peltata Roxb. ex Buch.-Ham.

Licuala spinosa Wurmb

Livistona jenkinsiana Griff.

Phoenix acaulis Buch.-Ham. ex Roxb.

Phoenix humilis Royle ex Becc. & Hook. f.

Phoenix loureiroi Kunth

Phoenix rupicola T. Anderson

Phoenix sylvestris Roxb.

Pinanga baviensis Becc.

Pinanga gracilis Blume

Plectocomia himalayana Griff.

Trachycarpus fortunei (Hook.) H. Wendl.

Trachycarpus martianus (Wall. ex Mart.) H. Wendl.

Trachycarpus takil Becc.

Wallichia caryotoides Roxb.

Wallichia densiflora Mart.

Wallichia disticha T. Anderson

Wallichia nana Griff.

Wallichia oblongifolia Griff.

Wallichia triandra (J. Joseph) S. K. Basu

88. COMMELINACEAE

Amischophacelus axillaris (L.) R. S. Rao & Kammathy

Amischotolype hispida (A. Richard) D. Y. Hong

Amischotolype hookeri (Hassk.) Hara

Amischotolype mollissima (Blume) Hassk.

Belosynapsis ciliata (Blume) R. S. Rao

Commelina appendiculata C. B. Clarke

Commelina benghalensis L.

Commelina caroliniana Walter

Commelina communis L.

Commelina diffusa Burm. f.

Commelina hasskarlii C. B. Clarke

Commelina maculata Edgew.

Commelina paludosa Blume

Commelina sikkimensis C. B. Clarke

Commelina suffruticosa Blume

Cyanotis axillaris (L.) D. Don ex Sweet

Cyanotis barbata D. Don

Cyanotis cristata (L.) D. Don

Cyanotis fasciculata (B. Heyne ex Roth) Schult. & Schult. f.

Cyanotis papilionacea (Burm. f.) Schult. & Schult. f.

Cyanotis vaga (Lour.) Schult. & Schult. f.

Dictyospermum scaberrimum (Blume) J. K. Morton ex D. Y. Hong

Floscopa scandens Lour.

Murdannia blumei (Hassk.) Brenan

Murdannia divergens (C. B. Clarke) G. Brückn.

Murdannia edulis (Stokes) Faden

Murdannia gigantea (Vahl) G. Brückn.

Murdannia japonica (Thunb.) Faden

Murdannia keisak (Hassk.) Hand.-Mazz.

Murdannia loriformis (Hassk.) R. S. Rao & Kammathy

Murdannia nudiflora (L.) Brenan

Murdannia scapiflora (Roxb.) Royle

Murdannia simplex (Vahl) Brenan

Murdannia spirata (L.) G. Brückn.

Murdannia stenothyrsa (Diels) Hand.-Mazz.

Murdannia triquetra (Wall. ex C. B. Clarke) G. Brückn.

Pollia hasskarlii R. S. Rao

Pollia secundiflora (Blume) Bakh. f.

Pollia subumbellata C. B. Clarke

Rhopalephora scaberrima (Blume) Faden

Spatholirion longifolium (Gagnep.) Dunn

Streptolirion volubile Edgew.

Streptolirion volubile subsp. **khasianum** (C. B. Clarke) D. Y. Hong

Tricarpelema giganteum (Hassk.) Hara

Tricarpelema glanduliferum (J. Joseph & R. S. Rao) R. S. Rao

Tricarpelema xizangense D. Y. Hong

89. PONTEDERIACEAE

Eichhornia crassipes (Mart.) Solms

Monochoria hastata (L.) Solms

Monochoria vaginalis (Burm. f.) C. Presl ex Kunth

90. MUSACEAE

Ensete glaucum (Roxb.) Cheesman

Ensete wilsonii (Tutcher) Cheesman

Musa acuminata Colla

Musa argentii Gogoi & Borah

Musa arunachalensis A. Joe, Sreejith & M. Sabu

Musa aurantiaca G. Mann ex Baker

Musa balbisiana Colla

Musa griersonii Noltie

Musa kamengensis Gogoi & Häkkinen

Musa nagensium Prain

Musa paradisiaca L.

Musa puspanjaliae Gogoi & Häkkinen

Musa rubra Wall. ex Kurz

Musa sanguinea Hook. f.

Musa sikkimensis Kurz

Musa thomsonii (King ex Baker) A. M. Cowan & Cowan

Musa velutina H. Wendl. & Drude

Musella lasiocarpa (Franch.) C. Y. Wu ex H. W. Li

91. CANNACEAE

Canna indica L.

92. MARANTACEAE

Phrynium placentarium (Lour.) Merr.

Phrynium pubinerve Blume

Schumannianthus dichotomus (Roxb.) Gagnep.

93. COSTACEAE

Costus lacerus Gagnep.

Costus oblongus S. Q. Tong

Costus speciosus (J. Koenig) Sm.

94. ZINGIBERACEAE

Alpinia allughas (Retz.) Roscoe

Alpinia blepharocalyx K. Schum.

Alpinia bracteata Roscoe

Alpinia calcarata Roscoe

Alpinia malaccensis (Burm. f.) Roscoe

Alpinia nigra (Gaertn.) B. L. Burtt

Alpinia ovoidocarpa H. Dong & G. J. Xu

Amomum aromaticum Roxb.

Amomum dealbatum Roxb.

Amomum kingii Baker

Amomum maximum Roxb.

Amomum nimkeyense M. Sabu, Hareesh, Tatum & A. K. Das

Amomum riwatchii M. Sabu & Hareesh

Amomum sabuanum V. P. Thomas, Nissar & U. Gupta

Amomum sericeum Roxb.

Amomum subulatum Roxb.

Amomum villosum Lour.

Boesenbergia longiflora (Wall.) Kuntze

Caulokaempferia sikkimensis (King ex Baker) K. Larsen

Cautleya cathcartii Baker

Cautleya gracilis (Sm.) Dandy

Cautleya gracilis var. **gracilior** (K. Schum.) Sanjappa

Cautleya spicata (Sm.) Baker

Curcuma amada Roxb.

Curcuma angustifolia Roxb.

Curcuma aromatica Salisb.

Curcuma exigua N. Liu

Curcuma longa L.

Curcuma longiflora Salisb.

Curcuma montana Roxb.

Curcuma phaeocaulis Valeton

Curcuma rubescens Roxb.

Etlingera linguiformis (Roxb.) R. M. Sm.

Etlingera loroglossa (Gagnep.) R. M. Sm.

Globba andersonii C. B. Clarke ex Baker

Globba clarkei Baker

Globba hookeri C. B. Clarke ex Baker

Globba macroclada Gagnep.

Globba multiflora Wall. ex Baker

Globba orixensis Roxb.

Globba racemosa Sm.

Hedychium aurantiacum Roscoe

Hedychium aureum C. B. Clarke & H. Mann ex Baker

Hedychium bijiangense T. L. Wu & S. J. Chen

Hedychium coccineum Buch.-Ham. ex Sm.

Hedychium coronarium J. Koenig

Hedychium densiflorum Wall.

Hedychium efilamentosum Hand.-Mazz.

Hedychium elatum R. Br.

Hedychium ellipticum Buch.-Ham. ex Sm.

Hedychium flavescens Carey ex Roscoe

Hedychium flavum Roxb.

Hedychium forrestii Diels

Hedychium gardnerianum Sheppard ex Ker Gawl.

Hedychium glaucum Roscoe

Hedychium gomezianum Wall.

Hedychium gracile Roxb.

Hedychium gracillimum A. S. Rao & Verma

Hedychium greenii W. W. Sm.

Hedychium griersonianum R. M. Sm.

Hedychium longipedunculatum Sastry & Verma

Hedychium marginatum C. B. Clarke

Hedychium parvibracteatum T. L. Wu & S. J. Chen

Hedychium putaoense Y. H. Tan & H. B. Ding

Hedychium radiatum A. S. Rao & Hajra

Hedychium robustum A. S. Rao & Hajra

Hedychium sinoaureum Stapf

Hedychium spicatum Sm.

Hedychium spicatum var. **acuminatum** (Roscoe) Wall.

Hedychium spicatum var. **trilobum** (Roscoe) Wall.

Hedychium stenopetalum Lodd.

Hedychium tengchongense Y. B. Luo

Hedychium thyrsiforme Buch.-Ham. ex Sm.

Hedychium venustum Wight

Hedychium villosum Wall.

Hedychium wardii C. E. C. Fisch.

Hedychium yunnanense Gagnep.

Hemiorchis pantlingii King

Hornstedtia arunachalensis S. Tripathi & V. Prakash

Hornstedtia tibetica T. L. Wu & S. J. Chen

Kaempferia rotunda L.

Larsenianthus arunachalensis M. Sabu, Sanoj & Rajesh Kumar

Larsenianthus careyanus (Benth.) W. J. Kress & Mood

Monolophus arunachalensis Bhaumik, S. Dey & Langhu

Rhynchanthus beesianus W. W. Sm.

Roscoea alpina Royle

Roscoea auriculata K. Schum.

Roscoea bhutanica Ngamr.

Roscoea cangshanensis M. H. Luo, X. F. Gao & H. H. Lin

Roscoea capitata Sm.

Roscoea cautleyoides Gagnep.

Roscoea cautleoides var. **pubescens** (Z. Y. Zhu) T. L. Wu

Roscoea chamaeleon Gagnep.

Roscoea debilis Gagnep.

Roscoea debilis var. **limprichtii** (Loes.) Cowley

Roscoea forrestii Cowley

Roscoea ganeshensis Cowley & W. J. Baker

Roscoea humeana Balf. f. & W. W. Sm.

Roscoea megalantha Tosh. Yoshida & Yangzom

Roscoea nepalensis Cowley

Roscoea praecox K. Schum.

Roscoea purpurea Sm.

Roscoea purpurea var. **auriculata** Sm.

Roscoea purpurea var. **gigantea** Sm.

Roscoea schneideriana (Loes.) Cowley

Roscoea scillifolia (Gagnep.) Cowley

Roscoea tibetica Batalin

Roscoea tumjensis Cowley

Roscoea wardii Cowley

Stahlianthus involucratus (King ex Baker) Craib ex Loes.

Zingiber arunachalensis A. Joe, T. Jayakr., Hareesh & M. Sabu

Zingiber capitatum Roxb.

Zingiber chrysanthum Roscoe

Zingiber clarkei King ex Baker

Zingiber montanum (J. Koenig) Link ex A. Dietr.

Zingiber officinale Roscoe

Zingiber roseum (Roxb.) Roscoe

Zingiber rubens Roxb.

Zingiber striolatum Diels

Zingiber yunnanense S. Q. Tong & X. Z. Liu

Zingiber zerumbet (L.) Roscoe ex Sm.

95. TYPHACEAE

Sparganium confertum Y. D. Chen

Sparganium fallax Graebn.

Sparganium glomeratum (Laest. ex Beurl.) Beurl.

Sparganium limosum Y. D. Chen

Sparganium stoloniferum (Buch.-Ham. ex Graebn.) Buch.-Ham. ex Juz.

Typha angustifolia L.

Typha domingensis Pers.

Typha elephantina Roxb.

Typha latifolia L.

Typha laxmannii Lep.

Typha minima Hoppe

Typha orientalis C. Presl

96. BROMELIACEAE

Ananas comosus (L.) Merr.

97. XYRIDACEAE

Xyris capensis Thunb.

Xyris indica L.

Xyris pauciflora Willd.

98. ERIOCAULACEAE

Eriocaulon alpestre Hook. f. & Thomson ex Körn.

Eriocaulon atrum Nakai

Eriocaulon bhutanicum Noltie

Eriocaulon brownianum Mart.

Eriocaulon cinereum R. Br.

Eriocaulon cristatum Mart.

Eriocaulon edwardii Fyson

Eriocaulon exsertum Satake

Eriocaulon hamiltonianum Mart.

Eriocaulon kathmanduense Satake

Eriocaulon leianthum W. L. Ma

Eriocaulon luzulifolium Mart.

Eriocaulon minusculum Moldenke

Eriocaulon nepalense J. D. Prescott ex Bong.

Eriocaulon obclavatum Satake

Eriocaulon oryzetorum Mart.

Eriocaulon quinquangulare L.

Eriocaulon rockianum Hand.-Mazz.

Eriocaulon rockianum var. **latifolium** W. L. Ma

Eriocaulon sollyanum Royle

Eriocaulon staintonii Satake

Eriocaulon trisectoides Satake

Eriocaulon viride Körn.

Eriocaulon xeranthemum Mart.

99. JUNCACEAE

Juncus alatus Franch. & Sav.

Juncus allioides Franch.

Juncus amplifolius A. Camus

Juncus articulatus L.

Juncus auritus K. F. Wu

Juncus benghalensis Kunth

Juncus bhutanensis Satake

Juncus biglumoides Hara

Juncus brachystigma Sam.

Juncus bracteatus Buchenau

Juncus bryophilus Noltie

Juncus bufonius L.

Juncus bufonius var. **rechingeri** (Snogerup) Jafri

Juncus castaneus Sm.

Juncus cephalostigma Sam.

Juncus cephalostigma var. **dingjieensis** K. F. Wu

Juncus chrysocarpus Buchenau

Juncus clarkei Buchenau

Juncus clarkei var. **marginatus** A. Camus

Juncus compressus Jacq.

Juncus concinnus D. Don

Juncus concolor Sam.

Juncus crassistylus A. Camus

Juncus dongchuanensis K. F. Wu

Juncus duthiei (C. B. Clarke) Noltie

Juncus effusus L.

Juncus elegans Royle ex D. Don

Juncus fugongensis S. Y. Bao

Juncus ganeshii Miyam. & H. Ohba

Juncus giganteus Sam.

Juncus glaucoturgidus Noltie

Juncus gracilicaulis A. Camus

Juncus grisebachii Buchenau

Juncus himalensis Klotzsch

Juncus hydrophilus Noltie

Juncus inflexus L.

Juncus kangdingensis K. F. Wu

Juncus kangpuensis K. F. Wu

Juncus khasiensis Buchenau

Juncus kingii Rendle

Juncus lanpinguensis Novikov

Juncus leptospermus Buchenau

Juncus leschenaultii Gay ex Laharpe

Juncus leucanthus Royle ex D. Don

Juncus leucomelas Royle ex D. Don

Juncus luteocarpus Satake

Juncus longiflorus (A. Camus) Noltie

Juncus longistamineus A. Camus

Juncus megalophyllus S. Y. Bao

Juncus meiguensis K. F. Wu

Juncus membranaceus Royle ex D. Don

Juncus milashanensis A. M. Lu & Zhi Y. Zhang

Juncus minimus Buchenau

Juncus miyiensis K. F. Wu

Juncus modicus N. E. Br.

Juncus mustagensis Miyam. & H. Ohba

Juncus nepalicus Miyam. & H. Ohba

Juncus nigroviolaceus K. F. Wu

Juncus ochraceus Buchenau

Juncus perparvus K. F. Wu

Juncus perpusillus Sam.

Juncus potaninii Buchenau

Juncus prismatocarpus R. Br.

Juncus prismatocarpus subsp. **teretifolius** K. F. Wu

Juncus przewalskii Buchenau

Juncus przewalskii var. **discolor** Sam.

Juncus pseudocastaneus (Lingelsh.) Sam.

Juncus ranarius Songeon & E. P. Perrier

Juncus rohatngensis Goel & Aswal

Juncus setchuensis Buchenau ex Diels

Juncus sherei Miyam. & H. Ohba

Juncus sikkimensis Hook. f.

Juncus sikkimensis var. **monocephalus** Hook. f.

Juncus sphacelatus Decne.

Juncus sphacelatus var. **himalensis** (Klotzsch) Jafri

Juncus spumosus Noltie

Juncus tanguticus Sam.

Juncus thomsonii Buchenau

Juncus tibeticus Egorova

Juncus trachyphyllus Miyam. & H. Ohba

Juncus trichophyllus W. W. Sm.

Juncus triglumis L.

Juncus triglumis subsp. **wakhaniensis** Snog.

Juncus turkestanicus V. I. Krecz. & Gontsch.

Juncus uniflorus W. W. Sm.

Juncus unifolius A. M. Lu & Z. Y. Zhang

Juncus wallichianus J. Gay ex Laharpe

Juncus yüi S. Y. Bao

Juncus yunnanensis A. Camus

Luzula bomiensis K. F. Wu

Luzula campestris (L.) DC.

Luzula effusa Buchenau

Luzula effusa var. **chinensis** (N. E. Br.) K. F. Wu

Luzula jilongensis K. F. Wu

Luzula multiflora (Ehrh.) Lej.

Luzula oligantha Samuelsson

Luzula plumosa E. Mey.

Luzula sichuanensis K. F. Wu

Luzula spicata (L.) DC.

100. CYPERACEAE

Actinoschoenus yunnanensis (C. B. Clarke) Y. C. Tang

Actinoscirpus grossus (L. f.) Goetgh. & D. A. Simpson

Actinoscirpus grossus var. **kysoor** (Roxb.) Noltie

Anosporum cephalotes (Vahl) Kurz

Blysmus compressus (L.) Panz. ex Link

Blysmus compressus subsp. **brevifolius** (Decne.) Kukkonen

Blysmus rufus (Huds.) Link

Blysmus sinocompressus Tang & F. T. Wang

Bulbostylis barbata (Rottb.) C. B. Clarke

Bulbostylis densa (Wall.) Hand.-Mazz.

Carex agglomerata C. B. Clarke

Carex alatauensis S. R. Zhang

Carex alta Boott

Carex aristatisquamata Tang & F. T. Wang ex L. K. Dai

Carex aristulifera P. C. Li

Carex atrata L.

Carex atrata subsp. **pullata** (Boott) Kük.

Carex atrofusca Schkuhr

Carex atrofusca subsp. **minor** (Boott) T. Koyama

Carex atrofuscoides K. T. Fu

Carex baccans Nees

Carex bhutanensis S. R. Zhang

Carex bigelowii Torr. ex Schwein.

Carex bijiangensis S. Yun Liang & S. R. Zhang

Carex bistaminata (W. Z. Di & M. J. Zhong) S. R. Zhang

Carex blinii H. Lév. & Vaniot

Carex bonatiana (Kük.) Ivanova

Carex borii Nelmes

Carex breviculmis R. Br.

Carex breviprophylla O. Yano

Carex brunnea Thunb.

Carex burangensis (Y. C. Yang) S. R. Zhang

Carex burttii Noltie

Carex caespititia Nees

Carex canescens L.

Carex capillifolia (Decne.) S. R. Zhang

Carex capilliformis Franch.

Carex cardiolepis Nees

Carex caudispicata F. T. Wang & Tang ex P. C. Li

Carex chinensis Retz.

Carex chitralensis Nelmes

Carex chiwuana F. T. Wang & Tang ex P. C. Li

Carex chlorocephalula F. T. Wang & Tang ex P. C. Li

Carex chui Nelmes

Carex clavispica S. R. Zhang

Carex composita Boott

Carex condensata Nees

Carex coninux (F. T. Wang & Tang) S. R. Zhang

Carex continua C. B. Clarke

Carex coriophora Fisch. & C. A. Mey. ex Kunth

Carex courtallensis Nees ex Boott

Carex crebra V. I. Krecz.

Carex cremostachys Franch.

Carex cruciata Wahlenb.

Carex cruenta Nees

Carex curticeps C. B. Clarke

Carex cylindrostachys Franch.

Carex daltonii Boott

Carex deasyi (C. B. Clarke) O. Yano & S. R. Zhang

Carex decaulescens subsp. **alsia** (Raymond) Kukkonen

Carex decora Boott

Carex delavayi Franch.

Carex deqinensis L. K. Dai

Carex desponsa Boott

Carex diandra Schrank

Carex dielsiana Kük.

Carex diluta M. Bieb.

Carex dimorpholepis Steud.

Carex dispalata Boott ex A. Gray

Carex dispalata subsp. **laxiflorens** Koyama

Carex divulsa subsp. **leersii** (Kneucker) W. Koch

Carex doniana Spreng.

Carex drepanorhyncha Franch.

Carex drukyulensis (Noltie) Jim. Mejías & Noltie

Carex echinata Murray

Carex enervis C. A. Mey.

Carex esbirajbhandari (Rajbh. & H. Ohba) O. Yano

Carex esenbeckii Kunth

Carex fastigiata Franch.

Carex fedia Nees

Carex filicina Nees

Carex filispica S. R. Zhang

Carex fissiglumis (C. B. Clarke) S. R. Zhang & O. Yano

Carex fluviatilis Boott

Carex foliosa D. Don

Carex forrestii Kük.

Carex fucata Boott ex C. B. Clarke

Carex fusiformis Nees

Carex gammiei (C. B. Clarke) S. R. Zhang & O. Yano

Carex gandakiensis Katsuy.

Carex gentilis Franch.

Carex gibba Wahlenb.

Carex goligongshanensis P. C. Li

Carex gonggaensis P. C. Li

Carex gongshanensis Tang & F. T. Wang ex Y. C. Yang

Carex haematostoma Nees

Carex halleriana Asso

Carex handelii Kük.

Carex handel-mazzettii (Ivanova) S. R. Zhang

Carex harae (Rajbh. & H. Ohba) O. Yano

Carex harrysmithii Kük.

Carex hebecarpa C. A. Mey.

Carex hebecarpa subsp. **ligulata** (Nees) Koyama

Carex hemineuros Koyama

Carex henryi (C. B. Clarke) L. K. Dai

Carex herbacoeli Jim. Mejías & Roalson

Carex heterostachya Bunge

Carex heudesii H. Lév. & Vaniot

Carex himalaica Koyama

Carex hirtelloides (Kük.) F. T. Wang & Tang ex P. C. Li

Carex hongyuanensis Y. C. Tang & S. Yun Liang

Carex huashanica Tang & F. T. Wang ex L. K. Dai

Carex hughii S. R. Zhang

Carex hymenolepis Nees

Carex inanis Kunth

Carex inclinis Boott ex C. B. Clarke

Carex infusctata Nees

Carex insignis Boott

Carex ivanoviae T. V. Egorova

Carex jackiana Boott

Carex kanaii (Rajbh. & H. Ohba) S. R. Zhang & O. Yano

Carex kangdingensis S. R. Zhang

Carex kansuensis Nelmes

Carex karlongensis Kük.

Carex kashmirensis C. B. Clarke

Carex kokanica (Regel) S. R. Zhang

Carex lanceolata Boott

Carex ledebouriana C. A. Mey. ex Trevir.

Carex lehmannii Drejer

Carex lenta D. Don

Carex lepidochlamys (F. T. Wang & Tang ex P. C. Li) S. R. Zhang

Carex liangshanensis S. R. Zhang

Carex ligulata Nees

Carex litorhyncha Franch.

Carex littledalei (C. B. Clarke) S. R. Zhang

Carex longicruris Nees

Carex longipes D. Don ex Tilloch & Taylor

Carex longipes var. **nepalensis** (Spreng.) Boott

Carex longispiculata Y. C. Yang

Carex macroprophylla (Y. C. Yang) S. R. Zhang

Carex magnoutriculata Tang & F. T. Wang ex L. K. Dai

Carex makuensis P. C. Li

Carex mallae (Rajbh. & H. Ohba) O. Yano

Carex maritima Gunnerus

Carex maubertiana Boott

Carex melanantha C. A. Mey.

Carex microglochin Wahlenb.

Carex montis-everesti Kük.

Carex moorcroftii Falc. ex Boott

Carex mosoynensis Franch.

Carex motuoensis Y. C. Yang

Carex moupinensis Franch.

Carex muliensis Hand.-Mazz.

Carex munda Boott

Carex myosurus Nees

Carex neesii S. R. Zhang

Carex nemostachys Steud.

Carex nigra (L.) Reichard

Carex nivalis Boott

Carex noltiei S. R. Zhang

Carex notha Kunth

Carex nubigena D. Don ex Tilloch & Taylor

Carex nudicarpa (Y. C. Yang) S. R. Zhang

Carex obliquitruncata Y. C. Tang & S. Yun Liang

Carex obscura Nees

Carex obscura var. **brachycarpa** C. B. Clarke

Carex obscuriceps Kük.

Carex oedorrhampha Nelmes

Carex oligocarya C. B. Clarke

Carex olivacea Boott

Carex omeiensis Tang
Carex orbicularis Boott
Carex ovoidispica O. Yano
Carex oxyphylla Franch.
Carex pamirensis C. B. Clarke
Carex parva Nees
Carex parvula O. Yano
Carex peichuniana S. R. Zhang
Carex perakensis C. B. Clarke
Carex pergracilis Nelmes
Carex phacota Spreng.
Carex plectobasis V. I. Krecz.
Carex polycephala Boott
Carex polymascula P. C. Li
Carex polyphylla Kar. & Kir.
Carex praeclara Nelmes
Carex prainii Kük.
Carex prescottiana Boott
Carex prolongata Kük.
Carex pruinosa Boott
Carex przewalskii T. V. Egorova
Carex pseudobicolor Boeckeler
Carex pseudocyperus L.
Carex pseudofoetida Kük.
Carex pseudogammiei S. R. Zhang
Carex pseudohumilis F. T. Wang & Y. L. Chang ex P. C. Li
Carex pseudolaxa (C. B. Clarke) O. Yano & S. R. Zhang
Carex pseudoligulata L. K. Dai
Carex pseudosupina Y. C. Tang ex L. K. Dai
Carex pseuduncinoides (Noltie) O. Yano & S. R. Zhang
Carex psychrophila Nees
Carex pterocaulos Nelmes
Carex pulchra Boott
Carex purpureosquamata L. K. Dai
Carex pycnostachya Kar. & Kir.
Carex radicalis Boott
Carex rara Boott
Carex remota subsp. **stewartii** Kukk.
Carex rhizopoda Maxim.
Carex rhynchophora Franch.
Carex ridongensis P. C. Li
Carex rochebrunii Franch. & Sav.
Carex rostrata Stokes
Carex rubrobrunnea C. B. Clarke
Carex sagaensis Y. C. Yang
Carex sanguinea Boott
Carex sargentiana (Hemsl.) S. R. Zhang
Carex satakeana Koyama
Carex scabrirostris Kük.
Carex schneideri Nelmes
Carex scolopendriformis F. T. Wang & Tang ex P. C. Li
Carex sendaica Franch.
Carex setigera D. Don
Carex setosa Boott
Carex setschwanensis (Hand.-Mazz.) S. R. Zhang
Carex siamensis (Ohwi) S. R. Zhang
Carex simulans C. B. Clarke
Carex speciosa Kunth
Carex stenophylla Wahlenb.
Carex stenophylloides V. I. Krecz.
Carex stracheyi Boott ex C. B. Clarke
Carex stramentitia Boott ex Boeckeler
Carex subfilicinoides Kük.
Carex supina Willd. ex Wahlenb.
Carex tangulashanensis Y. C. Yang
Carex tapintzensis Franch.
Carex tatsiensis (Franch.) Kük.
Carex teinogyna Boott
Carex teres Boott
Carex thibetica Franch.
Carex thomsonii Boott

Carex tibetikobresia S. R. Zhang
Carex tunicata (Hand.-Mazz.) S. R. Zhang
Carex uncinioides Boott
Carex unisexualis C. B. Clarke
Carex vaginosa (C. B. Clarke) S. R. Zhang
Carex vesiculosa Boott
Carex vidua Boott ex C. B. Clarke
Carex viridula Michx.
Carex winterbottomii C. B. Clarke
Carex wutuensis K. T. Fu
Carex yadongensis (Y. C. Yang) S. R. Zhang
Carex yajiangensis Tang & F. T. Wang ex S. Yun Liang
Carex yanchengii S. R. Zhang, *nom. nov*. (≡ *Kobresia stolonifera* Y. C. Tang ex P. C. Li)
Carex yangii (S. R. Zhang) S. R. Zhang
Carex yulungshanensis P. C. Li
Carex yunlingensis P. C. Li
Carex yunnanensis Franch.
Carex zhenkangensis Tang & F. T. Wang ex S. Yun Liang
Cladium jamaicence subsp. **chinense** (Nees) T. Koyama
Cladium jamaicense Crantz
Cladium mariscus (L.) Pohl
Courtoisina cyperoides (Roxb.) Soják
Cyperus alulatus J. Kern
Cyperus amuricus Maxim.
Cyperus atkinsoni C. B. Clarke
Cyperus babakan Steud.
Cyperus castaneus Willd.
Cyperus compactus Retz.
Cyperus compressus L.
Cyperus conglomeratus Rottb.
Cyperus corymbosus Rottb.
Cyperus cuspidatus Kunth
Cyperus cyperinus (Retz.) J. V. Suringar
Cyperus cyperoides (L.) Kuntze
Cyperus difformis L.
Cyperus diffusus Vahl
Cyperus digitatus Roxb.
Cyperus distans L. f.
Cyperus eleusinoides Kunth
Cyperus esculentus L.
Cyperus exaltatus Retz.
Cyperus fuscus L.
Cyperus glomeratus L.
Cyperus haspan L.
Cyperus haspan subsp. **juncoides** (Lam.) Kük.
Cyperus imbricatus Retz.
Cyperus involucratus Rott.
Cyperus iria L.
Cyperus laxus Lam.
Cyperus longus L.
Cyperus malaccensis Lam.
Cyperus michelianus (L.) Link
Cyperus nigrofuscus L. K. Dai
Cyperus niveus Retz.
Cyperus nutans Vahl
Cyperus pangorei Rott.
Cyperus paniceus Boeckeler
Cyperus pilosus Vahl
Cyperus platystylis R. Br.
Cyperus procerus Rott.
Cyperus pygmaeus Rott.
Cyperus rotundus L.
Cyperus sanguinolentus Vahal
Cyperus scariosus R. Br.
Cyperus serotinus Rott.
Cyperus sikkimensis Kük.
Cyperus silletensis Nees
Cyperus tenuiculmis Boeckeler
Cyperus tenuispica Steud.
Cyperus thomsonii Boeckeler
Cyperus trisulcus D. Don

Cyperus tuberosus Rottb.

Cyperus wallichianus Spreng.

Diplacrum caricinum R. Br.

Eleocharis acicularis (L.) Roem. & Schult.

Eleocharis acutangula (Roxb.) Schult.

Eleocharis atropurpurea (Retz.) J. Presl & C. Presl

Eleocharis congesta D. Don

Eleocharis dulcis (Burm. f.) Trin. ex Hensch.

Eleocharis mitracarpa Steud.

Eleocharis palustris (L.) Roem. & Schult.

Eleocharis parishii Britton

Eleocharis quinqueflora (Hartmann) O. Schwarz

Eleocharis retroflexa (Poiret) Urb.

Eleocharis retroflexa subsp. **chaetaria** (Roem. & Schult.) Koyama

Eleocharis setulosa P. C. Li

Eleocharis tetraquetra Nees

Eleocharis thomsonii Boeckeler

Eleocharis tuberculosa (Michx.) Roem. & Schult.

Eleocharis uniglumis (Link) Schult.

Eleocharis valleculosa var. **setosa** Ohwi

Eleocharis yokoscensis (Franch. & Savat.) Tang & F. T. Wang

Eriophorum comosum (Wall.) Nees

Eriophorum gracile W. D. J. Koch ex Roth

Eriophorum microstachyum Boeckeler

Eriophorum scheuchzeri Hoppe

Erioscirpus comosus (Wall.) Palla

Erioscirpus microstachyus (Boeckeler) Palla

Fimbristylis acuminata Vahl

Fimbristylis aestivalis (Retz.) Vahl

Fimbristylis albo-viridis C. B. Clarke

Fimbristylis bisumbellata (Forssk.) Bubani

Fimbristylis complanata (Retz.) Link

Fimbristylis dichotoma (L.) Vahl

Fimbristylis dichotoma subsp. **podocarpa** (Nees) T. Koyama

Fimbristylis falcata (Vahl) Kunth

Fimbristylis filifolia Boeckeler

Fimbristylis fimbristyloides (F. Muell.) Druce

Fimbristylis fusca (Nees) C. B. Clarke

Fimbristylis fuscinux C. B. Clarke

Fimbristylis littoralis Gaudich.

Fimbristylis miliacea (L.) Vahl

Fimbristylis nigrobrunnea Thwait.

Fimbristylis ovata (Burm. f.) J. Kern

Fimbristylis pierotii Miq.

Fimbristylis quinquangularis (Vahl) Kunth

Fimbristylis rigidula Nees

Fimbristylis schoenoides (Retz.) Vahl

Fimbristylis squarrosa Vahl

Fimbristylis stolonifera C. B. Clarke

Fimbristylis tenera Schult.

Fimbristylis tetragona R. Br.

Fimbristylis thomsonii Boeckeler

Fimbristylis tomentosa Vahl

Fimbristylis umbellaris (Lam.) Vahl

Fuirena ciliaris (L.) Roxb.

Fuirena umbellata Rottb.

Heleocharis yunnanensis Svenson

Hypolytrum nemorum (Vahl) Spreng.

Iodocourtoisia cyperoides (Roxb.) Bennet & Raizada

Isolepis setacea (L.) R. Br.

Kyllinga brevifolia Rott.

Kyllinga brevifolia subsp. **leiolepis** (Franch. & Sav.) T. Koyama

Kyllinga brevifolia var. **stellulata** (Valck. Sur.) S. S. Hooper ex Karthik.

Kyllinga cylindrica Nees

Kyllinga nemoralis (J. R. Forst. & G. Forst.) Dandy ex Hutch. & Dalziel

Kyllinga odorata subsp. **cylindrica** (Nees) T. Koyama

Kyllinga odorata Vahal

Kyllinga squamulata Vahl

Kyllinga tenuifolia Steud.

Kyllinga triceps Rottb.

Lipocarpha chinensis (Osbeck) J. Kern

Lipocarpha sphacelata (Vahl) Kunth

Lipocarpha squarrosa (L.) Goetgh.

Machaerina rubiginosa (Solander ex G. Forst.) T. Koyama

Mapania arunachalensis G. D. Pal

Mapania palustris (Hassk. ex Steud.) Fern.-Vill.

Pycreus delavayi C. B. Clarke

Pycreus diaphanus (Schrad. ex Roem. & Schult.) S. S. Hooper & T. Koyama

Pycreus flavidus (Retz.) T. Koyama

Pycreus lijiangensis L. K. Dai

Pycreus polystachyos (Rottb.) P. Beauv.

Pycreus pumilus (L.) Nees

Pycreus sanguin (Vahl) Nees

Pycreus sanguinolentus (Vahl) Nees ex C. B. Clarke

Pycreus stramineus C. B. Clarke

Pycreus sulcinux (C. B. Clarke) C. B. Clarke

Pycreus unioloides (R. Br.) Urb.

Queenslandiella hyalina (Vahl) F. Ballard

Rhynchospora corymbosa (L.) Britton

Rhynchospora rubra (Lour.) Makino

Rhynchospora rugosa (Vahl) Gale

Rhynchospora rugosa subsp. **brownii** (Roem. & Schult.) T. Koyama

Schoenoplectus articulatus (L.) Palla

Schoenoplectus chen-moui (Tang & F. T. Wang) Hayas.

Schoenoplectus fuscorubens (T. Koyama) T. Koyama

Schoenoplectus grossus (L. f.) Palla

Schoenoplectus juncoides (Roxb.) Palla

Schoenoplectus lacustris (L.) Palla

Schoenoplectus lacustris subsp. **hippolytii** (V. Krecz.) Kukkonen

Schoenoplectus litoralis (Schrad.) Palla

Schoenoplectus lupulinus (Nees) V. I. Krecz.

Schoenoplectus maritimus (L.) Lye

Schoenoplectus mucronatus (L.) Palla

Schoenoplectus mucronatus subsp. **robustus** (Miq.) T. Koyama

Schoenoplectus rechingeri Kukkonen

Schoenoplectus roylei (Nees) Ovcz. & Czukav.

Schoenoplectus supinus (L.) Palla

Schoenoplectus supinus subsp. **lateriflorus** (Gmel.) Soják

Schoenoplectus tabernaemontani (C. C. Gmel.) Palla

Schoenoplectus triqueter (L.) Palla

Schoenoplectus wallichii (Nees) T. Koyama

Scirpus dialgamensis Majeed Kak & Javeid

Scirpus lushanensis Ohwi

Scirpus paniculatocorymbosus Kük.

Scirpus quadrangulus D. Don

Scirpus rosthornii Diels

Scirpus ternatanus Reinw. ex Miq.

Scirpus wichurae Boeckeler

Scleria benthamii C. B. Clarke

Scleria biflora Roxb.

Scleria dulungensis P. C. Li

Scleria hookeriana Boeckeler

Scleria laevis Willd.

Scleria levis Retz.

Scleria lithosperma (L.) Sw.

Scleria parvula Steud.

Scleria pergracilis (Nees) Kunth

Scleria terrestris (L.) Fassett

Scleria tessellata Willd.

Trichophorum distigmaticum (Kük.) T. V. Egorova

Trichophorum pumilum (Vahl) Schinz & Thell.

101. POACEAE

Achnatherum brandisii (Mez) Z. L. Wu

Achnatherum chingii (Hitchc.) Keng

Achnatherum duthiei (Hook. f.) P. C. Kuo & S. L. Lu

Achnatherum inaequiglume Keng

Achnatherum inebrians (Hance) Keng ex Tzvelev

Achnatherum jacquemontii (Jaub. & Spach) P. C. Kuo & S. L. Lu

Achnatherum psilantherum Keng ex Tzvelev

Achnatherum sibiricum (L.) Keng ex Tzvelev

Achnatherum splendens (Trin.) Nevski

Acrachne racemosa (Heyne ex Roem. & Schult.) Ohwi

Acroceras zizanioides (Kunth) Dandy

Aegilops tauschii Coss.

Agrostis brachiata Munro ex Hook. f.

Agrostis canina L.

Agrostis capillaris L.

Agrostis clavata Trin.

Agrostis filipes Hook. f.

Agrostis gigantea Roth

Agrostis griffithiana (Hook. f.) Bor

Agrostis hissarica Roshev.

Agrostis hookeriana C. B. Clarke ex Hook. f.

Agrostis hugoniana Rendle

Agrostis inaequiglumis Griseb.

Agrostis kunmingensis B. S. Sun & Y. Cai Wang

Agrostis limprichtii Pilg.

Agrostis mackliniae Bor

Agrostis micrantha Steud.

Agrostis munroana Aitch. & Hemsl.

Agrostis myriantha Hook. f.

Agrostis nervosa Nees ex Trin.

Agrostis pendryi Paszko

Agrostis petelotii (Hitchc.) Noltie

Agrostis pilosula Trin.

Agrostis sikkimensis Bor

Agrostis sinocontracta S. M. Phillips & S. L. Lu

Agrostis sinorupestris L. Liu ex S. M. Phillips & S. L. Lu

Agrostis stolonifera L.

Agrostis tenuis Sibth.

Agrostis triaristata (Hook. f.) Bor

Agrostis ushae Noltie

Agrostis vinealis Schreb.

Agrostis viridis Gouan

Agrostis zenkeri Trin.

Aira caryophyllea L.

Alloteropsis cimicina (L.) Stapf

Alloteropsis semialata (R. Br.) Hitchc.

Alloteropsis semialata var. **eckloniana** (Nees) Pilger

Alopecurus aequalis Sobol.

Alopecurus apiatus Ovcz.

Alopecurus arundinaceus Poir.

Alopecurus himalaicus Hook. f.

Alopecurus mucronatus Paulsen

Alopecurus myosuroides Huds.

Alopecurus nepalensis Trin. ex Steud.

Alopecurus pratensis L.

Ampelocalamus mianningensis (Q. Li & X. Jiang) D. Z. Li & Stapleton

Ampelocalamus patellaris (Gamble) Stapleton

Andropogon ascinodis C. B. Clarke

Andropogon chinensis (Nees) Merr.

Andropogon munroi C. B. Clarke

Aniselytron treutleri (Kuntze) Soják

Anthoxanthum flexuosum (Hook. f.) Veldkamp

Anthoxanthum glabrum (Trin.) Veldkamp

Anthoxanthum hookeri (Griseb.) Rendle

Anthoxanthum nitens (Weber) Y. Schouten & Veldkamp

Anthoxanthum odoratum L.

Anthoxanthum potaninii (Tzvelev) S. M. Phillips & Z. L. Wu

Anthoxanthum sikkimense (Maxim.) Ohwi

Anthoxanthum tibeticum (Bor) Veldkamp

Apluda mutica L.

Apocopis paleacea (Trin.) Hochr.

Aristida adscensionis L.

Aristida alpina L. Liu

Aristida batangensis Z. X. Tang & H. X. Liu

Aristida brevissima L. Liu

Aristida cumingiana Trin. & Rupr.

Aristida cyanantha Nees ex Steud.

Aristida funiculata Trin. & Rupr.

Aristida mutabilis Trin. & Rupr.

Aristida scabrescens L. Liu

Aristida triseta Keng

Aristida tsangpoensis L. Liu

Arthraxon castratus (Griff.) V. Naray. ex Bor

Arthraxon echinatus (Nees) Hochst.

Arthraxon epectinatus B. S. Sun & H. Peng

Arthraxon hispidus (Thunb.) Makino

Arthraxon lancifolius (Trin.) Hochst.

Arthraxon microphyllus (Trin.) Hochst.

Arthraxon nudus (Nees ex Steud.) Hochst.

Arthraxon prionodes (Steud.) Dandy

Arthraxon submuticus (Nees ex Steud.) Hochst.

Arthraxon typicus (Buse) Koord.

Arundinaria faberi Rendle

Arundinaria falcata Nees

Arundinaria racemosa Munro

Arundinaria spanostachya (T. P. Yi) D. Z. Li

Arundinella bengalensis (Spreng.) Druce

Arundinella birmanica Hook. f.

Arundinella dagana Noltie

Arundinella decempedalis (Kuntze) Janowski

Arundinella hirta (Thunb.) Tanaka

Arundinella hookeri Munro ex Keng

Arundinella intricata Hughes

Arundinella khaseana Nees ex Steud.

Arundinella nepalensis Trin.

Arundinella setosa Trin.

Arundinella tricholepis B. S. Sun & Z. H. Hu

Arundinella yunnanensis Keng ex B. S. Sun & Z. H. Hu

Arundo donax L.

Avena barbata Pott ex Link

Avena byzantina K. Koch

Avena fatua L.

Avena fatua subsp. **meridionalis** Malz.

Avena sterilis subsp. **ludoviciana** (Durieu) M. Gillet & Magne

Axonopus compressus (Sw.) P. Beauv.

Axonopus fissifolius (Raddi) Kuhlm.

Bambusa alamii Stapleton

Bambusa arundinacea (Retz.) Willd.

Bambusa balcooa Roxb.

Bambusa clavata Stapleton

Bambusa jaintiana R. B. Majumdar

Bambusa longispiculata Gamble

Bambusa mompaeana Naithani

Bambusa multiplex (Lour.) Raeusch. ex Schult.

Bambusa nepalensis Stapleton

Bambusa nutans Wall. ex Munro

Bambusa pallida Munro

Bambusa teres Buch.-Ham. ex Munro

Bambusa tulda Roxb.

Bambusa vulgaris Schrad.

Bambusa xueana Ohrnb.

Beckmannia syzigachne (Steud.) Fernald

Bothriochloa bladhii (Retz.) S. T. Blake

Bothriochloa glabra (Roxb.) A. Camus

Bothriochloa intermedia (R. Br.) A. Camus

Bothriochloa intermedia var. **punctata** (R. Br.) A. Camus

Bothriochloa ischaemum (L.) Keng

Bothriochloa kuntzeana (Hack.) Henrard

Bothriochloa pertusa (L.) A. Camus

Brachiaria brizantha (A. Rich.) Stapf

Brachiaria deflexa (Schumach.) C. E. Hubbard ex Robyns

Brachiaria dictyoneura (Fig. & De Not.) Stapf

Brachiaria distachya (L.) Stapf

Brachiaria eruciformis (Sm.) Griseb.

Brachiaria kurzii (Hook. f.) E. G. Camus & A. Camus

Brachiaria mutica (Forssk.) Stapf

Brachiaria ramosa (L.) Stapf

Brachiaria reptans (L.) C. A. Gardner & C. E. Hubb.

Brachiaria setigera (Retz.) C. E. Hubb.

Brachiaria subquadripara (Trin.) Hitchc.

Brachiaria villosa (Lam.) A. Camus

Brachypodium distachyon (L.) P. Beauv.

Brachypodium pinnatum (L.) P. Beauv.

Brachypodium pratense Keng ex Keng f.

Brachypodium sylvaticum (Huds.) P. Beauv.

Briza media L.

Bromus catharticus Vahl

Bromus confinis Nees ex Steud.

Bromus danthoniae Trin. ex C. A. Mey.

Bromus epilis Keng f.

Bromus erectus Huds.

Bromus gedrosianus Pénzes

Bromus gracillimus Bunge

Bromus grossus Desf. ex DC.

Bromus himalaicus Stapf

Bromus hordeaceus L.

Bromus intermedius Guss.

Bromus inermis Leyss.

Bromus japonicus Thunb.

Bromus koeieanus Melderis

Bromus lanceolatus Roth

Bromus madritensis L.

Bromus magnus Keng

Bromus mairei Hack. ex Hand.-Mazz.

Bromus nepalensis Melderis

Bromus oxyodon Schrenk

Bromus paulsenii Hackel ex Paulsen

Bromus pectinatus Thunb.

Bromus plurinodis Keng

Bromus porphyranthos Cope

Bromus pseudoramosus Keng ex P. C. Keng

Bromus racemosus L.

Bromus ramosus Huds.

Bromus remotiflorus (Steud.) Ohwi

Bromus scoparius L.

Bromus secalinus L.

Bromus sericeus Drobov

Bromus sinensis Keng ex P. C. Keng

Bromus staintonii Melderis

Bromus stenostachyus Boiss.

Bromus sterilis L.

Bromus tectorum L.

Bromus tomentosus Trin.

Bromus variegatus M. Bieb.

Bromus willdenowii Kunth

Brylkinia caudata (Munro) F. Schmidt

Calamagrostis decora Hook. f.

Calamagrostis elatior (Griseb.) A. Camus

Calamagrostis emodensis Griseb.

Calamagrostis epigeios (L.) Roth

Calamagrostis hedinii Pilg.

Calamagrostis lahulensis G. Singh

Calamagrostis nagarum (Bor) G. Singh

Calamagrostis pseudophragmites (Hall. f.) Koeler

Calamagrostis stoliczkai Hook. f.

Capillipedium assimile (Steud.) A. Camus

Capillipedium kuoi L. B. Cai

Capillipedium parviflorum (R. Br.) Stapf

Capillipedium pteropechys (C. B. Clarke) Stapf

Catabrosa aquatica (L.) P. Beauv.

Catabrosa aquatica var. **angusta** Stapf

Catabrosa capusii Franch.

Catabrosa sikkimensis Stapf

Catabrosella himalaica (Hook. f.) Tzvelev

Catapodium rigidum (L.) C. E. Hubb.

Cenchrus biflorus Roxb.

Cenchrus orientalis (Rich.) Morrone

Cenchrus setigerus Vahl

Centotheca lappacea (L.) Desv.

Cephalostachyum capitatum Munro

Cephalostachyum latifolium Munro

Cephalostachyum mannii (Gamble) Stapleton & D. Z. Li

Cephalostachyum pallidum Munro

Cephalostachyum scandens Bor

Chimonobambusa armata (Gamble) Hsueh & T. P. Yi

Chimonobambusa arunachalensis T. P. Sharma & Borthakur

Chimonobambusa callosa (Murno) Nakai

Chimonobambusa griffithianus (Munro) Hsuch & Yi

Chimonobambusa metuoensis Hsueh & T. P. Yi

Chimonocalamus griffithianus (Munro) Hsueh & T. P. Yi

Chimonocalamus montanus Hsueh & T. P. Yi

Chionachne koenigii (Spreng.) Thwaites

Chloris dolichostachya Lag.

Chloris gayana Kunth

Chloris virgata Sw.

Chrysopogon aciculatus (Retz.) Trin.

Chrysopogon aucheri (Boiss.) Stapf

Chrysopogon fulvus (Spreng.) Chiov.

Chrysopogon gryllus (L.) Trin.

Chrysopogon gryllus subsp. **echinulatus** (Nees) Cope

Chrysopogon lancearius (Hook. f.) Haines

Chrysopogon serrulatus Trin.

Chrysopogon zizanioides (L.) Roberty

Cleistogenes songorica (Roshev.) Ohwi

Coelachne simpliciuscula (Steud.) Munro ex Benth.

Coelorachis khasiana (Hack.) Stapf ex Bor

Coelorachis striata (Steud.) A. Camus

Coix aquatica Roxb.

Coix lacryma-jobi L.

Colpodium altaicum Trin. ex Ledeb.

Colpodium tibeticum Bor

Colpodium wallichii (Stapf) Bor

Crypsis aculeata (L.) Aiton

Crypsis schoenoides (L.) Lam.

Cyathopus sikkimensis Stapf

Cymbopogon bhutanicus Noltie

Cymbopogon citratus (DC.) Stapf

Cymbopogon distans (Nees ex Steud.) Will. Watson

Cymbopogon flexuosus (Nees ex Steud.) Will. Watson

Cymbopogon flexuosus var. **sikkimensis** Bor

Cymbopogon jwarancusa (Jones) Schult.

Cymbopogon jwarancusa subsp. **olivieri** (Boiss.) S. Soenarko

Cymbopogon khasianus (Hack.) Stapf ex Bor

Cymbopogon liangshanensis S. M. Phillips & S. L. Chen

Cymbopogon martini (Roxb.) Will. Watson

Cymbopogon microthecus (Hook. f.) A. Camus

Cymbopogon nardus (L.) Rendle

Cymbopogon pendulus (Nees ex Steud.) Will. Watson

Cymbopogon pospischilii (K. Schum.) C. E. Hubb.

Cymbopogon schoenanthus (L.) Spreng.

Cymbopogon tungmaiensis L. Liu

Cymbopogon xichangensis R. S. Zhang & B. S. Sun

Cynodon dactylon (L.) Pers.

Cynodon plectostachyus (K. Schum.) Pilg.

Cynodon radiatus Roth

Cyrtococcum accrescens (Trin.) Stapf

Cyrtococcum oxyphyllum (Steud.) Stapf

Cyrtococcum patens (L.) A. Camus

Dactylis glomerata L.

Dactylis glomerata var. **himalayensis** Domin.

Dactyloctenium aegyptium (L.) Willd.

Danthonia cachemyriana Jaub. & Spach

Danthonia cumminsii Hook. f.

Danthonia schneideri Pilg.

Dendrocalamus fugongensis Hsueh & D. Z. Li

Dendrocalamus giganteus Munro

Dendrocalamus hamiltonii Nees & Arn. ex Munro

Dendrocalamus hookeri Munro

Dendrocalamus parishii Munro

Dendrocalamus sahnii H. B. Naithani & Bahadur

Dendrocalamus sikkimensis Gamble ex Oliv.

Dendrocalamus strictus (Roxb.) Nees

Dendrocalamus tibeticus Hsueh & T. P. Yi

Deschampsia cespitosa (L.) P. Beauv.

Deschampsia cespitosa subsp. **macrothyrsa** (Tatew. & Ohwi) Tzvelev

Deschampsia cespitosa subsp. **sikkimensis** Noltie

Deschampsia flexuosa (L.) Trin.

Deschampsia koelerioides Regel

Deyeuxia abnormis Hook. f.

Deyeuxia anthoxanthoides Munro ex Hook. f.

Deyeuxia debilis (Hook. f.) Veldkamp

Deyeuxia diffusa Keng

Deyeuxia effusiflora Rendle

Deyeuxia elatior (Griseb.) Hook. f.

Deyeuxia flaccida (P. C. Keng) Keng ex S. L. Lu

Deyeuxia flavens Keng

Deyeuxia himalaica L. Liu ex Wen L. Chen

Deyeuxia holciformis (Jaub. & Spach) Bor

Deyeuxia lapponica (Wahlenb.) Kunth

Deyeuxia mazzettii Veldkamp

Deyeuxia moupinensis (Franch.) Pilger

Deyeuxia neglecta (Ehrhart) Kunth

Deyeuxia nepalensis Bor

Deyeuxia nivicola Hook. f.

Deyeuxia nyingchiensis P. C. Kuo & S. L. Lu

Deyeuxia petelotii (Hitchc.) S. M. Phillips & Wen L. Chen

Deyeuxia pulchella Hook. f.

Deyeuxia purpurea (Trin.) Kunth

Deyeuxia pyramidalis (Host) Veldkamp

Deyeuxia rosea Bor

Deyeuxia scabrescens (Griseb.) Munro ex Duthie

Deyeuxia sichuanensis (J. L. Yang) S. M. Phillips & Wen L. Chen

Deyeuxia tianschanica (Rupr.) Bor

Deyeuxia tibetica Bor

Deyeuxia tibetica var. **przevalskyi** (Tzvelev) P. C. Kuo & S. L. Lu

Deyeuxia yanyuanensis (J. L. Yang) L. Liu

Deyeuxia zangxiensis P. C. Kuo & S. L. Lu

Dichanthium annulatum (Forssk.) Stapf

Dichanthium caricosum (L.) A. Camus

Dichanthium foveolatum (Del.) Roberty

Digitaria abludens (Roem. & Schult.) Veldkamp

Digitaria bicornis (Lam.) Roem. & Schult.

Digitaria ciliaris (Retz.) Koeler

Digitaria compacta (Roth ex Roem. & Schult.) Veldkemp

Digitaria cruciata (Nees) A. Camus

Digitaria fuscescens (J. Presl) Henrard

Digitaria hengduanensis L. Liu

Digitaria ischaemum (Schreb.) Muhl.

Digitaria longiflora (Retz.) Pers.

Digitaria radicosa (J. Presl) Miq.

Digitaria sanguinalis (L.) Scop.

Digitaria setigera Roth ex Roem. & Schult.

Digitaria stewartiana Bor

Digitaria stricta Roth ex Roem. & Schult.

Digitaria stricta var. **denudata** (Link) Henrard

Digitaria stricta var. **glabrescens** Bor

Digitaria ternata (A. Rich.) Stapf

Digitaria violascens Link

Dimeria fuscescens Trin.

Dimeria ornithopoda Trin.

Dinochloa macclellandii (Munro) Kurz

Drepanostachyum ampullare (T. P. Yi) Demoly

Drepanostachyum annulatum Stapleton

Drepanostachyum falcatum (Nees) Keng f.

Drepanostachyum fractiflexum (T. P. Yi) D. Z. Li

Drepanostachyum intermedium (Munro) Keng f.

Drepanostachyum khasianum (Munro) Keng f.

Drepanostachyum membranaceum (T. P. Yi) D. Z. Li

Drepanostachyum polystachyum (Kurz ex Gamble) R. B. Majumdar

Drepanostachyum semiorbiculatum (T. P. Yi) Stapleton

Duthiea brachypodium (P. Candargy) Keng & Keng f.

Duthiea nepalensis Bor

Echinochloa colona (L.) Link

Echinochloa crus-galli (L.) P. Beauv.

Echinochloa crus-pavonis (Kunth) Schult.

Echinochloa frumentacea Link

Echinochloa glabrescens Munro ex Hook. f.

Echinochloa oryzoides (Ard.) Fritsch

Echinochloa picta (K. D. Koenig) P. W. Michael

Echinochloa pyramidalis (Lam.) Hitchc. & Chase

Echinochloa stagnina (Retz.) P. Beauv.

Eleusine coracana (L.) Gaertn.

Eleusine indica (L.) Gaertn.

Elymus altissimus (Keng) Á. Löve ex B. Rong Lu

Elymus anthosachnoides (Keng) Á. Löve ex B. Rong Lu

Elymus antiquus (Nevski) Tzvelev

Elymus aristiglumis (Keng & S. L. Chen) S. L. Chen

Elymus barystachyus L. B. Cai

Elymus batalinii (Krasn.) Á. Löve

Elymus brevipes (Keng) S. L. Chen

Elymus burchan-buddae (Nevski) Tzvelev

Elymus cacuminis B. Rong Lu & B. Salomon

Elymus caesifolius Á. Löve ex S. L. Chen

Elymus caianus S. L. Chen & G. Zhu

Elymus canaliculatus (Nevski) Tzvelev

Elymus caninus (L.) L.

Elymus ciliaris (Trin.) Tzvelev

Elymus cognata (Hack.) T. A. Cope

Elymus curtiaristatus (L. B. Cai) S. L. Chen & G. Zhu

Elymus dahuricus Griseb.

Elymus dentatus (Hook. f.) T. A. Cope

Elymus dolichatherus (Keng) S. L. Chen

Elymus durus (Keng) S. L. Chen

Elymus duthiei (Melderis) G. Singh

Elymus elytrigioides (C. Yen & J. L. Yang) S. L. Chen

Elymus excelsus Turcz. ex Griseb.

Elymus fedtschenkoi Tzvelev

Elymus himalayanus (Nevski) Tzvelev

Elymus hongyuanensis (L. B. Cai) S. L. Chen & G. Zhu

Elymus jacquemontii (Hook. f.) Cope

Elymus kamoji (Ohwi) S. L. Chen

Elymus laxinodis (L. B. Cai) S. L. Chen & G. H. Zhu

Elymus leianthus (Keng) S. L. Chen

Elymus longearistatus (Boiss.) Tzvelev

Elymus mutabilis (Drobov) Tzvelev

Elymus nepalensis (Melderis) Melderis

Elymus nutans Griseb.

Elymus pulanensis (H. L. Yang) S. L. Chen

Elymus purpurascens (Keng) S. L. Chen

Elymus repens (L.) Gould

Elymus retroflexus B. Rong Lu & B. Salomon

Elymus russellii (Melderis) Cope

Elymus schrenkianus (Fisch. & C. A. Mey.) Tzvelev

Elymus schugnanicus (Nevski) Tzvelev

Elymus semicostatus (Nees ex Steud.) Meld.

Elymus shouliangiae (L. B. Cai) S. L. Chen & G. H. Zhu

Elymus sibiricus L.

Elymus sikkimensis (Melderis) Melderis

Elymus sinicus (Keng) S. L. Chen

Elymus sinosubmuticus S. L. Chen

Elymus stewartii (Melderis) Cope

Elymus strictus (Keng) S. L. Chen

Elymus tangutorum (Nevski) Hand.-Mazz.

Elymus tenuispicus (J. L. Yang & Y. H. Zhou) S. L. Chen

Elymus thoroldianus (Oliv.) G. Singh

Elymus tibeticus (Melderis) G. Singh

Elymus yangiae B. Rong Lu

Elymus yushuensis (L. B. Cai) S. L. Chen & G. H. Zhu

Elytrigia repens (L.) Desvaux ex B. D. Jackson

Elytrophorus spicatus (Willd.) A. Camus

Enneapogon desvauxii P. Beauv.

Enneapogon persicus Boiss.

Enteropogon dolichostachyus (Lag.) Keng ex Lazarides

Eragrostiella nardoides (Trin.) Bor

Eragrostis atrovirens (Desf.) Trin. ex Steud.

Eragrostis cilianensis (All.) Janch.

Eragrostis coarctata Stapf

Eragrostis cylindrica (Roxb.) Arn.

Eragrostis ferruginea (Thunb.) P. Beauv.

Eragrostis gangetica (Roxb.) Steud.

Eragrostis japonica (Thunb.) Trin.

Eragrostis minor Host

Eragrostis multicaulis Steud.

Eragrostis nardoides (Trinius) Bor

Eragrostis nigra Nees ex Steud.

Eragrostis nutans (Retz.) Nees ex Steud.

Eragrostis papposa (Desf. ex Roem. & Schult.) Steud.

Eragrostis pilosa (L.) P. Beauv.

Eragrostis tenella (L.) P. Beauv. ex Roem. & Schult.

Eragrostis tremula Hochst. ex Steud.

Eragrostis unioloides (Retz.) Nees ex Steud.

Eragrostis viscosa (Retz.) Trin.

Eremochloa zeylanica (Hack. ex Trimen) Hack.

Eremopoa altaica (Trin.) Roshev.

Erianthus filifolius (Steud.) Hack.

Erianthus longesetosus Anderss.

Erianthus longesetosus var. **hookeri** Anderss.

Erianthus ravennae (L.) P. Beauv.

Erianthus rufipilus (Steud.) Griseb.

Erianthus sikkimensis Hook. f.

Eriochloa villosa (Thunb.) Kunth

Eulalia contorta (Brongn.) Kuntze

Eulalia fastigiata (Nees ex Steud.) Haines

Eulalia leschenaultiana (Decne.) Ohwi

Eulalia manipurensis Bor
Eulalia mollis (Griseb.) Kuntze
Eulalia pallens (Hack.) Kuntze
Eulalia quadrinervis (Hack.) Kuntze
Eulalia staintonii Bor
Eulalia trispicata (Schult.) Henrard
Eulalia yunnanensis Keng & S. L. Chen
Eulaliopsis binata (Retz.) C. E. Hubb.
Eulaliopsis sykesii Bor
Fargesia acuticontracta T. P. Yi
Fargesia adpressa T. P. Yi
Fargesia albocerea Hsueh & T. P. Yi
Fargesia angustissima T. P. Yi
Fargesia canaliculata T. P. Yi
Fargesia communis T. P. Yi
Fargesia contracta T. P. Yi
Fargesia declivis T. P. Yi
Fargesia dulcicula T. P. Yi
Fargesia dura T. P. Yi
Fargesia edulis Hsueh & T. P. Yi
Fargesia elegans T. P. Yi
Fargesia emaculata T. P. Yi
Fargesia exposita T. P. Yi
Fargesia extensa T. P. Yi
Fargesia farcta T. P. Yi
Fargesia ferax (Keng) T. P. Yi
Fargesia frigidis T. P. Yi
Fargesia glabrifolia T. P. Yi
Fargesia gongshanensis T. P. Yi
Fargesia grossa T. P. Yi
Fargesia jiulongensis T. P. Yi
Fargesia lushuiensis Hsueh & T. P. Yi
Fargesia macclureana (Bor) Stapleton
Fargesia macrophylla Hsueh & C. M. Hui
Fargesia mali T. P. Yi
Fargesia melanostachys (Hand.-Mazz.) T. P. Yi
Fargesia nitida (Mitford) Keng f. ex T. P. Yi
Fargesia nujiangensis Hsueh & C. M. Hui
Fargesia orbiculata T. P. Yi
Fargesia papyrifera T. P. Yi
Fargesia pauciflora (Keng) T. P. Yi
Fargesia pleniculmis (Hand.-Mazz.) T. P. Yi
Fargesia praecipua T. P. Yi
Fargesia robusta T. P. Yi
Fargesia sagittatinea T. P. Yi
Fargesia solida T. P. Yi
Fargesia stenoclada T. P. Yi
Fargesia stricta Hsueh & C. M. Hui
Fargesia subflexuosa T. P. Yi
Fargesia sylvestris T. P. Yi
Fargesia yulongshanensis T. P. Yi
Fargesia yunnanensis Hsueh & T. P. Yi
Fargesia zayuensis T. P. Yi
Festuca alaica Drobow
Festuca alatavica (St.-Yves) Roshev.
Festuca arundinacea Schreb.
Festuca asthenica Hook. f.
Festuca bhutanica E. B. Alexeev
Festuca boriana E. B. Alexeev
Festuca changduensis L. Liu
Festuca chayuensis L. Liu
Festuca chumbiensis E. B. Alexeev
Festuca coelestis (St.-Yves) V. I. Krecz. & Bobr.
Festuca cumminsii Stapf
Festuca debilis (Stapf) E. B. Alexeev
Festuca dolichantha Keng ex Keng f.
Festuca eriobasis H. Scholz
Festuca fascinata Keng ex S. L. Lu
Festuca forrestii St.-Yves
Festuca forrestii subsp. **nepalensis** H. Scholz
Festuca georgii E. B. Alexeev
Festuca gigantea (L.) Vill.

Festuca handelii (St.-Yves) E. B. Alexeev

Festuca hartmannii (Markgr.-Dann.) E. B. Alexeev

Festuca japonica Makino

Festuca kashmiriana Stapf

Festuca leptopogon Stapf

Festuca levingei Stapf

Festuca liangshanica L. Liu

Festuca longiglumis S. L. Lu

Festuca modesta Nees ex Steud.

Festuca nepalica E. B. Alexeev

Festuca nitidula Stapf

Festuca olgae (Regel) Krivot.

Festuca ovina L.

Festuca pamirica Tzvelev

Festuca parvigluma Steud.

Festuca poluninii E. B. Alexeev

Festuca polycolea Stapf

Festuca pubiglumis S. L. Lu

Festuca rubra L.

Festuca rubra subsp. **arctica** (Hack.) Govor.

Festuca rubra subsp. **clarkei** (Stapf) Saint-Yves

Festuca scabriflora L. Liu

Festuca simlensis (Stapf) E. B. Alexeev

Festuca sinensis Keng ex E. B. Alexeev

Festuca sinomutica Xiang Chen & S. M. Phillips

Festuca stapfii E. B. Alexeev

Festuca tibetica (Stapf) E. B. Alexeev

Festuca undata Stapf

Festuca valesiaca Schleich. ex Gaudin

Festuca valesiaca subsp. **hypsophila** (St.-Yves) Tzvelev

Festuca valesiaca subsp. **pseudovina** (Hack. ex Wiesb.) Hegi

Festuca valesiaca subsp. **sulcata** (Hack.) Schinz & R. Keller

Festuca vierhapperi Hand.-Mazz.

Festuca wallichiana E. B. Alexeev

Festuca yulungschanica E. B. Alexeev

Festuca yunnanensis St.-Yves

Festuca yunnanensis var. **villosa** St.-Yves

Gaoligongshania megalothyrsa (Hand.-Mazz.) D. Z. Li, Hsueh & N. H. Xia

Garnotia acutigluma (Steud.) Ohwi

Garnotia emodi (Arn. & Nees) Janowski

Garnotia polypogonoides Munro ex Oliv.

Garnotia tenella (Arn. ex Miq.) Janowski

Gigantochloan macrostachya Kurz

Glyceria declinata Bréb.

Glyceria tonglensis C. B. Clarke

Hackelochloa granularis (L.) Kuntze

Hackelochloa porifera (Hack.) D. Rhind

Harpachne harpachnoides (Hack.) B. S. Sun & S. Wang

Helictotrichon asperum (Munro) Bor

Helictotrichon delavayi (Hack.) Henrard

Helictotrichon junghuhnii (Buse) Henrard

Helictotrichon parviflorum (Hook. f.) Bor

Helictotrichon potaninii Tzvelev

Helictotrichon pratense (L.) Pilger

Helictotrichon tibeticum (Roshev.) J. Holub

Helictotrichon tibeticum var. **laxiflorum** Keng ex Z. L. Wu

Helictotrichon virescens (Nees ex Steud.) Henrard

Helictotrichon yunnanense B. S. Sun & S. Wang

Hemarthria compressa (L. f.) R. Br.

Hemarthria protensa Steud.

Hemarthria vaginata Buse

Heteropogon contortus (L.) P. Beauv. ex Roem. & Schult.

Hierochloe flexuosa Hook. f.

Hierochloe laxa R. Br. ex Hook f.

Hierochloe tibetica Bor
Himalayacalamus asper Stapleton
Himalayacalamus brevinodus Stapleton
Himalayacalamus collaris (T. P. Yi) Ohrnb.
Himalayacalamus cupreus Stapleton
Himalayacalamus falconeri (Hook. f. ex Munro) Keng f.
Himalayacalamus fimbriatus Stapleton
Himalayacalamus hookerianus (Munro) Stapleton
Himalayacalamus porcatus Stapleton
Holcus lanatus L.
Hordeum agriocrithon Åberg
Hordeum bogdanii Wilensky
Hordeum brevisubulatum (Trin.) Link
Hordeum brevisubulatum subsp. **nevskianum** (Bowden) Tzvelev
Hordeum brevisubulatum subsp. **turkestanicum** Tzvelev
Hordeum distichon L.
Hordeum marinum subsp. **gussoneanum** (Parl.) Thell.
Hordeum murinum subsp. **glaucum** (Steud.) Tzvelev
Hordeum murinum subsp. **leporinum** (Link) Arcang.
Hordeum spontaneum K. Koch
Hordeum spontaneum var. **ischnatherum** (Cosson) Thell.
Hordeum spontaneum var. **proskowetzii** Nábělek
Hordeum turkestanicum Nevski
Hordeum vulgare L.
Hordeum vulgare var. **coeleste** L.
Hordeum vulgare var. **trifurcatum** (Schltdl.) Alef.
Hyalopoa nutans (Stapf) E. B. Alexeev ex T. A. Cope
Hygroryza aristata (Retz.) Nees ex Wight & Arn.
Hymenachne amplexicaulis (Rudge) Nees
Hymenachne assamica (Hook. f.) Hitchc.
Hymenachne pseudointerrupta Müll. Hal.
Hystrix duthiei (Stapf) Bor
Ichnanthus pallens (Sw.) Munro ex Benth.
Ichnanthus vicinus (Bailey) Merr.
Imperata cylindrica (L.) Raeusch.
Imperata latifolia (Hook. f.) L. Liu
Indocalamus longiauritus Hand.-Mazz.
Isachne albens Trin.
Isachne clarkei Hook. f.
Isachne dimyloides Bor
Isachne globosa (Thunb.) Kuntze
Isachne himalaica Hook. f.
Isachne miliacea Roth
Isachne pulchella Roth
Isachne scabrosa Hook. f.
Isachne sikkimensis Bor
Ischaemum polystachyum J. Presl
Ischaemum rugosum Salisb.
Ischaemum thomsonianum Stapf ex C. E. C. Fisch.
Kengyilia kokonorica (Keng) J. L. Yang, C. Yen & B. R. Baum
Kengyilia laxiflora (Keng) J. L. Yang, C. Yen & B. R. Baum
Kengyilia rigidula (Keng) J. L. Yang, C. Yen & B. R. Baum
Kengyilia thoroldiana (Oliv.) J. L. Yang, C. Yen & B. R. Baum
Kengyilia thoroldiana var. **laxiuscula** (Melderis) S. L. Chen
Kengyilia zadoiensis S. L. Lu & Y. H. Wu
Koeleria argentea Griseb.
Koeleria atroviolacea Domin
Koeleria cristata (L.) Pers.
Koeleria glaucovirens Domin
Koeleria litvinowii Domin

Koeleria litvinowii subsp. **argentea** (Griseb.) S. M. Phillips & Z. L. Wu

Koeleria macrantha (Ledeb.) Schult.

Lagurus ovatus L.

Leersia hexandra Sw.

Leersia oryzoides (L.) Sw.

Leptochloa chinensis (L.) Nees

Leptochloa panicea (Retz.) Ohwi

Leucopoa albida (Turcz.) V. I. Krecz. & Bobr.

Leymus lanatus (Korsh.) Tzvelev

Leymus secalinus (Georgi) Tzvelev

Littledalea alaica (Korsh.) Petrov ex Nevski

Littledalea przevalskyi Tzvelev

Littledalea racemosa Keng

Littledalea tibetica Hemsl.

Lolium × **hybridum** Hausskn.

Lolium multiflorum Lam.

Lolium perenne L.

Lolium persicum Boiss. & Hoh.

Lolium remotum Schrank

Lolium rigidum Gaudin

Lolium temulentum L.

Lophatherum gracile Brongn.

Lygeum spartum Loa. ex L.

Melica jacquemontii Decne.

Melica longiligulata Z. L. Wu

Melica nutans L.

Melica onoei Franch. & Sav.

Melica persica Kunth

Melica przewalskyi Roshev.

Melica radula Franch.

Melica scaberrima (Nees ex Steud.) Hook. f.

Melica scabrosa Trin.

Melica schuetzeana W. Hempel

Melica secunda Regel

Melica tangutorum Tzvelev

Melica taylorii W. Hempel

Melica tibetica Roshev.

Melica virgata Turcz. ex Trin.

Melica yajiangensis Z. L. Wu

Melinis minutiflora P. Beauv.

Melocalamus elevatissimus Hsueh & T. P. Yi

Melocanna baccifera (Roxb.) Kurz

Microchloa indica (L. f.) P. Beauv.

Microchloa kunthii Desv.

Microstegium batangense (S. L. Zhong) S. M. Phillips & S. L. Chen

Microstegium ciliatum (Trin.) A. Camus

Microstegium falconeri (Hook. f.) Clayton

Microstegium fasciculatum (L.) Henrard

Microstegium multiciliatum B. S. Sun

Microstegium nudum (Trin.) A. Camus

Microstegium petiolare (Trin.) Bor

Microstegium reticulatum B. S. Sun ex H. Peng & X. Yang

Microstegium vagans (Nees ex Steud.) A. Camus

Microstegium vimineum (Trin.) A. Camus

Milium effusum L.

Miscanthus nepalensis (Trin.) Hack.

Miscanthus nudipes (Griseb.) Hack.

Miscanthus paniculatus (B. S. Sun) Renvoize & S. L. Chen

Miscanthus wardii Bor

Mnesithea laevis (Retz.) Kunth

Muhlenbergia duthieana Hack.

Muhlenbergia himalayensis Hack. ex Hook. f.

Muhlenbergia huegelii Trin.

Neomicrocalamus andropogonifolius (Griff.) Stapleton

Neomicrocalamus mannii (Gamble) R. B. Majumdar

Neomicrocalamus microphyllus Hsueh & T. P. Yi

Neomicrocalamus prainii (Gamble) Keng f.

Neyraudia arundinacea (L.) Henrard

Neyraudia curvipes Ohwi

Neyraudia reynaudiana (Kunth) Keng ex Hitchc.

Ophiuros exaltatus (L.) Kunrtze

Ophiuros megaphyllus Stapf ex Haines

Oplismenus burmannii (Retz.) P. Beauv.

Oplismenus compositus (L.) P. Beauv.

Oplismenus compositus var. **rariflorus** (C. Presl) U. Scholz

Oplismenus undulatifolius (Ard.) Roem. & Schult.

Orinus anomala Keng ex Keng f. & L. Liu

Orinus thoroldii (Stapf ex Hemsl.) Bor

Orinus tibetica N. X. Zhao

Orthoraphium roylei Nees

Oryza meyeriana (Zoll. & Moritzi) Baill.

Oryza minuta J. Presl

Oryza nivara S. D. Sharma & Shastry

Oryza officinalis Wall. ex Watt

Oryza rufipogon Griff.

Oryza sativa L.

Panicum antidotale Retz.

Panicum atrosanguineum Hochst. ex A. Rich.

Panicum auritum J. Presl ex Nees

Panicum bisulcatum Thunb.

Panicum brevifolium L.

Panicum curviflorum Hornem.

Panicum elegantissimum Hook. f.

Panicum humidorum Buch.-Ham. ex Hook. f.

Panicum humile Nees ex Steud.

Panicum incisum Munro ex C. B. Clarke

Panicum incomtum Trin.

Panicum khasianum Munro ex Hook. f.

Panicum maximum Jacq.

Panicum miliaceum L.

Panicum notatum Retz.

Panicum paludosum Roxb.

Panicum psilopodium Trin.

Panicum repens L.

Panicum sarmentosum Roxb.

Panicum sumatrense Roth ex Roem. & Schult.

Panicum trypheron Schult.

Panicum virgatum L.

Panicum walense Mez

Paspalidium flavidum (Retz.) A. Camus

Paspalidium punctatum (Burm.) A. Camus

Paspalum conjugatum P. J. Bergius

Paspalum delavayi Henrard

Paspalum dilatatum Poir.

Paspalum distichum L.

Paspalum longifolium Roxb.

Paspalum notatum Flugge

Paspalum paspalodes (Michx.) Scribn.

Paspalum scrobiculatum L.

Paspalum thunbergii Kunth ex Steud.

Pennisetum alopecuroides (L.) Spreng.

Pennisetum clandestinum Hochst. ex Chiov.

Pennisetum flaccidum Griseb.

Pennisetum glaucum (L.) R. Br.

Pennisetum hohenackeri Hochst. ex Steud.

Pennisetum hordeoides (Lam.) Steud.

Pennisetum lanatum Klotzsch

Pennisetum longissimum S. L. Chen & Y. X. Jin

Pennisetum pedicellatum subsp. **unispiculum** Brunken

Pennisetum pedicellatum Trin.

Pennisetum polystachion (L.) Schult.

Pennisetum purpureum Schumach.

Pennisetum qianningense S. L. Zhong

Pennisetum sichuanense S. L. Chen & Y. X. Jin

Perotis hordeiformis Nees

Perotis indica (L.) Kuntze

Phacelurus speciosus (Steud.) C. E. Hubb.

Phacelurus zea (C. B. Clarke) Clayton

Phaenosperma globosa Munro ex Benth.

Phalaris arundinacea L.
Phalaris minor Retz.
Phalaris minor var. **nepalensis** (Trin.) Bor
Phleum alpinum L.
Phleum bertolonii DC.
Phleum himalaicum Mez
Phleum paniculatum Huds.
Phleum pratense L.
Phragmites australis (Cav.) Trin. ex Steud.
Phragmites karka (Retz.) Trin. ex Steud.
Phyllostachys arcana McClure
Phyllostachys assamica Gamble ex Brandis
Phyllostachys heteroclada Oliv.
Phyllostachys mannii Gamble
Piptatherum aequiglume (Duthie ex Hook. f.) Roshev.
Piptatherum gracile Mez
Piptatherum grandispiculum (P. C. Kuo & Z. L. Wu) S. M. Phillips & Z. L. Wu
Piptatherum hilariae Pazij
Piptatherum laterale (Regel) Munro ex Nevski
Piptatherum munroi (Stapf ex Hook. f.) Mez
Piptatherum pamiralaicum (Grig.) Roshev.
Piptatherum platyanthum Nevski
Piptatherum purpurascens (Hack.) Roshev.
Piptatherum tibeticum Roshev.
Piptatherum vicarium (Grigorj.) Roshev.
Pleioblastus simonii (Carr.) Nakai
Poa aitchisonii Boiss.
Poa albertii Regel
Poa alpigena (Blytt) Lindm.
Poa alpina (Blytt) Lindm.
Poa angustifolia L.
Poa annua L.
Poa arnoldii Melderis
Poa asperifolia Bor
Poa attenuata Trin.
Poa bactriana Roshev.
Poa bomiensis C. Ling
Poa bulbosa L.
Poa burmanica Bor
Poa calliopsis Litvinov
Poa chumbiensis Noltie
Poa compressa L.
Poa cooperi Noltie
Poa crymophila Keng ex C. Ling
Poa debilior Hitchc.
Poa digena Melderis
Poa dolichachyra G. Q. Song
Poa dzongicola Noltie
Poa eleonorae Bor
Poa faberi Rendle
Poa falconeri Stapf
Poa gammieana Stapf
Poa garhwalensis D. C. Nautiyal & R. D. Gaur
Poa gilgitica Dickoré
Poa glauca Vahl
Poa grandis Hand.-Mazz.
Poa harae Rajbh.
Poa himalayana Nees ex Steud.
Poa hirtiglumis Stapf
Poa hylobates Bor
Poa ianthina Keng ex Shan Chen
Poa imperialis Bor
Poa indattenuata G. Q. Song
Poa infirma Kunth
Poa kanaii Rajbh.
Poa khasiana Stapf
Poa koelzii Bor
Poa lachenensis Noltie
Poa ladakhensis H. Hartm.
Poa lahulensis Bor
Poa langtangensis Melderis
Poa lhasaensis Bor
Poa litvinoviana Ovcz.

Poa longii Noltie
Poa mairei Hack.
Poa mustangensis Rajbh.
Poa nemoraliformis Roshev.
Poa nemoralis L.
Poa nepalensis (Wall. ex Griseb.) Duthie
Poa nimuana C. Ling
Poa nitidespiculata Bor
Poa nubigeana Keng ex L. Liu
Poa orinosa Keng
Poa pagophila Bor
Poa palustris L.
Poa perennis Keng ex Keng f.
Poa polycolea Stapf
Poa polyneuron Bor
Poa poophagorum Bor
Poa pratensis L.
Poa pseudoamoena Bor
Poa rajbahandarii Noltie
Poa rhadiana Bor
Poa roemeri Bor
Poa rossbergiana K. S. Hao
Poa royleana Nees ex Steud.
Poa setulosa Bor
Poa sikkimensis (Stapf) Bor
Poa spontanea Bor
Poa staphiana Bor
Poa sunbisinii Soreng & G. H. Zhu
Poa supina Schrad.
Poa suruana H. Hartmann
Poa szechuensis Rendle
Poa tibetica Munro ex Stapf
Poa tibeticola Bor
Poa trivialis L.
Poa varia Keng ex L. Liu
Poa wardiana Bor
Poa yakiangensis L. Liu
Poa zhongbaensis C. Ling
Pogonatherum crinitum (Thunb.) Kunth
Pogonatherum paniceum (Lam.) Hack.
Polypogon fugax Nees ex Steud.
Polypogon hissaricus (Roshev.) Bor
Polypogon monspeliensis (L.) Desf.
Polypogon viridis (Gouan) Breistr.
Polytoca digitata (L. f.) Druce
Polytoca wallichiana (Steud.) Benth.
Pseudechinolaena polystachya (Humb., Bonpl. & Kunth) Stapf
Pseudodanthonia himalaica (Hook. f.) Bor & C. E. Hubb.
Pseudopogonatherum contortum (Brongn.) A. Camus
Pseudoraphis minuta (Mez) Pilg.
Pseudosorghum fasciculare (Roxb.) A. Camus
Pseudostachyum polymorphum Munro
Ptilagrostis concinna (Hook. f.) Roshev.
Ptilagrostis dichotoma Keng ex Tzvelev
Ptilagrostis junatovii Grubov
Ptilagrostis luquensis P. M. Peterson, Soreng & Z. L. Wu
Ptilagrostis mongholica (Turcz. ex Trin.) Griseb.
Ptilagrostis yadongensis P. C. Keng & J. S. Tang
Puccinellia degeensis L. Liu
Puccinellia distans (Jacq.) Parl.
Puccinellia hackeliana (V. I. Krecz.) V. I. Krecz.
Puccinellia himalaica Tzvelev
Puccinellia humilis (Litv. ex V. I. Krecz.) Bor
Puccinellia kashmiriana Bor
Puccinellia koeieana Melderis
Puccinellia ladakhensis (H. Hartmann) Dickoré
Puccinellia leiolepis L. Liu
Puccinellia micrandra (Keng) Keng
Puccinellia multiflora L. Liu

Puccinellia nudiflora (Hack.) Tzvelev

Puccinellia pamirica (Roshev.) V. I. Krecz. ex Ovcz. & Czukav.

Puccinellia pauciramea (Hack.) V. I. Krecz. ex Ovcz. & Czukav.

Puccinellia stapfiana R. R. Stewart

Puccinellia tenuiflora (Griseb.) Scribn. & Merr.

Puccinellia thomsonii (Stapf) R. R. Stewart

Puccinellia vachanica Ovcz. & Czukav.

Rostraria clarkeana (Domin) Holub

Rostraria cristata (L.) Tzvelev

Rottboellia cochinchinensis (Lour.) Clayton

Saccharum arundinaceum Retz.

Saccharum bengalense Retz.

Saccharum filifolium Nees ex Steud.

Saccharum longisetosum (Andersson) V. Naray. ex Bor

Saccharum narenga (Nees ex Steud.) Hack.

Saccharum officinarum L.

Saccharum procerum Roxb.

Saccharum ravennae (L.) L.

Saccharum rufipilum Steud.

Saccharum sikkimense (Hook. f.) V. Naray. ex Bor

Saccharum spontaneum L.

Saccharum wardii (Bor) Bor

Saccharum williamsii (Bor) Bor ex Cope

Sacciolepis indica (L.) Chase

Sacciolepis interrupta (Willd.) Stapf

Sacciolepis myosuroides (R. Br.) A. Camus

Schismus arabicus Nees

Schismus barbatus (L.) Thell.

Schizachne purpurascens subsp. **callosa** (Turcz. ex Griseb.) T. Koyama & Kawano

Schizachyrium brevifolium (Sw.) Buse

Schizachyrium delavayi (Hack.) Bor

Schizachyrium impressum (Hack.) A. Calms

Schizachyrium sanguineum (Retz.) Alston

Schizostachyum arunachalensis Naithani

Schizostachyum capitatum (Munro) R. Majumdar

Schizostachyum dullooa (Gamble) R. Majumdar

Schizostachyum fuchsianum (Gamble) R. Majumdar

Schizostachyum latifolium (Munro) R. Majumdar

Schizostachyum pallidum (Munro) R. Majunmdar

Schizostachyum pergracile (Munro) R. Majumdar

Schizostachyum polymorphum (Munro) R. Majumdar

Sclerochloa dura (L.) P. Beauv.

Secale cereale L.

Secale segetale (Zhuk.) Rozhev.

Semiarundinaria pantlingii (Gamble) Nakai

Setaria barbata (Lam.) Kunth

Setaria forbesiana (Nees ex Steud.) Hook. f.

Setaria geniculata (Lam.) P. Beauv.

Setaria glauca (L.) P. Beauv.

Setaria homonyma (Steud.) Chiov.

Setaria intermedia Roem. & Schult.

Setaria italica (L.) P. Beauv.

Setaria palmifolia (J. Koenig) Stapf

Setaria plicata (Lam.) T. Cooke

Setaria pumila (Poir.) Roem. & Schult.

Setaria sphacelata (Schumach.) Stapf & C. E. Hubb. ex Moss

Setaria tomentosa (Roxb.) Kunth

Setaria verticillata (L.) P. Beauv.

Setaria viridis (L.) P. Beauv.

Setaria yunnanensis Keng & G. D. Yu ex P. C. Keng & Y. K. Ma

Sinochasea trigyna Keng

Sorghum arundinaceum (Desv.) Stapf

Sorghum bicolor (L.) Moench

Sorghum halepense (L.) Pers.

Sorghum miliaceum (Roxb.) Snowden

Sorghum nitidum (Vahl) Pers.

Sorghum roxburghii Stapf

Sorghum vulgare Pers.

Spodiopogon cotulifer (Thunb.) Hack.

Spodiopogon dubius Hack.

Spodiopogon duclouxii A. Camus

Spodiopogon lacei Hole

Spodiopogon yuexiensis S. L. Zhong

Sporobolus diandrus (Retz.) P. Beauv.

Sporobolus fertilis (Steud.) Clayton

Sporobolus helvolus (Trin.) T. Dur. & Schinz

Sporobolus indicus (L.) R. Br.

Sporobolus indicus var. **fertilis** (Steud.) Jovet & Guedes

Sporobolus piliferus (Trin.) Kunth

Stipa aliena Keng

Stipa arabica Trin. & Rupr.

Stipa arabica var. **pamirica** (Roshev.) Freitag

Stipa baicalensis Roshev.

Stipa bhutanica Noltie

Stipa brandisii Mez

Stipa breviflora Griseb.

Stipa bungeana Trin.

Stipa capensis Thunb.

Stipa capillacea Keng

Stipa capillata L.

Stipa caucasica Schmalh.

Stipa chitralensis Bor

Stipa concinna Hook. f.

Stipa consanguinea Trin. & Rupr.

Stipa duthiei Hook. f.

Stipa himalaica Rozhev.

Stipa jacquemontii Jaub. & Spach

Stipa kirghisorum Smirnow

Stipa macroglossa Smirnow

Stipa mongholica Turcz. ex Trin.

Stipa orientalis Trin.

Stipa penicillata Hand.-Mazz.

Stipa przewalskyi Roshev.

Stipa purpurea Griseb.

Stipa regeliana Hack.

Stipa roborowskyi Roshev.

Stipa rohmooianan Nooltie

Stipa roylei (Nees) Duthie

Stipa sareptana A. K. Becker

Stipa sibirica (L.) Lam.

Stipa splendens Trin.

Stipa staintonii Bor

Stipa subsessiliflora (Rupr.) Roshev.

Stipa tianschanica Roshev.

Stipa trichoides Smirnow

Stipa turkestanica Hack.

Stipagrostis pennata (Trin.) De Winter

Stipagrostis plumosa (L.) Munro ex T. Anderss.

Tetrapogon villosus Desf.

Thamnocalamus arunachalensis Naithani

Thamnocalamus chigar (Stapleton) Stapleton

Thamnocalamus spathiflorus (Trin.) Munro

Thamnocalamus spathiflorus subsp. **nepalensis** Stapleton

Thamnocalamus spathiflorus var. **bhutanensis** Stapleton

Thamnocalamus spathiflorus var. **crassinodus** (T. P. Yi) Stapleton

Themeda anathera (Nees ex Steud.) Hack.

Themeda arundinacea (Roxb.) A. Camus

Themeda caudata (Nees) A. Camus

Themeda hookeri (Griseb.) A. Camus

Themeda intermedia (Hack.) Bor

Themeda minor L. Liu

Themeda quadrivalvis (L.) Kuntze

Themeda triandra Forssk.

Themeda villosa (Lam.) A. Camus

Thysanolaena latifolia (Roxb. ex Hornem.) Honda

Thysanolaena maxima (Roxb.) Kuntze

Tragus berteronianus Schult.

Tragus mongolorum Ohwi

Tragus roxburghii Panigrahi

Trikeraia hookeri (Stapf) Bor

Trikeraia oreophila Cope

Trikeraia pappiformis (Keng) P. C. Kuo & S. L. Lu

Tripogon bromoides Roth

Tripogon chinensis (Franch.) Hack.

Tripogon debilis L. B. Cai

Tripogon filiformis Nees ex Steud.

Tripogon humilis H. L. Yang

Tripogon liouae S. M. Phillips & S. L. Chen

Tripogon purpurascens Duthie

Tripogon rupestris S. M. Phillips & S. L. Chen

Tripogon sichuanicus S. M. Phillips & S. L. Chen

Tripogon trifidus Munro ex Hook. f.

Tripogon wardii Bor

Tripogon yunnanensis J. L. Yang ex S. M. Phillips & S. L. Chen

Trisetaria loeflingiana (L.) Paunero

Trisetum aeneum (Hook. f.) R. R. Stewart

Trisetum bifidum (Thunb.) Ohwi

Trisetum clarkei (Hook. f.) R. R. Stewart

Trisetum debile Chrtek

Trisetum kangdingense (Z. L. Wu) S. M. Phillips & Z. L. Wu

Trisetum scitulum Bor

Trisetum sibiricum Rupr.

Trisetum sikkimense (Hook. f.) Chrtek

Trisetum spicatum (L.) K. Richt.

Trisetum spicatum subsp. **alascanum** (Nash) Hultén

Trisetum spicatum subsp. **himalaicum** Hultén ex Veldkamp

Trisetum spicatum subsp. **mongolicum** Hultén ex Veldkamp

Trisetum spicatum var. **pubiflorum** (Hack.) L. Liou

Trisetum yunnanense Chrtek

Triticum aestivum L.

Triticum boeoticum Boiss.

Triticum sphaerococcum Percival

Triticum turgidum L.

Urochloa panicoides P. Beauv.

Vulpia bromoides (L.) S. F. Gray

Vulpia ciliata Dumort.

Vulpia megalura (L.) C. C. Gmelin

Vulpia myuros (L.) C. C. Gmelin

Vulpia unilateralis (L.) Stace

Yushania brevipaniculata (Hand.-Mazz.) T. P. Yi

Yushania canoviridis G. H. Ye & Z. P. Wang

Yushania crassicollis T. P. Yi

Yushania crispata T. P. Yi

Yushania emeryi (Stapleton) Demoly

Yushania farcticaulis T. P. Yi

Yushania hirsuta (Munro) R. B. Majumdar

Yushania levigata T. P. Yi

Yushania lineolata T. P. Yi

Yushania maculata T. P. Yi

Yushania maling (Gamble) R. B. Majumdar & Karthik.

Yushania microphylla (Munro) R. B. Majumdar

Yushania multiramea T. P. Yi

Yushania pantlingii (Gamble) R. B. Majumdar

Yushania polytricha Hsueh & T. P. Yi

Yushania vigens T. P. Yi

Yushania violascens (Keng) T. P. Yi

Yushania weixiensis T. P. Yi

Yushania xizangensis T. P. Yi

Yushania yadongensis T. P. Yi

Zea mays L.

102. CERATOPHYLLACEAE

Ceratophyllum demersum L.

103. EUPTELEACEAE

Euptelea pleiosperma Hook. f. & Thomson

104. PAPAVERACEAE

Argemone mexicana L.

Cathcartia chelidonifolia (Bureau & Franch.) Grey-Wilson

Cathcartia smithiana Hand.-Mazz.

Cathcartia villosa Hook. f. ex Hook.

Corydalis acropteryx Fedde

Corydalis adiantifolia Hook. f. & Thomson

Corydalis adoxifolia C. Y. Wu

Corydalis adunca Maxim.

Corydalis adiantifolia Hook. f. & Thomson

Corydalis aeaeae X. F. Gao, Lidén, Y. W. Wang & Y. L. Peng

Corydalis alburyi Ludlow & Stearn

Corydalis alpestris C. A. Mey.

Corydalis ampelos Lidén & Z. Y. Su

Corydalis amphipogon Lidén

Corydalis anaginova Lidén & Z. Y. Su

Corydalis ananke Lidén

Corydalis angusta Z. Y. Su & Lidén

Corydalis anthocrene Lidén & Van De Veire

Corydalis appendiculata Hand.-Mazz.

Corydalis aquilegioides Z. Y. Su

Corydalis atuntsuensis W. W. Sm.

Corydalis aurantiaca Ludlow & Stearn

Corydalis auricilla Lidén & Z. Y. Su

Corydalis auriculata Lidén & Z. Y. Su

Corydalis balsamiflora Prain

Corydalis barbisepala Hand.-Mazz. & Fedde

Corydalis benecincta W. W. Sm.

Corydalis bijiangensis C. Y. Wu & H. Chuang

Corydalis bimaculata C. Y. Wu & Z. Y. Su

Corydalis binderae Fedde

Corydalis bowes-lyonii D. G. Long

Corydalis brachyceras Lidén & Van De Veire

Corydalis brevipedunculata (Z. Y. Su) Z. Y. Su & Lidén

Corydalis brevirostrata C. Y. Wu & Z. Y. Su

Corydalis brevirostrata subsp. **tibetica** (Maxim.) Lidén

Corydalis brunneovaginata Fedde

Corydalis bulbifera C. Y. Wu

Corydalis bulleyana Diels

Corydalis bulleyana subsp. **muliensis** Lidén & Z. Y. Su

Corydalis calcicola W. W. Sm.

Corydalis calliantha D. G. Long

Corydalis calycina Lidén

Corydalis calycosa H. Chuang

Corydalis capitata X. F. Gao, Lidén, Y. W. Wang & Y. L. Peng

Corydalis carinata Lidén & Z. Y. Su

Corydalis cashmeriana Royle

Corydalis cashmeriana subsp. **longicalcarata** (D. G. Long) Lidén

Corydalis casimiriana Duthie & Prain ex Prain

Corydalis casimiriana subsp. **brachycarpa** Lidén

Corydalis cataractarum Lidén

Corydalis cavei D. G. Long

Corydalis chaerophylla DC.

Corydalis chamdoensis C. Y. Wu & H. Chuang

Corydalis changuensis D. G. Long

Corydalis cheirifolia Franch.

Corydalis chrysosphaera C. Marquand & Airy Shaw

Corydalis clarkei Prain

Corydalis clavibracteata Ludlow & Stearn

Corydalis conspersa Maxim.

Corydalis cornuta Royle

Corydalis corymbosa C. Y. Wu & Z. Y. Su

Corydalis crassifolia Royle

Corydalis crispa Prain

Corydalis crispa subsp. **laeviangula** (C. Y. Wu & H. Chuang) Lidén & Z. Y. Su

Corydalis cristata Maxim.

Corydalis crithmifolia Royle

Corydalis curviflora Maxim.

Corydalis curviflora subsp. **rosthornii** (Fedde) C. Y. Wu

Corydalis cytisiflora (Fedde) Lidén ex C. Y. Wu, H. Chuang & Z. Y. Su

Corydalis cytisiflora subsp. **altecristata** (C. Y. Wu & H. Chuang) Lidén

Corydalis cytisiflora subsp. **minuticristata** (Fedde) Lidén

Corydalis cytisiflora subsp. **pseudosmithii** (Fedde) Lidén

Corydalis dajingensis C. Y. Wu & Z. Y. Su

Corydalis dasyptera Maxim.

Corydalis davidii Franch.

Corydalis degensis C. Y. Wu & H. Chuang

Corydalis delavayi Franch.

Corydalis delicatula D. G. Long

Corydalis delphinioides Fedde

Corydalis densispica C. Y. Wu

Corydalis denticulatobracteata Fedde

Corydalis diffusa Lidén

Corydalis diphylla Wall.

Corydalis diphylla var. **cyrtocentra** (Prain) Jafri

Corydalis dolichocentra Z. Y. Su & Lidén

Corydalis dorjii D. G. Long

Corydalis drakeana Prain

Corydalis drepanantha D. G. Long

Corydalis dubia Prain

Corydalis dulongjiangensis H. Chuang

Corydalis duthiei Maxim.

Corydalis ecristata (Prain) D. G. Long

Corydalis elata Bureau & Franch.

Corydalis elegans Hook. f. & Thomson

Corydalis elegans subsp. **robusta** Lidén

Corydalis ellipticarpa C. Y. Wu & Z. Y. Su

Corydalis enantiophylla Lidén

Corydalis eugeniae Fedde

Corydalis falconeri Hook. f. & Thomson

Corydalis feddeana H. Lév.

Corydalis fedtschenkoana subsp. **metallica** (Wendelbo) Wendelbo

Corydalis filicina Prain

Corydalis filiformis Royle

Corydalis filisecta C. Y. Wu

Corydalis fimbrillifera Korsh.

Corydalis flabellata Edgew.

Corydalis flaccida Hook. f. & Thomson

Corydalis flexuosa Franch.

Corydalis flexuosa subsp. **pseudoheterocentra** (Fedde) Lidén ex C. Y. Wu, H. Chuang & Z. Y. Su

Corydalis foetida C. Y. Wu & Z. Y. Su

Corydalis foliaceobracteata C. Y. Wu & Z. Y. Su

Corydalis franchetiana Prain

Corydalis gaoxinfeniae Lidén

Corydalis geraniifolia Hook. f. & Thomson

Corydalis gerdae Fedde

Corydalis giraldii Fedde

Corydalis glaucissima Lidén & Z. Y. Su

Corydalis glycyphyllos Fedde

Corydalis gortschakovii Schrenk

Corydalis govaniana Wall.

Corydalis govaniana var. **malukiana** Jafri

Corydalis govaniana var. **swatensis** (Kitam.) Jafri

Corydalis gracillima C. Y. Wu ex Govaerts

Corydalis grandiflora C. Y. Wu & Z. Y. Su

Corydalis gymnopoda Z. Y. Su & Lidén

Corydalis gyrophylla Lidén

Corydalis hamata Franch.

Corydalis harrysmithii Lidén & Z. Y. Su

Corydalis hebephylla C. Y. Wu & Z. Y. Su

Corydalis helodes Lidén & Van De Veire

Corydalis hemidicentra Hand.-Mazz.

Corydalis hendersonii Hemsl.

Corydalis hendersonii var. **altocristata** C. Y. Wu & Z. Y. Su

Corydalis hepaticifolia C. Y. Wu & Z. Y. Su

Corydalis heterocentra Diels

Corydalis heterothylax C. Y. Wu ex Z. Y. Su & Lidén

Corydalis hindukushensis Wendelbo & Grey-Wilson

Corydalis homopetala Diels

Corydalis hongbashanensis Lidén & Y. W. Wang

Corydalis hookeri Prain

Corydalis humicola Hand.-Mazz.

Corydalis imbricata Z. Y. Su & Lidén

Corydalis inopinata Prain ex Fedde

Corydalis iochanensis H. Lév.

Corydalis jigmei C. E. C. Fisch. & Kaul

Corydalis juncea Wall.

Corydalis kingdonis Airy Shaw

Corydalis kingii Prain

Corydalis kokiana Hand.-Mazz.

Corydalis kuruchuensis Lidén

Corydalis laelia Prain

Corydalis laelia subsp. **bhutanica** D. G. Long

Corydalis lagochila Lidén & Z. Y. Su

Corydalis lasiocarpa Lidén & Z. Y. Su

Corydalis lathyroides Prain

Corydalis lathyrophylla C. Y. Wu

Corydalis lathyrophylla subsp. **dawuensis** Lidén

Corydalis latiflora Hook. f. & Thomson

Corydalis latiloba (Franch.) Hand.-Mazz.

Corydalis latiloba var. **tibetica** Z. Y. Su & Lidén

Corydalis laucheana Fedde

Corydalis laxiflora Lidén

Corydalis ledebouriana Kar. & Kir.

Corydalis leptocarpa Hook. f. & Thomson

Corydalis lhasaensis C. Y. Wu & Z. Y. Su

Corydalis lhorongensis C. Y. Wu & H. Chuang

Corydalis liana Lidén & Z. Y. Su

Corydalis lidenii Z. Y. Su

Corydalis linarioides Maxim.

Corydalis linearis C. Y. Wu

Corydalis linstowiana Fedde

Corydalis livida Maxim.

Corydalis livida var. **denticulatocristata** Z. Y. Su

Corydalis longibracteata Ludlow & Stearn

Corydalis longipes DC.

Corydalis longistyla Z. Y. Su & Lidén

Corydalis longkiensis C. Y. Wu, Lidén & Z. Y. Su

Corydalis lopinensis Franch.

Corydalis lowndesii Lidén

Corydalis ludlowii Stearn

Corydalis lupinoides C. Marquand & Airy Shaw

Corydalis madida Lidén & Z. Y. Su

Corydalis magni Pusalkar

Corydalis mairei H. Lév.

Corydalis mayae Hand.-Mazz.

Corydalis megacalyx Ludlow & Stearn

Corydalis megalosperma Z. Y. Su

Corydalis meifolia Wall.

Corydalis melanochlora Maxim.

Corydalis microflora (C. Y. Wu & H. Chuang) Z. Y. Su & Lidén

Corydalis milarepa Lidén & Z. Y. Su

Corydalis minutiflora C. Y. Wu

Corydalis mira (Batalin) C. Y. Wu & H. Chuang

Corydalis moorcroftiana Wall. ex Hook. f. & Thomson

Corydalis moupinensis Franch.

Corydalis mucronata Franch.

Corydalis mucronifera Maxim.

Corydalis mucronipetala (C. Y. Wu & H. Chuang) Lidén & Z. Y. Su

Corydalis muliensis C. Y. Wu & Z. Y. Su

Corydalis murreeana Jafri

Corydalis myriophylla Lidén

Corydalis nana Royle

Corydalis nemoralis C. Y. Wu & H. Chuang

Corydalis nigroapiculata C. Y. Wu

Corydalis nubicola Z. Y. Su & Lidén

Corydalis oligantha Ludlow & Stearn

Corydalis oligosperma C. Y. Wu & Z. Y. Su

Corydalis onobrychis Fedde

Corydalis ophiocarpa Hook. f. & Thomson

Corydalis oreocoma Lidén & Z. Y. Su

Corydalis oxalidifolia Ludlow & Stearn

Corydalis oxypetala Franch.

Corydalis oxypetala subsp. **balfouriana** (Diels) Lidén

Corydalis pachycentra Franch.

Corydalis pachypoda (Franch.) Hand.-Mazz.

Corydalis pakistanica Jafri

Corydalis panda Lidén & Y. W. Wang

Corydalis papillosa Z. Y. Su & Lidén

Corydalis peltata Lidén & Z. Y. Su

Corydalis petrodoxa Lidén & Z. Y. Su

Corydalis petrophila Franch.

Corydalis pingwuensis C. Y. Wu

Corydalis polygalina Hook. f. & Thomson

Corydalis polyphylla Hand.-Mazz.

Corydalis porphyrantha C. Y. Wu

Corydalis potaninii Maxim.

Corydalis prattii Franch.

Corydalis procera Lidén & Z. Y. Su

Corydalis pseudoadoxa (C. Y. Wu & H. Chuang) C. Y. Wu & H. Chuang

Corydalis pseudobalfouriana Lidén & Z. Y. Su

Corydalis pseudobarbisepala Fedde

Corydalis pseudocristata Fedde

Corydalis pseudocrithmifolia Jafri

Corydalis pseudodensispica Z. Y. Su & Lidén

Corydalis pseudodrakeana Lidén

Corydalis pseudofargesii H. Chuang

Corydalis pseudofilisecta Lidén & Z. Y. Su

Corydalis pseudofluminicola Fedde

Corydalis pseudoimpatiens Fedde

Corydalis pseudoincisa C. Y. Wu, Z. Y. Su & Lidén

Corydalis pseudojuncea Ludlow & Stearn

Corydalis pseudolongipes Lidén

Corydalis pseudomucronata C. Y. Wu, Z. Y. Su & Lidén

Corydalis pseudorupestris Lidén & Z. Y. Su

Corydalis pseudosibirica Lidén & Z. Y. Su

Corydalis pseudotongolensis Lidén

Corydalis pseudoweigoldii Z. Y. Su

Corydalis pterygopetala Hand.-Mazz.

Corydalis pterygopetala var. **divaricata** Z. Y. Su & Lidén

Corydalis pterygopetala var. **megalantha** (Diels) Lidén & Z. Y. Su

Corydalis pterygopetala var. **parviflora** Lidén

Corydalis pubicaulis C. Y. Wu & H. Chuang

Corydalis pygmaea C. Y. Wu & Z. Y. Su

Corydalis quantmeyeriana Fedde

Corydalis quinquefoliolata Ludlow & Stearn

Corydalis racemosa (Thunb.) Pers.

Corydalis radicans Hand.-Mazz.

Corydalis regia Z. Y. Su & Lidén

Corydalis retingensis Ludlow

Corydalis rheinbabeniana Fedde

Corydalis rorida H. Chuang

Corydalis rostellata Lidén

Corydalis rubrisepala Lidén

Corydalis rubrisepala subsp. **zhuangiana** Lidén

Corydalis rutifolia (Sm.) DC.

Corydalis saltatoria W. W. Sm.

Corydalis sarcolepis Lidén & Z. Y. Su

Corydalis scaberula Maxim.

Corydalis schelesnowiana Regel & Schmalh.

Corydalis schistostigma X. F. Gao, Lidén, Y. W. Wang & Y. L. Peng

Corydalis schweriniana Fedde

Corydalis shakyae Lidén

Corydalis sheareri S. Moore

Corydalis shimienensis C. Y. Wu & Z. Y. Su

Corydalis sigmantha Z. Y. Su & C. Y. Wu

Corydalis sigmoides C. Y. Wu & H. Chuang

Corydalis sikkimensis (Prain) Fedde

Corydalis simplex Lidén

Corydalis smithiana Fedde

Corydalis sophronitis Z. Y. Su & Lidén

Corydalis spathulata Prain ex Craib

Corydalis spicata Lidén

Corydalis staintonii Ludlow & Stearn

Corydalis stenantha Franch.

Corydalis stipulata Lidén

Corydalis stolonifera Lidén

Corydalis stewartii Fedde

Corydalis stracheyi Duthie ex Prain

Corydalis stracheyi var. **ecristata** Prain

Corydalis stracheyioides Fedde

Corydalis straminea Maxim. ex Hemsl.

Corydalis stramineoides C. Y. Wu & Z. Y. Su

Corydalis striatocarpa H. Chuang

Corydalis stricta Stephan ex Fisch.

Corydalis stricta subsp. **pamirica** Michajlova

Corydalis susannae Lidén

Corydalis taliensis Franch.

Corydalis tangutica Peschkova

Corydalis tangutica subsp. **bullata** (Lidén) Z. Y. Su

Corydalis temolana C. Y. Wu & H. Chuang

Corydalis temulifolia Franch.

Corydalis temulifolia subsp. **aegopodioides** (H. Lév. & Vaniot) C. Y. Wu

Corydalis tenerrima C. Y. Wu, Z. Y. Su & Lidén

Corydalis tenuipes Lidén & Z. Y. Su

Corydalis terracina Lidén

Corydalis thyrsiflora Prain

Corydalis tibetica Hook. f. & Thomson

Corydalis tibetoalpina C. Y. Wu & Z. Y. Su

Corydalis tibeto-oppositifolia C. Y. Wu & Z. Y. Su

Corydalis tongolensis Franch.

Corydalis trachycarpa Maxim.

Corydalis trifoliata Franch.

Corydalis trilobipetala Hand.-Mazz.

Corydalis triternatifolia C. Y. Wu

Corydalis tsangensis Lidén & Z. Y. Su

Corydalis tsayulensis C. Y. Wu & H. Chuang

Corydalis uncinata Lidén

Corydalis uncinatella Lidén

Corydalis urosepala Fedde

Corydalis uvaria Lidén & Z. Y. Su

Corydalis vaginans Royle

Corydalis verna Z. Y. Su & Lidén

Corydalis vivipara Fedde

Corydalis weigoldii Fedde

Corydalis wuzhengyiana Z. Y. Su & Lidén

Corydalis yaoi Lidén & Z. Y. Su

Corydalis yargongensis C. Y. Wu

Corydalis yui Lidén

Corydalis yunnanensis Franch.

Corydalis zadoiensis L. H. Zhou

Corydalis zhongdianensis Z. Y. Su & Lidén

Dactylicapnos burmanica (K. R. Stern) Lidén

Dactylicapnos cordata Lidén

Dactylicapnos gaoligongshanensis Lidén

Dactylicapnos grandifoliolata Merr.

Dactylicapnos leiosperma Lidén

Dactylicapnos lichiangensis (Fedde) Hand.-Mazz.

Dactylicapnos macrocapnos (Prain) Hutch.

Dactylicapnos macrocapnos subsp. **echinosperma** Lidén

Dactylicapnos odontocarpa Lidén

Dactylicapnos roylei (Hook. f. & Thomson) Hutch.

Dactylicapnos scandens (D. Don) Hutch.

Dactylicapnos schneideri (Fedde) Lidén

Dactylicapnos torulosa (Hook. f. & Thomson) Hutch.

Dicranostigma lactucoides Hook. f. & Thomson

Dicranostigma leptopodum (Maxim.) Fedde

Dicranostigma platycarpum C. Y. Wu & H. Chuang

Eschscholzia californica Cham.

Fumaria indica (Hausskn.) Pugsley

Fumaria parviflora Lam.

Fumaria vaillantii Loisel.

Glaucium elegans Fisch. & C. A. Mey.

Glaucium fimbrilligerum (Trautv.) Boiss.

Hypecoum erectum L.

Hypecoum leptocarpum Hook. f. & Thomson

Hypecoum parviflorum Kar. & Kir.

Hypecoum zhukanum Lidén

Ichtyoselmis macrantha (Oliv.) Lidén

Meconopsis aculeata Royle

Meconopsis argemonantha Prain

Meconopsis auriculata Stapf

Meconopsis autumnalis P. A. Egan

Meconopsis baileyi Prain

Meconopsis baileyi subsp. **multidentata** Grey-Wilson

Meconopsis baileyi subsp. **pratensis** Kingdon-Ward ex Grey-Wilson

Meconopsis balangensis Tosh. Yoshida, H. Sun & Boufford

Meconopsis balangensis var. **atrata** Tosh. Yoshida, H. Sun & Boufford

Meconopsis bella Prain

Meconopsis bella subsp. **grandifolia** Grey-Wilson

Meconopsis betonicifolia Franch.

Meconopsis bhutanica Tosh. Yoshida & Grey-Wilson

Meconopsis bijiangensis H. Ohba, Tosh. Yoshida & H. Sun

Meconopsis bijiangensis subsp. **chimiliensis** Grey-Wilson

Meconopsis chankheliensis Grey-Wilson

Meconopsis compta Prain

Meconopsis concinna Prain

Meconopsis delavayi (Franch.) Franch. ex Prain

Meconopsis dhwojii G. Taylor ex Hay

Meconopsis discigera Prain

Meconopsis exilis Tosh. Yoshida, H. Sun & Grey-Wilson

Meconopsis florindae Kingdon-Ward

Meconopsis forrestii Prain

Meconopsis ganeshensis Grey-Wilson

Meconopsis georgei G. Taylor

Meconopsis gracilipes G. Taylor

Meconopsis grandis Prain

Meconopsis grandis subsp. **jumlaensis** Grey-Wilson

Meconopsis grandis subsp. **orientalis** Grey-Wilson

Meconopsis henrici Bureau & Franch.

Meconopsis horridula Hook. f. & Thomson

Meconopsis horridula subsp. **drukyulensis** Grey-Wilson

Meconopsis impedita Prain

Meconopsis impedita subsp. **rubra** (Kingdon-Ward) Grey-Wilson

Meconopsis integrifolia (Maxim.) Franch.

Meconopsis integrifolia subsp. **souliei** (Fedde) Grey-Wilson

Meconopsis lamjungensis Tosh. Yoshida, H. Sun & Grey-Wilosn

Meconopsis lancifolia (Franch.) Franch. ex Prain

Meconopsis lancifolia subsp. **eximia** (Prain) Grey-Wilson

Meconopsis latifolia (Prain) Prain

Meconopsis lhasaensis Grey-Wilson

Meconopsis lijiangensis (Grey-Wilson) Grey-Wilson

Meconopsis longipetiolata G. Taylor

Meconopsis ludlowii Grey-Wilson

Meconopsis lyrata (H. A. Cummins & Prain) Fedde ex Prain

Meconopsis manasluensis P. A. Egan

Meconopsis muscicola Tosh. Yoshida, H. Sun & Boufford

Meconopsis napaulensis DC.

Meconopsis neglecta G. Taylor

Meconopsis nyingchiensis L. H. Zhou

Meconopsis paniculata (D. Don) Prain

Meconopsis paniculata subsp. **paniculata** var. **rubra** Grey-Wilson

Meconopsis paniculata subsp. **pseudoregia** Grey-Wilson

Meconopsis pinnatifolia C. Y. Wu & X. Zhuang

Meconopsis polygonoides (Prain) Prain

Meconopsis prainiana Kingdon-Ward

Meconopsis prainiana var. **lutea** (G. Taylor) Grey-Wilosn

Meconopsis prattii Prain

Meconopsis primulina Prain

Meconopsis pseudointegrifolia Prain

Meconopsis pseudointegrifolia subsp. **daliensis** Grey-Wilson

Meconopsis pseudovenusta G. Taylor

Meconopsis psilonomma Farrer

Meconopsis pulchella Tosh. Yoshida, H. Sun & Boufford

Meconopsis punicea Maxim.

Meconopsis quintuplinervia Regel

Meconopsis racemosa Maxim.

Meconopsis regia G. Taylor

Meconopsis robusta Hook. f. & Thomson

Meconopsis rudis (Prain) Prain

Meconopsis sherriffii G. Taylor

Meconopsis simikotensis Grey-Wilson

Meconopsis simplicifolia (D. Don) Walp.

Meconopsis simplicifolia subsp. **grandiflora** Grey-Wilosn

Meconopsis sinomaculata Grey-Wilson

Meconopsis sinuata Prain

Meconopsis speciosa Prain

Meconopsis speciosa subsp. **cawdoriana** (Kingdon-Ward) Grey-Wilson

Meconopsis staintonii Grey-Wilson

Meconopsis sulphurea Grey-Wilson

Meconopsis sulphurea subsp. **gracilifolia** Grey-Wilson

Meconopsis superba King ex Prain

Meconopsis taylorii L. H. J. Williams

Meconopsis tibetica Grey-Wilson

Meconopsis torquata Prain

Meconopsis venusta Prain

Meconopsis violacea Kingdon-Ward

Meconopsis wallichii Hook.

Meconopsis wilsonii Grey-Wilson

Meconopsis wilsonii subsp. **australis** Grey-Wilson

Meconopsis xiangchengensis R. Li & Z. L. Dao

Meconopsis zhongdianensis Grey-Wilson

Papaver decaisnei Hochst. & Steud. ex Elkan

Papaver dubium L.

Papaver hybridum L.

Papaver involucratum Popov

Papaver macrostomum Boiss. & Huet ex Boiss.

Papaver nudicaule L.

Papaver orientale L.

Papaver pavoninum C. A. Mey.

Papaver rhoeas L.

Papaver somniferum L.

Roemeria hybrida (L.) DC.

Roemeria refracta DC.

105. CIRCAEASTERACEAE

Circaeaster agrestis Maxim.

Kingdonia uniflora Balf. f. & W. W. Sm.

106. LARDIZABALACEAE

Akebia apetala (Quan Xia, J. Z. Sun & Z. X. Peng) Christenh.

Decaisnea insignis (Griff.) Hook. f. & Thomson

Sinofranchetia chinensis (Franch.) Hemsl.

Stauntonia angustifolia (Wall.) Christenh.

Stauntonia brunoniana Wall. ex Hemsl.

Stauntonia elliptica Hemsl.

Stauntonia grandiflora (Réaub.) Christenh.

Stauntonia latifolia (Wall.) Christenh.

Stauntonia libera H. N. Qin

Stauntonia linearifolia (T. Chen & H. N. Qin) Christenh.

Stauntonia medogensis (H. N. Qin) Christenh.

107. MENISPERMACEAE

Aspidocarya uvifera Hook. f. & Thomson

Cissampelos pareira L.

Cissampelos pareira var. **hirsuta** (Buch.-Ham. ex DC.) Forman

Cocculus hirsutus (L.) Diels

Cocculus laurifolius DC.

Cocculus orbiculatus (L.) DC.

Cocculus orbiculatus var. **mollis** (Wall. ex Hook. f. & Thomson) Hara

Cyclea barbata Miers

Cyclea bicristata (Griff.) Diels

Cyclea peltata (Lam.) Hook. f. & Thomson

Diploclisia glaucescens (Blume) Diels

Eleutharrhena macrocarpa (Diels) Forman

Haematocarpus validus (Miers) Bakh. f. ex Forman

Hypserpa nitida Miers ex Benth.

Pachygone yunnanensis H. S. Lo

Parabaena sagittata Miers

Pericampylus aduncus Miers

Pericampylus glaucus (Lam.) Merr.

Pycnarrhena pleniflora Miers

Sinomenium acutum (Thunb.) Rehder & E. H. Wilson

Stephania brevipedunculata C. Y. Wu & D. D. Tao

Stephania delavayi Diels

Stephania ebracteata S. Y. Zhao & H. S. Lo

Stephania elegans Hook. f. & Thomson

Stephania epigaea H. S. Lo

Stephania glabra (Roxb.) Miers

Stephania glandulifera Miers

Stephania gracilenta Miers

Stephania japonica (Thunb.) Miers

Stephania japonica var. **discolor** (Blume) Forman

Stephania miyiensis S. Y. Zhao & H. S. Lo

Stephania subpeltata H. S. Lo

Tiliacora acuminata (Lam.) Miers

Tinomiscium petiolare Hook. f. & Thomson

Tinospora cordifolia (Willd.) Hook. f. & Thoms.

Tinospora crispa (L.) Hook. f. & Thomson

Tinospora sagittata (Oliver) Gagnep.

Tinospora sinensis (Lour.) Merr.

108. BERBERIDACEAE

Berberis aggregata C. K. Schneid.

Berberis agricola Ahrendt

Berberis ahrendtii R. R. Rao & Uniyal

Berberis amabilis C. K. Schneid.

Berberis amoena Dunn

Berberis angulosa Wall. ex Hook. f. & Thomson

Berberis approximata Sprague

Berberis aristata DC.

Berberis asiatica Roxb. ex DC.

Berberis asmyana C. K. Schneid.

Berberis atrocarpa C. K. Schneid.

Berberis atroviridiana T. S. Ying

Berberis batangensis T. S. Ying

Berberis beaniana C. K. Schneid.

Berberis beesiana Ahrendt

Berberis bergmanniae C. K. Schneid.

Berberis bhutanensis Ahrendt

Berberis brachypoda Maxim.

Berberis bracteata (Ahrendt) Ahrendt

Berberis brandisiana Ahrendt

Berberis calcipratorum Ahrendt

Berberis calliobotrys Aitch. ex Koehne

Berberis campylotropa T. S. Ying

Berberis candidula (C. K. Schneid.) C. K. Schneid.

Berberis centiflora Diels

Berberis ceratophylla G. Don

Berberis chitria Buch.-Ham. ex Lindl.

Berberis chrysosphaera Mulligan

Berberis collettii C. K. Schneid.

Berberis concinna Hook. f.

Berberis concolor W. W. Sm.

Berberis contracta T. S. Ying

Berberis cooperi Ahrendt

Berberis coriaria Royle ex Lindl.

Berberis coryi Veitch

Berberis crassilimba C. Y. Wu ex S. Y. Bao

Berberis daochengensis T. S. Ying

Berberis dasyclada Ahrendt

Berberis dasystachya Maxim.

Berberis davidii Ahrendt

Berberis dawoensis K. Mey.

Berberis deinacantha C. K. Schneid.

Berberis delavayi C. K. Schneid.

Berberis derongensis T. S. Ying

Berberis dictyoneura C. K. Schneid.

Berberis dictyophylla Franch.

Berberis dictyophylla var. **epruinosa** C. K. Schneid.

Berberis dumicola C. K. Schneid.

Berberis duthieana C. K. Schneid.

Berberis edgeworthiana C. K. Schneid.

Berberis erythroclada Ahrendt

Berberis everestiana Ahrendt

Berberis fallax C. K. Schneid.

Berberis feddeana C. K. Schneid.

Berberis fengii S. Y. Bao

Berberis ferdinandi-coburgii C. K. Schneid.

Berberis forrestii Ahrendt

Berberis franchetiana C. K. Schneid.

Berberis francisci-ferdinandi C. K. Schneid.

Berberis gagnepainii C. K. Schneid.

Berberis gilungensis T. S. Ying

Berberis glaucocarpa Stapf

Berberis graminea Ahrendt

Berberis grandibracteata Ahrendt

Berberis griffithiana C. K. Schneid.

Berberis grodtmanniana C. K. Schneid.

Berberis grodtmanniana var. **flavoramea** C. K. Schneid.

Berberis gyalaica Ahrendt

Berberis hamiltoniana Ahrendt

Berberis hemsleyana Ahrendt

Berberis hobsonii Ahrendt

Berberis holocraspedon Ahrendt

Berberis hookeri Lem.

Berberis hookeri var. **microcarpa** Ahrendt

Berberis hookeri var. **viridis** C. K. Schneid.

Berberis huegeliana C. K. Schneid.

Berberis humidoumbrosa Ahrendt

Berberis hypericifolia T. S. Ying

Berberis ignorata C. K. Schneid.

Berberis insignis Hook. f. & Thomson

Berberis insignis subsp. **incrassata** (Ahrendt) D. F. Chamb. & C. M. Hu

Berberis insignis var. **gouldii** Ahrendt

Berberis insignis var. **shergaonensis** Ahrendt

Berberis insignis var. **zelaica** Ahrendt

Berberis insolita C. K. Schneid.

Berberis jaeschkeana C. K. Schneid.

Berberis jaeschkeana var. **apiculata** (Ahrendt) Naithani & S. Biswas

Berberis jaeschkeana var. **bimbilaica** Ahrendt

Berberis jaeschkeana var. **usteriana** C. K. Schneid.

Berberis jamesiana Forrest & W. W. Sm.

Berberis jiulongensis T. S. Ying

Berberis johannis Ahrendt

Berberis kangdingensis T. S. Ying

Berberis kashmirana Ahrendt

Berberis khasiana Ahrendt

Berberis koehneana C. K. Schneid.

Berberis koehneana var. **auramea** Ahrendt

Berberis kongboensis Ahrendt

Berberis kumaonensis C. K. Schneid.

Berberis kunawurensis Royle

Berberis leboensis T. S. Ying

Berberis lecomtei C. K. Schneid.

Berberis lepidifolia Ahrendt

Berberis levis Franch.

Berberis lijiangensis C. Y. Wu ex S. Y. Bao

Berberis liophylla C. K. Schneid.

Berberis longispina T. S. Ying

Berberis ludlowii Ahrendt

Berberis luhuoensis T. S. Ying

Berberis lycium Royle

Berberis macrosepala Hook. f. & Thomson

Berberis macrosepala var. **setifolia** Ahrendt

Berberis medogensis T. S. Ying
Berberis mekongensis W. W. Sm.
Berberis metapolyantha Ahrendt
Berberis mianningensis T. S. Ying
Berberis microtricha C. K. Schneid.
Berberis minutiflora C. K. Schneid.
Berberis mouillacana C. K. Schneid.
Berberis mucrifolia Ahrendt
Berberis muliensis Ahrendt
Berberis muliensis var. **atuntzeana** Ahrendt
Berberis multicaulis T. S. Ying
Berberis multiovula T. S. Ying
Berberis multiserrata T. S. Ying
Berberis nullinervis T. S. Ying
Berberis nutanticarpa C. Y. Wu ex S. Y. Bao
Berberis obovatifolia T. S. Ying
Berberis orthobotrys Bien. ex Aitch.
Berberis orthobotrys var. **canescens** Ahrendt
Berberis orthobotrys var. **capitata** Jafri
Berberis orthobotrys var. **conwayi** Ahrendt
Berberis orthobotrys var. **rupicunda** Ahrendt
Berberis orthobotrys var. **sinthanensis** Ahrendt
Berberis pachyacantha Koehne
Berberis pallens Franch.
Berberis papillifera (Franch.) Koehne
Berberis parapruinosa T. S. Ying
Berberis paraspecta Ahrendt
Berberis parisepala Ahrendt
Berberis parkeriana C. K. Schneid.
Berberis petiolaris Wall. ex G. Don
Berberis pingwuensis T. S. Ying
Berberis platyphylla (Ahrendt) Ahrendt
Berberis poluninii Ahrendt
Berberis polyantha Hemsl.
Berberis potaninii Maxim.
Berberis praecipua C. K. Schneid.
Berberis prattii C. K. Schneid.
Berberis pruinocarpa C. Y. Wu ex S. Y. Bao
Berberis pruinosa Franch.
Berberis pseudoamoena T. S. Ying
Berberis pseudotibetica C. Y. Wu ex S. Y. Bao
Berberis pseudoumbellata Parker
Berberis pseudoumbellata var. **gilgitica** Jafri
Berberis pulangensis T. S. Ying
Berberis racemulosa T. S. Ying
Berberis replicata W. W. Sm.
Berberis reticulinervis T. S. Ying
Berberis retusa T. S. Ying
Berberis royleana Ahrendt
Berberis sabulicola T. S. Ying
Berberis sherriffii Ahrendt
Berberis sichuanica T. S. Ying
Berberis sikkimensis (C. K. Schneid.) Ahrendt
Berberis silva-taroucana C. K. Schneid.
Berberis stewartiana Jafri
Berberis sublevis W. W. Sm.
Berberis taliensis C. K. Schneid.
Berberis taronensis Ahrendt
Berberis temolaica Ahrendt
Berberis tenuipedicellata T. S. Ying
Berberis thomsoniana C. K. Schneid.
Berberis tischleri C. K. Schneid.
Berberis tomentulosa Ahrendt
Berberis trichiata T. S. Ying
Berberis tsangpoensis Ahrendt
Berberis tsarica Ahrendt
Berberis tsarica var. **ritangensis** Ahrendt
Berberis tsarongensis Stapf
Berberis ulicina Hook. f. & Thomson
Berberis umbratica T. S. Ying
Berberis valida (C. K. Schneid.) C. K. Schneid.
Berberis vernae C. K. Schneid.
Berberis verruculosa Hemsl. & E. H. Wilson

Berberis vinifera T. S. Ying

Berberis virescens Hook. f.

Berberis wallichiana DC.

Berberis walteriana Ahrendt

Berberis weisiensis C. Y. Wu ex S. Y. Bao

Berberis wilsoniae Hemsl.

Berberis wilsoniae var. **guhtzunica** (Ahrendt) Ahrendt

Berberis xanthophlaea Ahrendt

Berberis yui T. S. Ying

Berberis yunnanensis Franch.

Caulophyllum robustum Maxim.

Diphylleia sinensis H. L. Li

Dysosma aurantiocaulis (Hand.-Mazz.) Hu

Dysosma delavayi (Franch.) Hu

Dysosma tsayuensis T. S. Ying

Dysosma versipellis (Hance) M. Cheng

Epimedium chlorandrum Stearn

Epimedium davidii Franch.

Epimedium elatum Morr. & Decne.

Epimedium elongatum Kom.

Epimedium flavum Stearn

Epimedium latisepalum Stearn

Epimedium platypetalum K. Mey.

Mahonia acanthifolia G. Don

Mahonia borealis Takeda

Mahonia bracteolata Takeda

Mahonia calamicaulis subsp. **kingdon-wardiana** (Ahrendt) T. S. Ying & Boufford

Mahonia duclouxiana Gagnep.

Mahonia dulongensis H. Li

Mahonia lushuiensis T. S. Ying & H. Li

Mahonia monyulensis Ahrendt

Mahonia napaulensis DC.

Mahonia polyodonta Fedde

Mahonia taronensis Hand.-Mazz.

Sinopodophyllum hexandrum (Royle) T. S. Ying

109. RANUNCULACEAE

Aconitum abietetorum W. T. Wang & L. Q. Li

Aconitum alpinonepalense Tamura

Aconitum amplexicaule Lauener

Aconitum angulatum Tamura

Aconitum assamicum Lauener

Aconitum balangrense Lauener

Aconitum balfourii Stapf

Aconitum balfourii var. **rhombilobum** Stapf

Aconitum bhedingense Lauener

Aconitum bisma (Buch.-Ham.) Rapaics

Aconitum brachypodum Diels

Aconitum brachypodum var. **laxiflorum** Fletcher & Lauener

Aconitum bracteolatum Lauener

Aconitum brevicalcaratum (Finet & Gagnep.) Diels

Aconitum brevicalcaratum var. **parviflorum** Chen & Liu

Aconitum brevilimbum Lauener

Aconitum brunneum Hand.-Mazz.

Aconitum bulbilliferum Hand.-Mazz.

Aconitum bulleyanum Diels

Aconitum calthifolium H. F. Comber

Aconitum campylorrhynchum Hand.-Mazz.

Aconitum changianum W. T. Wang

Aconitum chasmanthum Stapf

Aconitum chayuense W. T. Wang

Aconitum chiachaense var. **glandulosum** W. T. Wang

Aconitum chiachaense W. T. Wang

Aconitum chrysotrichum W. T. Wang

Aconitum contortum Finet & Gagnep.

Aconitum crassiflorum Hand.-Mazz.

Aconitum deinorrhizum Stapf

Aconitum delavayi Franch.

Aconitum dhwojii Lauener

Aconitum elliotii Lauener

Aconitum elliotii var. **doshongense** (Lauener) W. T. Wang

Aconitum elliotii var. **glabrescens** W. T. Wang & L. Q. Li

Aconitum elliotii var. **pilopetalum** W. T. Wang & L. Q. Li

Aconitum elwesii Stapf

Aconitum episcopale H. Lév.

Aconitum falciforme Hand.-Mazz.

Aconitum falconeri Stapf

Aconitum fengii W. T. Wang

Aconitum fengii var. **crispulum** Q. E. Yang

Aconitum ferox Wall. ex Seringue

Aconitum flavum Hand.-Mazz.

Aconitum fletcherianum G. Taylor

Aconitum franchetii Finet & Gagnep.

Aconitum franchetii var. **villosulum** W. T. Wang

Aconitum funiculare Stapf

Aconitum gammiei Stapf

Aconitum geniculatum Fletcher & Lauener

Aconitum georgei Comber

Aconitum glabrisepalum W. T. Wang

Aconitum habaense W. T. Wang

Aconitum hemsleyanum E. Pritz.

Aconitum hemsleyanum var. **atropurpureum** (Hand.-Mazz.) W. T. Wang

Aconitum henryi E. Pritz.

Aconitum heterophylloides Stapf

Aconitum heterophylloides var. **leucanthum** (Bruhl) Lauener

Aconitum heterophyllum Wall. ex Royle

Aconitum heterophyllum var. **bracteatum** Stapf

Aconitum hicksii Lauener

Aconitum hookeri Stapf

Aconitum huiliense Hand.-Mazz.

Aconitum incisofidum W. T. Wang

Aconitum kashimicum Stapf ex Cov.

Aconitum kongboense Lauener

Aconitum kongboense var. **villosum** W. T. Wang

Aconitum laciniatum Stapf

Aconitum laeve Royle

Aconitum legendrei Hand-Mazz.

Aconitum legendrei var. **albovillosum** (Chen & Liu) Y. Luo & Q. E. Yang

Aconitum leiwuqiense W. T. Wang

Aconitum lethale Griff.

Aconitum leucanthum Stapf

Aconitum liangshanicum W. T. Wang

Aconitum liljestrandii Hand-Mazz.

Aconitum liljestrandii var. **fangianum** (W. T. Wang) Y. Luo & Q. E. Yang

Aconitum longilobum W. T. Wang

Aconitum longipedicellatum Lauener

Aconitum longipetiolatum Lauener

Aconitum ludlowii Exell

Aconitum milinense W. T. Wang

Aconitum moschatum Stapf

Aconitum nagarum Stapf

Aconitum nagarum var. **acaule** (Finet & Gagnep.) Q. E. Yang

Aconitum namlaense W. T. Wang

Aconitum naviculare (Brühl) Stapf

Aconitum nepalense Lauener

Aconitum nielamuense W. T. Wang

Aconitum novoluridum Munz

Aconitum nutantiflorum Chang ex W. T. Wang

Aconitum orochryseum Stapf

Aconitum ouvrardianum Hand.-Mazz.

Aconitum palmatum D. Don

Aconitum parapolyanthum Lauener

Aconitum patulum Tamura

Aconitum pendulicarpum Chang ex W. T. Wang

Aconitum pendulum Busch

Aconitum phyllostegium Hand.-Mazz.

Aconitum piepunense Hand.-Mazz.

Aconitum piepunense var. **pilosum** Comber

Aconitum poluninii Lauener

Aconitum polyanthum (Fin. & Gagnep.) Hand.-Mazz.

Aconitum polyanthum var. **puberulum** Finet & Gagnep.

Aconitum polyschistum Hand.-Mazz.

Aconitum pomeense W. T. Wang

Aconitum prominens Lauener

Aconitum pseudobrunneum W. T. Wang

Aconitum pseudodivaricatum W. T. Wang

Aconitum pseudokongboense W. T. Wang & L. Q. Li

Aconitum pulchellum Hand.-Mazz.

Aconitum pulchellum var. **hispidum** Hand.-Mazz.

Aconitum racemulosum Franch.

Aconitum racemulosum var. **grandibracteolatum** W. T. Wang

Aconitum refractum (Finet & Gagnep.) Hand.-Mazz.

Aconitum rhombifolium Chen

Aconitum richardsonianum Lauener

Aconitum richardsonianum var. **pseudosessiliflorum** Lauener

Aconitum rilongense Kadota

Aconitum rotundifolium Kar. & Kir.

Aconitum scaposum Franch.

Aconitum secundiflorum W. T. Wang

Aconitum sherriffii Lauener

Aconitum sinoaxillare W. T. Wang

Aconitum sinomontanum Nakai

Aconitum sinomontanum var. **pilocarpum** W. T. Wang

Aconitum soongoricum Stapf

Aconitum souliei Finet & Gagnep.

Aconitum spathulatum W. T. Wang

Aconitum spicatum (Bruehl) Stapf

Aconitum spiripetalum Hand.-Mazz.

Aconitum staintonii Lauener

Aconitum stapfianum Hand.-Mazz.

Aconitum stylosoides W. T. Wang

Aconitum stylosum Stapf

Aconitum subspathulatum W. T. Wang

Aconitum tamuranum Lauener

Aconitum tangense Marquand & Airy Shaw

Aconitum tanguticum (Maxim.) Stapf

Aconitum taronense (Hand.-Mazz.) Fletcher & Lauener

Aconitum tatsienense Finet & Gagnep.

Aconitum tongolense Ulbr.

Aconitum transsectum Diels

Aconitum tsaii W. T. Wang

Aconitum vilmorinianum Kom.

Aconitum violaceum Jacquem. ex Stapf

Actaea acuminata Wall. ex Royle

Actaea asiatica Hara

Actaea brachycarpa (P. K. Hsiao) J. Compton

Actaea dahurica (Turcz. ex Fisch. & C. A. Mey.) Franch.

Actaea lancifoliolata (X. F. Pu & M. R. Jia) J. P. Luo, Q. Yuan & Q. E. Yang

Actaea muliensis J. P. Luo, Q. E. Yang & Q. Yuan

Actaea spicata L.

Actaea spicata var. **acuminata** (Wall. ex Royle) Hara

Adonis aestivalis L.

Adonis aestivalis subsp. **parviflora** (Fisch. ex DC.) N. Busch

Adonis aestivalis subsp. **squarrosa** (Steven) Nyman

Adonis chrysocyathus Hook. f. & Thomson

Adonis coerulea Maxim.

Adonis davidii Franch.

Adonis nepalensis Simonov.

Adonis sutchuenensis Franch.

Anemoclema glaucifolium (Franch.) W. T. Wang

Anemone baicalensis Turcz.

Anemone biflora DC.

Anemone brachystema W. T. Wang

Anemone brevistyla C. C. Chang ex W. T. Wang

Anemone coelestina Franch.

Anemone coelestina var. **holophylla** (Diels) Ziman & B. E. Dutton

Anemone coelestina var. **linearis** (Brühl) Ziman & B. E. Dutton

Anemone davidii Franch.

Anemone delavayi Franch.

Anemone delavayi var. **oligocarpa** (C. Pei) Ziman & B. E. Dutton

Anemone demissa Hook. f. & Thomson

Anemone demissa var. **major** W. T. Wang

Anemone demissa var. **villosissima** Brühl

Anemone demissa var. **yunnanensis** Franch.

Anemone elongata D. Don

Anemone erythrophylla Finet & Gagnep.

Anemone exigua Maxim.

Anemone falconeri Thomson

Anemone flaccida Fr. Schmidt

Anemone fuscopurpurea Hara

Anemone geum H. Lév.

Anemone geum subsp. **ovalifolia** (Brühl) R. P. Chaudhary

Anemone griffithii Hook. f. & Thomson

Anemone howellii Jeffrey & W. W. Sm.

Anemone hupehensis (Lemoine) Lemoine

Anemone imbricata Maxim.

Anemone jiachaensis W. T. Wang

Anemone laceratoincisa W. T. Wang

Anemone milinensis W. T. Wang

Anemone motuoensis W. T. Wang

Anemone narcissiflora L.

Anemone obtusiloba D. Don

Anemone obtusiloba subsp. **leiophylla** W. T. Wang

Anemone obtusiloba subsp. **nepalensis** R. P. Chaudhary

Anemone obtusiloba subsp. **omalocarpella** Bruh

Anemone obtusiloba subsp. **potentilloides** (Cambess. ex Lauener) R. P. Chaudhary

Anemone patula C. C. Chang ex W. T. Wang

Anemone patula var. **minor** W. T. Wang

Anemone polyanthes D. Don

Anemone polycarpa W. E. Evans

Anemone prattii Huth ex Ulbr.

Anemone raui Goel & U. C. Bhattach

Anemone rivularis Buch.-Ham. ex DC.

Anemone rivularis var. **pilosisepala** W. T. Wang

Anemone robusta W. T. Wang

Anemone rockii Ulbr.

Anemone rockii var. **multicaulis** W. T. Wang

Anemone rupestris subsp. **gelida** (Maxim.) Lauener

Anemone rupestris Wall. ex Hook. f. & Thomson

Anemone rupicola Cambess.

Anemone scabriuscula W. T. Wang

Anemone smithiana Lauener & Panigrahi

Anemone subindivisa W. T. Wang

Anemone subpinnata W. T. Wang

Anemone tetrasepala Royle

Anemone tibetica W. T. Wang

Anemone tomentosa (Maxim.) C. Pei

Anemone trullifolia Hook. f. & Thomson

Anemone trullifolia var. **liangshanica** (W. T. Wang) Ziman & B. E. Dutton

Anemone trullifolia var. **lutienensis** (W. T. Wang) Ziman & B. E. Dutton

Anemone tschernjaewii Regel

Anemone vitifolia Buch.-Ham. ex DC.

Anemone yulongshanica var. **truncata** (H. F. Comber) W. T. Wang

Anemone yulongshanica W. T. Wang

Aquilegia ecalcarata Maxim.

Aquilegia fragrans Benth.

Aquilegia fragrans var. **kanawarensis** (Jacquem. ex Camb.) H. Riedl

Aquilegia moorcroftiana var. **afghanica** (Bruehl) H. Riedl

Aquilegia moorcroftiana Wall. ex Royle

Aquilegia nakaoi Tamura

Aquilegia nivalis Falc. ex Jackson

Aquilegia pubiflora var. **mussooriensis** Royle

Aquilegia pubiflora Wall. ex Royle

Aquilegia rockii Munz

Aquilegia yangii Y. Luo & L. Li

Asteropyrum cavaleriei (H. Lév. & Vaniot) J. R. Drumm. & Hutch.

Asteropyrum peltatum (Franch.) J. R. Drumm & Hutch.

Batrachium bungei (Steudel) L. Liou

Batrachium foeniculaceum (Gilib.) V. I. Krecz.

Batrachium rionii (Lagger) Nym.

Batrachium trichophyllum (Chaix ex Vill.) Bosch

Beesia calthifolia Ulbr.

Beesia deltophylla C. Y. Wu ex P. K. Hsiao

Calathodes oxycarpa Sprague

Calathodes palmata Hook. f. & Thomson

Callianthemum alatavicum Freyn

Callianthemum anemonoides (Zahlbr.) Endl.

Callianthemum farreri W. W. Sm.

Callianthemum pimpinelloides (D. Don) Hook. f. & Thomson

Caltha govaniana Wall. ex Royle

Caltha lobulatifolia W. T. Wang

Caltha palustris L.

Caltha palustris var. **alba** (Cambess.) Hook. f. & Thomson

Caltha palustris var. **barthei** Hance

Caltha palustris var. **himalaica** Tamura

Caltha palustris var. **purpurea** Sparre & Fisch.

Caltha palustris var. **umbrosa** Diels

Caltha rubriflora B. L. Burtt & Lauener

Caltha scaposa Hook. f. & Thomson

Caltha sinogracilis W. T. Wang

Cimicifuga foetida L.

Cimicifuga foetida var. **bifida** W. T. Wang & P. K. Hsiao

Cimicifuga foetida var. **foliolosa** P. K. Hsiao

Cimicifuga foetida var. **longibracteata** P. K. Hsiao

Cimicifuga foetida var. **velutina** Franch. ex Finet & Gagnep.

Cimicifuga simplex (DC.) Wormsk. ex Turcz.

Cimicifuga yunnanensis P. K. Hsiao

Clematis acuminata DC.

Clematis acuminata subsp. **sikkimensis** (Hook. f. & Thomson) Bruehl

Clematis acuminata var. **longicaudata** W. T. Wang

Clematis acutangula Hook. f. & Thomson

Clematis akebioides (Maxim.) Veitch

Clematis alternata Kitam. & Tamura

Clematis andersonii (C. B. Clarke ex Kuntze) H. Eichler

Clematis apiculata Hook. f. & Thomson

Clematis armandii Franch.

Clematis aspleniifola Schrenk

Clematis austrogansuensis W. T. Wang & L. Q. Li

Clematis barbellata Edgew.

Clematis barbellata var. **obtusa** Kitamura & Tamura

Clematis bracteolata Tamura

Clematis brevicaudata DC.

Clematis buchananiana DC.

Clematis cadmia Buch.-Ham. ex Hook. f. & Thomson

Clematis chrysocoma Franch.

Clematis confusa Grey-Wilson

Clematis connata DC.

Clematis connata var. **lanceolata** S. N. Biswas

Clematis connata var. **pseudoconnata** (Kuntze) W. T. Wang

Clematis connata var. **trullifera** (Franch.) W. T. Wang

Clematis delavayi Franch.

Clematis delavayi var. **calvescens** C. K. Schneid.

Clematis delavayi var. **limprichtii** (Ulbr.) M. C. Chang

Clematis delavayi var. **spinescens** Balfour ex Diels

Clematis diebuensis W. T. Wang

Clematis fasciculiflora Franch.

Clematis glabrifolia K. Sun & M. S. Yan

Clematis gouriana Roxb. ex DC.

Clematis gracilifolia Rehder & E. H. Wilson

Clematis gracilifolia var. **dissectifolia** W. T. Wang & M. C. Chang

Clematis gracilifolia var. **lasiocarpa** W. T. Wang

Clematis gracilifolia var. **macrantha** W. T. Wang & M. C. Chang

Clematis grandidentata (Rehder & E. H. Wilson) W. T. Wang

Clematis grandidentata var. **likiangensis** (Rehder) W. T. Wang

Clematis grata Wall.

Clematis graveolens Lindl.

Clematis grewiiflora DC.

Clematis hilariae Kovalevsk.

Clematis jialasaensis W. T. Wang

Clematis kockiana C. K. Schneid.

Clematis ladakhiana Grey-Wilson

Clematis lancifolia Bureau & Franch.

Clematis lancifolia var. **ternata** W. T. Wang & M. C. Chang

Clematis lasiandra Maxim.

Clematis lushuiensis W. T. Wang

Clematis manipurensis (Brühl) W. T. Wang

Clematis manipurensis var. **lasioclada** W. T. Wang

Clematis metuoensis M. Y. Fang

Clematis montana Buch.-Ham. ex DC.

Clematis montana subsp. **praecox** Kuntze

Clematis montana var. **brevifoliola** Kuntze

Clematis montana var. **glabrescens** (H. F. Comber) W. T. Wang & M. C. Chang

Clematis montana var. **longipes** W. T. Wang

Clematis montana var. **sterilis** Hand.-Mazz.

Clematis nagaensis W. T. Wang

Clematis nainitalensis W. T. Wang

Clematis napaulensis DC.

Clematis ningjingshanica W. T. Wang

Clematis nukiangensis M. Y. Fang

Clematis nutans Royle

Clematis orientalis L.

Clematis orientalis var. **uniflora** Tamura

Clematis parviloba Gardner & Champ.

Clematis peterae Hand.-Mazz.

Clematis phlebantha L. H. J. Williams

Clematis pianmaensis W. T. Wang

Clematis pinchuanensis W. T. Wang & M. Y. Fang

Clematis pogonandra Maxim.

Clematis pogonandra var. **pilosula** Rehder & E. H. Wilson

Clematis potaninii Maxim.

Clematis pseudopogonandra Finet & Gagnep.

Clematis puberula Hook. f. & Thomson

Clematis puberula var. **ganpiniana** (H. Lév. & Vaniot) W. T. Wang

Clematis puberula var. **subsericea** (Rehder & E. H. Wilson) W. T. Wang

Clematis ranunculoides Franch.

Clematis rehderiana Craib

Clematis roylei Rehder

Clematis siamensis J. R. Drumm. & Craib

Clematis siamensis var. **clarkei** (Kuntze) W. T. Wang

Clematis siamensis var. **monantha** (W. T. Wang & L. Q. Li) W. T. Wang & L. Q. Li

Clematis sikkimensis (Hook. f. & Thomson) J. R. Drumm. ex Burkill

Clematis smilacifolia Wall.

Clematis staintonii W. T. Wang

Clematis tangutica (Maxim.) Korsh.

Clematis tangutica var. **obtusiuscula** Rehder & E. H. Wilson

Clematis tangutica var. **pubescens** M. C. Chang & P. P. Ling

Clematis tengchongensis W. T. Wang

Clematis terniflora DC.

Clematis tibetana Kuntze

Clematis tibetana var. **laciniifolia** Grey-Wilson

Clematis tibetana var. **vernayi** (C. E. C. Fischer) W. T. Wang

Clematis tongluensis (Bruhl) Tamura

Clematis tongluensis var. **mollisepala** W. T. Wang

Clematis tortuosa Wall. ex C. E. C. Fisch.

Clematis tripartita W. T. Wang

Clematis tsaii W. T. Wang

Clematis urophylla Franch.

Clematis venusta M. C. Chang

Clematis viridis (W. T. Wang & M. C. Chang) W. T. Wang

Clematis wallichii W. T. Wang

Clematis wenxianensis W. T. Wang

Clematis yui W. T. Wang

Clematis yunnanensis Franch.

Clematis zandaensis W. T. Wang

Clematis zemuensis W. W. Sm.

Consolida ajacis (L.) Schur

Consolida ambigua (L.) Ball & Heywood

Consolida orientalis (Gay) Schröd.

Consolida schlagintweitii (Huth) Munz

Coptis chinensis Franch.

Coptis teeta Wall.

Delphinium afghanicum Rech. f.

Delphinium albocoeruleum Maxim.

Delphinium altissimum Wall.

Delphinium angustirhombicum W. T. Wang

Delphinium anthriscifolium Hance

Delphinium autumnale Hand.-Mazz.

Delphinium batangense Finet & Gagnep.

Delphinium beesianum var. **latisectum** W. T. Wang

Delphinium beesianum var. **radiatifolium** (Hand.-Mazz.) W. T. Wang

Delphinium beesianum W. W. Sm.

Delphinium bhutanicum Munz

Delphinium brevisepalum W. T. Wang

Delphinium brunonianum Royle

Delphinium bulleyanum Forrest ex Diels

Delphinium caeruleum Jacquem.

Delphinium caeruleum var. **crassicalcaratum** W. T. Wang & M. J. Warnock

Delphinium caeruleum var. **obtusilobum** Brühl ex Huth

Delphinium campylocentrum Maxim.

Delphinium candelabrum Ostenf.

Delphinium candelabrum var. **monanthum** (Hand.-Mazz.) W. T. Wang

Delphinium cashmerianum Royle

Delphinium caudatolobum W. T. Wang

Delphinium ceratophoroides W. T. Wang

Delphinium ceratophorum Franch.

Delphinium ceratophorum var. **brevicorniculatum** W. T. Wang

Delphinium ceratophorum var. **hirsutum** W. T. Wang

Delphinium ceratophorum var. **robustum** W. T. Wang

Delphinium chayuense W. T. Wang

Delphinium cheilanthum Fisch. ex DC.

Delphinium chenii W. T. Wang

Delphinium chitralense Riedl

Delphinium chrysotrichum Finet & Gagnep.

Delphinium chrysotrichum var. **tsarongense** (Hand.-Mazz.) W. T. Wang

Delphinium chumulangmaense W. T. Wang

Delphinium chungbaense W. T. Wang

Delphinium coleopodum Hand.-Mazz.

Delphinium conaense W. T. Wang

Delphinium cooperi Munz

Delphinium davidii Franch.

Delphinium delavayi Franch.

Delphinium delavayi var. **lasiandrum** W. T. Wang

Delphinium delavayi var. **pogonanthum** (Hand.-Mazz.) W. T. Wang

Delphinium densiflorum Duthie ex Huth

Delphinium denudatum Wall. ex Hook. f. & Thomson

Delphinium dolichocentroides W. T. Wang

Delphinium dolichocentroides var. **leiogynum** W. T. Wang

Delphinium drepanocentrum (Bruehl ex Huth) Munz

Delphinium erlangshanicum W. T. Wang

Delphinium forrestii Diels

Delphinium forrestii var. **viride** (W. T. Wang) W. T. Wang

Delphinium giraldii Diels

Delphinium glabricaule W. T. Wang

Delphinium glaciale Hook. f. & Thomson

Delphinium gonggaense W. T. Wang

Delphinium grandiflorum L.

Delphinium grandiflorum var. **mosoynense** (Franch.) Huth

Delphinium grandilimbum W. T. Wang & M. J. Warnock

Delphinium gyalanum C. Marquand & Airy Shaw

Delphinium hamatum Franch.

Delphinium handelianum W. T. Wang

Delphinium hillcoatiae Munz

Delphinium himalayae Munz

Delphinium hirtifolium W. T. Wang

Delphinium hui F. H. Chen

Delphinium humilius (W. T. Wang) W. T. Wang

Delphinium incisolobulatum W. T. Wang

Delphinium incisum (Hook. f. & Thomson) Wall. ex Munz

Delphinium kamaonense Huth

Delphinium kamaonense var. **glabrescens** (W. T. Wang) W. T. Wang

Delphinium kantzeense W. T. Wang

Delphinium kingianum Brühl ex Huth

Delphinium kingianum var. **acuminatissimum** (W. T. Wang) W. T. Wang

Delphinium kingianum var. **eglandulosum** W. T. Wang

Delphinium kingianum var. **leiocarpum** Brühl ex Huth

Delphinium koelzii Munz

Delphinium kohatense (Brühl) Munz

Delphinium lacostei Danguy

Delphinium lasiantherum W. T. Wang

Delphinium latirhombicum W. T. Wang

Delphinium laxicymosum W. T. Wang

Delphinium leiophyllum (W. T. Wang) W. T. Wang

Delphinium leiostachyum W. T. Wang

Delphinium liangshanense W. T. Wang

Delphinium likiangense Franch.

Delphinium longipedicellatum W. T. Wang

Delphinium ludlowii Munz

Delphinium majus (W. T. Wang) W. T. Wang

Delphinium malacophyllum Hand.-Mazz.

Delphinium maoxianense W. T. Wang

Delphinium maximowiczii Franch.

Delphinium medogense W. T. Wang

Delphinium micropetalum Finet & Gagnep.

Delphinium motingshanicum W. T. Wang & M. J. Warnock

Delphinium muliense W. T. Wang

Delphinium muliense var. **minutibracteolatum** W. T. Wang

Delphinium muscosum Exell & Hillc.

Delphinium nangchienense W. T. Wang

Delphinium nangziense W. T. Wang

Delphinium nepalense Kitam. & Tamura

Delphinium ninglangshanicum W. T. Wang

Delphinium nordhagenii Wendelbo

Delphinium nordhagenii var. **acutidentatum** W. T. Wang

Delphinium nortonii Dunn

Delphinium obcordatilimbum W. T. Wang

Delphinium omeiense W. T. Wang

Delphinium omeiense var. **pubescens** W. T. Wang

Delphinium orthocentrum Franch.

Delphinium oxycentrum W. T. Wang

Delphinium pachycentrum Hemsl.

Delphinium pachycentrum var. **lancisepalum** (Hand.-Mazz.) W. T. Wang

Delphinium pergameneum W. T. Wang

Delphinium pomeense W. T. Wang

Delphinium potaninii Huth

Delphinium potaninii var. **bonvalotii** (Franch.) W. T. Wang

Delphinium potaninii var. **latibracteolatum** W. T. Wang

Delphinium pseudocampylocentrum W. T. Wang

Delphinium pseudocandelabrum W. T. Wang

Delphinium pseudoglaciale W. T. Wang

Delphinium pseudohamatum W. T. Wang

Delphinium pseudomosoynense W. T. Wang

Delphinium pseudopulcherrimum W. T. Wang

Delphinium pseudothibeticum W. T. Wang & M. J. Warnock

Delphinium pseudotongolense W. T. Wang

Delphinium pseudoyunnanense W. T. Wang & M. J. Warnock

Delphinium pulanense W. T. Wang

Delphinium pumilum W. T. Wang

Delphinium purpurascens W. T. Wang

Delphinium pycnocentrum Franch.

Delphinium pylzowii var. **trigynum** W. T. Wang

Delphinium pyramidale Royle

Delphinium rangtangense W. T. Wang

Delphinium roylei Munz

Delphinium saxatile W. T. Wang

Delphinium scabriflorum D. Don

Delphinium sherriffii Munz

Delphinium sinoscaposum W. T. Wang

Delphinium sinovitifolium W. T. Wang

Delphinium smithianum Hand.-Mazz.

Delphinium souliei Franch.

Delphinium spirocentrum Hand.-Mazz.

Delphinium stapeliosmum Brühl

Delphinium stapeliosum Bruehl ex Huth

Delphinium subspathulatum W. T. Wang

Delphinium sutchuenense Franch.

Delphinium tabatae Tamura

Delphinium taliense Franch.

Delphinium tangkulaense W. T. Wang

Delphinium tatsienense Franch.

Delphinium tenii H. Lév.

Delphinium thibeticum Finet & Gagnep.

Delphinium thibeticum var. **laceratilobum** W. T. Wang

Delphinium tongolense Franch.

Delphinium trichophorum Franch.

Delphinium trichophorum var. **playcentrum** W. T. Wang

Delphinium trichophorum var. **subglaberrimum** Hand.-Mazz.

Delphinium umbrosum Hand.-Mazz.

Delphinium umbrosum var. **drepanocentrum** (Brühl ex Huth) W. T. Wang & M. J. Warnock

Delphinium umbrosum var. **hispidum** W. T. Wang

Delphinium uncinatum Hook. f. & Thomson

Delphinium unifolium Tamura

Delphinium vestitum Wall. ex Royle

Delphinium viscosum Hook. f. & Thomson

Delphinium wardii C. Marquand & Airy Shaw

Delphinium wenchuanense W. T. Wang

Delphinium williamsii Munz

Delphinium wrightii F. H. Chen

Delphinium xichangense W. T. Wang

Delphinium yajiangense W. T. Wang

Delphinium yangii W. T. Wang

Delphinium yanwaense W. T. Wang

Delphinium yongningense W. T. Wang & M. J. Warnock

Delphinium yuanum F. H. Chen

Delphinium yulungshanicum W. T. Wang

Delphinium yunnanense (Franch.) Franch.

Dichocarpum adiantifolium (Hook. f. & Thomson) W. T. Wang & P. K. Hsiao

Dichocarpum auriculatum (Franch.) W. T. Wang & P. K. Hsiao

Dichocarpum fargesii (Franch.) W. T. Wang & P. K. Hsiao

Dichocarpum franchetii (Franch.) W. T. Wang & P. K. Hsiao

Dichocarpum sutchuenense (Franch.) W. T. Wang & P. K. Hsiao

Eranthis albiflora Franch.

Eranthis lobulata W. T. Wang

Eranthis lobulata var. **elatior** W. T. Wang

Gymnaconitum gymnandrum (Maxim.) Wei Wang & Z. D. Chen

Halerpestes lancifolia (Bertol.) Hand.-Mazz.

Halerpestes salsuginosa (Pall.) Greene

Halerpestes sarmentosa (Adams) Kom.

Halerpestes tricuspis (Maxim.) Hand.-Mazz.

Helleborus thibetanus Franch.

Hepatica falconeri (Thomson) Juz.

Isopyrum anemonoides Kar. & Kir.

Isopyrum ludlowii Tamura & Lauener

Metanemone ranunculoides W. T. Wang

Naravelia zeylanica (L.) DC.

Nigella damascena L.

Nigella glandulifera Freyn & Sint.

Nigella sativa L.

Oxygraphis delavayi Franch.

Oxygraphis endlicheri (Walp.) Bennet & Sum. Chandra

Oxygraphis glacialis (Fisch. ex DC.) Bunge

Oxygraphis kumaonensis I. D. Rai & G. S. Rawat

Oxygraphis nepalensis Tamura

Oxygraphis shaftoanus Aitch. & Hemsl.

Oxygraphis tenuifolia W. E. Evans

Paraquilegia anemonoides (Willd.) Ulbr.

Paraquilegia caespitosa (Boiss. & Hohen.) J. R. Drumm. & Hutch.

Paraquilegia microphylla (Royle) J. R. Drumm. & Hutch.

Paroxygraphis sikkimensis W. W. Sm.

Pulsatilla campanella Fisch. ex Krylov

Pulsatilla millefolium (Hemsl. & E. H. Wilson) Ulbr.

Pulsatilla wallichiana (Royle) Ulbr.

Ranunculus adoxifolius Hand.-Mazz.

Ranunculus albertii Regel & Schmalh.

Ranunculus alpigenus Kom.

Ranunculus arvensis L.

Ranunculus aucheri Boiss.

Ranunculus balangshanicus W. T. Wang

Ranunculus banguoensis L. Liu

Ranunculus banguoensis var. **grandiflorus** W. T. Wang

Ranunculus bikramii Aswal & Mehrotra

Ranunculus brotherusii Freyn

Ranunculus brotherusii var. **latisectus** Riedl

Ranunculus cantoniensis DC.

Ranunculus chinensis Bunge

Ranunculus chuanchingensis L. Liu

Ranunculus decandrus W. T. Wang

Ranunculus densiciliatus W. T. Wang

Ranunculus dielsianus Ulbr.

Ranunculus dielsianus var. **suprasericeus** Hand.-Mazz.

Ranunculus diffusus DC.

Ranunculus dingjieensis L. Liou

Ranunculus distans Wall. ex Royle

Ranunculus dongrergensis Hand.-Mazz.

Ranunculus duoxionglashanicus W. T. Wang

Ranunculus felixii H. Lév.

Ranunculus felixii var. **forrestii** Hand.-Mazz.

Ranunculus ficariifolius A. Lév. & Vaniot

Ranunculus flavidus (Hand.-Mazz.) R. R. Stewart

Ranunculus furcatifidus W. T. Wang

Ranunculus glabricaulis (Hand.-Mazz.) L. Liu

Ranunculus glacialiformis Hand.-Mazz.

Ranunculus glareosus Hand.-Mazz.

Ranunculus gongheensis W. T. Wang

Ranunculus himalaicus Y. Kadota

Ranunculus hirtellus Royle

Ranunculus hirtellus var. **humilis** W. T. Wang

Ranunculus hirtellus var. **orientalis** W. T. Wang

Ranunculus humillimus W. T. Wang

Ranunculus hyperboreus Rottb.

Ranunculus indivisus (Maxim.) Hand.-Mazz.

Ranunculus indivisus var. **abaensis** (W. T. Wang) W. T. Wang

Ranunculus limprichtii Ulbr.

Ranunculus limprichtii var. **flavus** Hand.-Mazz.

Ranunculus lingua L.

Ranunculus longicaulis C. A. Mey.

Ranunculus mainlingensis W. T. Wang

Ranunculus makaluensis Kadota

Ranunculus melanogynus W. T. Wang

Ranunculus membranaceus Royle

Ranunculus membranaceus var. **pubescens** (W. T. Wang) W. T. Wang

Ranunculus micronivalis Hand.-Mazz.

Ranunculus minjanensis Rech. f.

Ranunculus minor (L. Liu) W. T. Wang

Ranunculus munroanus J. R. Drumm. ex Dunn

Ranunculus muricatus L.

Ranunculus muscigenus W. T. Wang

Ranunculus napahaiensis W. T. Wang & L. Liao

Ranunculus napaulensis DC.

Ranunculus natans C. A. Mey.

Ranunculus nematolobus Hand.-Mazz.

Ranunculus nephelogenes Edgew.

Ranunculus nephelogenes var. **geniculatus** (Hand.-Mazz.) W. T. Wang

Ranunculus nyalamensis var. **angustipetalus** W. T. Wang

Ranunculus nyalamensis W. T. Wang

Ranunculus oreionannos C. Marquand & Airy Shaw

Ranunculus pagaeus Hand.-Mazz.

Ranunculus palmatifidus Riedl

Ranunculus pamiri Korsh.

Ranunculus pedicellatus Hand.-Mazz.

Ranunculus pegaeus Hand.-Mazz.

Ranunculus petrogeiton Ulbr.

Ranunculus pianmaensis W. T. Wang

Ranunculus platypetalus (Hand.-Mazz.) Hand.-Mazz.

Ranunculus platypetalus var. **macranthus** W. T. Wang

Ranunculus popovii Ovcz.

Ranunculus popovii var. **stracheyanus** (Maxim.) W. T. Wang

Ranunculus potaninii Kom.

Ranunculus pseudohirculus Schrenk

Ranunculus pseudolobatus L. Liu

Ranunculus pseudopygmaeus Hand.-Mazz.

Ranunculus pulchellus C. A. Mey.

Ranunculus repens L.

Ranunculus rionii Lagger

Ranunculus rubrocalyx Kom.

Ranunculus rufosepalus Franch.

Ranunculus sceleratus L.

Ranunculus shaftoanus (Aitch. & Hemsl.) Boiss.

Ranunculus siamensis Tamura

Ranunculus sikkimensis Hand.-Mazz.

Ranunculus silerifolius H. Lév.

Ranunculus similis Hemsl.

Ranunculus sinovaginatus W. T. Wang

Ranunculus sphaerospermus Boiss. & C. I. Blanche

Ranunculus stenorhynchus Franch.

Ranunculus stewartii Riedl

Ranunculus tanguticus (Maxim.) Ovcz.

Ranunculus tanguticus var. **dasycarpus** (Maxim.) L. Liu

Ranunculus tengchongensis W. T. Wang

Ranunculus tetrandrus W. T. Wang

Ranunculus tongrenensis W. T. Wang

Ranunculus trautvetterianus Regel ex Ovcz.

Ranunculus trichophyllus Chaix

Ranunculus trigonus Hand.-Mazz.

Ranunculus trilobus Desf.

Ranunculus trivedii B. S. Aswal & B. N. Mehrotra

Ranunculus wangianus Q. E. Yang

Ranunculus yaoanus W. T. Wang

Ranunculus yunnanensis Franch.

Ranunculus zhungdianensis W. T. Wang

Ranunculus zhouquensis W. T. Wang

Souliea vaginata (Maxim.) Franch.

Thalictrum alpinum L.

Thalictrum alpinum var. **elatum** O. E. Ulbr.

Thalictrum alpinum var. **microphyllum** (Royle) Hand.-Mazz.

Thalictrum atriplex Finet & Gagnep.

Thalictrum baicalense Turcz.

Thalictrum brevisericeum W. T. Wang & S. H. Wang

Thalictrum chayuense W. T. Wang

Thalictrum chelidonii DC.

Thalictrum cultratum Wall.

Thalictrum delavayi Franch.

Thalictrum delavayi var. **acuminatum** Franch.

Thalictrum delavayi var. **decorum** Franch.

Thalictrum diffusiflorum C. Marquand & Airy Shaw

Thalictrum elegans Wall. ex Royle

Thalictrum falconeri Lecoy.

Thalictrum fargesii Franch. ex Finet & Gagnep.

Thalictrum finetii B. Boivin

Thalictrum foetidum L.

Thalictrum foliolosum DC.

Thalictrum fusiforme W. T. Wang

Thalictrum glareosum Hand.-Mazz.

Thalictrum grandiflorum Maxim.

Thalictrum javanicum Blume

Thalictrum leuconotum Franch.

Thalictrum leve (Franch.) W. T. Wang

Thalictrum microgynum Lecoyer ex Oliv.

Thalictrum minus L.

Thalictrum neurocarpum Royle

Thalictrum oligandrum Maxim.

Thalictrum pedunculatum Edgew.

Thalictrum petaloideum L.

Thalictrum platycarpum Hook. f. & Thomson

Thalictrum przewalskii Maxim.

Thalictrum punduanum Wall.

Thalictrum reniforme Wall.

Thalictrum reticulatum Franch.

Thalictrum rostellatum Hook. f. & Thomson

Thalictrum rotundifolium DC.

Thalictrum rutifolium Hook. f. & Thomson

Thalictrum saniculaeforme DC.

Thalictrum saniculiforme DC.

Thalictrum scabrifolium Franch.

Thalictrum secundum Edgew.

Thalictrum smithii B. Boivin

Thalictrum squamiferum Lecoy.

Thalictrum squarrosum Stephen ex Willd.

Thalictrum tenuisubulatum W. T. Wang

Thalictrum trichopus Franch.

Thalictrum tsawarungense W. T. Wang & S. H. Wang

Thalictrum uncatum Maxim.

Thalictrum uncinulatum Franch. ex Lecoy.

Thalictrum virgatum Hook. f. & Thomson

Thalictrum viscosum W. T. Wang & S. H. Wang

Thalictrum wangii B. Boivin

Trollius acaulis Lindl.

Trollius bhotanicus (King ex Brühl) Mukerjee

Trollius farreri Stapf

Trollius micranthus Hand.-Mazz.

Trollius pumilus D. Don

Trollius pumilus var. **tanguticus** Brühl

Trollius pumilus var. **tehkehensis** (W. T. Wang) W. T. Wang

Trollius ranunculoides Hemsl.

Trollius vaginatus Hand.-Mazz.

Trollius yunnanensis (Franch.) Ulbr.

Trollius yunnanensis var. **anemonifolius** (Brühl) W. T. Wang

Trollius yunnanensis var. **eupetalus** (Stapf) W. T. Wang

110. SABIACEAE

Meliosma alba (Schltdl.) Walp.

Meliosma arnottiana (Wight) Walp.

Meliosma cuneifolia Franch.

Meliosma cuneifolia var. **glabriuscula** Cufod.

Meliosma dilleniifolia (Wall. ex Wight & Arn.) Walp.

Meliosma dumicola W. W. Sm.

Meliosma parviflora Lecomte

Meliosma pinnata (Roxb.) Maxim.

Meliosma simplicifolia (Roxb.) Walp.

Meliosma thomsonii King ex Brandis

Meliosma veitchiorum Hemsl.

Meliosma yunnanensis Franch.

Sabia campanulata Wall.

Sabia campanulata var. **kingiana** M. P. Nayar & S. C. Majumdar

Sabia fasciculata Lecomte ex L. Chen

Sabia lanceolata Colebr.

Sabia lanceolata var. **siamensis** Water

Sabia limoniacea Wall. ex Hook. f. & Thomson

Sabia paniculata Edgew. ex Hook. f. & Thomson

Sabia parviflora Wall.

Sabia purpurea Hook. f. & Thomson

Sabia purpurea subsp. **dumicola** (W. W. Sm.) Water

Sabia wardii W. W. Sm.

Sabia yunnanensis Franch.

111. NELUMBONACEAE

Nelumbo nucifera Gaertn.

112. PLATANACEAE

Platanus orientalis L.

113. PROTEACEAE

Grevillea robusta A. Cunn. ex R. Br.

Helicia clivicola W. W. Sm.

Helicia excelsa (Roxb.) Blume

Helicia nilagirica Bedd.

Helicia robusta (Roxb.) R. Br. ex Blume

Helicia tibetensis H. S. Kiu

Helicia vestita var. **longipes** W. T. Wang

Heliciopsis terminalis (Kurz) Sleumer

114. TROCHODENDRACEAE

Tetracentron sinense Oliv.

115. BUXACEAE

Buxus hebecarpa Hatus.

Buxus microphylla Siebold & Zucc.

Buxus mollicula W. W. Sm.

Buxus mollicula var. **glabra** Hand.-Mazz.

Buxus papillosa C. K. Schneid.

Buxus rugulosa Hatus.

Buxus rugulosa var. **rupicola** (W. W. Sm.) P. Brückner & T. L. Ming

Buxus rugulosa var. **intermedia** Hatus.

Buxus rugulosa var. **prostrata** (W. W. Sm.) M. Cheng

Buxus wallichiana Baill.

Pachysandra axillaris Franch.

Sarcococca coriacea (Hook.) Sweet

Sarcococca hookeriana Baill.

Sarcococca hookeriana var. **digyna** Franch.

Sarcococca ruscifolia Stapf

Sarcococca saligna (D. Don) Müll. Arg.

Sarcococca vagans Stapf

Sarcococca wallichii Stapf

116. DILLENIACEAE

Dillenia aurea Sm.

Dillenia indica L.

Dillenia pentagyna Roxb.

Dillenia scabrella Roxb. ex Wall.

Tetracera sarmentosa (L.) Vahl

Tetracera scandens (L.) Merr.

117. PAEONIACEAE

Paeonia anomala subsp. **veitchii** (Lynch) D. Y. Hong & K. Y. Pan

Paeonia decomposita Hand.-Mazz.

Paeonia decomposita subsp. **rotundiloba** D. Y. Hong

Paeonia delavayi Franch.

Paeonia emodi Wall. ex Royle

Paeonia lactiflora Pall.

Paeonia ludlowii (Stern & G. Taylor) D. Y. Hong

Paeonia mairei H. Lév.

Paeonia obovata Maxim.

Paeonia obovata subsp. **willmottiae** (Stapf) D. Y. Hong & K. Y. Pan

Paeonia rockii (S. G. Haw & Lauener) T. Hong & J. J. Li ex D. Y. Hong

Paeonia sterniana H. R. Fletcher

118. ALTINGIACEAE

Altingia excelsa Noronha

119. HAMAMELIDACEAE

Corylopsis brevistyla H. T. Chang

Corylopsis glaucescens Hand.-Mazz.

Corylopsis himalayana Griff.

Corylopsis microcarpa H. T. Chang

Corylopsis trabeculosa Hu & Cheng

Corylopsis velutina Hand.-Mazz.

Corylopsis willmottiae Rehder & E. H. Wilson

Corylopsis yui Hu & Cheng

Corylopsis yunnanensis Diels

Exbucklandia populnea (R. Br. ex Griff.) R. W. Br.

Loropetalum chinense (R. Br.) Oliv.

Parrotiopsis jacquemontiana (Decne.) Rehder

Rhodoleia forrestii Chun ex Exell

120. CERCIDIPHYLLACEAE

Cercidiphyllum japonicum Siebold & Zucc.

121. DAPHNIPHYLLACEAE

Daphniphyllum chartaceum Rosenth.

Daphniphyllum himalense (Benth.) Müll. Arg.

122. ITEACEAE

Itea chinensis Hook. & Arn.

Itea indochinensis var. **pubinervia** (H. T. Chang) C. Y. Wu

Itea kiukiangensis C. C. Huang & S. C. Huang

Itea macrophylla Wall.

Itea yunnanensis Franch.

123. GROSSULARIACEAE

Ribes acuminatum Wall. ex G. Don

Ribes alpestre Decne.

Ribes alpestre var. **eglandulosum** L. T. Lu

Ribes alpestre var. **giganteum** Jancz.

Ribes davidii Franch.

Ribes davidii var. **ciliatum** L. T. Lu

Ribes davidii var. **lobatum** L. T. Lu

Ribes glabricalycinum L. T. Lu

Ribes glaciale Wall. ex Roxb.

Ribes griffithii Hook. f. & Thomson

Ribes griffithii var. **gongshanense** (T. C. Ku) L. T. Lu

Ribes heterotrichum C. A. Mey.

Ribes himalense Royle ex Decne.

Ribes humile Jancz.

Ribes janczewskii Pojark.

Ribes kialanum Jancz.

Ribes laurifolium Jancz.

Ribes laurifolium var. **yunnanense** L. T. Lu

Ribes longeracemosum Franch.

Ribes longeracemosum var. **davidii** Jancz.

Ribes longeracemosum var. **pilosum** T. C. Ku

Ribes luridum Hook. f. & Thomson

Ribes meyeri Maxim.

Ribes moupinense Franch.

Ribes moupinense var. **muliense** S. H. Yu & J. M. Xu

Ribes moupinense var. **pubicarpum** L. T. Lu

Ribes moupinense var. **tripartitum** (Batalin) Jancz.

Ribes orientale Desf.

Ribes pseudofasciculatum K. S. Hao

Ribes rubrisepalum L. T. Lu

Ribes setchuense Jancz.

Ribes soulieanum Jancz.

Ribes stenocarpum Maxim.

Ribes tenue Jancz.

Ribes tenue var. **incisum** L. T. Lu

Ribes tianquanense S. H. Yu & J. M. Xu

Ribes vilmorinii Jancz.

Ribes vilmorinii var. **pubicarpum** L. T. Lu

Ribes xizangense L. T. Lu

124. SAXIFRAGACEAE

Astilbe chinensis (Maxim.) Franch. & Sav.

Astilbe rivularis Buch.-Ham. ex D. Don

Astilbe rivularis var. **angustifoliolata** Hara

Astilbe rivularis var. **myriantha** (Diels) J. T. Pan

Astilbe rubra Hook. f. & Thomson

Bergenia ciliata (Haw.) Sternb.

Bergenia emeiensis C. Y. Wu

Bergenia emeiensis var. **rubellina** J. T. Pan

Bergenia pacumbis (Buch.-Ham. ex D. Don) C. Y. Wu & J. T. Pan

Bergenia purpurascens (Hook. f. & Thomson) Engl.

Bergenia stracheyi (Hook. f. & Thomson) Engl.

Bergenia tianquanensis J. T. Pan

Chrysosplenium absconditicapsulum J. T. Pan

Chrysosplenium axillare Maxim.

Chrysosplenium carnosum Hook. f. & Thomson

Chrysosplenium cavaleriei H. Lév. & Vaniot

Chrysosplenium davidianum Decne. ex Maxim.

Chrysosplenium delavayi Franch.

Chrysosplenium forrestii Diels

Chrysosplenium fuscopuncticulosum Z. P. Jien

Chrysosplenium griffithii Hook. f. & Thomson

Chrysosplenium griffithii var. **intermedium** (Hara) J. T. Pan

Chrysosplenium lanuginosum Hook. f. & Thomson

Chrysosplenium lanuginosum var. **ciliatum** (Franch.) J. T. Pan

Chrysosplenium lanuginosum var. **gracile** (Franch.) Hara

Chrysosplenium lixianense Z. P. Jien ex J. T. Pan

Chrysosplenium nepalense D. Don

Chrysosplenium nudicaule Bunge

Chrysosplenium nudicaule var. **intermedium** Hara

Chrysosplenium oxygraphoides Hand.-Mazz.

Chrysosplenium sikangense Hara

Chrysosplenium sinicum Maxim.

Chrysosplenium tenellum Hook. f. & Thomson

Chrysosplenium trichospermum Edgew. ex Hook. f. & Thomson

Chrysosplenium uniflorum Maxim.

Rodgersia aesculifolia Batalin

Rodgersia aesculifolia var. **henrici** (Franch.) C. Y. Wu ex J. T. Pan

Rodgersia nepalensis Cope ex Cullen

Rodgersia pinnata Franch.

Rodgersia pinnata var. **strigosa** J. T. Pan

Rodgersia sambucifolia Hemsl.

Rodgersia sambucifolia var. **estrigosa** J. T. Pan

Saxifraga afghanica Aitc. & Hemsl.

Saxifraga alpigena Harry Sm.

Saxifraga amabilis H. Ohba & Wakab.

Saxifraga andersonii Engl.

Saxifraga androsacea L.

Saxifraga androsacea var. **uniflora** Wulfen

Saxifraga angustata Harry Sm.

Saxifraga aristulata Hook. f. & Thomson

Saxifraga aristulata var. **longipila** (Engl. & Irmsch.) J. T. Pan

Saxifraga asarifolia Sternb.

Saxifraga atrata Engl.

Saxifraga atuntsiensis W. W. Sm.

Saxifraga aurantiaca Franch.

Saxifraga auriculata Engl. & Irmsch.

Saxifraga auriculata var. **conaensis** J. T. Pan

Saxifraga baimashanensis C. Y. Wu

Saxifraga balfourii Engl. & Irmsch.

Saxifraga benzilanensis H. Chuang

Saxifraga bergenioides C. Marquand

Saxifraga bijiangensis H. Chuang

Saxifraga brachyphylla Franch.

Saxifraga brachypoda D. Don

Saxifraga brachypodoidea J. T. Pan

Saxifraga brevicaulis Harry Sm.

Saxifraga brunneopunctata Harry Sm.

Saxifraga brunonis Wall. ex Ser.

Saxifraga bulleyana Engl. & Irmsch.

Saxifraga cacuminum Harry Sm.

Saxifraga candelabrum Franch.

Saxifraga cardiophylla Franch.

Saxifraga carnosula Mattf.

Saxifraga caveana var. **lanceolata** J. T. Pan

Saxifraga caveana W. W. Sm.

Saxifraga cernua L.

Saxifraga champutungensis H. Chuang

Saxifraga chionophila Franch.

Saxifraga chrysanthoides Engl. & Irmsch.

Saxifraga chumbiensis Engl. & Irmsch.

Saxifraga ciliatopetala (Engl. & Irmsch.) J. T. Pan

Saxifraga cinerascens Engl. & Irmsch.

Saxifraga cinerea Harry Sm.

Saxifraga clavistaminea Engl. & Irmsch.

Saxifraga clivorum Harry Sm.

Saxifraga coarctata W. W. Sm.

Saxifraga congestiflora Engl. & Irmsch.

Saxifraga consanguinea W. W. Sm.

Saxifraga contraria Harry Sm.

Saxifraga cordigera Hook. f. & Thomson

Saxifraga culcitosa Mattf.

Saxifraga daliensis H. Chuang

Saxifraga daochengensis J. T. Pan

Saxifraga davidii Franch.

Saxifraga decora Harry Sm.

Saxifraga decussata J. Anthony

Saxifraga densifoliata Engl. & Irmsch.

Saxifraga densifoliata var. **nedongensis** J. T. Pan

Saxifraga deqenensis C. Y. Wu

Saxifraga dhwojii (Harry Sm.) S. Akiyama & H. Ohba

Saxifraga dianxibeiensis J. T. Pan

Saxifraga diapensia Harry Sm.

Saxifraga dielsiana Engl. & Irmsch.

Saxifraga diffusicallosa C. Y. Wu

Saxifraga divaricata Engl. & Irmsch.

Saxifraga diversifolia Wall. ex Ser.

Saxifraga dongwanensis H. Chuang

Saxifraga doyalana Harry Sm.

Saxifraga drabiformis Franch.

Saxifraga draboides C. Y. Wu

Saxifraga dshagalensis Engl.

Saxifraga dungbooi Engl. & Irmsch.

Saxifraga eglandulosa Engl.

Saxifraga egregia Engl.

Saxifraga egregia var. **eciliata** J. T. Pan

Saxifraga egregia var. **xiaojinensis** J. T. Pan

Saxifraga egregioides J. T. Pan

Saxifraga elatinoides Hand.-Mazz.

Saxifraga elliotii Harry Sm.

Saxifraga elliptica Engl. & Irmsch.

Saxifraga engleriana Harry Sm.

Saxifraga epiphylla Gornall & H. Ohba

Saxifraga erectisepala J. T. Pan

Saxifraga erinacea Harry Sm.

Saxifraga excellens Harry Sm.

Saxifraga filicaulis Wall. ex Ser.

Saxifraga filifolia J. Anthony

Saxifraga finitima W. W. Sm.

Saxifraga flaccida J. T. Pan

Saxifraga flagellaris Willd. ex Sternb.

Saxifraga flagellaris subsp. **crassiflagellata** Hulten

Saxifraga flexilis W. W. Sm.

Saxifraga forrestii Engl. & Irmsch.

Saxifraga gageana W. W. Sm.

Saxifraga ganeshii H. Ohba & S. Akiyama

Saxifraga gedangensis J. T. Pan

Saxifraga gemmigera Engl.

Saxifraga gemmigera var. **gemmuligera** (Engl.) J. T. Pan & Gornall

Saxifraga gemmipara Franch.

Saxifraga georgei J. Anthony

Saxifraga giraldiana Engl.

Saxifraga glabricaulis Harry Sm.

Saxifraga glacialis Harry Sm.

Saxifraga glaucophylla Franch.

Saxifraga gonggashanensis J. T. Pan

Saxifraga gongshanensis T. C. Ku

Saxifraga gouldii C. E. C. Fisch.

Saxifraga gouldii var. **eglandulosa** Harry Sm.

Saxifraga granulifera Harry Sm.

Saxifraga gyalana C. Marquand & Airy Shaw

Saxifraga habaensis C. Y. Wu

Saxifraga haplophylloides Franch.

Saxifraga harai H. Ohba & Wakab.

Saxifraga heleonastes Harry Sm.

Saxifraga hemisphaerica Hook. f. & Thomson

Saxifraga hengduanensis H. Chuang

Saxifraga heteroclada var. **aurantia** Harry Sm.

Saxifraga heterocladoides J. T. Pan

Saxifraga heterotricha C. Marquand & Airy Shaw

Saxifraga heterotricha var. **anadena** (Harry Sm.) J. T. Pan & Gornall

Saxifraga hirculoides Decne.

Saxifraga hirculus L.

Saxifraga hirculus subsp. **alpina** (Engl.) Podlech

Saxifraga hispidula D. Don

Saxifraga hookeri Engl. & Irmsch.

Saxifraga hypericoides Franch.

Saxifraga hypericoides var. **aurantiascens** (Engl. & Irmsch.) J. T. Pan & Gornall

Saxifraga hypericoides var. **rockii** (Mattf.) J. T. Pan & Gornall

Saxifraga hypostoma Harry Sm.

Saxifraga imparilis Balf. f.

Saxifraga implicans Harry Sm.

Saxifraga insolens Irmsch.

Saxifraga isophylla Harry Sm.

Saxifraga jacquemontiana Decne.

Saxifraga jainzhuglaensis J. T. Pan

Saxifraga jaljalensis H. Ohba & S. Akiyama

Saxifraga kingdonii C. Marquand

Saxifraga kingiana Engl. & Irmsch.

Saxifraga komarovii Los.-Losinsk.

Saxifraga kongboensis Harry Sm.

Saxifraga kumaunensis Engl.

Saxifraga lamninamensis H. Ohba

Saxifraga latiflora Hook. f. & Thomson

Saxifraga lepida Harry Sm.

Saxifraga lepidostolonosa Harry Sm.

Saxifraga likiangensis Franch.

Saxifraga lilacina Duthie

Saxifraga linearifolia Engl. & Irmsch.

Saxifraga litangensis Engl.

Saxifraga lixianensis T. C. Ku

Saxifraga llonakhensis W. W. Sm.

Saxifraga loripes J. Anthony

Saxifraga lowndesii Harry Sm.

Saxifraga ludingensis J. T. Pan

Saxifraga ludlowii Harry Sm.

Saxifraga lumpuensis Engl.

Saxifraga lushuiensis H. Chuang

Saxifraga lychnitis Hook. f. & Thomson

Saxifraga macrostigmatoides Engl.

Saxifraga mallae H. Ohba & Wakab.

Saxifraga matta-viridis Harry Sm.

Saxifraga maxionggouensis J. T. Pan

Saxifraga medogensis J. T. Pan

Saxifraga meeboldii Engl. & Irmsch.

Saxifraga megacordia C. Y. Wu

Saxifraga melanocentra Franch.

Saxifraga micans Harry Sm.

Saxifraga microgyna Engl. & Irmsch.

Saxifraga microphylla Royle ex Hook. f. & Thomson

Saxifraga microviridis Hara

Saxifraga minutifoliosa C. Y. Wu

Saxifraga mira Harry Sm.

Saxifraga miralana Harry Sm.

Saxifraga monantha Harry Sm.

Saxifraga montanella Harry Sm.

Saxifraga montanella var. **retusa** J. T. Pan

Saxifraga moorcroftiana (Ser.) Wall. ex Sternb.

Saxifraga mucronulata Royle

Saxifraga mucronulatoides J. T. Pan

Saxifraga nakaoi Kitam.

Saxifraga nakaoides J. T. Pan
Saxifraga nambulana Harry Sm.
Saxifraga namdoensis Harry Sm.
Saxifraga nana Engler
Saxifraga nanella Engl. & Irmsch.
Saxifraga nanella var. **glabrisepala** J. T. Pan
Saxifraga nanelloides C. Y. Wu
Saxifraga nangqenica J. T. Pan
Saxifraga nangxianensis J. T. Pan
Saxifraga neopropagulifera Hara
Saxifraga nigroglandulifera N. P. Balakr.
Saxifraga nigroglandulosa Engl. & Irmsch.
Saxifraga omphalodifolia Hand.-Mazz.
Saxifraga oppositifolia L.
Saxifraga oreophila Franch.
Saxifraga oresbia J. Anthony
Saxifraga paiquensis J. T. Pan
Saxifraga pallida Wall. ex Ser.
Saxifraga palpebrata Hook. f. & Thomson
Saxifraga pardanthina Hand.-Mazz.
Saxifraga parkaensis J. T. Pan
Saxifraga parnassiifolia D. Don
Saxifraga parva Hemsl.
Saxifraga parvula Engl. & Irmsch.
Saxifraga pellucida C. Y. Wu
Saxifraga peplidifolia Franch.
Saxifraga peraristulata Mattf.
Saxifraga perpusilla Hook. f. & Thomson
Saxifraga pilifera Hook. f. & Thomson
Saxifraga poluninana Harry Sm.
Saxifraga pratensis Engl. & Irmsch.
Saxifraga prattii Engl. & Irmsch.
Saxifraga prattii var. **obtusata** Engl.
Saxifraga przewalskii Engl.
Saxifraga pseudohirculus Engl.
Saxifraga pseudo-pallida Engl. & Irmsch.
Saxifraga pseudoparvula H. Chuang
Saxifraga pulchra Engl. & Irmsch.
Saxifraga pulvinaria Harry Sm.
Saxifraga punctulata Engl.
Saxifraga punctulata var. **minuta** J. T. Pan
Saxifraga punctulatoides J. T. Pan
Saxifraga quadrifaria Engl. & Irmsch.
Saxifraga ramulosa Wall. ex Ser.
Saxifraga rhodopetala Harry Sm.
Saxifraga rizhaoshanensis J. T. Pan
Saxifraga rolwalingensis H. Ohba
Saxifraga rotundipetala J. T. Pan
Saxifraga roylei Harry Sm.
Saxifraga rubriflora Harry Sm.
Saxifraga rufescens Balf. f.
Saxifraga rupicola Franch.
Saxifraga saginoides Hook. f. & Thomson
Saxifraga sanguinea Franch.
Saxifraga saxatilis Harry Sm.
Saxifraga saxorum Harry Sm.
Saxifraga sediformis Engl. & Irmsch.
Saxifraga serrula Harry Sm.
Saxifraga sessiliflora Harry Sm.
Saxifraga sheqilaensis J. T. Pan
Saxifraga sherriffii Harry Sm.
Saxifraga sibirica L.
Saxifraga signata Engl. & Irmsch.
Saxifraga signatella C. Marquand
Saxifraga sikkimensis Engl.
Saxifraga sinomontana J. T. Pan & Gornall
Saxifraga sinomontana var. **amabilis** Harry Sm. ex J. T. Pan
Saxifraga smithiana Irmsch.
Saxifraga sphaeradena Harry Sm.
Saxifraga staintonii Harry Sm.
Saxifraga stella-aurea Hook. f. & Thomson
Saxifraga stella-aurea var. **macrostellata** H. Chuang
Saxifraga stellariifolia Franch.

Saxifraga stenophylla Royle

Saxifraga stolitzkae Duthie ex Engl. & Irmsch.

Saxifraga strigosa Wall. ex Ser.

Saxifraga strigosa var. **ramosa** (Engl. & Irmsch.) H. Chuang

Saxifraga subaequifoliata Irmsch.

Saxifraga subamplexicaulis Engl. & Irmsch.

Saxifraga sublinearifolia J. T. Pan

Saxifraga subomphalodifolia J. T. Pan

Saxifraga subsessiliflora Engl. & Irmsch.

Saxifraga subspathulata Engl. & Irmsch.

Saxifraga substrigosa J. T. Pan

Saxifraga subternata Harry Sm.

Saxifraga subtsangchanensis J. T. Pan

Saxifraga tangutica Engl.

Saxifraga tangutica var. **platyphylla** (Harry Sm.) J. T. Pan

Saxifraga taraktophylla C. Marquand & Airy Shaw

Saxifraga tatsienluensis Engl.

Saxifraga tentaculata C. E. C. Fisch.

Saxifraga tibetica Losinsk.

Saxifraga tigrina Harry Sm.

Saxifraga triaristulata Hand.-Mazz.

Saxifraga tsangchanensis Franch.

Saxifraga umbellulata Hook. f. & Thomson

Saxifraga umbellulata var. **muricola** (C. Marquand & Airy Shaw) J. T. Pan

Saxifraga umbellulata var. **pectinata** (C. Marquand & Airy Shaw) J. T. Pan

Saxifraga unguiculata Engl.

Saxifraga uninervia J. Anthony

Saxifraga valleculosa H. Chuang

Saxifraga versicallosa C. Y. Wu

Saxifraga viridipetala Z. X. Zhang & Gornall

Saxifraga wallichiana Sternb.

Saxifraga wardii W. W. Sm.

Saxifraga wardii var. **glabripedicellata** J. T. Pan

Saxifraga wenchuanensis T. C. Ku

Saxifraga williamsii Harry Sm.

Saxifraga xiaozhongdianensis J. T. Pan

Saxifraga yarlungzangboensis J. T. Pan

Saxifraga yezhiensis C. Y. Wu

Saxifraga yushuensis J. T. Pan

Saxifraga zayuensis T. C. Ku

Saxifraga zhidoensis J. T. Pan

Saxifraga zimmermannii Baehni

Schizophragma crassum Hand.-Mazz.

Schizophragma crassum var. **hsitaoanum** (Chun) C. F. Wei

Schizophragma integrifolium Oliver

Schizophragma integrifolium var. **glaucescens** Rehder

Tiarella polyphylla D. Don

125. CRASSULACEAE

Hylotelephium angustum (Maxim.) H. Ohba

Hylotelephium erythrostictum (Miq.) H. Ohba

Hylotelephium ewersii (Ledeb.) H. Ohba

Hylotelephium pakistanicum (G. R. Sarwar) Holub

Hylotelephium verticillatum (L.) H. Ohba

Kungia aliciae (Raym.-Hamet) K. T. Fu

Kungia aliciae var. **komarovii** (Raym.-Hamet) K. T. Fu

Kungia schoenlandii (Raym.-Hamet) K. T. Fu

Kungia schoenlandii var. **stenostachya** (Fröd.) K. T. Fu

Ohbaea balfourii (Raym.-Hamet) V. V. Byalt & I. V. Sokolova

Orostachys chanetii (H. Lév.) A. Berger

Orostachys fimbriata (Turcz.) A. Berger

Orostachys spinosa (L.) Sweet

Phedimus aizoon (L.) 't Hart

Phedimus aizoon var. **latifolius** (Maxim.) H. Ohba, K. T. Fu & B. M. Barthol.

Phedimus aizoon var. **scabrus** (Maxim.) H. Ohba, K. T. Fu & B. M. Barthol.

Rhodiola alsia (Fröd.) S. H. Fu

Rhodiola alterna S. H. Fu

Rhodiola amabilis (H. Ohba) H. Ohba

Rhodiola atropurpurea (Turcz.) Trautv. & Mey.

Rhodiola atsaensis (Fröd.) H. Ohba

Rhodiola atuntsuensis (Praeger) S. H. Fu

Rhodiola bupleuroides (Wall. ex Hook. f. & Thomson) S. H. Fu

Rhodiola calliantha (H. Ohba) H. Ohba

Rhodiola chrysanthemifolia (H. Lév.) S. H. Fu

Rhodiola coccinea (Royle) Boriss.

Rhodiola coccinea subsp. **scabrida** (Franch.) H. Ohba

Rhodiola crenulata (Hook. f. & Thomson) H. Ohba

Rhodiola cretinii (Raym.-Hamet) H. Ohba

Rhodiola cretinii subsp. **sinoalpina** (Fröd. ex Hand.-Mazz.) H. Ohba

Rhodiola discolor (Franch.) S. H. Fu

Rhodiola dumulosa (Franch.) S. H. Fu

Rhodiola fastigiata (Hook. f. & Thomson) S. H. Fu

Rhodiola forrestii (Raym.-Hamet) S. H. Fu

Rhodiola gannanica K. T. Fu

Rhodiola gelida Schrenk

Rhodiola handelii H. Ohba

Rhodiola heterodonta (Hook. f. & Thomson) Boriss.

Rhodiola himalensis (D. Don) S. H. Fu

Rhodiola himalensis var. **bouvieri** (Raym.-Hamet) Jacobsen

Rhodiola himalensis subsp. **taohoensis** (S. H. Fu) H. Ohba

Rhodiola hobsonii (Prain ex Raym.-Hamet) S. H. Fu

Rhodiola humilis (Hook. f. & Thomson) S. H. Fu

Rhodiola imbricata Edgew.

Rhodiola junggarica Chang Y. Yang & N. R. Cui

Rhodiola kansuensis (Fröd.) S. H. Fu

Rhodiola kaschgarica Boriss.

Rhodiola kirilowii (Regel) Maxim.

Rhodiola liciae (Raym.-Hamet) S. H. Fu

Rhodiola linearifolia (Royle) Fu

Rhodiola litwinowii Boriss.

Rhodiola macrocarpa (Praeger) S. H. Fu

Rhodiola multibracteata H. Chuang

Rhodiola nobilis (Franch.) S. H. Fu

Rhodiola ovatisepala (Raym.-Hamet) S. H. Fu

Rhodiola pachyclados (Aitch. & Hemsl.) H. Ohba

Rhodiola pamiroalaica Boriss.

Rhodiola prainii (Raym.-Hamet) H. Ohba

Rhodiola primuloides (Franch.) S. H. Fu

Rhodiola purpureoviridis (Praeger) S. H. Fu

Rhodiola purpureoviridis subsp. **phariensis** (H. Ohba) H. Ohba

Rhodiola quadrifida (Pall.) Schrenk

Rhodiola sacra (Prain ex Raym.-Hamet) S. H. Fu

Rhodiola sacra var. **tsuiana** (S. H. Fu) S. H. Fu

Rhodiola sedoides Lidén & Bharali

Rhodiola semenovii (Regel & Herder) Boriss.

Rhodiola serrata H. Ohba

Rhodiola sexifolia S. H. Fu

Rhodiola sherriffii H. Ohba

Rhodiola sinuata (Royle ex Edgew.) S. H. Fu

Rhodiola smithii (Raym.-Hamet) S. H. Fu

Rhodiola stapfii (Raym.-Hamet) S. H. Fu

Rhodiola subopposita (Maxim.) Jacobsen

Rhodiola tangutica (Maxim.) S. H. Fu

Rhodiola tibetica (Hook. f. & Thomson) S. H. Fu

Rhodiola wallichiana (Hook.) S. H. Fu

Rhodiola wallichiana var. **cholaensis** (Praeger) S. H. Fu

Rhodiola yunnanensis (Franch.) S. H. Fu

Rosularia adenotricha (Wall. ex Edgew.) Jansson & Rech

Rosularia adenotricha subsp. **chitralica** G. R. Sarwar

Rosularia adenotricha subsp. **viguieri** (Wall. ex Edgew.) Jansson

Rosularia alpestris (Kar. & Kir.) Borissova

Rosularia alpestris subsp. marnieri (R.-Hamet ex H. Ohba) Eggli

Rosularia glabra (Regel & Winkler) Berger

Rosularia modesta (Bornm.) Parsa

Rosularia modesta var. **linearifolia** (C.-A. Jansson) Eggli

Rosularia platyphylla (Schrenk) Berger

Rosularia rosulata (Edgew.) H. Ohba

Rosularia sedoides (Decne.) H. Ohba

Sedum alfredii Hance

Sedum barbeyi Raym.-Hamet

Sedum beauverdii Raym.-Hamet

Sedum blepharophyllum Fröd.

Sedum celatum Fröd.

Sedum celiae Raym.-Hamet

Sedum chauveaudii Raym.-Hamet

Sedum chauveaudii var. **margaritae** (Raym.-Hamet) Fröd.

Sedum concarpum Fröd.

Sedum correptum Fröd.

Sedum costantinii Raym.-Hamet

Sedum daigremontianum Raym.-Hamet

Sedum dielsii Raym.-Hamet

Sedum dimorphophyllum K. T. Fu & G. Y. Rao

Sedum dugueyi Raym.-Hamet

Sedum elatinoides Franch.

Sedum emarginatum Migo

Sedum engleri Raym.-Hamet

Sedum engleri var. **dentatum** S. H. Fu

Sedum erici-magnusii Fröd.

Sedum erici-magnusii subsp. **chilianense** K. T. Fu

Sedum fedtschenkoi Raym.-Hamet

Sedum filipes Hemsl.

Sedum fischeri Raym.-Hamet

Sedum forrestii Raym.-Hamet

Sedum franchetii Grande

Sedum fui Rowley

Sedum fui var. **longisepalum** (K. T. Fu) S. H. Fu

Sedum gagei Raym.-Hamet

Sedum giajae Raym.-Hamet

Sedum glaebosum Fröd.

Sedum heckelii Raym.-Hamet

Sedum hengduanense K. T. Fu

Sedum hispanicum L.

Sedum holei Raym.-Hamet

Sedum jacquemontii Fröd.

Sedum japonicum Siebold ex Miq.

Sedum kingdonii H. Ohba

Sedum leblancae Raym.-Hamet

Sedum leucocarpum Franch.

Sedum longifuniculatum K. T. Fu

Sedum longyanense K. T. Fu

Sedum luchuanicum K. T. Fu

Sedum lutzii Raym.-Hamet

Sedum lutzii var. **viridiflavum** K. T. Fu

Sedum magae Raym.-Hamet

Sedum magniflorum K. T. Fu

Sedum majus (Hemsl.) Migo

Sedum morotii Raym.-Hamet

Sedum morotii var. **pinoyi** (Raym.-Hamet) Fröd.

Sedum multicaule Wall. ex Lindl.

Sedum multicaule subsp. **rugosum** K. T. Fu

Sedum muyaicum K. T. Fu

Sedum nothodugueyi K. T. Fu

Sedum obtusipetalum Franch.

Sedum oligospermum Maire

Sedum oreades (Decne.) Raym.-Hamet

Sedum pagetodes Fröd.

Sedum perpusillum Hook f. & Thomson

Sedum platysepalum Franch.

Sedum prasinopetalum Fröd.

Sedum przewalskii Maxim.

Sedum pseudo-multicaule H. Ohba

Sedum purdomii W. W. Sm.

Sedum ramentaceum K. T. Fu

Sedum raymondii Fröd.

Sedum roborowskii Maxim.

Sedum rosei Raym.-Hamet

Sedum rosei var. **magniflorum** Fröd.

Sedum sagittipetalum Fröd.

Sedum sarmentosum Bunge

Sedum semilunatum K. T. Fu

Sedum sinoglaciale K. T. Fu

Sedum somenii Raym.-Hamet

Sedum stellariifolium Franch.

Sedum stimulosum K. T. Fu

Sedum subgaleatum K. T. Fu

Sedum susannae Raym.-Hamet

Sedum susannae var. **macrosepalum** K. T. Fu

Sedum triactina A. Berger

Sedum triactina subsp. **leptum** Fröd.

Sedum trichospermum K. T. Fu

Sedum trullipetalum Hook. f. & Thomson

Sedum tsinghaicum K. T. Fu

Sedum tsonanum K. T. Fu

Sedum ulricae Fröd.

Sedum victorianum C.-A. Jansson

Sedum woronowii Raym.-Hamet

Sinocrassula ambigua (Praeger) A. Berger

Sinocrassula densirosulata (Praeger) A. Berger

Sinocrassula diversifolia H. Chuang

Sinocrassula indica (Decne.) A. Berger

Sinocrassula indica var. **forrestii** (Raym.-Hamet) S. H. Fu

Sinocrassula indica var. **luteorubra** (Praeger) S. H. Fu

Sinocrassula indica var. **obtusifolia** (Fröd.) S. H. Fu

Sinocrassula indica var. **serrata** (Raym.-Hamet) S. H. Fu

Sinocrassula indica var. **viridiflora** K. T. Fu

Sinocrassula longistyla (Praeger) S. H. Fu

Sinocrassula techinensis (S. H. Fu) S. H. Fu

Sinocrassula yunnanensis (Franch.) A. Berger

Tillaea alata Viviani

Tillaea likiangensis H. Chuang

Tillaea schimperi (Fisch. & C. A. Mey.) M. G. Gilbert, H. Ohba & K. T. Fu

126. PENTHORACEAE

Penthorum chinense Pursh

127. HALORAGACEAE

Gonocarpus micranthus Thunb.

Myriophyllum spicatum L.

Myriophyllum verticillatum L.

128. VITACEAE

Ampelocissus artemisiifolia Planch.

Ampelocissus barbata (Wall.) Planch.

Ampelocissus butoensis C. L. Li

Ampelocissus divaricata (M. A. Lawson) Planch.

Ampelocissus latifolia (Roxb.) Planch.

Ampelocissus rugosa (Wall.) Planch.

Ampelocissus sikkimensis (M. A. Lawson) Planch.

Ampelocissus xizangensis C. L. Li

Ampelopsis aconitifolia Bunge

Ampelopsis acutidentata W. T. Wang

Ampelopsis bodinieri (H. Lév. & Vaniot) Rehder

Ampelopsis delavayana Planch. ex Franch.

Ampelopsis glandulosa (Wall.) Momiy.

Ampelopsis mollifolia W. T. Wang

Ampelopsis tomentosa Planch. ex Franch.

Ampelopsis vitifolia (Boiss.) Planch.

Causonis japonica (Thunb.) Raf.

Causonis trifolia (L.) Mabb. & J. Wen

Cayratia cardiospermoides (Planch. ex Franch.) Gagnep.

Cayratia daliensis C. L. Li

Cayratia fugongensis C. L. Li

Cayratia geniculata (Blume) Gegnep.

Cayratia medogensis C. L. Li

Cayratia mollissima (Planch.) Gagnep.

Cayratia pedata (Lam.) Gagnep.

Cissus adnata Roxb.

Cissus aristata Blume

Cissus assamica (M. A. Lawson) Craib

Cissus discolor Blume

Cissus elongata Roxb.

Cissus heyneana Planch.

Cissus javana DC.

Cissus quadrangularis L.

Cissus repanda Vahl

Cissus repens Lam.

Cissus spectabilis (Kurz) Planch.

Cyphostemma auriculatum (Roxb.) P. Singh & B. V. Shetty

Nekemias gongshanensis (C. L. Li) J. Wen & Z. L. Nie

Nekemias rubifolia (Wall.) J. Wen & Z. L. Nie

Parthenocissus chinensis C. L. Li

Parthenocissus semicordata (Roxb.) Planch.

Tetrastigma affine (Gagnep. ex Osmaston) Raizada & H. O. Saxena

Tetrastigma angustifolium (Roxb.) Planch.

Tetrastigma aplinianum (Collett & Hemsl.) Momiy.

Tetrastigma bracteolatum (Wall.) Planch.

Tetrastigma campylocarpum (Kurz) Planch.

Tetrastigma delavayi Gagnep.

Tetrastigma dubium (M. A. Lawson) Planch.

Tetrastigma hemsleyanum Diels & Gilg

Tetrastigma henryi Gagnep.

Tetrastigma hookeri (M. A. Lawson) Planch.

Tetrastigma hypoglaucum Planch.

Tetrastigma lanceolarium (Roxb.) Planch.

Tetrastigma leucostaphylum (Dennst.) Alston

Tetrastigma obovatum Gagnep.

Tetrastigma obtectum (Wall. ex M. A. Lawson) Planch. ex Franch.

Tetrastigma obtectum var. **glabrum** (H. Lév.) Gagnep.

Tetrastigma planicaule (Hook. f.) Gagnep.

Tetrastigma rumicispermum (M. A. Lawson) Planch.

Tetrastigma serrulatum (Roxb.) Planch.

Tetrastigma thomsonianum Planch.

Tetrastigma triphyllum (Gagnep.) W. T. Wang

Tetrastigma triphyllum var. **hirtum** (Gagnep.) W. T. Wang

Tetrastigma xizangense C. L. Li

Tetrastigma yunnanense Gagnep.

Tetrastigma yunnanense var. **mollissimum** C. Y. Wu ex W. T. Wang

Vitis betulifolia Diels & Gilg

Vitis davidii (Rom. Caill.) Foex

Vitis flexuosa Thunb.

Vitis heyneana Roem. & Schult.

Vitis jacquemontii R. Parker

Vitis pedicellata M. A. Lawson

Vitis vinifera L.

Vitis wilsoniae H. J. Veitch

Yua thomsonii (M. A. Lawson) C. L. Li

129. LEEACEAE

Leea aequata L.

Leea alata Edgew.

Leea asiatica (L.) Ridsdale

Leea compactiflora Kurz

Leea erecta Voll & Brade

Leea guineensis G. Don

Leea indica (Burm. f.) Merr.

Leea macrophylla Roxb. ex Hornem.

130. ZYGOPHYLLACEAE

Balanites aegyptiaca Delile

Balanites roxburghii Planch.

Fagonia bruguieri DC.

Fagonia bruguieri var. **laxa** DC.

Fagonia bruguieri var. **rechingeri** DC.

Fagonia olivieri DC.

Tribulus longipetalus subsp. **macropterus** Viv.

Tribulus terrestris L.

Zygophyllum atriplicoides Fisch. & C. A. Mey.

Zygophyllum bucharicum B. Fedtsch.

Zygophyllum eurypterum subsp. **gontscharowii** Boiss. & Buhse

Zygophyllum fabago L.

Zygophyllum fabago subsp. **dolichocarpum** L.

Zygophyllum fabago subsp. **orientale** L.

Zygophyllum heterocladum Rech. f. & Patzak.

Zygophyllum miniatum Cham & Schlechtend

Zygophyllum miniatum var. **englerianum** Cham & Schlechtend

Zygophyllum obliquum Popov

131. FABACEAE

Abrus precatorius L.

Abrus pulchellus Wall. ex Thwaites

Acacia auriculiformis A. Cunn. ex Benth.

Acacia confusa Merr.

Acacia dealbata Link

Acacia decurrens Willd.

Acacia mearnsii De Wild.

Acacia microcephala Graham ex Wall.

Acacia pruinescens Kurz

Acrocarpus fraxinifolius Arn.

Adenanthera pavonina L.

Aeschynomene indica L.

Albizia arunachalensis K. C. Sahni & H. B. Naithani

Albizia chinensis (Osbeck) Merr.

Albizia crassiramea Lace

Albizia duclouxii Gagnep.

Albizia garrettii I. C. Nielsen

Albizia julibrissin Durazz.

Albizia kalkora (Roxb.) Prain

Albizia lebbeck (L.) Benth.

Albizia lucidior (Steud.) I. C. Nielsen ex Hara

Albizia mollis (Wall.) Boivin

Albizia myriophylla Benth.

Albizia odoratissima (L. f.) Benth.

Albizia procera (Roxb.) Benth.

Albizia sherriffii Baker f.

Alhagi maurorum Medik.

Alhagi nepalensis (D. Don) Shap.

Alysicarpus heyneanus var. **meeboldii** (Schindl.) A. Pramanik & Thoth.

Alysicarpus rugosus (Willd.) DC.

Alysicarpus vaginalis (L.) DC.

Alysicarpus yunnanensis Y. C. Yang & P. H. Huang

Amphicarpaea bracteata (L.) Fernald

Amphicarpaea edgeworthii Benth.

Amphicarpaea ferruginea Benth.

Amphicarpaea ferruginea var. **bracteosa** (Prain) H. Ohashi & Tateishi

Apios carnea (Wall.) Benth. ex Baker

Apios dclavayi Franch.

Apios macrantha Oliver

Arachis hypogaea L.

Archidendron alternifoliolatum (T. L. Wu) I. C. Nielsen

Archidendron arunachalense S. S. Dash & Sanjappa

Archidendron bigeminum (L.) I. C. Nielsen

Archidendron chevalieri (Kosterm.) I. C. Nielsen

Archidendron clypearia (Jack) I. C. Nielsen

Archidendron ellipticum (Blume) I. C. Nielsen

Archidendron kerrii (Gagnep.) I. C. Nielsen

Archidendron laoticum (Gagnep.) I. C. Nielsen

Archidendron lucidum (Benth.) I. C. Nielsen

Archidendron nielsenianum S. S. Dash & Sanjappa

Archidendron robinsonii (Gagnep.) I. C. Nielsen

Archidendron xichouense (C. Chen & H. Sun) X. Y. Zhu

Argyrolobium flaccidum (Royle) Jaub. & Spach

Argyrolobium roseum (Camb.) Jaub. & Spach

Astragalus affghanus Boiss.

Astragalus alitschuri O. Fedtsch.

Astragalus amherstianus Royle ex Benth.

Astragalus ammophilus var. **persepolitanus** (Boiss.) Ali

Astragalus arnoldii Hemsl. & H. Pearson

Astragalus austrotibetanus Podlech & L. R. Xu

Astragalus bakeri Ali

Astragalus batangensis E. Peter

Astragalus baxoiensis Podlech & L. R. Xu

Astragalus bhotanensis Baker

Astragalus bicuspis Fisch.

Astragalus bouffordii Podlech

Astragalus brachytrichus Podlech & L. R. Xu

Astragalus brevialatus H. T. Tsai & T. T. Yu

Astragalus breviscapus B. Fedtsch.

Astragalus caeruleopetalinus Y. C. Ho

Astragalus candolleanus Royle ex Benth.

Astragalus chadjanensis Franch.

Astragalus chagyabensis P. C. Li & C. C. Ni

Astragalus changduensis Y. C. Ho

Astragalus changmuicus C. C. Ni & P. C. Li

Astragalus charguschanus Freyn

Astragalus chateri Vassilcz.

Astragalus chilienshanensis Y. C. Ho

Astragalus chitralensis Ali

Astragalus chiukiangensis H. T. Tsai & T. T. Yu

Astragalus chlorostachys Lindl.

Astragalus chrysopterus Bunge

Astragalus clarkeanus Ali

Astragalus cobresiiphilus Podlech & L. R. Xu

Astragalus coluteocarpus Boiss.

Astragalus complanatus subsp. **eutrichus** (Hand.-Mazz.) Wenn.

Astragalus compositus Pavlov

Astragalus conaensis Podlech & L. R. Xu

Astragalus concretus Benth.

Astragalus confertus Benth. ex Bunge

Astragalus craibianus N. D. Simpson

Astragalus crassifolius Ulbr.

Astragalus damxungensis Podlech & L. R. Xu

Astragalus davidii Franch.

Astragalus degensis Ulbr.

Astragalus densiflorus Kar. & Kir.

Astragalus despectus Podlech & L. R. Xu

Astragalus dickorei Podlech & L. R. Xu

Astragalus dignus Boriss.

Astragalus dingjiensis C. C. Ni & P. C. Li

Astragalus droshensis Ali

Astragalus dulungkiangensis P. C. Li

Astragalus dumetorum Hand.-Mazz.

Astragalus edelbergianus Širj. & Rech. f.

Astragalus emodi Steud.

Astragalus ernestii Comber

Astragalus euchlorus K. T. Fu

Astragalus falconeri Bunge

Astragalus floridulus Podlech

Astragalus floridus Bunge

Astragalus forrestii N. D. Simpson

Astragalus frigidus (L.) A. Gray

Astragalus gahiratensis Ali

Astragalus gilgitensis Ali

Astragalus gongshanensis Podlech & L. R. Xu

Astragalus gracilipes Benth. ex Bunge

Astragalus grahamianus Royle ex Benth.

Astragalus graveolens Buch.-Ham. ex Benth.

Astragalus handelii H. T. Tsai & T. T. Yu

Astragalus himalayanus Klotzsch

Astragalus hoffmeisteri (Klotzsch) Ali

Astragalus hosackioides (Royle ex Benth.) Benth.

Astragalus hysophilus Podlech & L. R. Xu

Astragalus imitensis Ali

Astragalus iskanderi Lipsky

Astragalus jabbor-khailii Kitam.

Astragalus jiazaensis Podlech & L. R. Xu

Astragalus kashmirensis Bunge

Astragalus khasianus Bunge

Astragalus kialensis N. D. Simpson

Astragalus kongrensis Baker

Astragalus kuschakewiczii B. Fedtsch.

Astragalus laetabilis Podlech & L. R. Xu

Astragalus lamalaensis C. C. Ni

Astragalus lasiosemius Boiss.

Astragalus laspurensis Ali

Astragalus latiunguiculatus Y. C. Ho

Astragalus laxmannii Jacq.

Astragalus lessertioides Bunge

Astragalus leucocephalus Grah. ex Benth.

Astragalus lhorongensis P. C. Li & C. C. Ni

Astragalus licentianus Hand.-Mazz.

Astragalus longilobus E. Peter

Astragalus longiscapus C. C. Ni & P. C. Li

Astragalus lowarensis Ali

Astragalus lucidus H. T. Tsai & T. T. Yu

Astragalus luteolus H. T. Tsai & T. T. Yu

Astragalus macropterus DC.

Astragalus maddenianus Benth. ex Baker

Astragalus mahoschanicus Hand.-Mazz.

Astragalus malacophyllus Benth. ex Bunge

Astragalus malcolmii Hemsl. & H. Pearson

Astragalus maowensis Podlech & L. R. Xu
Astragalus matiensis P. C. Li
Astragalus maxwellii Royle ex Benth.
Astragalus mekongensis Podlech
Astragalus melanostachys Bunge
Astragalus melilotoides Pallas
Astragalus minutidentatus Y. C. Ho
Astragalus mirabilis Lipsky
Astragalus miyalomontis P. C. Li
Astragalus monadelphus Bunge
Astragalus monbeigii N. D. Simpson
Astragalus mongholicus Bunge
Astragalus montivagus Podlech & L. R. Xu
Astragalus moupinensis Franch.
Astragalus muliensis Hand.-Mazz.
Astragalus multiceps Wall. ex Benth.
Astragalus munroi Benth. ex Bunge
Astragalus nakaoi Kitam.
Astragalus nanellus H. T. Tsai & T. T. Yu
Astragalus nanfengensis C. C. Ni
Astragalus nangxianensis P. C. Li & C. C. Ni
Astragalus neomonodelphus H. T. Tsai & T. T. Yu
Astragalus nivalis Kar. & Kir.
Astragalus ophiocarpus Benth. ex Bunge
Astragalus oplites Benth. ex R. Parker
Astragalus orbicularifolius P. C. Li & C. C. Ni
Astragalus orbiculatus Ledeb.
Astragalus oreocharis Podlech & L. R. Xu
Astragalus oxyodon Baker
Astragalus peduncularis Royle ex Benth.
Astragalus persepolitanus Boiss.
Astragalus peterae H. T. Tsai & T. T. Yu
Astragalus pindreensis (Baker) Ali
Astragalus polyacanthus Benth.
Astragalus polycladus Bureau & Franch.
Astragalus porphyrocalyx Y. C. Ho
Astragalus praeteritus Podlech & L. R. Xu
Astragalus przewalskii Bunge
Astragalus pseudochlorostachys Ali
Astragalus pseudomahoschanicus Podlech
Astragalus pseudoversicolor Y. C. Ho
Astragalus psilocentros Fisch.
Astragalus psilocentros var. **pilosus** Parker
Astragalus pullus N. D. Simpson
Astragalus retamocarpus Boiss. & Hohen.
Astragalus retusifoliatus Y. C. Ho
Astragalus rhizanthus Royle ex Benth.
Astragalus rhizocephalus subsp. **hindukushensis** (Wendelbo) Podlech
Astragalus rhododendrophilus Podlech & L. R. Xu
Astragalus rigidulus Bunge
Astragalus schachdarinus Lipsky
Astragalus scheremetevianus B. Fedtsch.
Astragalus sedaensis Y. C. Ho
Astragalus semicircularis P. C. Li
Astragalus sichuanensis L. Meng, X. Y. Zhu & P. K. Hsiao
Astragalus sikkimensis Bunge
Astragalus skorniakovii B. Fedtsch.
Astragalus skythropos Bunge
Astragalus smithianus E. Peter
Astragalus souliei N. D. Simpson
Astragalus staintonianus Ali
Astragalus stewartii Baker
Astragalus stipulatus D. Don ex Sims
Astragalus strictus Grah. ex Benth.
Astragalus subscaposus M. Popov ex A.G. Borisova
Astragalus subuliformis DC.
Astragalus sungpanensis E. Peter
Astragalus supralaevis Podlech & L. R. Xu
Astragalus sutchuenensis Franch.
Astragalus tecti-mundi Freyn

Astragalus thomsonii Podlech
Astragalus tibetanus Bunge
Astragalus tibeticola Podlech & L. R. Xu
Astragalus tongolensis Ulbr.
Astragalus toppinianus Ali
Astragalus trichocarpus Graham
Astragalus tsangpoensis Podlech & L. R. Xu
Astragalus tumbatsicus C. Marquand & Airy Shaw
Astragalus verus Olivier
Astragalus webbianus Graham ex Benth.
Astragalus weixinensis Y. C. Ho
Astragalus wolungensis P. C. Li
Astragalus yangii C. Chen & Zi G. Qian
Astragalus yunnanensis Franch.
Astragalus yunningensis H. T. Tsai & T. T. Yu
Astragalus zadaensis Podlech & L. R. Xu
Astragalus zanskarensis Benth. ex Bunge
Astragalus zayuensis C. C. Ni & P. C. Li
Astragalus zemuensis W. W. Sm.
Bauhinia acuminata L.
Bauhinia bohniana L. Chen
Bauhinia brachycarpa Wall. ex Benth.
Bauhinia comosa Craib
Bauhinia delavayi Franch.
Bauhinia divergens Baker
Bauhinia glauca (Benth.) Benth.
Bauhinia khasiana Baker
Bauhinia malabarica Roxb.
Bauhinia ornata var. **kerrii** (Gagnep.) K. Larsen & S. S. Larsen
Bauhinia ovatifolia T. Chen
Bauhinia purpurea L.
Bauhinia scandens L.
Bauhinia semla Wunderlin
Bauhinia vahlii Wight & Arn.
Bauhinia variegata L.
Bauhinia wallichii J. F. Macbr.
Bauhinia yunnanensis Franch.
Butea buteiformis (Voigt) Grierson & D. G. Long
Butea minor Baker
Butea monosperma (Lam.) Taub.
Butea xizangensis Z. Y. Zhu & Y. F. Du
Caesalpinia bonduc (L.) Roxb.
Caesalpinia cucullata Roxb.
Caesalpinia decapetala (Roth) Alston
Caesalpinia digyna Rottler
Caesalpinia enneaphylla Roxb.
Caesalpinia minax Hance
Caesalpinia pulcherrima (L.) Sw.
Caesalpinia tortuosa Roxb.
Cajanus cajan (L.) Millsp.
Cajanus elongatus (Benth.) Maesen
Cajanus goensis Dalzell
Cajanus grandiflorus (Benth. ex Baker) Maesen
Cajanus mollis (Benth.) Maesen
Cajanus platycarpus (Benth.) Maesen
Cajanus scarabaeoides (L.) Thouars
Cajanus villosus (Baker) Maesen
Cajanus volubilis (Blanco) Blanco
Callerya cinerea (Benth.) Schot
Calliandra umbrosa (Wall.) Benth.
Campylotropis alopochroa H. Ohashi
Campylotropis brevifolia Ricker
Campylotropis capillipes (Franch.) Schindl.
Campylotropis delavayi (Franch.) Schindl.
Campylotropis drummondii Schindl.
Campylotropis eriocarpa (DC.) Schindl.
Campylotropis hirtella (Franch.) Schindl.
Campylotropis howellii Schindl.
Campylotropis luhitensis H. Ohashi
Campylotropis macrostyla (D. Don) Schindl.
Campylotropis meeboldii (Schindl.) Schindl.
Campylotropis pauciflora C. J. Chen

Campylotropis polyantha (Franch.) Schindl.

Campylotropis sargentiana Schindl.

Campylotropis speciosa subsp. **eriocarpa** (Schindl.) Iokawa & H. Ohashi

Campylotropis stenocarpa (Klotzsch) Schindl.

Campylotropis teretiracemosa P. C. Li & C. J. Chen

Campylotropis trigonoclada (Franch.) Schindl.

Campylotropis wilsonii Schindl.

Campylotropis yunnanensis (Franch.) Schindl.

Canavalia ensiformis (L.) DC.

Caragana aegacanthoides (R. Parker) L. B. Chaudhary & S. K. Srivastava

Caragana aliensis Y. Z. Zhao

Caragana aurantiaca Koehne

Caragana bicolor Kom.

Caragana boisii C. K. Schneid.

Caragana brevifolia Kom.

Caragana brevispina Royle ex Benth.

Caragana campanulata Vassilcz.

Caragana changduensis Y. X. Liou

Caragana chinghaiensis Y. X. Liou

Caragana chumbica Prain

Caragana conferta Benth. ex Baker

Caragana crassispina C. Marquand

Caragana cuneatoalata Y. X. Liou

Caragana decorticans Hemsl.

Caragana densa Kom.

Caragana erinacea Kom.

Caragana franchetiana Kom.

Caragana franchetiana var. **gyirongensis** (C. C. Ni) Y. X. Liou

Caragana gerardiana Benth.

Caragana jubata (Pallas) Poir.

Caragana jubata var. **czetyrkininii** (Sanchir) Y. X. Liou

Caragana jubata var. **recurva** Y. X. Liou

Caragana junatovii Gorbunova

Caragana kozlowii Kom.

Caragana limprichtii Harms

Caragana opulens Kom.

Caragana polyacantha Royle

Caragana rosea Turcz. ex Maxim.

Caragana sukiensis C. K. Schneid.

Caragana tangutica Maxim.

Caragana tibetica (Maxim. ex C. K. Schneid.) Kom.

Caragana tragacanthoides var. **himalaica** Kom.

Caragana versicolor Benth.

Cassia fistula L.

Cassia grandis L. f.

Cassia javanica L.

Cassia javanica subsp. **nodosa** (Roxb.) K. Larsen & S. S. Larsen

Ceratonia siliqua L.

Cercis glabra Pamp.

Cercis griffithii Boiss.

Chamaecrista auricoma (Benth.) V. Singh

Chamaecrista leschenaultiana (DC.) O. Deg.

Chamaecrista mimosoides (L.) Greene

Chamaecrista nictitans subsp. **patellaria** (Collad.) H. S. Irwin & Barneby

Chamaecrista nomame (Sieber) H. Ohashi

Chamaecrista pumila (Lam.) K. Larsen

Chesneya acaulis (Baker) Popov

Chesneya crassipes Boriss.

Chesneya cuneata (Benth.) Ali

Chesneya depressa (Oliver) Popow

Chesneya nubigena (D. Don) Ali

Chesneya nubigena subsp. **purpurea** (P. C. Li) X. Y. Zhu

Chesneya polystichoides (Hand.-Mazz.) Ali

Chesneya polystichoides subsp. **bhutanica** H. Ohashi

Chesneya spinosa P. C. Li

Cicer arietinum L.

Cicer macracanthum Popow

Cicer microphyllum Royle ex Benth.

Cicer nuristanicum Kitam.

Cladrastis delavayi (Franch.) Prain

Clitoria mariana L.

Clitoria ternatea L.

Cochlianthus gracilis Benth.

Cochlianthus gracilis var. **brevipes** C. F. Wei

Codariocalyx gyroides (Link) Hassk.

Codariocalyx microphyllus (Thunb.) H. Ohashi

Codariocalyx motorius (Houtt.) H. Ohashi

Codariocalyx motorius var. **glaber** X. Y. Zhu & Y. F. Du

Colutea delavayi Franch.

Colutea multiflora Ali

Colutea nepalensis Sims

Colutea paulsenii Freyn

Colutea paulsenii subsp. **mesantha** (Shap. ex Ali) Ali

Corethrodendron multijugum (Maxim.) B. H. Choi & H. Ohashi

Craspedolobium schochii Harms

Crotalaria acicularis Buch.-Ham. ex Benth.

Crotalaria alata Buch.-Ham. ex D. Don

Crotalaria albida Heyne ex Roth

Crotalaria bracteata Roxb. ex DC.

Crotalaria calycina Schrank

Crotalaria capitata Benth. ex Lam.

Crotalaria chinensis L.

Crotalaria cytisoides DC.

Crotalaria ferruginea Graham ex Benth.

Crotalaria humifusa Graham ex Benth.

Crotalaria juncea L.

Crotalaria kanaii H. Ohashi

Crotalaria khasiana Thoth. & A. A. Ansari

Crotalaria laburnifolia L.

Crotalaria linifolia L. f.

Crotalaria mairei H. Lév.

Crotalaria medicaginea Lam.

Crotalaria medicaginea var. **luxurians** (Benth.) Baker

Crotalaria mysorensis Roth

Crotalaria nana var. **patula** Baker

Crotalaria occulta Graham ex Benth.

Crotalaria pallida Aiton

Crotalaria prostrata Rottler ex Willd.

Crotalaria psoraleoides D. Don

Crotalaria quinquefolia L.

Crotalaria retusa L.

Crotalaria sessiliflora L.

Crotalaria spectabilis Roth

Crotalaria tetragona Roxb. ex Andrews

Crotalaria verrucosa L.

Crotalaria yunnanensis Franch.

Crotalaria yunnanensis var. **heqingensis** (C. Y. Yang) C. Chen & J. Q. Li

Cullen corylifolium (L.) Medik.

Cyamopsis tetragonoloba (L.) Taub.

Dalbergia assamica Benth.

Dalbergia clarkei Thoth.

Dalbergia lanceolaria L. f.

Dalbergia latifolia Roxb.

Dalbergia mimosoides Franch.

Dalbergia oliveri Gamble ex Prain

Dalbergia pinnata (Lour.) Prain

Dalbergia rimosa Roxb.

Dalbergia sericea G. Don

Dalbergia sissoo Roxb. ex DC.

Dalbergia stipulacea Roxb.

Dalbergia volubilis Roxb.

Dalhousiea bracteata (Roxb.) Benth.

Delonix regia (Bojer) Raf.

Dendrolobium triangulare (Retz.) Schindl.

Derris acuminata Benth.

Derris cuneifolia Benth.

Derris ferruginea (Roxb.) Benth.

Derris hainesiana Thoth.

Derris harrowiana (Diels) Z. Wei

Derris marginata (Roxb.) Benth.

Derris monticola (Kurz) Prain

Derris polystachya Benth.

Derris pseudorobusta Thoth.

Derris robusta (DC.) Benth.

Derris scabricaulis (Franch.) Gagnep.

Derris scandens (Roxb.) Benth.

Derris secunda Baker

Desmanthus pernambucanus (L.) Thell.

Desmodium concinnum DC.

Desmodium diffusum DC.

Desmodium gangeticum (L.) DC.

Desmodium griffithianum Benth.

Desmodium heterocarpon (L.) DC.

Desmodium heterocarpon var. **strigosum** Meeuwen

Desmodium heterophyllum (Willd.) DC.

Desmodium hispidum Franch.

Desmodium laxiflorum DC.

Desmodium oblongum Wall. ex Benth.

Desmodium oojeinense (Roxb.) H. Ohashi

Desmodium renifolium (L.) Schindl.

Desmodium tiliaceum Wall.

Desmodium triflorum (L.) DC.

Desmodium velutinum (Willd.) DC.

Desmodium villosum (Mill.) DC.

Desmodium williamsii H. Ohashi

Dichrostachys cinerea (L.) Wight & Arn.

Dolichos staintonii H. Ohashi & Tateishi

Dolichos tenuicaulis (Baker) Craib

Dumasia cordifolia Benth. ex Baker

Dumasia forrestii Diels

Dumasia villosa DC.

Dumasia yunnanensis Y. T. Wei & S. K. Lee

Dunbaria rotundifolia (Lour.) Merr.

Dysolobium grande (Benth.) Prain

Dysolobium lucens (Benth.) Prain

Dysolobium pilosum (Klein ex Willd.) Maréchal

Entada phaseoloides (L.) Merr.

Entada rheedii Spreng.

Entada rheedii subsp. **sinohimalensis** (Grierson & D. G. Long) Panigrahi

Eriosema chinense Vogel

Erythrina arborescens Roxb.

Erythrina stricta Roxb.

Erythrina suberosa Roxb.

Euchresta horsfieldii (Lesch.) Benn.

Faba vulgaris Moench

Flemingia bhutanica Grierson

Flemingia bracteata (Roxb.) Wight

Flemingia chappar Buch.-Ham. ex Benth.

Flemingia congesta Roxb. ex W. T. Aiton

Flemingia fluminalis C. B. Clarke ex Prain

Flemingia fruticulosa Wall. ex Benth.

Flemingia macrophylla (Willd.) Merr.

Flemingia paniculata Wall. ex Benth.

Flemingia semialata Roxb. ex W. T. Aiton

Flemingia strobilifera (L.) W. T. Aiton

Galactia tenuiflora (Willd.) Wight & Arn.

Geissaspis cristata Wight & Arn.

Gleditsia assamica Bor

Gleditsia japonica var. **delavayi** (Franch.) L. C. Li

Gleditsia sinensis Lam.

Glycine max (L.) Merr.

Glycine soja Siebold & Zucc.

Glycyrrhiza glabra L.

Glycyrrhiza uralensis Fisch.

Glycyrrhiza yunnanensis S. H. Cheng & L. K. Dai ex P. C. Li

Gueldenstaedtia verna (Georgi) Boriss.

Gymnocladus assamicus P. C. Kanjilal

Gymnocladus chinensis Baill.

Hedysarum algidum L. Z. Shue

Hedysarum algidum var. **speciosum** (Hand.-Mazz.) Y. H. Wu

Hedysarum alii Haidar Ali & Qaiser

Hedysarum alpinum L.

Hedysarum astragaloides Benth. ex Baker

Hedysarum brahuicum Boiss.

Hedysarum cachemirianum Benth. ex Baker

Hedysarum campylocarpon H. Ohashi

Hedysarum citrinum Baker f.

Hedysarum cuonanum P. L. Liu, J. Wen & Zhao Y. Chang

Hedysarum falconeri Baker

Hedysarum ferganense var. **minjanense** (K. H. Rechinger) L. Z. Shue

Hedysarum fistulosum Hand.-Mazz.

Hedysarum jinchuanense L. Z. Shue

Hedysarum kumaonense Benth. ex Baker

Hedysarum laxiflorum Benth. ex Baker

Hedysarum limitaneum Hand.-Mazz.

Hedysarum longigynophorum C. C. Ni

Hedysarum manaslense (Kitam.) H. Ohashi

Hedysarum manaslense var. **nepalense** (H. Ohashi) H. Ohashi

Hedysarum microcalyx Baker

Hedysarum minjanense Rech. f.

Hedysarum nagarzense C. C. Ni

Hedysarum polybotrys Hand.-Mazz.

Hedysarum pseudoastragalus Ulbr.

Hedysarum sericeum M. Bieb.

Hedysarum shahjinalense Haidar Ali & Qaiser

Hedysarum sikkimense Benth. ex Baker

Hedysarum tanguticum B. Fedtsch.

Hedysarum thiochroum Hand.-Mazz.

Hedysarum tibeticum (Benth.) B. H. Choi & H. Ohashi

Hedysarum volkii Rech. f.

Hedysarum wakhanicum Podlech & Anders

Hedysarum xizangense C. C. Ni

Hylodesmum laxum (DC.) H. Ohashi & R. R. Mill

Hylodesmum longipes (Franch.) H. Ohashi & R. R. Mill

Hylodesmum podocarpum (DC.) H. Ohashi & R. R. Mill

Hylodesmum podocarpum subsp. **oxyphyllum** (DC.) H. Ohashi & R. R. Mill

Hylodesmum podocarpum subsp. **szechuenense** (Craib) H. Ohashi & R. R. Mill

Hylodesmum repandum (Vahl) H. Ohashi & R. R. Mill

Hylodesmum williamsii (H. Ohashi) H. Ohashi & R. R. Mill

Indigofera acutipetala Y. Y. Fang & C. Z. Zheng

Indigofera amblyantha Craib

Indigofera argentea Burm. f.

Indigofera argutidens Craib

Indigofera arrecta Hochst. ex A. Rich.

Indigofera astragalina DC.

Indigofera atropurpurea Hornem.

Indigofera balfouriana Craib

Indigofera bella Prain

Indigofera bracteata Baker

Indigofera bracteata var. **khasiana** Sanjappa

Indigofera bungeana Walp.

Indigofera calcicola Craib

Indigofera cassioides Rottler ex DC.

Indigofera caudata Dunn

Indigofera cedrorum Dunn

Indigofera chaetodonta Franch.

Indigofera colutea (Burm. f.) Merr.

Indigofera constricta (Thwaites) Trimen

Indigofera cordifolia B. Heyne ex Roth

Indigofera cylindracea Graham ex Baker

Indigofera delavayi Franch.

Indigofera dichroa Craib

Indigofera dolichochaete Craib

Indigofera dosua Buch.-Ham. ex D. Don

Indigofera dosua var. **simlensis** Sanjappa

Indigofera dumetorum Craib

Indigofera exilis Grierson & D. G. Long

Indigofera franchetii X. F. Gao & Schrire

Indigofera gangetica Sanjappa

Indigofera glabra L.

Indigofera hamiltonii Graham ex Duthie & Prain

Indigofera hancockii Craib

Indigofera hebepetala Benth. ex Baker

Indigofera hebepetala var. **glabra** Ali

Indigofera hendecaphylla Jacq.

Indigofera henryi Craib

Indigofera heterantha Wall. ex Brandis

Indigofera himachalensis V. D. Chauhan & A. K. Pandey

Indigofera himalayensis Ali

Indigofera hirsuta L.

Indigofera howellii Craib & W. W. Sm.

Indigofera lenticellata Craib

Indigofera linifolia (L. f.) Retz.

Indigofera linnaei Ali

Indigofera mairei Pamp.

Indigofera mengtzeana Craib

Indigofera muliensis Y. Y. Fang & C. Z. Zheng

Indigofera myosurus Craib

Indigofera neosericopetala P. C. Li

Indigofera nigrescens Kurz ex King & Prain

Indigofera pampaniniana Craib

Indigofera pendula Franch.

Indigofera penduloides Y. Y. Fang & C. Z. Zheng

Indigofera prostrata Willd.

Indigofera pseudonigrescens X. F. Gao and Xue Li Zhao

Indigofera pseudoreticulata Grierson & D. G. Long

Indigofera reticulata Franch.

Indigofera rigioclada Craib

Indigofera sensitiva Franch.

Indigofera sesquipedalis Sanjappa

Indigofera silvestrii Pamp.

Indigofera squalida Prain

Indigofera stachyodes Lindl.

Indigofera sticta Craib

Indigofera suffruticosa Mill.

Indigofera szechuensis Craib

Indigofera tinctoria L.

Indigofera trifoliata L.

Indigofera trifoliata var. **nepalensis** H. Ohashi

Indigofera trita L. f.

Indigofera wightii Graham ex Wight & Arn.

Indigofera wilsonii Craib

Indopiptadenia oudhensis (Brandis) Brenan

Kummerowia striata (Thunb.) Schindl.

Lablab purpureus (L.) Sweet

Lathyrus aphaca L.

Lathyrus davidii Hance

Lathyrus dielsianus Harms

Lathyrus emodii (Fritsch.) Ali

Lathyrus erectus Lag.

Lathyrus humilis (Ser.) Spreng.

Lathyrus palustris var. **exalatus** (H. P. Tsui) X. Y. Zhu

Lathyrus palustris var. **linearifolius** Ser.

Lathyrus pratensis L.

Lathyrus sativus L.

Lathyrus sphaericus Retz.

Lens culinaris Medik.

Lespedeza cuneata (Dum. Cours.) G. Don

Lespedeza davidii Franch.

Lespedeza elegans Cambess.

Lespedeza elliptica Benth.

Lespedeza fasciculiflora Franch.

Lespedeza fasciculiflora var. **hengduanshanensis** C. J. Chen

Lespedeza floribunda Bunge

Lespedeza forrestii Schindler

Lespedeza gerardiana Wall. ex Maxim.

Lespedeza griffithii (Schindl.) H. Ohashi

Lespedeza hispida (Franch.) T. Nemoto & H. Ohashi

Lespedeza inschanica (Maxim.) Schindl.

Lespedeza juncea (L. f.) Pers.

Lespedeza pilosa (Thunb.) Siebold & Zucc.

Lespedeza potaninii V. N. Vassil.

Lespedeza speciosa Royle ex Schindl.

Lespedeza tomentosa (Thunb.) Siebold ex Maxim.

Leucaena leucocephala (Lam.) de Wit

Lotus corniculatus L.

Lotus corniculatus var. **japonicus** Regel

Lotus krylovii Schischk. & Serg.

Macrotyloma uniflorum (Lam.) Verdc.

Maniltoa polyandra (Roxb.) Harms

Mastersia assamica Benth.

Medicago archiducis-nicolaii Širjaev

Medicago falcata L.

Medicago lupulina L.

Medicago minima (L.) L.

Medicago monantha (C. A. Mey.) Trautv.

Medicago orbicularis (L.) Bartal.

Medicago polymorpha L.

Medicago ruthenica (L.) Trautv.

Medicago sativa L.

Medicago varia Martyn

Melilotus albus Medik.

Melilotus indicus (L.) All.

Melilotus officinalis (L.) Pall.

Millettia caudata (Benth.) Baker

Millettia extensa (Benth.) Baker

Millettia fruticosa (DC.) Baker

Millettia glaucescens Kurz

Millettia nepalensis R. Parker

Millettia pachycarpa Benth.

Millettia piscidia (Roxb.) Wight

Millettia pulchra Kurz

Millettia vellutina Dunn

Mimosa pudica L.

Mimosa rubicaulis Lam.

Mimosa rubicaulis subsp. **himalayana** (Gamble) H. Ohashi

Mucuna calophylla W. W. Sm.

Mucuna gigantea (Willd.) DC.

Mucuna macrocarpa Wall.

Mucuna nigricans (Lour.) Steud.

Mucuna pruriens (L.) DC.

Mucuna pruriens var. **utilis** (Wall. ex Wight) L. H. Bailey

Mucuna sempervirens Hemsl.

Ohwia caudata (Thunb.) H. Ohashi

Onobrychis cornuta (L.) Desv.

Onobrychis dasycephala Baker ex Aitch.

Onobrychis dealbata Stocks

Onobrychis laxioflora Baker

Onobrychis spinosissima Baker ex Aitch.

Onobrychis stewartii Baker

Onobrychis tanaitica Spreng.

Onobrychis viciifolia Scop.

Ononis arvensis L.

Ormosia assamica Yakovlev

Ormosia robusta Baker

Ototropis calliantha (Franch.) H. Ohashi & K. Ohashi

Ototropis conferta (DC.) H. Ohashi & K. Ohashi

Ototropis elegans (DC.) H. Ohashi & K. Ohashi

Ototropis elegans subsp. **wolohoensis** (Schindl.) H. Ohashi & K. Ohashi

Ototropis elegans var. **handelii** (Schindl.) H. Ohashi & K. Ohashi

Ototropis khasiana (Prain) H. Ohashi & K. Ohashi

Ototropis kulhaitensis (C. B. Clarke ex Prain) H. Ohashi & K. Ohashi

Ototropis likabalia (Bennet & Sum. Chandra) H. Ohashi & K. Ohashi

Ototropis multiflora (DC.) H. Ohashi & K. Ohashi

Ototropis sequax (Wall.) H. Ohashi & K. Ohashi

Ototropis stenophylla (Pamp.) H. Ohashi & K. Ohashi

Ototropis yunnanensis (Franch.) H. Ohashi & K. Ohashi

Oxytropis arenae-ripariae Vassilcz.

Oxytropis assiensis Vassilcz.

Oxytropis auriculata C. W. Chang

Oxytropis avisoides P. C. Li

Oxytropis baxoiensis P. C. Li

Oxytropis bella B. Fedtsch.

Oxytropis biflora P. C. Li

Oxytropis brevipedunculata P. C. Li

Oxytropis cachemiriana Cambess.

Oxytropis chiliophylla Royle ex Benth.

Oxytropis chinglingensis C. W. Chang

Oxytropis chitralensis Ali

Oxytropis cinerascens Bunge

Oxytropis crassiuscula Boriss.

Oxytropis deflexa (Pall.) DC.

Oxytropis densa Benth. ex Bunge

Oxytropis densiflora P. C. Li

Oxytropis densiflora var. **multiramosa** (P. C. Li) X. Y. Zhu & H. Ohashi

Oxytropis duthieana Ali

Oxytropis falcata Bunge

Oxytropis fasciculiflorum Vassilcz.

Oxytropis gerzeensis P. C. Li

Oxytropis gilgitensis Vassilcz.

Oxytropis giraldii Ulbr.

Oxytropis glabra DC.

Oxytropis globiflora Bunge

Oxytropis gorbunovii Boriss.

Oxytropis graminetorum Vassilcz.

Oxytropis hirsutiuscula Freyn

Oxytropis humifusa Kar. & Kir.

Oxytropis hypoglottoides (Baker) Ali

Oxytropis imbricata Kom.

Oxytropis immersa (Aitch.) B. Fedtsch.

Oxytropis kansuensis Bunge

Oxytropis kazidanica Vassilcz.

Oxytropis lapponica (Wahlenb.) Gay

Oxytropis latibracteata Jurtzev

Oxytropis lehmannii Bunge

Oxytropis lhasaensis X. Y. Zhu

Oxytropis longialata P. C. Li

Oxytropis malloryana Dunn

Oxytropis meinshausenii Schrenk

Oxytropis melanocalyx Bunge

Oxytropis merkensis Bunge

Oxytropis microphylla (Pall.) DC.

Oxytropis mollis Royle ex Benth.

Oxytropis morenarum Vassilcz.

Oxytropis nepalensis Vassilcz.

Oxytropis ochrantha Turcz.

Oxytropis ochrocephala Bunge

Oxytropis ochrolongibracteata X. Y. Zhu & H. Ohashi

Oxytropis oreophila A. Gray

Oxytropis pagobia Bunge

Oxytropis parasericeopetala P. C. Li

Oxytropis pauciflora Bunge

Oxytropis platonychia Bunge

Oxytropis platysema Schrenk

Oxytropis podoloba Kar. & Kir.

Oxytropis poincinsii Franch.

Oxytropis proboscidea Bunge

Oxytropis pseudocoerulea P. C. Li

Oxytropis pusilla Bunge

Oxytropis qamdoensis X. Y. Zhu, Y. F. Du & H. Ohashi

Oxytropis qilianshanica C. W. Chang & C. L. Zhang ex X. Y. Zhu & H. Ohashi

Oxytropis qingnanensis Y. H. Wu

Oxytropis reniformis P. C. Li

Oxytropis riparia Litv.

Oxytropis salicetorum Vassilcz.

Oxytropis savellanica Boiss.

Oxytropis sericopetala Prain ex C. E. C. Fisch.

Oxytropis sichuanica C. W. Chang

Oxytropis stracheyana Bunge

Oxytropis subpodoloba P. C. Li

Oxytropis taochensis Kom.

Oxytropis tatarica Baker

Oxytropis tianschanica Bunge

Oxytropis torrentium Vassilcz.

Oxytropis trichosphaera Freyn

Oxytropis tudanensis X. Y. Zhu, H. Ohashi & Si Feng Li

Oxytropis vakhdzhiri Vassilcz.

Oxytropis williamsii Vassilcz.

Oxytropis yunnanensis Franch.

Pachyrhizus erosus (L.) Urb.

Paracalyx scariosus (Roxb.) Ali

Parkia timoriana (DC.) Merr.

Parochetus communis Buch.-Ham. ex D. Don

Peltophorum pterocarpum (DC.) K. Heyne

Phaseolus lunatus L.

Phaseolus vulgaris L.

Phyllodium pulchellum (L.) Desv.

Phyllolobium balfourianum (N. D. Simpson) M. L. Zhang & Podlech

Phyllolobium camptodontum (Franch.) M. L. Zhang & Podlech

Phyllolobium chapmanianum (Wenn.) M. L. Zhang & Podlech

Phyllolobium dolichochaete (Diels) M. L. Zhang & Podlech

Phyllolobium donianum (DC.) M. L. Zhang & Podlech

Phyllolobium enneaphyllum (P. C. Li) M. L. Zhang & Podlech

Phyllolobium eutrichus (Hand.-Mazz.) M. L. Zhang & Podlech

Phyllolobium flavovirens (K. T. Fu) M. L. Zhang & Podlech

Phyllolobium heydei (Baker) M. L. Zhang & Podlech

Phyllolobium lachungense (L. B. Chaudhary) Podlech

Phyllolobium lasaense (C. C. Ni & P. C. Li) M. L. Zhang & Podlech

Phyllolobium lineariauriferum (P. C. Li) M. L. Zhang & Podlech

Phyllolobium milingense (C. C. Ni & P. C. Li) M. L. Zhang & Podlech

Phyllolobium pastorium (H. T. Tsai & T. T. Yu) M. L. Zhang & Podlech

Phyllolobium prodigiosum (K. T. Fu) M. L. Zhang & Podlech

Phyllolobium sanbilingense (H. T. Tsai & T. T. Yu) M. L. Zhang & Podlech

Phyllolobium siccaneum (P. C. Li) M. L. Zhang & Podlech

Phyllolobium sichuanense Podlech

Phyllolobium tanguticum (Batalin) X. Y. Zhu

Phyllolobium tingriense (C. C. Ni & P. C. Li) M. L. Zhang & Podlech

Phyllolobium tribulifolium (Benth. ex Bunge) M. L. Zhang & Podlech

Phyllolobium turgidocarpum (K. T. Fu) M. L. Zhang & Podlech

Piptanthus nepalensis (Hook.) Sweet

Piptanthus tomentosus Franch.

Pisum sativum L.

Pithecellobium dulce L.

Podolotus hosackioides Benth.

Pongamia pinnata (L.) Pierre

Prosopis juliflora (Sw.) DC.

Pterocarpus marsupium Roxb.

Pterolobium hexapetalum Kurz

Pterolobium punctatum Hemsl.

Pueraria bella Prain

Pueraria edulis Pamp.

Pueraria montana var. **lobata** (Willd.) Maesen & S. M. Almeida ex Sanjappa & Predeep

Pueraria montana var. **thomsonii** (Benth.) M. R. Almeida

Pueraria peduncularia (Benth.) Benth.

Pueraria phaseoloides (Roxb.) Benth.

Pueraria sikkimensis Prain

Pueraria tuberosa (Willd.) DC.

Pueraria wallichii DC.

Rhynchosia falconeri Baker

Rhynchosia harae H. Ohashi & Tateishi

Rhynchosia himalensis Benth. ex Baker

Rhynchosia himalensis var. **craibiana** (Rehder) E. Peter

Rhynchosia minima (L.) DC.

Rhynchosia nepalensis H. Ohashi & Tateishi

Rhynchosia pseudo-cajan Camb.

Rhynchosia sericea Span.

Rhynchosia viscosa (Roth) DC.

Rhynchosia yunnanensis Franch.

Salweenia wardii Baker f.

Saraca asoca (Roxb.) Willd.

Senegalia caesia (L.) Maslin, Seigler & Ebinger

Senegalia catechu (L. f.) P. J. H. Hurter & Mabb.

Senegalia delavayi (Franch.) Maslin, Seigler & Ebinger

Senegalia gageana (Craib) Maslin, Seigler & Ebinger

Senegalia intsia (L.) Maslin, Seigler & Ebinger

Senegalia lenticularis (Buch.-Ham ex Benth.) Ragup., Seigler, Ebinger & Maslin

Senegalia megaladena (Desv.) Maslin, Seigler & Ebinger

Senegalia modesta (Wall.) P. J. H. Hurter

Senegalia pennata (L.) Maslin

Senegalia pennata subsp. **hainanensis** (Hayata) Maslin, Seigler & Ebinger

Senegalia pruinescens (Kurz) Maslin, Seigler & Ebinger

Senegalia rugata (Lam.) Britton & Rose

Senegalia teniana (Harms) Maslin, Seigler & Ebinger

Senegalia torta (Roxb.) Maslin, Seigler & Ebinger

Senegalia yunnanensis (Franch.) Maslin, Seigler & Ebinger

Senna alata (L.) Roxb.

Senna floribunda (Cav.) H. S. Irwin & Barneby

Senna hirsuta (L.) H. S. Irwin & Barneby

Senna italica subsp. **micrantha** (Brenan) Lock

Senna occidentalis (L.) Link

Senna spectabilis (DC.) H. S. Irwin & Barneby

Senna tora (L.) Roxb.

Sesbania bispinosa (Jacq.) W. Wight

Sesbania cannabina (Retz.) Pers.

Sesbania grandiflora (L.) Pers.

Sesbania rostrata Bremek. & Oberm.

Sesbania sesban (L.) Merr.

Shuteria ferruginea (Kurz) Baker

Shuteria hirsuta Baker

Shuteria involucrata (Wall.) Wight & Arn.

Shuteria involucrata var. **glabrata** (Wight & Arn.) H. Ohashi

Shuteria vestita Wight & Arn.

Smithia ciliata Royle

Smithia conferta Sm.

Smithia grandis Baker

Smithia sensitiva Aiton

Sophora albescens (Rehder) C. Y. Ma
Sophora alopecuroides L.
Sophora benthamii Steenis
Sophora davidii (Franch.) Skeels
Sophora davidii var. **chuansiensis** (C. Y. Ma) C. Y. Ma ex B. J. Bao & Vincent
Sophora dunnii Prain
Sophora flavescens Aiton
Sophora mollis (Royle) Baker
Sophora moorcroftiana (Benth.) Benth. ex Baker
Sophora velutina Lindl.
Sophora wightii Baker
Sophora wilsonii Craib
Spatholobus parviflorus (Roxb. ex DC.) Kuntze
Spongiocarpella nubigena (D. Don) Yakovlev
Tadehagi pseudotriquetrum (DC.) H. Ohashi
Tadehagi triquetrum (L.) H. Ohashi
Tamarindus indicus L.
Tephrosia candida (Roxb.) DC.
Tephrosia purpurea (L.) Pers.
Thermopsis alpina (Pallas) Ledeb.
Thermopsis barbata Benth.
Thermopsis gyirongensis S. Q. Wei
Thermopsis inflata Cambess.
Thermopsis lanceolata R. Br.
Thermopsis przewalskii Czefr.
Thermopsis smithiana E. Peter
Thermopsis yushuensis S. Q. Wei
Tibetia forrestii (Ali) P. C. Li
Tibetia himalaica (Baker) H. P. Tsui
Tibetia tongolensis (Ulbr.) H. P. Tsui
Tibetia yadongensis H. P. Tsui
Tibetia yunnanensis (Franch.) H. P. Tsui
Tibetia yunnanensis var. **coelestis** (Diels) X. Y. Zhu
Trifolium alexandrinum L.
Trifolium dubium Sibth.
Trifolium fragiferum L.
Trifolium pratense L.
Trifolium repens L.
Trifolium resupinatum L.
Trigonella cachemiriana Cambess.
Trigonella corniculata (L.) L.
Trigonella emodi Benth.
Trigonella foenum-graecum L.
Trigonella gracilis Benth.
Trigonella pubescens Edgew. ex Baker
Uraria crinita (L.) DC.
Uraria lagopodoides (L.) DC.
Uraria lagopus DC.
Uraria picta (Jacq.) Desv. ex DC.
Uraria rufescens (DC.) Schindl.
Uraria sinensis (Hemsl.) Franch.
Vachellia eburnea (L. f.) P. J. H. Hurter & Mabb.
Vachellia farnesiana (L.) Wight & Arn.
Vachellia leucophloea (Roxb.) Maslin, Seigler & Ebinger
Vachellia nilotica subsp. **indica** (Benth.) Kyal. & Boatwr.
Vachellia pseudoeburnea (J. R. Drumm. ex Dunn) Ragup., Seigler, Ebinger & Maslin
Vicia amoena Fisch. ex Ser.
Vicia angustifolia L.
Vicia bakeri Ali
Vicia benthamiana Ali
Vicia bungei Ohwi
Vicia chinensis Franch.
Vicia costata Ledeb.
Vicia cracca L.
Vicia dichroantha Diels
Vicia faba L.
Vicia hirsuta (L.) Gray
Vicia megalotropis Ledeb.

Vicia monantha Retz.

Vicia multicaulis Ledeb.

Vicia narbonensis L.

Vicia nummularia Hand.-Mazz.

Vicia perelegans K. T. Fu

Vicia pseudo-orobus Fisch. & C. A. Mey.

Vicia rigidula Royle

Vicia sativa L.

Vicia sepinum L.

Vicia sinogigantea B. J. Bao & Turland

Vicia subvillosa (Ledeb.) Boiss.

Vicia tenuifolia Roth

Vicia ternata Z. D. Xia

Vicia tetrasperma (L.) Schreb.

Vicia tibetica Prain ex C. E. C. Fisch.

Vicia unijuga A. Braun

Vicia variegata subsp. **himalaica** Radzhi

Vicia villosa Roth

Vicia wushanica Z. D. Xia

Vigna angularis (Willd.) Ohwi & H. Ohashi

Vigna angularis var. **nipponensis** (Ohwi) Ohwi & H. Ohashi

Vigna clarkei Prain

Vigna mungo (L.) Hepper

Vigna nepalensis Tateishi & Maxted

Vigna radiata (L.) R. Wilczek

Vigna umbellata (Thunb.) Ohwi & H. Ohashi

Vigna unguiculata (L.) Walp.

Vigna unguiculata subsp. **cylindrica** (L.) Verdc.

Vigna unguiculata subsp. **sesquipedalis** (L.) Verdc.

Vigna vexillata var. **angustifolia** (Schumach. & Thonn.) Baker

Zornia diphylla (L.) Pers.

Zornia gibbosa Span.

132. POLYGALACEAE

Epirixanthes elongata Blume

Polygala abyssinica R. Br. ex Fresen.

Polygala arillata Buch.-Ham. ex D. Don

Polygala bomiensis S. K. Chen & J. Parnell

Polygala chinensis L.

Polygala crotalarioides Buch.-Ham. ex DC.

Polygala didyma C. Y. Wu

Polygala erioptera DC.

Polygala furcata Royle

Polygala globulifera Dunn

Polygala globulifera var. **kachinensis** (Mukerjee) R. N. Banerjee

Polygala globulifera var. **longiracemosa** S. K. Chen

Polygala irregularis Boiss.

Polygala karensium Kurz

Polygala lhunzeensis C. Y. Wu & S. K. Chen

Polygala lijiangensis C. Y. Wu & S. K. Chen

Polygala linariifolia Poir.

Polygala longifolia Poir.

Polygala mariesii Hemsl. ex Forb. & Hemsl.

Polygala persicariifolia DC.

Polygala sibirica L.

Polygala sibirica var. **elegans** (Wall. ex Royle) Hara

Polygala sibirica var. **megalopha** Franch.

Polygala stocksiana Boiss.

Polygala tatarinowii Regel

Polygala tenuifolia Willd.

Polygala tricholopha Chodat

Polygala triflora L.

Polygala wattersii Hance

Polygala wenxianensis Y. S. Zhou & Z. X. Peng

Salomonia cantoniensis Lour.

Salomonia ciliata (L.) DC.

Securidaca inappendiculata Hassk.

Xanthophyllum burkillii J. R. Drumm. & Dunn

Xanthophyllum flavescens Roxb.

Xanthophyllum griffithii Hook. f. ex A. W. Benn.

133. ROSACEAE

Acomastylis elata (Wall. ex G. Don) F. Bolle

Acomastylis elata var. **humilis** (Royle) F. Bolle

Acomastylis macrosepala (Ludlow) T. T. Yu & C. L. Li

Agrimonia pilosa Ledeb.

Agrimonia pilosa var. **nepalensis** (D. Don) Nakai

Alchemilla glabra Neygenf.

Alchemilla japonica Nakai & Hara

Argentina anserina (L.) Rydb.

Argentina aristata (Soják) Soják

Argentina cardotiana (Hand.-Mazz.) Soják

Argentina commutata (Lehm.) Soják

Argentina contigua (Soják) Y. H. Tong & N. H. Xia

Argentina fallens (Cardot) Soják

Argentina festiva (Soják) Soják

Argentina glabriuscula (T. T. Yu & C. L. Li) Soják

Argentina gombalana (Hand.-Mazz.) Soják

Argentina interrupta (T. T. Yu & C. L. Li) Soják

Argentina leuconota (D. Don) Soják

Argentina lineata (Trevir.) Soják

Argentina micropetala (D. Don) Soják

Argentina microphylla (D. Don) Soják

Argentina peduncularis (D. Don) Soják

Argentina phanerophlebia (Yu & Li) T. Feng & H. Wang

Argentina polyphylla (Wall. ex Lehm.) Soják

Argentina polyphylloides (H. Ikeda & H. Ohba) Y. H. Tong & N. H. Xia

Argentina sericophylla (R. Parker) Soják

Argentina smithiana (Hand.-Mazz.) Soják

Argentina songzhuensis T. Feng & Heng C. Wang

Argentina stenophylla (Franch.) Soják

Argentina tapetodes (Soják) Soják

Argentina tristis (Soják) Soják

Argentina turfosa (Hand.-Mazz.) Soják

Argentina wenchuensis (H. Ikeda & H. Ohba) Y. H. Tong & N. H. Xia

Aruncus dioicus subsp. **triternatus** (Maxim.) Hara

Aruncus gombalanus (Hand.-Mazz.) Hand.-Mazz.

Aruncus sylvester Kostel. ex Maxim.

Chaenomeles cathayensis (Hemsl.) C. K. Schneid.

Chaenomeles speciosa (Sweet) Nakai

Chaenomeles thibetica T. T. Yu

Coluria longifolia Maxim.

Coluria oligocarpa (J. Krause) F. Bolle

Cotoneaster acuminatus Lindl.

Cotoneaster acutifolius Turcz.

Cotoneaster adpressus Bois

Cotoneaster affinis Lindl.

Cotoneaster ambiguus Rehder & E. H. Wilson

Cotoneaster apiculatus Rehder & E. H. Wilson

Cotoneaster assamensis G. Klotz

Cotoneaster bullatus Bois

Cotoneaster buxifolius var. **rockii** (G. Klotz) L. T. Lu & A. R. Brach

Cotoneaster buxifolius Wall. ex Lindl.

Cotoneaster cashmiriensis G. Klotz

Cotoneaster chengkangensis T. T. Yu

Cotoneaster conspicuus Comber ex Marquand

Cotoneaster coriaceus Franch.

Cotoneaster dammeri C. K. Schneid.

Cotoneaster dammeri var. **radicans** (Dammer ex C. K. Schneid.) C. K. Schneid.

Cotoneaster dielsianus E. Pritz.

Cotoneaster divaricatus Rehder & E. H. Wilson

Cotoneaster foveolatus Rehder & E. H. Wilson

Cotoneaster franchetii Bois

Cotoneaster frigidus Wall. ex Lindl.

Cotoneaster garhwalensis G. Klotz

Cotoneaster gilgitensis G. Klotz

Cotoneaster glabratus Rehder & E. H. Wilson

Cotoneaster glaucophyllus Franch.

Cotoneaster glomerulatus W. W. Sm.

Cotoneaster gracilis Rehder & E. H. Wilson

Cotoneaster harrysmithii Flinck & B. Hylmö

Cotoneaster hebephyllus Diels

Cotoneaster hebephyllus var. **incanus** W. W. Sm.

Cotoneaster hebephyllus var. **majusculus** W. W. Sm.

Cotoneaster horizontalis Decne.

Cotoneaster integrifolius (Roxb.) G. Klotz

Cotoneaster langei G. Klotz

Cotoneaster microphyllus Wall. ex Lindl.

Cotoneaster moupinensis Franch.

Cotoneaster multiflorus Bunge

Cotoneaster multiflorus var. **atropurpureus** T. T. Yu

Cotoneaster nitens Rehder & E. H. Wilson

Cotoneaster nitidifolius Marquand

Cotoneaster nitidus Jacques

Cotoneaster nitidus var. **duthieanus** (C. K. Schneid.) T. T. Yu

Cotoneaster nitidus var. **parvifolius** (T. T. Yu) T. T. Yu

Cotoneaster nivalis (G. Klotz) Panigrahi & Arv. Kumar

Cotoneaster obscurus Rehder & E. H. Wilson

Cotoneaster pannosus Franch.

Cotoneaster pannosus var. **robustior** W. W. Sm.

Cotoneaster pojarkovae Zakirov

Cotoneaster racemiflorus (Desf.) K. Koch

Cotoneaster racemiflorus var. **royleanus** Dippel

Cotoneaster reticulatus Rehder & E. H. Wilson

Cotoneaster rotundifolius var. **lanatus** (Jacques) C. K. Schneid.

Cotoneaster rotundifolius Wall. ex Lindl.

Cotoneaster rubens var. **minimus** T. T. Yu

Cotoneaster rubens W. W. Sm.

Cotoneaster salicifolius Franch.

Cotoneaster soongoricus (Regel & Herder) Popov

Cotoneaster subadpressus T. T. Yu

Cotoneaster submultiflorus Popov

Cotoneaster symondsii T. Moore

Cotoneaster tenuipes Rehder & E. H. Wilson

Cotoneaster turbinatus Craib

Cotoneaster verruculosus Diels

Cotoneaster wardii W. W. Sm.

Crataegus oresbia W. W. Sm.

Crataegus wilsonii Sarg.

Cydonia oblonga Mill.

Dasiphora dryadanthoides Juz.

Dichotomanthes tristaniicarpa Kurz

Docynia delavayi (Franch.) C. K. Schneid.

Docynia indica (Wall.) Decne.

Duchesnea chrysantha (Zoll. & Moritzi) Miq.

Duchesnea indica (Andrews) Focke

Duchesnea indica var. **microphylla** T. T. Yu & T. C. Ku

Eriobotrya bengalensis (Roxb.) Hook. f.

Eriobotrya dubia (Lindl.) Decne.

Eriobotrya elliptica Lindl.

Eriobotrya fragrans Champ. ex Benth.

Eriobotrya hookeriana Decne.

Eriobotrya japonica (Thunb.) Lindl.

Eriobotrya petiolata Hook. f.

Eriobotrya salwinensis Hand.-Mazz.

Eriobotrya tengyuehensis W. W. Sm.

Filipendula vestita (Wall. ex G. Don) Maxim.

Fragaria daltoniana J. Gay

Fragaria gracilis Losinsk.

Fragaria moupinensis (Franch.) Cardot

Fragaria nilgerrensis Schltdl. ex J. Gay

Fragaria nubicola (Hook. f.) Lindl. ex Lacaita

Fragaria vesca L.

Geum aleppicum Jacq.

Geum elatum Wall. ex G. Don

Geum japonicum var. **chinense** F. Bolle

Geum roylei Wall. ex F. Bolle

Geum sikkimense Prain

Kerria japonica (L.) DC.

Malus baccata (L.) Borkh.

Malus chitralensis Vassilcz.

Malus daochengensis C. L. Li

Malus halliana Koehne

Malus kansuensis (Batalin) C. K. Schneid.

Malus muliensis T. C. Ku

Malus ombrophila Hand.-Mazz.

Malus prattii (Hemsl.) C. K. Schneid.

Malus pumila Mill.

Malus sikkimensis (Wenzig) Koehne

Malus toringoides (Rehder) Hughes

Malus transitoria (Batalin) C. K. Schneid.

Malus transitoria var. **glabrescens** T. T. Yu & T. C. Ku

Malus yunnanensis (Franch.) C. K. Schneid.

Malus yunnanensis var. **veitchii** (Osborn) Rehder

Neillia affinis Hemsl.

Neillia affinis var. **polygyna** Cardot ex J. E. Vidal

Neillia breviracemosa T. C. Ku

Neillia densiflora T. T. Yu & T. C. Ku

Neillia fugongensis T. C. Ku

Neillia gracilis Franch.

Neillia ribesioides Rehder

Neillia rubiflora D. Don

Neillia sinensis Oliv.

Neillia sparsiflora Rehder

Neillia thibetica Bureau & Franch.

Neillia thrysiflora D. Don

Neillia thyrsiflora var. **tunkinensis** (Vidal) Vidal

Neillia velutina Pendry

Osteomeles schwerinae C. K. Schneid.

Osteomeles schwerinae var. **microphylla** Rehder & E. H. Wilson

Photinia beauverdiana C. K. Schneid.

Photinia berberidifolia Rehder & E. H. Wilson

Photinia cuspidata (Bertol.) N. P. Balakr.

Photinia glomerata Rehder & E. H. Wilson

Photinia hookeri (Decne.) Merr.

Photinia integrifolia Lindl.

Photinia megaphylla T. T. Yu & T. C. Ku

Photinia prionophylla (Franch.) C. K. Schneid.

Photinia tsaii Rehder

Photinia wardii C. E. C. Fisch.

Potentilla agrimonioides M. Bieb.

Potentilla arbuscula D. Don

Potentilla argentea L.

Potentilla argyrophylla var. **atrosanguinea** (Loddiges, G. Loddiges & W. Loddiges) Hook. f.

Potentilla argyrophylla Wall. ex Lehm.

Potentilla articulata Franch.

Potentilla articulata var. **latipetiolata** (C. E. C. Fisch.) T. T. Yü & C. L. Li

Potentilla atrosanguinea Lodd., G. Lodd. & W. Lodd.

Potentilla bhutanica Ludlow

Potentilla biflora Willd. ex Schlecht.

Potentilla biflora var. **lahulensis** Th. Wolf

Potentilla bifurca L.

Potentilla bifurca subsp. **orientalis** (Juz.) Soják

Potentilla bifurca var. **humilior** Osten-Sacken & Rupr.

Potentilla blanda J. Soják

Potentilla centigrana Maxim.

Potentilla chrysantha Trevir.

Potentilla collettiana Aitch. & Hemsl.

Potentilla conferta Bunge

Potentilla conferta var. **trijuga** T. T. Yu & C. L. Li

Potentilla coriandrifolia D. Don

Potentilla coriandrifolia var. **dumosa** Franch.

Potentilla crenulata T. T. Yu & C. L. Li

Potentilla cuneata Wall. ex Lehm.

Potentilla curviseta Hook. f.

Potentilla delavayi Franch.

Potentilla desertorum Bunge

Potentilla doubjouneana Cambess.

Potentilla dryadanthoides (Juz.) Viroshilov

Potentilla eriocarpa Wall. ex Lehm.

Potentilla eriocarpa var. **major** Kitam.

Potentilla eriocarpa var. **tsarongensis** W. E. Evans

Potentilla evestita Th. Wolf

Potentilla flabellata Regel & Schmalh. ex Regel

Potentilla flagellaris Willd. ex Schltdl.

Potentilla fragarioides L.

Potentilla freyniana Bornm.

Potentilla fruticosa L.

Potentilla fruticosa var. **albicans** Rehder & E. H. Wilson

Potentilla fruticosa var. **arbuscula** (D. Don) Maxim.

Potentilla fruticosa var. **ochreata** Lindl. ex Lehm.

Potentilla fruticosa var. **pumila** Hook. f.

Potentilla fruticosa var. **rigida** (Wall. ex Lehm.) Th. Wolf

Potentilla gelida C. A. Mey.

Potentilla gilgitica M. Shah & Wilcock

Potentilla glabra G. Lodd.

Potentilla glabra var. **mandshurica** (Maxim.) Hand.-Mazz.

Potentilla granulosa T. T. Yu & C. L. Li

Potentilla griffithii Hook. f.

Potentilla griffithii var. **velutina** Cardot

Potentilla grisea Juz.

Potentilla grisea var. **vilijuga** (Soják) M. Shah & Wilcock

Potentilla hololeuca Boiss.

Potentilla hypargyrea Hand.-Mazz.

Potentilla hypargyrea var. **subpinnata** T. T. Yu & C. L. Li

Potentilla indica (Andrews) Th. Wolf

Potentilla kashmirica Hook. f.

Potentilla lancinata Cardot

Potentilla leschenaultiana Ser.

Potentilla longifolia Willd. ex Schltdl.

Potentilla macrosepala Cardot

Potentilla makaluensis H. Ikeda & H. Ohba

Potentilla monanthes Lindl. ex Lehm.

Potentilla monanthes var. **sibthorpioides** Hook. f.

Potentilla multicaulis Bunge

Potentilla multiceps T. T. Yu & C. L. Li

Potentilla multifida L.

Potentilla multifida var. **nubigena** Th. Wolf

Potentilla multifida var. **ornithopoda** (Tausch) Th. Wolf

Potentilla nepalensis Hook.

Potentilla nervosa Juz.

Potentilla pamirica Th. Wolf

Potentilla pamiroalaica Juz.

Potentilla parvifolia Fisch. ex Lehm.

Potentilla parvifolia var. **hypoleuca** Hand.-Mazz.

Potentilla potaninii Th. Wolf

Potentilla potaninii var. **compsophylla** (Hand.-Mazz.) T. T. Yu & C. L. Li

Potentilla pteropoda Royle

Potentilla salesoviana Steph.

Potentilla saundersiana Royle

Potentilla saundersiana var. **caespitosa** Lehm. ex Th. Wolf

Potentilla saundersiana var. **jacquemontii** Franch.

Potentilla saundersiana var. **subpinnata** Hand.-Mazz.

Potentilla sischanensis Bunge ex Lehm.

Potentilla spectabilis Businsky & Sojak

Potentilla sprengeliana Lehm.

Potentilla stewartiana M. Shah & Wilcock

Potentilla sundaica (Blume) O. Kuntze

Potentilla supina L.

Potentilla supina subsp. **costata** Soják

Potentilla supina subsp. **paradoxa** (Nutt. ex Torr. & A. Gray) Soják

Potentilla taronensis C. Y. Wu ex T. T. Yu & C. L. Li

Potentilla tephroleuca (Th. Wolf) B. Fedtsch.

Potentilla turczaninowiana Stschegl.

Potentilla turfosoides H. Ikeda & H. Ohba

Potentilla veitchii E. H. Wilson

Potentilla virgata Lehm.

Potentilla xizangensis T. T. Yu & C. L. Li

Prinsepia uniflora Batalin

Prinsepia utilis Royle

Prunus arborea var. **montana** (Hook. f.) Kalkman

Prunus armeniaca L.

Prunus armeniaca var. **holosericea** Batalin

Prunus brachypoda Batalin

Prunus brahuica (Boiss.) Aitch. & Hemsl.

Prunus brunnescens (T. T. Yu & T. C. Ku) J. R. He

Prunus buergeriana Miq.

Prunus caudata Franch.

Prunus cerasoides Buch.-Ham. ex D. Don

Prunus ceylanica (Wight) Miq.

Prunus clarofolia C. K. Schneid.

Prunus conadenia Koehne

Prunus cornuta (Wall. ex Royle) Steud.

Prunus crataegifolia Hand.-Mazz.

Prunus davidiana (Carrière) Franch.

Prunus domestica L.

Prunus grayana Maxim.

Prunus gyirongensis Y. H. Tong & N. H. Xia

Prunus himalayana J. Wen

Prunus hypotricha Rehder

Prunus hypoxantha (Koehne) J. Wen

Prunus jajarkotensis Hara

Prunus jenkinsii Hook. f.

Prunus kansuensis Rehder

Prunus mira Koehne

Prunus mugus Hand.-Mazz.

Prunus napaulensis (Ser.) Steud.

Prunus obtusata Koehne

Prunus patentipila Hand.-Mazz.

Prunus persica (L.) Batsch

Prunus perulata Koehne

Prunus phaeosticta (Hance) Maxim.

Prunus pleiocerasus Koehne

Prunus rufa Wall. ex Hook. f.

Prunus salicina Lindl.

Prunus salicina var. **pubipes** (Koehne) L. H. Bailey

Prunus serrula Franch.

Prunus spinulosa Siebold & Zucc.

Prunus stipulacea Maxim.

Prunus tangutica (Batalin) Koehne

Prunus taplejunica H. Ohba & S. Akiyama

Prunus topkegolensis H. Ohba & S. Akiyama

Prunus tatsienensis Batalin

Prunus tomentosa Thunb.

Prunus trichantha Koehne

Prunus trichostoma Koehne

Prunus undulata Buch.-Ham. ex D. Don

Prunus wilsonii (C. K. Schneid.) Koehne

Prunus yaoiana (W. L. Zheng) Y. H. Tong & N. H. Xia

Prunus yunnanensis Franch.

Prunus zippeliana Miq.

Pygeum henryi Dunn

Pygeum wilsonii Koehne

Pygeum wilsonii var. **macrophyllum** L. T. Lu

Pygeum zeylanicum Gaertn.

Pyracantha angustifolia C. K. Schneid.

Pyracantha crenulata (D. Don) M. Roem.

Pyracantha crenulata var. **kansuensis** Rehder

Pyracantha fortuneana (Maxim.) H. L. Li

Pyrus betulifolia Bunge

Pyrus griffithii (Decne.) Hook. f.

Pyrus pashia Buch.-Ham. ex D. Don

Pyrus polycarpa Hook. f.

Pyrus pseudopashia T. T. Yu

Pyrus xerophila T. T. Yu

Rosa alba L.

Rosa banksiae W. T. Aiton

Rosa banksiopsis Baker

Rosa brunonii Lindl.

Rosa calyptopoda Cardot

Rosa caudata Baker

Rosa chinensis Jacq.

Rosa clinophylla Thory

Rosa cymosa Tratt.

Rosa davidii Crépin

Rosa deqenensis T. C. Ku

Rosa derongensis T. C. Ku

Rosa duplicata T. T. Yu & T. C. Ku

Rosa farreri Stapf ex Cox

Rosa forrestiana Boulenger

Rosa giraldii Crép.

Rosa glomerata Rehder & E. H. Wilson

Rosa graciliflora Rehder & E. H. Wilson

Rosa helenae Rehder & E. H. Wilson

Rosa henryi Boulenger

Rosa hugonis Hemsl.

Rosa indica L.

Rosa laevigata Michx.

Rosa leschenaultiana (Redout & Thory) Wight & Arn.

Rosa lichiangensis T. T. Yu & T. C. Ku

Rosa longicuspis Bertol.

Rosa longicuspis var. **sinowilsonii** (Hemsl.) T. T. Yu & T. C. Ku

Rosa ludingensis T. C. Ku

Rosa macrophylla Lindl.

Rosa macrophylla var. **glandulifera** T. T. Yu & T. C. Ku

Rosa mairei H. Lév.

Rosa miyiensis T. C. Ku

Rosa moyesii Hemsl. & E. H. Wilson

Rosa multibracteata Hemsl. & E. H. Wilson

Rosa murielae Rehder & E. H. Wilson

Rosa odorata (Andrews) Sweet

Rosa odorata var. **erubescens** (Focke) T. T. Yu & T. C. Ku

Rosa odorata var. **gigantea** (Crépin) Rehder & E. H. Wilson

Rosa odorata var. **pseudoindica** (Lindl.) Rehder

Rosa omeiensis Rolfe

Rosa persetosa Rolfe

Rosa pinnatisepala T. C. Ku

Rosa praelucens Bijh.

Rosa prattii Hemsl.

Rosa primula Boulenger

Rosa roxburghii Tratt.

Rosa rubus H. Lév. & Vaniot

Rosa sericea Lindl.

Rosa sertata Rolfe

Rosa setipoda Hemsl. & E. H. Wilson

Rosa sikangensis T. T. Yu & T. C. Ku

Rosa sinobiflora T. C. Ku

Rosa soulieana Crép.

Rosa soulieana var. **sungpanensis** Rehder

Rosa soulieana var. **yunnanensis** C. K. Schneid.

Rosa sweginzowii Koehne

Rosa sweginzowii var. **glandulosa** Cardot

Rosa sweginzowii var. **stevensii** (Rehder) T. C. Ku

Rosa taronensis T. T. Yu

Rosa tibetica T. T. Yu & T. C. Ku

Rosa webbiana Royle

Rosa weisiensis T. T. Yu & T. C. Ku

Rosa willmottiae Hemsl.

Rosa willmottiae var. **glandulifera** T. T. Yu & T. C. Ku

Rosa zhongdianensis T. C. Ku

Rubus acuminatus Sm.

Rubus alceifolius Poir.

Rubus alexeterius Focke

Rubus alexeterius var. **acaenocalyx** (Hara) T. T. Yu & L. T. Lu

Rubus amabilis Focke

Rubus angustibracteatus T. T. Yu & L. T. Lu

Rubus assamensis Focke

Rubus aurantiacus Focke

Rubus austrotibetanus T. T. Yu & L. T. Lu

Rubus biflorus Buch.-Ham. ex Sm.

Rubus biflorus var. **adenophorus** Franch.

Rubus biflorus var. **pubescens** T. T. Yu & L. T. Lu

Rubus birmanicus Hook. f.

Rubus bonatianus Focke

Rubus burkillii Rolfe

Rubus calycinoides Kuntze

Rubus calycinus Wall. ex D. Don

Rubus chrysobotrys Hand.-Mazz.

Rubus cinclidodictyus Cardot

Rubus clivicola E. Walker

Rubus cochinchinensis Tratt.

Rubus cockburnianus Hemsl.

Rubus columellaris Tutcher

Rubus corchorifolius L. f.

Rubus coreanus Miq.

Rubus delavayi Franch.

Rubus doyonensis Hand.-Mazz.

Rubus duthieanus N. P. Balakr.

Rubus efferatus Craib

Rubus ellipticus Sm.

Rubus ellipticus var. **obcordatus** (Franch.) Focke

Rubus erythrocarpus T. T. Yu & L. T. Lu

Rubus erythrocarpus var. **weixiensis** T. T. Yu & L. T. Lu

Rubus eucalyptus Focke

Rubus eucalyptus var. **etomentosus** T. T. Yu & L. T. Lu

Rubus faberi Focke

Rubus fairholmianus Gardner

Rubus flosculosus Focke

Rubus fockeanus Kurz

Rubus foliaceistipulatus T. T. Yu & L. T. Lu

Rubus foliolosus D. Don

Rubus forrestianus Hand.-Mazz.

Rubus franchetianus H. Lév.

Rubus fuscifolius T. T. Yu & L. T. Lu

Rubus ghanakantae R. S. Rao & Joseph

Rubus glandulifer N. P. Balakr.

Rubus gongshanensis T. T. Yu & L. T. Lu

Rubus griffithii Hook. f.

Rubus gyamdaensis L. T. Lu & Boufford

Rubus gyamdaensis var. **glabriusculus** (T. T. Yu & L. T. Lu) L. T. Lu & Boufford

Rubus hamiltonii Hook. f.

Rubus haridasanii Chand. Gupta & S. S. Dash

Rubus hemithyrsus Hand.-Mazz.

Rubus hexagynus Roxb.

Rubus hibiscifolius Focke

Rubus hoffmeisterianus Kunth & C. D. Bouché

Rubus hypargyrus Edgew.

Rubus hypopitys Focke

Rubus idaeopsis Focke

Rubus inopertus (Focke) Focke

Rubus insignis Hook. f.

Rubus irritans Focke

Rubus kumaonensis N. P. Balakr.

Rubus lambertianus Ser.

Rubus lasiostylus Focke

Rubus laxus Focke

Rubus lineatus Reinw. ex Blume

Rubus lineatus var. **angustifolius** Hook. f.

Rubus lucens Focke

Rubus macilentus Cambess.

Rubus macilentus var. **angulatus** Franch.

Rubus maershanensis Huang C. Wang & H. Sun

Rubus malifolius Focke

Rubus mesogaeus Focke ex Diels

Rubus metoensis T. T. Yu & L. T. Lu

Rubus moluccanus var. **macrocarpa** Gardn.

Rubus multisetosus T. T. Yu & L. T. Lu

Rubus niveus Thunb.

Rubus nyalamensis T. T. Yu & L. T. Lu

Rubus ourosepalus Cardot

Rubus paniculatus Sm.

Rubus pectinarioides Hara

Rubus pectinaris Focke

Rubus pentagonus Wall. ex Focke

Rubus pentagonus var. **eglandulosus** T. T. Yu & L. T. Lu

Rubus pentagonus var. **longisepalus** T. T. Yu & L. T. Lu

Rubus pentagonus var. **modestus** (Focke) T. T. Yu & L. T. Lu

Rubus piluliferus Focke

Rubus playfairianus Hemsl. ex Focke

Rubus poliophyllus Kuntze

Rubus polyodontus Hand.-Mazz.

Rubus potentilloides W. E. Evans

Rubus pseudopileatus Cardot

Rubus pseudopileatus var. **glabratus** T. T. Yu & L. T. Lu

Rubus pseudopileatus var. **kangdingensis** T. T. Yu & L. T. Lu

Rubus ptilocarpus T. T. Yu & L. T. Lu

Rubus pungens Cambess.

Rubus pungens var. **linearisepalus** T. T. Yu & L. T. Lu

Rubus pungens var. **oldhamii** (Miq.) Maxim.

Rubus ramachandrae S. S. Dash & Chand. Gupta

Rubus rosifolius Sm.

Rubus rosifolius var. **coronarius** (Sims) Focke

Rubus rubrisetulosus Cardot

Rubus salwinensis Hand.-Mazz.

Rubus × **seminepalensis** Naruh.

Rubus setchuenensis Bureau & Franch.

Rubus sikkimensis Hook. f.

Rubus simplex Focke
Rubus splendidissimus Hara
Rubus stans Focke
Rubus stimulans Focke
Rubus subinopertus T. T. Yu & L. T. Lu
Rubus subornatus Focke
Rubus subornatus var. **melanadenus** Focke
Rubus subtibetanus Hand.-Mazz.
Rubus subtibetanus var. **glandulosus** T. T. Yu & L. T. Lu
Rubus sumatranus Miq.
Rubus taronensis C. Y. Wu ex T. T. Yu & L. T. Lu
Rubus thomsonii Focke
Rubus treutleri Hook. f.
Rubus tricolor Focke
Rubus trijugus Focke
Rubus tsangii Merr.
Rubus wardii Merr.
Rubus xanthocarpus Bureau & Franch.
Rubus yunanicus Kuntze
Sanguisorba diandra (Hook. f.) Nordb.
Sanguisorba filiformis (Hook. f.) Hand.-Mazz.
Sanguisorba officinalis L.
Sibbaldia adpressa Bunge
Sibbaldia cuneata Hornem. ex Edgew.
Sibbaldia cuneifolia (Bertol.) Paule & Soják
Sibbaldia emodi H. Ikeda & H. Ohba
Sibbaldia micropetala (D. Don) Hand.-Mazz.
Sibbaldia perpusilloides (W. W. Sm.) Hand.-Mazz.
Sibbaldia procumbens L.
Sibbaldia procumbens var. **aphanopetala** (Hand.-Mazz.) T. T. Yu & C. L. Li
Sibbaldia purpurea Royle
Sibbaldia purpurea var. **macropetala** (Murav.) T. T. Yu & C. L. Li
Sibbaldia sikkimensis (Prain) Chatterjee
Sibbaldia tenuis Hand.-Mazz.
Sibbaldia tetrandra Bunge
Sibiraea angustata (Rehder) Hand.-Mazz.
Sibiraea laevigata (L.) Maxim.
Sibiraea tomentosa Diels
Sorbaria arborea C. K. Schneid.
Sorbaria arborea var. **glabrata** Rehder
Sorbaria arborea var. **subtomentosa** Rehder
Sorbaria tomentosa (Lindl.) Rehder
Sorbus albopilosa T. T. Yu & L. T. Lu
Sorbus arachnoidea Koehne
Sorbus arguta T. T. Yu
Sorbus aronioides Rehder
Sorbus astateria (Cardot) Hand.-Mazz.
Sorbus bhutanica (W. W. Sm.) N. P. Balakr.
Sorbus cashmiriana Hedl.
Sorbus coronata (Cardot) T. T. Yu & H. T. Tsai
Sorbus coronata var. **glabrescens** T. T. Yu & L. T. Lu
Sorbus dunnii Rehder
Sorbus epidendron Hand.-Mazz.
Sorbus esserteauana Koehne
Sorbus ferruginea (Wenzig) Rehder
Sorbus filipes Hand.-Mazz.
Sorbus folgneri (C. K. Schneid.) Rehder
Sorbus foliolosa (Wall.) Spach
Sorbus glomerulata Koehne
Sorbus hedlundii C. K. Schneid.
Sorbus helenae Koehne
Sorbus hemsleyi (C. K. Schneid.) Rehder
Sorbus hupehensis C. K. Schneid.
Sorbus insignis (Hook. f.) Hedl.
Sorbus keissleri (C. K. Schneid.) Rehder
Sorbus kiukiangensis T. T. Yu
Sorbus kiukiangensis var. **glabrescens** T. T. Yu
Sorbus koehneana C. K. Schneid.
Sorbus kurzii (Watt ex Prain) C. K. Schneid.

Sorbus lanata (D. Don) Schauer

Sorbus lanpingensis L. T. Lu

Sorbus medogensis L. T. Lu & T. C. Ku

Sorbus megalocarpa Rehder

Sorbus meliosmifolia Rehder

Sorbus microphylla (Wall. ex Hook. f.) Wenz.

Sorbus monbeigii (Cardot) N. P. Balakr.

Sorbus multijuga Koehne

Sorbus obsoletidentata (Cardot) T. T. Yu

Sorbus ochracea (Hand.-Mazz.) J. E. Vidal

Sorbus oligodonta (Cardot) Hand.-Mazz.

Sorbus pallescens Rehder

Sorbus poteriifolia Hand.-Mazz.

Sorbus prattii Koehne

Sorbus prattii var. **aestivalis** (Koehne) T. T. Yu

Sorbus pteridophylla Hand.-Mazz.

Sorbus reducta Diels

Sorbus rehderiana Koehne

Sorbus rehderiana var. **cupreonitens** Hand.-Mazz.

Sorbus rehderiana var. **grosseserrata** Koehne

Sorbus rhamnoides (Decne.) Rehder

Sorbus rufopilosa C. K. Schneid.

Sorbus rufopilosa var. **stenophylla** Koehne

Sorbus salwinensis T. T. Yu & L. T. Lu

Sorbus sargentiana Koehne

Sorbus scalaris Koehne

Sorbus setschwanensis (C. K. Schneid.) Koehne

Sorbus sharmae M. F. Watson, V. Manandhar & Rushforth

Sorbus subochracea T. T. Yu & L. T. Lu

Sorbus thibetica (Cardot) Hand.-Mazz.

Sorbus thomsonii (King ex Hook. f.) Rehder

Sorbus tianschanica Rupr.

Sorbus ursina (Wall. ex G. Don) S. Schauer

Sorbus vestita (Wall. ex G. Don) Lodd.

Sorbus vilmorinii C. K. Schneid.

Sorbus wallichii (Hook. f.) T. T. Yu

Sorbus wilsoniana C. K. Schneid.

Sorbus zayuensis T. T. Yu & L. T. Lu

Spenceria ramalana Trimen

Spenceria ramalana var. **parviflora** (Stapf) Kitamura

Spiraea adiantoides Businský

Spiraea alpina Pall.

Spiraea arcuata Hook. f.

Spiraea arunachalensis Panigrahi & K. M. Purohit

Spiraea bella Sims

Spiraea bella var. **pubicarpa** T. T. Yu & L. T. Lu

Spiraea calcicola W. W. Sm.

Spiraea canescens D. Don

Spiraea compsophylla Hand.-Mazz.

Spiraea cantoniensis Lour.

Spiraea daochengensis L. T. Lu

Spiraea diversifolia Dunn

Spiraea hemicryptophyta Grierson

Spiraea henryi Hemsl.

Spiraea hypericifolia L.

Spiraea hypoleuca Dunn

Spiraea japonica L. f.

Spiraea japonica var. **acuminata** Franch.

Spiraea japonica var. **acuta** T. T. Yu

Spiraea japonica var. **fortunei** (Planch.) Rehder

Spiraea japonica var. **incisa** T. T. Yu

Spiraea japonica var. **ovalifolia** Franch.

Spiraea laeta Rehder

Spiraea laeta var. **tenuis** Rehder

Spiraea lasiocarpa var. **villosa** Businský

Spiraea lichiangensis W. W. Sm.

Spiraea lobulata T. T. Yu & L. T. Lu

Spiraea longigemmis Maxim.

Spiraea micrantha Hook. f.

Spiraea mollifolia Rehder

Spiraea mollifolia var. **glabrata** T. T. Yu & L. T. Lu

Spiraea mongolica Maxim.

Spiraea mongolica var. **tomentulosa** T. T. Yu

Spiraea muliensis T. T. Yu & L. T. Lu

Spiraea myrtilloides Rehder

Spiraea ovalis Rehder

Spiraea papillosa Rehder

Spiraea papillosa var. **yunnanensis** T. T. Yu

Spiraea prunifolia Siebold & Zucc.

Spiraea purpurea Hand.-Mazz.

Spiraea rosthornii E. Pritz. ex Diels

Spiraea sargentiana Rehder

Spiraea schneideriana Rehder

Spiraea siccanea (W. W. Sm.) Rehder

Spiraea subdioica K. M. Purohit & Panigrahi

Spiraea sublobata Hand.-Mazz.

Spiraea vaccinifolia D. Don

Spiraea veitchii Hemsl.

Spiraea velutina Franch.

Spiraea wilsonii Duthie

Spiraea xizangensis L. T. Lu

Spiraea yunnanensis Franch.

Stranvaesia amphidoxa C. K. Schneid.

Stranvaesia davidiana Decne.

Stranvaesia davidiana var. **undulata** (Decne.) Rehder & E. H. Wilson

Stranvaesia nussia (Buch.-Ham. ex D. Don) Decne.

134. ELAEAGNACEAE

Elaeagnus angustata (Rehder) C. Y. Chang

Elaeagnus angustifolia L.

Elaeagnus bockii Diels

Elaeagnus bockii var. **muliensis** C. Y. Chang

Elaeagnus caudata Schltdl.

Elaeagnus caudata var. **aliena** Momiy.

Elaeagnus conferta Roxb.

Elaeagnus davidii Franch.

Elaeagnus delavayi Lecomte

Elaeagnus infundibularis Momiy.

Elaeagnus kanaii Momiy.

Elaeagnus lanceolata Warb.

Elaeagnus lanpingensis C. Y. Chang

Elaeagnus latifolia L.

Elaeagnus pallidiflora C. Y. Chang

Elaeagnus parvifolia Wall.

Elaeagnus pyriformis Hook. f.

Elaeagnus taliensis C. Y. Chang

Elaeagnus tricholepis Momiy.

Elaeagnus umbellata Thunb.

Elaeagnus xizangensis C. Y. Chang

Hippophae goniocarpa Y. S. Lian et al. ex Swenson & Bartish

Hippophae gyantsensis (Rousi) Y. S. Lian

Hippophae litangensis Y. S. Lian & Xue L. Chen ex Swenson & Bartish

Hippophae neurocarpa S. W. Liu & T. N. Ho

Hippophae neurocarpa subsp. **stellatopilosa** Y. S. Lian et al. ex Swenson & Bartish

Hippophae rhamnoides L.

Hippophae rhamnoides subsp. **sinensis** Rousi

Hippophae rhamnoides subsp. **turkestanica** Rousi

Hippophae rhamnoides subsp. **wolongensis** Y. S. Lian, K. Sun & X. L. Chen

Hippophae rhamnoides subsp. **yunnanensis** Rousi

Hippophae salicifolia D. Don

Hippophae tibetana Schltdl.

135. RHAMNACEAE

Berchemia edgeworthii Lawson

Berchemia flavescens (Wall.) Brongn.

Berchemia floribunda (Wall.) Brongn.

Berchemia hirtella H. T. Tsai & K. M. Feng

Berchemia lineata (L.) DC.

Berchemia longipedicellata Y. L. Chen & P. K. Chou

Berchemia medogensis Y. L. Chen & Y. F. Du

Berchemia polyphylla Wall. ex M. A. Lawson

Berchemia sinica C. K. Schneid.

Berchemia yunnanensis Franch.

Gouania leptostachya DC.

Gouania microcarpa DC.

Gouania napalensis Wall.

Gouania tiliifolia Lam.

Helinus lanceolatus Brand.

Hovenia acerba Lindl.

Hovenia acerba var. **kiukiangensis** (Hu & Cheng) C. Y. Wu ex Y. L. Chen & P. K. Chou

Hovenia dulcis Thunb.

Paliurus orientalis (Franch.) Hemsl.

Paliurus spina-christi Mill.

Rhamnella forrestii W. W. Sm.

Rhamnella gilgitica Mansf. & Melch.

Rhamnella julianae C. K. Schneid.

Rhamnella martini (H. Lév.) C. K. Schneid.

Rhamnella wilsonii C. K. Schneid.

Rhamnus aurea Heppeler

Rhamnus bodinieri H. Lév.

Rhamnus collettii Bhandari & Bhansali

Rhamnus dumetorum C. K. Schneid.

Rhamnus dumetorum var. **crenoserrata** Rehder & E. H. Wilson

Rhamnus esquirolii H. Lév.

Rhamnus flavescens Y. L. Chen & P. K. Chou

Rhamnus gilgiana Heppeler

Rhamnus hemsleyana C. K. Schneid.

Rhamnus hemsleyana var. **yunnanensis** C. Y. Wu ex Y. L. Chen & P. K. Chou

Rhamnus henryi C. K. Schneid.

Rhamnus heterophylla Oliver

Rhamnus maximovicziana J. J. Vassil.

Rhamnus minutus Grubov

Rhamnus napalensis (Wall.) Lawson

Rhamnus nigricans Hand.-Mazz.

Rhamnus ninglangensis Y. L. Chen

Rhamnus pentapomica R. Parker

Rhamnus persicus Boiss.

Rhamnus procumbens Edgew.

Rhamnus prostrata R. N. Parker

Rhamnus purpureus Edgew.

Rhamnus rosthornii E. Pritz. ex Diels

Rhamnus sargentiana C. K. Schneid.

Rhamnus tangutica J. J. Vassil.

Rhamnus triqueter (Wall.) M. A. Lawson

Rhamnus utilis var. **szechuanensis** Y. L. Chen & P. K. Chou

Rhamnus virgata Roxb.

Rhamnus virgata var. **hirsuta** (Wight & Arn.) Y. L. Chen & P. K. Chou

Rhamnus wightii Wight & Arn.

Rhamnus xizangensis Y. L. Chen & P. K. Chou

Sageretia brandrethiana Aitch.

Sageretia filiformis (Roth ex Schult.) G. Don

Sageretia gracilis J. R. Drumm. & Sprague

Sageretia hamosa (Wall.) Brongn.

Sageretia henryi J. R. Drumm. & Sprague

Sageretia horrida Pax & K. Hoffm.

Sageretia kashmirensis Bhandari & Bhansali

Sageretia kishtwarensis Bhandari & Bhansali

Sageretia lijiangensis G. S. Fan & S. K. Chen

Sageretia lucida Merr.

Sageretia parviflora (Roem. & Schult.) G. Don

Sageretia paucicostata Maxim.

Sageretia pycnophylla C. K. Schneid.

Sageretia subcaudata C. K. Schneid.

Sageretia thea (Osbeck) M. C. Johnst.

Sageretia thea var. **bornmuelleri** (Osbeck) M. C. Johnst.

Sageretia yilinii G. S. Fan & S. K. Chen

Ventilago denticulata Willd.

Ventilago maderaspatana Gaertn.

Ziziphus apetala Hook. f. ex M. A. Lawson

Ziziphus funiculosa Buch.-Ham. ex M. A. Lawson

Ziziphus glabrata B. Heyne ex Roth

Ziziphus incurva Roxb.

Ziziphus jujuba Mill.

Ziziphus mairei Dode

Ziziphus mauritiana Lam.

Ziziphus montana W. W. Sm.

Ziziphus nummularia (Burm. f.) Wight & Arn.

Ziziphus oenoplia (L.) Mill.

Ziziphus oxyphylla Edgew.

Ziziphus rubiginosa D. G. Long & Rae

Ziziphus rugosa Lam.

Ziziphus xiangchengensis Y. L. Chen & P. K. Chou

Ziziphus xylopyrus Retz.

136. ULMACEAE

Holoptelea integrifolia (Roxb.) Planch.

Ulmus androssowii var. **subhirsuta** (C. K. Schneid.) P. H. Huang et al.

Ulmus bergmanniana C. K. Schneid.

Ulmus bergmanniana var. **lasiophylla** C. K. Schneid.

Ulmus chumlia Melville & Heybroek

Ulmus lanceifolia Roxb. ex Wall.

Ulmus microcarpa L. K. Fu

Ulmus pumila L.

Ulmus villosa Brandis ex Gamble

Ulmus wallichiana Planch.

Zelkova schneideriana Hand.-Mazz.

Zelkova sinica C. K. Schneid.

137. CANNABACEAE

Aphananthe cuspidata (Blume) Planch.

Cannabis sativa L.

Celtis australis L.

Celtis biondii Pamp.

Celtis bungeana Blume

Celtis caucasica Willd.

Celtis cerasifera C. K. Schneid.

Celtis eriocarpa Decne.

Celtis julianae C. K. Schneid.

Celtis sinensis Pers.

Celtis tetrandra Roxb.

Celtis timorensis Span.

Humulus scandens (Lour.) Merr.

Humulus yunnanensis Hu

Trema cannabina Lour.

Trema levigata Hand.-Mazz.

Trema orientalis (L.) Blume

Trema politoria (Planch.) Blume

Trema tomentosa (Roxb.) Hara

138. MORACEAE

Artocarpus chama Buch.-Ham.

Artocarpus gomezianus Wall. ex Trécul

Artocarpus gongshanensis S. K. Wu ex C. Y. Wu & S. S. Chang

Artocarpus heterophyllus Lam.

Artocarpus integra (Thunb.) Merr.

Artocarpus lacucha Buch.-Ham.

Artocarpus lakoocha Roxb.

Broussonetia kaempferi Siebold

Broussonetia kurzii (Hook. f.) Corner

Broussonetia papyrifera (L.) L'Herit. ex Vent.

Cudrania spinosa Koord.

Ficus abelii Miq.

Ficus altissima Blume

Ficus amplissima Sm.

Ficus arnottiana (Miq.) Miq.

Ficus arnottiana var. **subcostata** Corner

Ficus auriculata Lour.

Ficus benghalensis L.

Ficus benjamina L.

Ficus carica L.

Ficus chapaensis Gagnep.

Ficus concinna (Miq) Miq.

Ficus conglobata King

Ficus curtipes Corner

Ficus cyrtophylla (Wall. ex Miq.) Miq.

Ficus drupacea Thunb.

Ficus drupacea var. **pubescens** (Roth) Corner

Ficus elastica Roxb. ex Hornem.

Ficus esquiroliana H. Lév.

Ficus filicauda Hand.-Mazz.

Ficus filicauda var. **longipes** S. S. Chang

Ficus fistulosa Reinw. ex Blume

Ficus fulva Reinw. ex Blume

Ficus gasparriniana Miq.

Ficus gasparriniana var. **laceratifolia** (H. Lév. & Vaniot) Corner

Ficus geniculata Kurz

Ficus glaberrima Blume

Ficus hederacea Roxb.

Ficus henryi Warb. ex Diels

Ficus heterophylla L. f.

Ficus heterophylla var. **assamica** (Miq.) Corner ex Chater

Ficus heteropleura Blume

Ficus hirta Vahl

Ficus hispida L. f.

Ficus hookeriana Corner

Ficus ischnopoda Miq.

Ficus krishnae C. DC.

Ficus lacor Buch.-Ham.

Ficus laevis Blume

Ficus lamponga Miq.

Ficus langkokensis Drake

Ficus maclellandii King

Ficus microcarpa L. f.

Ficus nepalensis Spreng.

Ficus neriifolia Sm.

Ficus nervosa B. Heyne ex Roth

Ficus oligodon Miq.

Ficus palmata Forssk.

Ficus palmata subsp. **virgata** (Roxb.) Browicz

Ficus prostrata (Wall. ex Miq.) Miq.

Ficus pubigera (Wall. ex Miq.) Kurz

Ficus pumila L.

Ficus racemosa L.

Ficus religiosa L.

Ficus rigida Jack

Ficus rumphii Blume

Ficus sagittata Vahl

Ficus sarmentosa Buch.-Ham. ex Sm.

Ficus sarmentosa var. **duclouxii** (H. Lév. & Vaniot) Corner

Ficus sarmentosa var. **luducca** (Roxb.) Corner

Ficus sarmentosa var. **nipponica** (Franch. & Sav.) Corner

Ficus semicordata Buch.-Ham. ex Sm.

Ficus sinuata Thunb.

Ficus squamosa Roxb.

Ficus subincisa Buch.-Ham. ex Sm.

Ficus subulata Blume

Ficus tikoua Bureau

Ficus tinctoria G. Forst.

Ficus tinctoria subsp. **gibbosa** (Blume) Corner

Ficus tinctoria subsp. **parasitica** G. Forst.

Ficus virens Aiton

Ficus xiphias C. E. C. Fisch.

Maclura amboinensis Blume

Maclura cochinchinensis (Lour.) Corner

Morus albus L.

Morus australis Poir.

Morus cathayana Hemsl.

Morus cathayana var. **gongshanensis** (Z. Y. Cao) Z. Y. Cao

Morus macroura Miq.

Morus mongolica (Bureau) C. K. Schneid.

Morus notabilis C. K. Schneid.

Morus serrata Roxb.

Streblus asper Lour.

139. URTICACEAE

Boehmeria clidemioides Miq.

Boehmeria clidemioides var. **diffusa** (Wedd.) Hand.-Mazz.

Boehmeria conica C. J. Chen, Wilmot-Dear & Friis

Boehmeria glomerulifera Miq.

Boehmeria hamiltoniana Wedd.

Boehmeria kamley Acharya & Yonek.

Boehmeria macrophylla Hornem.

Boehmeria macrophylla var. **canescens** (Wedd.) D. G. Long

Boehmeria macrophylla var. **rotundifolia** (D. Don) W. T. Wang

Boehmeria macrophylla var. **scabrella** (Wedd.) D. G. Long

Boehmeria macrophylla var. **tomentosa** (Wedd.) D. G. Long

Boehmeria nivea (L.) Gaudich.

Boehmeria penduliflora Wedd. ex D. G. Long

Boehmeria platyphylla D. Don

Boehmeria platyphylla var. **canescens** Wedd.

Boehmeria platyphylla var. **rotundifolia** (Buch.-Ham. ex D. Don) Wedd.

Boehmeria platyphylla var. **scabrella** (Roxb.) Wedd.

Boehmeria platyphylla var. **zaylanica** Wedd.

Boehmeria polystachya Wedd.

Boehmeria rugulosa Wedd.

Boehmeria ternifolia D. Don

Boehmeria tirapensis Deb & R. M. Dutta

Boehmeria tomentosa Wedd.

Boehmeria umbrosa (Hand.-Mazz.) W. T. Wang

Chamabainia cuspidata Wight

Debregeasia longifolia (Burm. f.) Wedd.

Debregeasia orientalis C. J. Chen

Debregeasia saeneb (Forssk.) Hepper & J. R. I. Wood

Debregeasia salicifolia (D. Don) Rendle

Debregeasia wallichiana (Wedd.) Wedd.

Dendrocnide meyeniana (Walp.) Chew

Dendrocnide sinuata (Blume) Chew

Droguetia iners (Forssk.) Schweinf.

Elatostema acuminatum (Poir.) Brongn.

Elatostema acutitepalum W. T. Wang

Elatostema aliferum W. T. Wang

Elatostema angustitepalum W. T. Wang

Elatostema apicicrassum W. T. Wang

Elatostema arcuans Dunn

Elatostema auriculatum var. **strigosum** W. T. Wang

Elatostema auriculatum W. T. Wang

Elatostema balansae Gagnep.

Elatostema beibengense W. T. Wang

Elatostema biglomeratum W. T. Wang

Elatostema bijiangense var. **weixiense** W. T. Wang

Elatostema boehmerioides W. T. Wang

Elatostema bomiense W. T. Wang

Elatostema brevipedunculatum W. T. Wang

Elatostema caveanum Grierson & D. G. Long

Elatostema cikaiense W. T. Wang

Elatostema cornutum Wedd.

Elatostema cucullatonaviculare W. T. Wang

Elatostema cuneatum Wight

Elatostema cuneiforme var. **gracilipes** W. T. Wang

Elatostema cuneiforme W. T. Wang

Elatostema cuspidatum Wight

Elatostema cyrtandrifolium (Zoll. & Moritzi) Miq.

Elatostema decipiens Wedd.

Elatostema didymocephalum W. T. Wang

Elatostema dissectoides W. T. Wang

Elatostema dissectum Wedd.

Elatostema dulongense W. T. Wang

Elatostema ebracteatum W. T. Wang

Elatostema eriocephalum W. T. Wang

Elatostema ferrugineum W. T. Wang

Elatostema ficoides Wedd.

Elatostema fugongense W. T. Wang

Elatostema furcatibracteum W. T. Wang

Elatostema grandidentatum W. T. Wang

Elatostema gungshanense W. T. Wang

Elatostema heterogrammicum W. T. Wang

Elatostema heterolobum (Wedd.) Hallier f.

Elatostema hookerianum Wedd.

Elatostema imbricans Dunn

Elatostema integrifolium (D. Don) Wedd.

Elatostema laevissimum W. T. Wang

Elatostema laxicymosum W. T. Wang

Elatostema lihengianum W. T. Wang

Elatostema lineolatum var. **majus** Wedd.

Elatostema lineolatum Wight

Elatostema litseifolium W. T. Wang

Elatostema longicaudatum Grierson & D. G. Long

Elatostema macintyrei Dunn

Elatostema medogense W. T. Wang

Elatostema microtrichum W. T. Wang

Elatostema monandrum (Buch.-Ham. ex D. Don) Hara

Elatostema monandrum var. **pinnatifidum** (Hook. f.) Murti

Elatostema nasutum Hook. f.

Elatostema obtusum Wedd.

Elatostema oxyodontum W. T. Wang

Elatostema pachyceras W. T. Wang

Elatostema papillosum Wedd.

Elatostema paracuminatum W. T. Wang

Elatostema parvum (Blume) Miq.

Elatostema pauciflorum W. T. Wang

Elatostema pianmaense W. T. Wang

Elatostema platyceras W. T. Wang

Elatostema platyphyllum var. **polycephalum** Hara

Elatostema platyphyllum Wedd.

Elatostema procridioides Wedd.

Elatostema pseudocuspidatum W. T. Wang

Elatostema pseudonasutum W. T. Wang

Elatostema punctatum (Buch.-Ham. ex D. Don) Wedd.

Elatostema pusillum C. B. Clarke ex Hook. f.

Elatostema pycnodontum W. T. Wang

Elatostema recticaudatum W. T. Wang

Elatostema reptans Hook. f.

Elatostema rhombiforme W. T. Wang

Elatostema rupestre (Buch.-Ham. ex D. Don) Wedd.

Elatostema sessile J. R. Forst. & G. Forst.

Elatostema sessile var. **polycephalum** Wedd.

Elatostema sikkimense C. B. Clarke

Elatostema stellatum Hook. f.

Elatostema stracheyanum Wedd.

Elatostema subincisum Wedd.

Elatostema tenuicaudatoides W. T. Wang

Elatostema tenuicornutum W. T. Wang

Elatostema tetratepalum W. T. Wang

Elatostema treutleri Hook. f.

Elatostema wangii Q. Lin & L. D. Duan

Elatostema wenxianense W. T. Wang & Z. X. Peng

Elatostema yangbiense W. T. Wang

Elatostema yui W. T. Wang

Forsskaolea tenacissima L.

Girardinia diversifolia (Link) Friis

Gonostegia hirta (Blume) Miq.

Gonostegia pentandra (Roxb.) Miq.

Laportea bulbifera (Siebold & Zucc.) Wedd.

Laportea cuspidata (Wedd.) Friis

Laportea medogensis C. J. Chen

Laportea stolonifera B. L. Bhellum & B. Singh

Laportea terminalis Wight

Lecanthus peduncularis (Royle) Wedd.

Lecanthus petelotii (Gagnep.) C. J. Chen

Lecanthus petelotii var. **corniculata** C. J. Chen

Lecanthus petelotii var. **yunnanensis** C. J. Chen

Maoutia puya (Hook.) Wedd.

Oreocnide frutescens (Thunb.) Miq.

Oreocnide frutescens subsp. **occidentalis** C. J. Chen

Oreocnide integrifolia (Gaudich.) Miq.

Parietaria judaica L.

Parietaria micrantha Ledeb.

Pellionia heteroloba Wedd.

Pellionia macroceras Gagnep.

Pellionia repens (Lour.) Merr.

Pilea angulata (Blume) Blume

Pilea anisophylla Wedd.

Pilea approximata C. B. Clarke

Pilea approximata var. **incisoserrata** C. J. Chen

Pilea auricularis C. J. Chen

Pilea bracteosa var. **oxyodon** (Wedd.) Hara

Pilea bracteosa Wedd.

Pilea cavaleriei H. Lév.

Pilea clarkei Hook. f.

Pilea cordifolia Hook. f.

Pilea elegantissima C. J. Chen

Pilea gansuensis C. J. Chen & Z. X. Peng

Pilea glaberrima (Blume) Blume

Pilea hilliana Hand.-Mazz.

Pilea hookeriana Wedd.

Pilea howelliana Hand.-Mazz.

Pilea insolens Wedd.

Pilea kanaii Hara

Pilea linearifolia C. J. Chen

Pilea longipedunculata S. S. Chien & C. J. Chen

Pilea macrocarpa C. J. Chen

Pilea martinii (H. Lév.) Hand.-Mazz.

Pilea medogensis C. J. Chen

Pilea melastomoides (Poir.) Wedd.

Pilea monilifera Hand.-Mazz.

Pilea multicellularis C. J. Chen

Pilea myriantha (Dunn) C. J. Chen

Pilea obesa (Wedd.) Wedd. ex Hara

Pilea oxyodon Wedd.

Pilea pauciflora C. J. Chen

Pilea pellionioides C. J. Chen

Pilea peperomioides Diels

Pilea peploides (Gaudich.) Hook. & Arn.

Pilea pumila (L.) A. Gray

Pilea racemosa (Royle) Tuyama

Pilea salwinensis (Hand.-Mazz.) C. J. Chen

Pilea scripta (D. Don) Wedd.

Pilea semisessilis Hand.-Mazz.

Pilea squamosa C. J. Chen

Pilea symmeria Wedd.

Pilea ternifolia Wedd.

Pilea umbrosa Blume

Pilea umbrosa var. **obesa** Wedd.

Poikilospermum lanceolatum (Trècul) Merr.

Poikilospermum naucleiflorum (Roxb. ex Lindl.) Chew

Poikilospermum suaveolens (Blume) Merr.

Pouzolzia bennettiana Wight

Pouzolzia calophylla W. T. Wang & C. J. Chen

Pouzolzia frondosa Kuntze

Pouzolzia sanguinea (Blume) Merr.

Pouzolzia sanguinea var. **elegans** (Wedd.) Friis, Wilmot-Dear & C. J. Chen

Pouzolzia sanguinea var. **fulgens** (Wedd.) Hara

Pouzolzia sanguinea var. **nepalensis** (Wedd.) Hara

Pouzolzia zeylanica (L.) Benn.

Procris crenata C. B. Rob.

Sarcochlamys pulcherrima Gaudich.

Urtica ardens Link

Urtica atrichocaulis (Hand.-Mazz.) C. J. Chen

Urtica cannabina L.

Urtica dioica L.

Urtica dioica subsp. **afghanica** Chrtek

Urtica dioica subsp. **gansuensis** C. J. Chen

Urtica hyperborea Jacquem. ex Wedd.

Urtica laetevirens Maxim.

Urtica mairei H. Lév.

Urtica parviflora Roxb.

Urtica triangularis Hand.-Mazz.

Urtica triangularis subsp. **pinnatifida** (Hand.-Mazz.) C. J. Chen

Urtica triangularis subsp. **trichocarpa** C. J. Chen

Urtica urens L.

140. FAGACEAE

Castanea mollissima Blume

Castanea sativa Mill.

Castanopsis argyrophylla King ex Hook. f.

Castanopsis armata (Roxb.) Spach

Castanopsis birmanica A. Camus

Castanopsis calathiformis (Skan) Rehder & E. H. Wilson

Castanopsis castanicarpa (Roxb.) Spach

Castanopsis ceratacantha Rehder & E. H. Wilson

Castanopsis clarkei King ex Hook. f.

Castanopsis delavayi Franch.

Castanopsis echinocarpa Hook. f. & Thomson ex Miq.

Castanopsis eyrei (Champ. ex Benth.) Tutch.

Castanopsis ferox (Roxb.) Spach

Castanopsis fissa (Champ. ex Benth.) Rehder & E. H. Wilson

Castanopsis griffithii A. Camus

Castanopsis hystrix Hook. f. & Thomson ex A. DC.

Castanopsis indica (Roxb. ex Lindl.) A. DC.

Castanopsis inermis (Lindl.) Benth. & Hook. f.

Castanopsis lanceifolia (Roxb.) Hickel ex A. Camus

Castanopsis longispina (King ex Hook. f.) C. C. Huang & Y. T. Chang

Castanopsis orthacantha Franch.

Castanopsis platyacantha Rehder & E. H. Wilson

Castanopsis purpurella (Miq.) N. P. Balakr.

Castanopsis tibetana Hance

Castanopsis tribuloides (Sm.) A. DC.

Castanopsis wattii (King ex Hook. f.) A. Camus

Cyclobalanopsis annulata (Sm.) Oerst.

Cyclobalanopsis delavayi (Franch.) Schottky

Cyclobalanopsis gambleana (A. Camus) Y. C. Hsu & H. W. Jen

Cyclobalanopsis glauca (Thunb.) Oerst.

Cyclobalanopsis glaucoides Schottky

Cyclobalanopsis kiukiangensis Y. T. Chang ex Y. C. Hsu & H. W. Jen

Cyclobalanopsis lamellosa (Sm.) Oersted

Cyclobalanopsis lobbii (Hook. f. & Thomson ex Wenzig) Y. C. Hsu & H. W. Jen

Cyclobalanopsis motuoensis (C. C. Huang) Y. C. Hsu & H. W. Jen

Cyclobalanopsis myrsinifolia (Blume) Oerst.

Cyclobalanopsis oxyodon (Miq.) Oerst.

Cyclobalanopsis rex (Hemsl.) Schottky

Cyclobalanopsis semiserrata (Roxb.) Oerst.

Cyclobalanopsis stewardiana (A. Camus) Y. C. Hsu & H. W. Jen

Fagus longipetiolata Seemen

Lithocarpus arcaulus (Buch.-Ham. ex Spreng.) C. C. Huang & Y. T. Chang

Lithocarpus chittagongus (King ex Hook. f.) Merr.

Lithocarpus collettii (King ex Hook. f.) A. Camus

Lithocarpus confinis C. C. Huang ex Y. C. Hsu & H. W. Jen

Lithocarpus craibianus Barnett

Lithocarpus dealbatus (Hook. f. & Thomson ex Miquel) Rehder

Lithocarpus elegans (Blume) Hatus. ex Soepadmo

Lithocarpus fangii (Hu & W. C. Cheng) C. C. Huang & Y. T. Chang

Lithocarpus fenestratus (Roxb.) Rehder

Lithocarpus gaoligongensis C. C. Huang & Y. T. Chang

Lithocarpus grandifolius (D. Don) S. N. Biswas

Lithocarpus hancei (Benth.) Rehder

Lithocarpus hypoglaucus (Hu) C. C. Huang ex Y. C. Hsu & H. W. Jen

Lithocarpus jenkinsii (Benth.) C. C. Huang & Y. T. Chang

Lithocarpus kamengii K. C. Sahni & H. B. Naithani

Lithocarpus lappaceus (Roxb.) Rehder

Lithocarpus lindleyanus (Wall. ex A. DC.) A. Camus

Lithocarpus listeri (King) Grierson & D. G. Long

Lithocarpus lithocarpaeus (Oerst.) Hjelmq.

Lithocarpus litseifolius (Hance) Chun

Lithocarpus magnificus (Brandis) A. Camus

Lithocarpus megalophyllus Rehder & E. H. Wilson

Lithocarpus milroyi (C. S. Purkay.) Barnett

Lithocarpus obscurus C. C. Huang & Y. T. Chang

Lithocarpus pachyphyllus (Kurz) Rehder

Lithocarpus pseudoxizangensis Z. K. Zhou & H. Sun

Lithocarpus thomsonii (Miq.) Rehder

Lithocarpus truncatus (King ex Hook. f.) Rehder & E. H. Wilson

Lithocarpus variolosus (Franch.) Chun

Lithocarpus xizangensis C. C. Huang & Y. T. Chang

Lithocarpus xylocarpus (Kurz) Markgr.

Quercus acrodonta Seemen

Quercus acutissima Carruth.

Quercus aliena Blume

Quercus aquifolioides Rehder & E. H. Wilson

Quercus baloot Griff.

Quercus baronii Skan

Quercus brandisiana Kurz

Quercus chapensis Hickel & A. Camus

Quercus cocciferoides Hand.-Mazz.

Quercus cocciferoides var. **taliensis** (A. Camus) Y. C. Hsu & H. Wei Jen

Quercus delavayi Franch.

Quercus × deleiensis A. Camus

Quercus dentata Thunb.

Quercus dolicholepis A. Camus

Quercus edithiae Skan

Quercus engleriana Seemen

Quercus fabri Hance

Quercus fimbriata Chun & C. C. Huang ex Y. C. Hsu & H. W. Jen

Quercus floribunda Lindl. ex A. Camus

Quercus glauca Thunb.

Quercus griffithii Hook. f. & Thomson ex Miq.

Quercus guajavifolia H. Lév.

Quercus helferiana A. DC.

Quercus hypargyrea (Seemen ex Diels) C. C. Huang & Y. T. Chang

Quercus kerrii Craib

Quercus kiukiangensis (Y. T. Chang) Y. T. Chang

Quercus lamellosa Sm.

Quercus lanata Sm.

Quercus lineata Blume

Quercus lobbii Ettingsh.

Quercus lodicosa O. E. Warb. & E. F. Warb.

Quercus monimotricha (Hand.-Mazz.) Hand.-Mazz.

Quercus motuoensis C. C. Huang

Quercus myrsinifolia Blume

Quercus oblongata D. Don

Quercus oxyodon Miq.

Quercus oxyphylla (E. H. Wilson) Hand.-Mazz.

Quercus rehderiana Hand.-Mazz.

Quercus schottkyana Rehder & E. H. Wilson

Quercus semecarpifolia Sm.

Quercus semiserrata Roxb.

Quercus senescens Hand.-Mazz.

Quercus serrata Murray

Quercus spinosa David ex Franch.

Quercus stewardiana A. Camus

Quercus thomsoniana A. DC.

Quercus tomentosinervis (Y. C. Hsu & H. Wei Jen) C. C. Huang

Quercus tungmaiensis Y. T. Chang

Quercus variabilis Blume

Quercus yunnanensis Franch.

141. MYRICACEAE

Myrica esculenta D. Don

Myrica nana A. Chev.

142. JUGLANDACEAE

Carya tonkinensis Lecomte

Engelhardia roxburghiana Wall.

Engelhardia spicata Lechen. ex Blume

Engelhardia spicata var. **acerifolia** (Reinw.) Koorders & Valeton

Engelhardia spicata var. **colebrookeana** (Lindl.) Koorders & Valeton

Engelhardia spicata var. **integra** (Kurz) Manning ex Steenis

Juglans regia L.

Juglans sigillata Dode

Platycarya strobilacea Siebold & Zucc.

Pterocarya hupehensis Skan

Pterocarya macroptera var. **delavayi** (Franch.) Manning

Pterocarya macroptera var. **insignis** (Rehder & E. H. Wilson) Manning

Pterocarya stenoptera C. DC.

143. BETULACEAE

Alnus cremastogyne Burkill

Alnus ferdinandi-coburgii C. K. Schneid.

Alnus lanata Duthie ex Bean

Alnus nepalensis D. Don

Alnus nitida (Spach) Endl.

Betula alnoides Buch.-Ham. ex D. Don

Betula ashburneri McAll. & Rushforth

Betula calcicola (W. W. Sm.) P. C. Li

Betula chitralica Browicz

Betula cylindrostachya Lindl. ex Wall.

Betula delavayi Franch.

Betula delavayi var. **microstachya** P. C. Li

Betula delavayi var. **polyneura** Hu ex P. C. Li

Betula gynoterminalis Y. C. Hsu & C. J. Wang

Betula insignis Franch.

Betula jacquemontii Spach

Betula jarmolenkoana Golosk.

Betula jiulungensis Hu ex P. C. Li

Betula luminifera H. Winkl.

Betula platyphylla Sukaczev

Betula potaninii Batalin

Betula rhombibracteata P. C. Li

Betula skvorsovii McAllister & Ashburner

Betula tadzhikistanica V. N. Vassil.

Betula utilis D. Don

Carpinus cordata var. **chinensis** Franch.

Carpinus faginea Lindl.

Carpinus fangiana Hu

Carpinus mollicoma Hu

Carpinus monbeigiana Hand.-Mazz.

Carpinus polyneura Franch.

Carpinus pubescens Burkill

Carpinus sungpanensis W. Y. Hsia

Carpinus tschonoskii Maxim.

Carpinus viminea Lindl.

Carpinus viminea var. **chiukiangensis** Hu

Corylus chinensis Franch.

Corylus fargesii (Franch.) C. K. Schneid.

Corylus ferox Wall.

Corylus ferox var. **thibetica** (Batalin) Franch.

Corylus heterophylla var. **sutchuanensis** Franch.

Corylus jacquemontii Decne.

Corylus sieboldiana Blume

Corylus wangii Hu

Corylus yunnanensis (Franch.) A. Camus

Ostrya japonica Sarg.

Ostrya yunnanensis W. K. Hu

Ostryopsis davidiana Decne.

Ostryopsis intermedia B. Tian & J. Q. Liu

Ostryopsis nobilis Balf. f. & W. W. Sm.

144. CORIARIACEAE

Coriaria nepalensis Wall.

Coriaria terminalis Hemsl.

145. CUCURBITACEAE

Actinostemma tenerum Griff.

Actinostemma tenerum var. **yunnanense** A. M. Lu & Zhi Y. Zhang

Biswarea tonglensis (C. B. Clarke) Cogn.

Bolbostemma biglandulosum var. **sinuatolobulatum** C. Y. Wu

Bolbostemma paniculatum (Maxim.) Franch.

Diplocyclos palmatus (L.) C. Jeffrey

Edgaria darjeelingensis C. B. Clarke

Gomphogyne cissiformis Griff.

Gomphogyne cissiformis var. **villosa** Cogn., A. DC. & C. DC.

Gomphogyne nepalensis W. J. de Wilde & Duyfjes

Gymnopetalum chinense (Lour.) Merr.

Gynostemma aggregatum C. Y. Wu & S. K. Chen

Gynostemma laxum (Wallich) Cogn.

Gynostemma longipes C. Y. Wu

Gynostemma pentaphyllum (Thunb.) Makino

Hemsleya amabilis Diels

Hemsleya chinensis Cogn. ex F. B. Forbes & Hemsl.

Hemsleya delavayi (Gagnep.) C. Jeffrey ex C. Y. Wu & Z. L. Chen

Hemsleya delavayi var. **yalungensis** (Hand.-Mazz.) C. Y. Wu & Z. L. Chen

Hemsleya dulongjiangensis C. Y. Wu

Hemsleya endecaphylla C. Y. Wu

Hemsleya gigantha W. J. Chang

Hemsleya heterosperma (Wall.) C. Jeffrey

Hemsleya lijiangensis A. M. Lu ex C. Y. Wu & Z. L. Chen

Hemsleya macrocarpa (Cogn.) C. Y. Wu ex C. Jeffrey

Hemsleya macrocarpa var. **grandiflora** (C. Y. Wu) D. Z. Li

Hemsleya macrosperma C. Y. Wu

Herpetospermum pedunculosum (Ser.) C. B. Clarke

Hodgsonia heteroclita (Roxb.) Hook. f. & Thomson

Indofevillea khasiana Chatterjee

Momordica cochinchinensis (Lour.) Spreng.

Momordica dioica Roxb. ex Willd.

Momordica subangulata subsp. **renigera** (Wall. ex G. Don) W. J. de Wilde

Mukia maderaspatana (L.) M. Roem.

Neoalsomitra clavigera (Wall.) Hutch.

Schizopepon bicirrhosus (C. B. Clarke) C. Jeffrey

Schizopepon bomiensis A. M. Lu & Z. Y. Zhang

Schizopepon longipes Gagnep.

Schizopepon macranthus Hand.-Mazz.

Schizopepon xizangensis A. M. Lu & Z. Y. Zhang

Sinobaijiania decipiens C. Jeffrey & W. J. de Wilde

Siraitia sikkimensis (Chakrav.) C. Jeffrey

Solena heterophylla Lour.

Solena heterophylla subsp. **napaulensis** (Ser.) W. J. de Wilde & Duyfjes

Thladiantha capitata Cogn.

Thladiantha cordifolia (Blume) Cogn.

Thladiantha davidii Franch.

Thladiantha dentata Cogn.

Thladiantha dubia Bunge

Thladiantha grandisepala A. M. Lu & Zhi Y. Zhang

Thladiantha hookeri C. B. Clarke

Thladiantha lijiangensis A. M. Lu & Zhi Y. Zhang

Thladiantha longisepala C. Y. Wu, A. M. Lu & Zhi Y. Zhang

Thladiantha medogensis A. M. Lu & J. Q. Li

Thladiantha montana Cogn.

Thladiantha nudiflora Hemsl.

Thladiantha oliveri Cogn. ex Mottet

Thladiantha pustulata (H. Lév.) C. Jeffrey ex A. M. Lu & Zhi Y. Zhang

Thladiantha pustulata var. **jingfushanensis** A. M. Lu & J. Q. Li

Thladiantha sessilifolia Hand.-Mazz.

Thladiantha setispina A. M. Lu & Zhi Y. Zhang

Thladiantha villosula Cogn.

Trichosanthes cordata Roxb.

Trichosanthes cucumeroides var. **dicaelosperma** (C. B. Clarke) S. K. Chen

Trichosanthes dunniana H. Lév.

Trichosanthes lepiniana (Naudin) Cogn.

Trichosanthes pilosa Lour.

Trichosanthes rosthornii Harms

Trichosanthes rosthornii var. **multicirrata** (C. Y. Cheng & C. H. Yueh) S. K. Chen

Trichosanthes rubriflos Thorel ex Cayla

Trichosanthes smilacifolia C. Y. Wu ex C. H. Yueh & C. Y. Cheng

Trichosanthes tricuspidata Lour.

Trichosanthes truncata C. B. Clarke

Trichosanthes wallichiana (Ser.) Wight

Zanonia indica L.

Zehneria bodinieri (H. Lév.) W. J. de Wilde & Duyfjes

Zehneria japonica (Thunb.) H. Y. Liu

146. TETRAMELACEAE

Tetrameles nudiflora R. Br.

147. DATISCACEAE

Datisca cannabina L.

148. BEGONIACEAE

Begonia aborensis Dunn

Begonia acetosella Craib

Begonia annulata K. Koch

Begonia asperifolia Irmsch.

Begonia asperifolia var. **tomentosa** T. T. Yu

Begonia asperifolia var. **unialata** T. C. Ku

Begonia augustinei Hemsl.

Begonia brevisetulosa C. Y. Wu

Begonia burkillii Dunn

Begonia cathcartii Hook. f. & Thomson

Begonia clavicaulis Irmsch.

Begonia dentatobracteata C. Y. Wu

Begonia dioica Buch.-Ham. ex D. Don

Begonia discrepans Irmsch.

Begonia flagellaris Hara

Begonia flaviflora Hara

Begonia flaviflora var. **gamblei** (Irmsch.) Golding & Kareg.

Begonia flaviflora var. **vivida** Golding & Kareg.

Begonia forrestii Irmsch.

Begonia gemmipara Hook. f. & Thomson

Begonia grandis Dryand.

Begonia griffithiana (A. DC.) Warb.

Begonia gungshanensis C. Y. Wu

Begonia handelii Irmsch.

Begonia hatacoa Buch.-Ham. ex D. Don

Begonia henryi Hemsl.

Begonia hymenophylloides F. K. Ward ex L. B. Sm. & Wassh.

Begonia imitans Irmsch.

Begonia iridescens Dunn

Begonia josephii A. DC.

Begonia kachinensis Nob. Tanaka

Begonia kingdon-wardii Tebbitt

Begonia labordei H. Lév

Begonia leptoptera Hara

Begonia longifolia Blume

Begonia medogensis Jian W. Li, Y. H. Tan, X. H. Jin

Begonia megaptera A. DC.

Begonia minicarpa Hara

Begonia modestiflora Kurz

Begonia muliensis T. T. Yu

Begonia nepalensis (A. DC.) Warb.

Begonia nuwakotensis S. Rajbh.

Begonia ovatifolia A. DC.

Begonia palmata D. Don

Begonia palmata var. **bowringiana** (Champion ex Benth.) Golding & Kareg.

Begonia palmata var. **khasiana** (Irmsch.) Golding & Kareg.

Begonia panchtharensis S. Rajbh.

Begonia pedunculosa Wall.

Begonia picta Sm.

Begonia repenticaulis Irmsch.

Begonia rex Putz.

Begonia rockii Irmsch.

Begonia roxburghii (Miq.) A. DC.

Begonia rubella Buch.-Ham. ex D. Don

Begonia satrapis C. B. Clarke

Begonia scintillans Dunn

Begonia scutata Wall. ex A. DC.

Begonia semperflorens Link & Otto

Begonia shilendrae Rekha Morris & P. D. McMillan

Begonia sikkimensis A. DC.

Begonia sikkimensis var. **kamengensis** Rekha Morris, P. D. McMillan & Golding ex Golding

Begonia silhetensis (A. DC.) C. B. Clarke

Begonia taliensis Gagnep.

Begonia taligera S. Rajbh.

Begonia tenella D. Don

Begonia tessaricarpa C. B. Clarke

Begonia thomsonii A. DC.

Begonia tribenensis C. R. Rao

Begonia wilsonii Gagnep.

Begonia xanthina Hook.

149. CELASTRACEAE

Celastrus alnifolius D. Don

Celastrus angulatus Maxim.

Celastrus gemmatus Loes.

Celastrus glaucophyllus Rehder & E. H. Wilson

Celastrus hindsii Benth.

Celastrus hirsutus Comber

Celastrus homaliifolius P. S. Hsu

Celastrus hookeri Prain

Celastrus hypoleucus (Oliv.) Warb. ex Loes.

Celastrus membranifolius Prain

Celastrus microcarpus D. Don

Celastrus monospermoides Loes.

Celastrus monospermus Roxb.

Celastrus orbiculatus Thunb.

Celastrus paniculatus Willd.

Celastrus rosthornianus Loes.

Celastrus rugosus Rehder & E. H. Wilson

Celastrus stylosus Wall.

Celastrus vaniotii (H. Lév.) Rehder

Elaeodendron albens (Retz.) M. Hassler [≡ *Cassine albens* (Retz.) Kosterm.]

Elaeodendron glaucum (Rottb.) Pers.

Elaeodendron paniculatum Wight & Arn.

Euonymus acanthocarpus Franch.

Euonymus aculeolus C. Y. Cheng ex J. S. Ma

Euonymus attenuatus Wall. ex M. A. Lawson

Euonymus bullatus Wall. ex Lodd.

Euonymus chui Hand.-Mazz.

Euonymus cinereus M. A. Lawson

Euonymus clivicola W. W. Sm.

Euonymus cornutoides Loes.

Euonymus cornutus Hemsl.

Euonymus echinatus Wall.

Euonymus fimbriatus Wall.

Euonymus fortunei (Turcz.) Hand.-Mazz.

Euonymus frigidus Wall.

Euonymus giraldii Loes.

Euonymus glaber Roxb.

Euonymus grandiflorus Wall.

Euonymus griffithii Kurz

Euonymus hamiltonianus Wall.

Euonymus hui J. S. Ma

Euonymus jinyangensis C. Y. Chang

Euonymus kachinensis Prain

Euonymus lawsonii C. B. Clarke ex Prain

Euonymus laxiflorus Champ. ex Benth.

Euonymus lichiangensis W. W. Sm.

Euonymus lucidus D. Don

Euonymus maackii Rupr.

Euonymus macrocarpus Gamble ex Oliv.

Euonymus microcarpus (Oliv. ex Loes.) Sprague

Euonymus myrianthus Hemsl.

Euonymus nanoides Loes. & Rehder

Euonymus parasimilis C. Y. Cheng ex J. S. Ma

Euonymus pendulus Wall. ex M. A. Lawson

Euonymus phellomanus Loes.

Euonymus porphyreus Loes.

Euonymus pseudovagans Pit.

Euonymus pygmeus W. W. Sm.

Euonymus rehderianus Loes.

Euonymus sanguineus Loes.

Euonymus schensianus Maxim.

Euonymus semenovii Regel & Herder

Euonymus serratifolius Bedd.

Euonymus szechuanensis Chen H. Wang

Euonymus ternifolius Hand.-Mazz.

Euonymus theifolius Wall. ex M. A. Lawson

Euonymus tibeticus W. W. Sm.

Euonymus tingens Wall.

Euonymus vagans Wall.

Euonymus venosus Hemsl.

Euonymus verrucosoides Loes.

Euonymus viburnoides Prain

Euonymus wilsonii Sprague

Euonymus yunnanensis Franch.

Glyptopetalum angulatum (Griff.) Chakrab. & M. Gangop.

Glyptopetalum griffithii (Kurz) Prain

Glyptopetalum ilicifolium (Franch.) C. Y. Cheng & Q. S. Ma

Gymnosporia acuminata Hook. f.

Gymnosporia berberoides W. W. Sm.

Gymnosporia jinyangensis (C. Y. Chang) Q. R. Liu & Funston

Gymnosporia ovata Wall. ex M. Lawson

Gymnosporia royleana Wall. ex M. A. Lawson

Gymnosporia rufa (Wall.) M. A. Lawson

Gymnosporia variabilis (Hemsl.) Loes.

Lophopetalum wightianum Arn.

Maytenus hookeri Loes.

Maytenus sikkimensis (Prain) D. C. S. Raju & Babu

Microtropis densiflora Wight

Microtropis discolor (Wall.) Arn.

Microtropis longifolia Wall. ex Kurz

Microtropis macrophylla Merr. & Freeman

Microtropis sphaerocarpa C. Y. Cheng & T. C. Kao

Microtropis tetragona Merr. & Freeman

Parnassia angustipetala T. C. Ku

Parnassia cacuminum Hand.-Mazz.

Parnassia chinensis Franch.

Parnassia chinensis var. **ganeshii** S. Akiyama & H. Ohba

Parnassia chinensis var. **sechuanensis** Z. P. Jien

Parnassia cooperi W. E. Evans

Parnassia cordata (Drude) Z. P. Jien ex T. C. Ku

Parnassia crassifolia Franch.

Parnassia davidii Franch.

Parnassia degeensis T. C. Ku

Parnassia delavayi Franch.

Parnassia deqenensis T. C. Ku

Parnassia epunctulata J. T. Pan

Parnassia farreri W. E. Evans

Parnassia kangdingensis T. C. Ku

Parnassia kumaonica Nekrass.

Parnassia labiata Z. P. Jien

Parnassia lanceolata T. C. Ku

Parnassia lanccolata var. **oblongipetala** T. C. Ku

Parnassia laxmannii Pall. ex Schult.

Parnassia lijiangensis T. C. Ku

Parnassia longipetala Hand.-Mazz.

Parnassia longipetala var. **brevipetala** Z. P. Jien ex T. C. Ku

Parnassia longipetaloides J. T. Pan

Parnassia lutea Batal.

Parnassia mysorensis F. Heyne ex Wight & Arn.

Parnassia noemiae Franch.

Parnassia nubicola Wall. ex Royle

Parnassia palustris L.

Parnassia pusilla Wall. ex Arn.

Parnassia qinghaiensis J. T. Pan

Parnassia scaposa Mattf.

Parnassia submysorensis J. T. Pan

Parnassia subscaposa C. Y. Wu ex T. C. Ku

Parnassia tenella Hook. f. & Thomson

Parnassia tibetana Z. P. Jien ex T. C. Ku

Parnassia trinervis Drude

Parnassia venusta Z. P. Jien

Parnassia viridiflora Batalin

Parnassia wightiana Wall. ex Wight & Arn.

Parnassia yanyuanensis T. C. Ku

Parnassia yui Z. P. Jien

Parnassia yulongshanensis T. C. Ku

Parnassia yunnanensis Franch.

Parnassia yunnanensis var. **longistipitata** Z. P. Jien

Pristimera arborea (Roxb.) A. C. Sm.

Reissantia indica (Willd.) N. Hallé

Salacia chinensis L.

Salacia khasiana C. S. Purkay.

Salacia membranacea M. A. Lawson

Salacia roxburghii Wall. ex Wight & Arn.

Salacia salacioides (Roxb.) R. S. Rao & Hemadri

Siphonodon celastrineus Griff.

Tripterygium wilfordii Hook. f.

150. CONNARACEAE

Connarus paniculatus Roxb.

Rourea caudata Planch.

Rourea minor (Gaertn.) Alston

151. OXALIDACEAE

Averrhoa bilimbi L.

Biophytum petersianum Klotzsch

Biophytum reinwardtii (Zucc.) Klotzsch

Biophytum sensitivum (L.) DC.

Oxalis acetosella L.

Oxalis cernua Thunb.

Oxalis corniculata L.

Oxalis corymbosa DC.

Oxalis debilis var. **corymbosa** (DC.) Lourteig

Oxalis dehradunensis Raizada

Oxalis deppei Lodd.

Oxalis griffithii Edgew. & Hook. f.

Oxalis latifolia Kunth

Oxalis leucolepis Diels

Oxalis stricta L.

152. ELAEOCARPACEAE

Elaeocarpus acuminatus Wall. ex Mast.

Elaeocarpus aristatus Roxb.

Elaeocarpus balansae DC.

Elaeocarpus braceanus Watt ex C. B. Clarke

Elaeocarpus brachystachyus H. T. Chang

Elaeocarpus bracteatus Kurz

Elaeocarpus dubius DC.

Elaeocarpus floribundus Blume

Elaeocarpus gaoligongshanensis Y. Tang & Z. L. Dao

Elaeocarpus glandulosus Wall. ex Merr.

Elaeocarpus harmandii Pierre

Elaeocarpus japonicus Siebold

Elaeocarpus lacunosus Wall. ex Kurz

Elaeocarpus lanceifolius Roxb.

Elaeocarpus petiolatus (Jack) Wall. ex Steudel

Elaeocarpus prunifolius Wall. ex Müll. Berol.

Elaeocarpus rugosus Roxb. ex G. Don

Elaeocarpus serratus L.

Elaeocarpus sikkimensis Mast.

Elaeocarpus simplex Kurz

Elaeocarpus sphaericus (Gaertn.) K. Schum.

Elaeocarpus stapfianus Gagnep.

Elaeocarpus tectorius (Lour.) Poir.

Elaeocarpus tuberculatus Roxb.

Elaeocarpus varunua Buch.-Ham. ex Mast.

Elaeocarpus wallichii Kurz

Sloanea dasycarpa (Benth.) Hemsl.

Sloanea forrestii W. W. Sm.

Sloanea hemsleyana (T. Itô) Rehder & E. H. Wilson

Sloanea sterculiacea (Benth.) Rehder & E. H. Wilson

Sloanea sterculiacea var. **assamica** (Benth.) Coode

Sloanea tomentosa (Benth.) Rehder & E. H. Wilson

153. RHIZOPHORACEAE

Carallia brachiata (Lour.) Merr.

154. ERYTHROXYLACEAE

Erythroxylum sinense Y. C. Wu

155. RAFFLESIACEAE

Sapria himalayana Griff.

156. EUPHORBIACEAE

Acalypha australis L.

Acalypha brachystachya Hornem.

Acalypha ciliata Forssk.

Acalypha indica L.

Acalypha mairei (H. Lév) C. K. Schneid.

Acalypha schneideriana Pax & Hoffm.

Acalypha supera Forssk.

Alchornea mollis Benth. ex Müll. Arg.

Alchornea tiliifolia (Benth.) Müll. Arg.

Balakata baccata (Roxb.) Esser

Baliospermum angustifolium Y. T. Chang

Baliospermum bilobatum T. L. Chin

Baliospermum calycinum Müll. Arg.

Baliospermum calycinum var. **densiflorum** (D. G. Long) Chakrab. & N. P. Balakr.

Baliospermum calycinum var. **nepalense** (Hurus. & Yu. Tanaka) Chakrab. & N. P. Balakr.

Baliospermum corymbiferum Hook. f.

Baliospermum micranthum Müll. Arg.

Baliospermum montanum (Willd.) Müll. Arg.

Baliospermum solanifolium (Burm.) Suresh

Chrozophora rottleri (Geisel.) A. Juss. ex Spreng.

Chrozophora tinctoria (L.) A. Juss.

Claoxylon khasianum Hook. f.

Claoxylon longipetiolatum Kurz

Claoxylon mercurialis (L.) Thwaites

Cleidion javanicum Blume

Cleidion javanicum var. **alongense** (Bennet & Sum. Chandra) Chakrab. & N. P. Balakr.

Cleidion javanicum var. **longipedicellatum** Chakrab. & M. Gangop.

Cleidion spiciflorum (Burm. f.) Merr.

Croton bonplandianus Baill.

Croton caudatus Geiseler

Croton chlorocalyx Müll. Arg.

Croton himalaicus D. G. Long

Croton joufra Roxb.

Croton persimilis Müll. Arg.

Croton roxburghii N. P. Balakr.

Croton tiglium L.

Croton yunnanensis W. W. Sm.

Endospermum chinense Benth.

Euphorbia altotibetica Paulsen

Euphorbia antiquorum L.

Euphorbia boissieriana (Woronow) Prokh.

Euphorbia cashmeriana Royle

Euphorbia clarkeana Hook. f.

Euphorbia cognata (Klotzsch & Garcke) Boiss.
Euphorbia cornigera Boiss.
Euphorbia cristata B. Heyne ex Roth
Euphorbia cyathophora Murray
Euphorbia donii Oudejans
Euphorbia dracunculoides Lam.
Euphorbia esula L.
Euphorbia falcata L.
Euphorbia fusiformis Buch.-Ham. ex D. Don
Euphorbia granulata Forssk.
Euphorbia griffithii Hook. f.
Euphorbia heishuiensis W. T. Wang
Euphorbia heterophylla L.
Euphorbia heyneana Spreng.
Euphorbia himalayensis Klotzsch
Euphorbia hirta L.
Euphorbia hispida Boiss.
Euphorbia humifusa Willd.
Euphorbia hypericifolia L.
Euphorbia indica Lam.
Euphorbia jacquemontii Boiss.
Euphorbia jolkinii Boiss.
Euphorbia kanaorica Boiss.
Euphorbia kansuensis Prokh.
Euphorbia lathyris L.
Euphorbia ligularia Roxb.
Euphorbia luteo-viridis D. G. Long
Euphorbia maddeni Boiss.
Euphorbia micractina Boiss.
Euphorbia microsciadia Boiss.
Euphorbia milii Des Moul.
Euphorbia nana Royle
Euphorbia orbiculata Kunth
Euphorbia osyridea Boiss.
Euphorbia pamirica (Prokh.) Prokh.
Euphorbia parviflora L.
Euphorbia peplus L.
Euphorbia prolifera Buch.-Ham. ex D. Don
Euphorbia prostrata Aiton
Euphorbia pseudosikkimensis (Hurus. & Yu. Tanaka) Radcl.-Sm.
Euphorbia pulcherrima Willd. ex Klotzsch
Euphorbia rosea Retz.
Euphorbia royleana Boiss.
Euphorbia schillingii Radcl.-Sm.
Euphorbia schillingii var. **praecox** A. D. Schill. ex Radcl.-Sm.
Euphorbia schugnanica B. Fedtsch.
Euphorbia sharmae U. C. Bhattach.
Euphorbia sieboldiana C. Morren & Decne.
Euphorbia sikkimensis Boiss.
Euphorbia stracheyi Boiss.
Euphorbia thomsoniana Boiss.
Euphorbia thymifolia L.
Euphorbia tibetica Boiss.
Euphorbia tirucalli L.
Euphorbia wallichii Hook. f.
Excoecaria acerifolia Didrichsen
Excoecaria agallocha L.
Excoecaria oppositifolia Griff.
Falconeria insignis Royle
Homonoia riparia Lour.
Jatropha curcas L.
Jatropha gossypifolia L.
Lasiococca symphyllifolia (Kurz) Hook. f.
Macaranga denticulata (Blume) Müll. Arg.
Macaranga gamblei Hook. f.
Macaranga indica Wight
Macaranga peltata (Roxb.) Müll. Arg.
Macaranga pustulata King ex Hook. f.
Mallotus nepalensis Müll. Arg.
Mallotus oreophilus Müll. Arg.
Mallotus philippensis (Lam.) Müll. Arg.
Mallotus repandus (Willd.) Müll. Arg.

Mallotus roxburghianus Müll. Arg.

Mallotus tetracoccus (Roxb.) Kurz

Manihot esculenta Crantz

Mercurialis leiocarpa Siebold & Zucc.

Ostodes katharinae Pax

Ostodes paniculata Blume

Plukenetia corniculata Sm.

Ricinus communis L.

Sapium baccatum Roxb.

Sapium eugeniaefolium Buch.-Ham. ex Hook. f.

Sapium insigne (Royle) Benth. ex Hook. f.

Sapium sebiferum (L.) Roxb.

Sumbaviopsis albicans (Blume) J. J. Sm.

Suregada multiflora (A. Juss.) Baill.

Tragia hispida Willd.

Trevia nudiflora L.

Vernicia cordata (Thunb.) Airy Shaw

Vernicia montana Lour.

157. CENTROPLACACEAE

Bhesa robusta (Roxb.) Ding Hou

158. OCHNACEAE

Ochna integerrima (Lour.) Merr.

Ochna obtusata DC.

Ochna pumila Buch.-Ham. ex D. Don

159. PHYLLANTHACEAE

Actephila excelsa (Dalzell) Müll. Arg.

Andrachne telephioides L.

Antidesma acidum Retz.

Antidesma acuminatum Wall.

Antidesma bunius (L.) Spreng.

Antidesma ghaesembilla Gaertn.

Antidesma japonicum Siebold & Zucc.

Antidesma khasianum Hook. f.

Antidesma montanum Blume

Antidesma nigricans Tul.

Antidesma roxburghii Wall. ex Tul.

Antidesma velutinosum Blume

Aporosa dioica (Roxb.) Müll. Arg.

Aporosa octandra (Buch.-Ham. ex D. Don) Vickery

Baccaurea ramiflora Lour.

Bischofia javanica Blume

Breynia androgyna (L.) Chakrab. & N. P. Balakr.

Breynia bishnupadae (M. Gangop. & Chakrab.) Chakrab. & N. P. Balakr.

Breynia delavayi (Croizat) Welzen & Pruesapan

Breynia gynophora Welzen & Pruesapan

Breynia macrantha (Hassk.) Chakrab. & N. P. Balakr.

Breynia quadrangularis (Willd.) Chakrab. & N. P. Balakr.

Breynia repanda (Müll. Arg.) Chakrab. & N. P. Balakr.

Breynia retusa (Dennst.) Alston

Breynia trinervia (Hook. f. & Thomson ex Müll. Arg.) Chakrab. & N. P. Balakr.

Breynia vitis-idaea (Burm. f.) C. E. C. Fisch.

Bridelia assamica Hook. f.

Bridelia glauca Blume

Bridelia montana (Roxb.) Willd.

Bridelia pubescens Kurz

Bridelia retusa (L.) A. Juss.
Bridelia sikkimensis Gehrm.
Bridelia stipularis (L.) Blume
Bridelia tomentosa Blume
Bridelia verrucosa Haines
Cleistanthus nokrensis B. Sigh
Flueggea leucopyrus Willd.
Flueggea suffruticosa (Pallas) Baill.
Flueggea virosa (Roxb. ex Willd.) Royle
Flueggea virosa subsp. **himalaica** D. G. Long
Glochidion acuminatum Müll. Arg.
Glochidion assamicum (Müll. Arg.) Hook. f.
Glochidion bhutanicum D. G. Long
Glochidion ellipticum Wight
Glochidion helferi (Müll. Arg.) Hook. f.
Glochidion heyneanum (Wight & Arn.) Wight
Glochidion hirsutum (Roxb.) Voigt
Glochidion khasicum (Müll. Arg.) Hook. f.
Glochidion lanceolarium (Roxb.) Voigt
Glochidion medogense T. L. Chin
Glochidion metanubigenum Hurus. & Ya. Tanaka
Glochidion mishmiense Hook. f.
Glochidion multiloculare (Rottl. ex Willd.) Voigt
Glochidion nubigenum Hook. f.
Glochidion oblatum Hook. f.
Glochidion puberum (L.) Hutch.
Glochidion sphaerogynum (Müll. Arg.) Kurz
Glochidion thomsonii (Müll. Arg.) Hook. f.
Glochidion triandrum (Blanco) C. B. Robinson
Glochidion velutinum Wight
Glochidion zeylanicum (Gaertn.) A. Juss.
Glochidion zeylanicum var. **tomentosum** (Dalzell) Trimen
Hymenocardia punctata Wall. ex Lindl.
Leptopus chinensis (Bunge) Pojark.
Leptopus cordifolius Decne.
Leptopus emicans (Dunn) Pojark.
Meineckia macropus (Hook. f.) G. L. Webster
Phyllanthus acidus (L.) Skeels
Phyllanthus airy-shwii Brunel & J. P. Roux
Phyllanthus amarus Schumach. & Thonn.
Phyllanthus brevipes Hook. f.
Phyllanthus clarkei Hook. f.
Phyllanthus cochinchinensis (Lour.) Spreng.
Phyllanthus debilis Klein ex Willd.
Phyllanthus emblica L.
Phyllanthus forrestii W. W. Sm.
Phyllanthus fraternus G. L. Webster
Phyllanthus glaucus Wall. ex Müll. Arg.
Phyllanthus leschenaultii Müll. Arg.
Phyllanthus parvifolius Buch.-Ham. ex D. Don
Phyllanthus praetervisus Müll. Arg.
Phyllanthus pseudoparvifolius R. L. Mitra & Sanjappa
Phyllanthus pulcher Wall. ex Müll. Arg.
Phyllanthus reticulatus Poir.
Phyllanthus rheedei Wight
Phyllanthus roxburghii Müll. Arg.
Phyllanthus sikkimensis Müll. Arg.
Phyllanthus tsarongensis W. W. Sm.
Phyllanthus urinaria L.
Phyllanthus virgatus G. Forst.

160. ELATINACEAE

Bergia ammannioides Roxb. ex Roth
Elatine ambigua Wight
Elatine gracilis Mason
Elatine triandra Schkuhr

161. MALPIGHIACEAE

Aspidopterys elliptica (Blume) A. Juss.

Aspidopterys glabriuscula A. Juss.

Aspidopterys glabriuscula var. **lohitensis** R. C. Srivast.

Aspidopterys indica (Willd.) W. Theob.

Aspidopterys nutans (Roxb. ex DC.) A. Juss.

Aspidopterys orbiculata (Roxb. ex Wall.) Nied.

Aspidopterys wallichii Hook. f.

Hiptage acuminata Wall. ex A. Juss.

Hiptage benghalensis (L.) Kurz

Hiptage yunnanensis Huang ex S. K. Chen

162. DICHAPETALACEAE

Dichapetalum gelonioides (Roxb.) Engl.

163. PUTRANJIVACEAE

Drypetes assamica (Hook. f.) Pax & K. Hoffm.

Drypetes eglandulosa (Roxb.) Pax & K. Hoffm.

Drypetes indica (Müll. Arg.) Pax & Hoffm.

Drypetes venusta (Wight) Pax & K. Hoffm.

Putranjiva roxburghii Wall.

164. PASSIFLORACEAE

Adenia cardiophylla (Masters) Engl.

Adenia cordifolia (Blume) Engl.

Adenia trilobata (Roxb.) Engl.

Passiflora caerulea L.

Passiflora edulis Sims

Passiflora foetida L.

Passiflora leschenaultii DC.

Passiflora napalensis Wall.

Passiflora wilsonii Hemsl.

165. SALICACEAE

Bennettiodendron leprosipes (Clos) Merr.

Carrierea calycina Franch.

Casearia elliptica Willd.

Casearia glomerata Roxb.

Casearia glomerata var. **sikkimensis** (N. Mukh.) R. C. Srivast.

Casearia graveolens Dalzell

Casearia kurzii C. B. Clarke

Casearia tomentosa Roxb.

Casearia vareca Roxb.

Casearia zeylanica (Gaertn.) Thwaites

Flacourtia indica (Burm. f.) Merr.

Flacourtia jangomas (Lour.) Raeusch.

Homalium ceylanicum (Gardner) Benth.

Homalium napaulense (DC.) Benth.

Homalium zeylanicum Benth.

Idesia polycarpa Maxim.

Idesia polycarpa var. **vestita** Diels

Poliothyrsis sinensis Hook. f.

Populus adenopoda Maxim.

Populus alba L.

Populus cathayana Rehder

Populus ciliata Wall. ex Royle

Populus ciliata var. **aurea** C. Marquand & Airy Shaw

Populus ciliata var. **gyirongensis** C. Wang & S. L. Tung

Populus ciliata var. **weixii** C. Wang & S. L. Tung

Populus davidiana Dode

Populus euphratica Olivier

Populus gamblei Dode

Populus glauca H. H. Haines

Populus haoana W. C. Cheng & C. Wang

Populus haoana var. **macrocarpa** C. Wang & S. L. Tung

Populus haoana var. **microcarpa** C. Wang & S. L. Tung

Populus jacquemontiana var. **glauca** Dode

Populus kangdingensis C. Wang & S. L. Tung

Populus lancifolia N. Chao

Populus lasiocarpa Oliv.

Populus mainlingensis C. Wang & S. L. Tung

Populus morisetiana Dode

Populus nigra L.

Populus pamirica Kom.

Populus pseudoglauca C. Wang & P. Y. Fu

Populus purdomii Rehder

Populus qamdoensis C. Wang & S. L. Tung

Populus rotundifolia Griff.

Populus schneideri (Rehder) N. Chao

Populus simonii Carrière

Populus szechuanica C. K. Schneid.

Populus triloba Dode

Populus trinervis C. Wang & S. L. Tung

Populus wenxianica Z. C. Feng & J. L. Guo ex G. Zhu

Populus wilsonii C. K. Schneid.

Populus wuana C. Wang & S. L. Tung

Populus xiangchengensis C. Wang & S. L. Tung

Populus yatungensis (C. Wang & P. Y. Fu) C. Wang & S. L. Tung

Populus yatungensis var. **crenata** C. Wang & S. L. Tung

Populus yuana C. Wang & S. L. Tung

Populus yunnanensis Dode

Salix acmophylla Boiss.

Salix aegyptiaca L.

Salix alba L.

Salix alba L. × **Salix paraplesia** C. K. Schneid.

Salix alfredii Goerz

Salix annulifera C. Marquand & Airy Shaw

Salix argyrophegga C. K. Schneid.

Salix argyrotrichocarpa C. F. Fang

Salix atopantha C. K. Schneid.

Salix austrotibetica N. Chao

Salix babylonica L.

Salix balfouriana C. K. Schneid.

Salix bhutanensis Flod.

Salix bhutanensis var. **yadongensis** (N. Chao) N. Chao

Salix bistyla Hand.-Mazz.

Salix blakii Goerz

Salix brachista C. K. Schneid.

Salix caesia Vill.

Salix calyculata Hook. f. ex Andersson

Salix camusii H. Lév.

Salix capusii Franch.

Salix cavaleriei H. Lév.

Salix cheilophila C. K. Schneid.

Salix clathrata Hand.-Mazz.

Salix coggygria Hand.-Mazz.

Salix coggygria var. **gymnocarpa** (P. Y. Mao & W. Z. Li) L. He

Salix crenata K. S. Hao ex C. F. Fang & A. K. Skvortsov

Salix daltoniana Andersson

Salix delavayana Hand.-Mazz.

Salix delavayana var. **medogensis** (Y. L. Chou) N. Chao

Salix delavayana var. **pilososuturalis** Y. L. Chou & C. F. Fang

Salix denticulata Andersson

Salix dibapha C. K. Schneid.

Salix disperma Roxb. ex D. Don

Salix dissa C. K. Schneid.

Salix driophila C. K. Schneid.

Salix elongata L. He & Z. X. Zhang

Salix eriostachya Wall. ex Andersson

Salix ernestii C. K. Schneid.

Salix excelsa J. F. Gmel.

Salix fargesii Burkill

Salix faxonianoides C. Wang & P. Y. Fu

Salix flabellaris Andersson

Salix floccosa Burkill

Salix fruticulosa Andersson

Salix gilashanica C. Wang & P. Y. Fu

Salix gonggashanica C. F. Fang & A. K. Skvortsov

Salix gyamdaensis C. F. Fang

Salix gyirongensis S. D. Zhao & C. F. Fang

Salix heishuiensis N. Chao

Salix heterochroma Seemen

Salix heterostemon Flod.

Salix hirticaulis Hand.-Mazz.

Salix hylonoma C. K. Schneid.

Salix hypoleuca Seemen

Salix iliensis Regel

Salix jinchuanica N. Chao

Salix jingdongensis C. F. Fang

Salix kamanica C. Wang & P. Y. Fu

Salix karelinii Turcz. ex Stschegl.

Salix lanifera C. F. Fang & S. D. Zhao

Salix lindleyana Wall. ex Andersson

Salix longiflora Andersson

Salix luctuosa H. Lév.

Salix ludlowiana A. K. Skvortsov

Salix maerkangensis N. Chao

Salix magnifica Hemsl.

Salix magnifica var. **ulotricha** (C. K. Schneid.) N. Chao

Salix maizhokunggarensis N. Chao

Salix matsudana Koidz.

Salix mazzettiana N. Chao

Salix microphyta Franch.

Salix minjiangensis N. Chao

Salix miyabeana Seemen

Salix moupinensis Franch.

Salix muliensis Goerz

Salix myrtillacea Andersson

Salix neoamnematchinensis T. Y. Ding & C. F. Fang

Salix nujiangensis N. Chao

Salix nuristanica A. K. Skvortsov

Salix obscura Andersson

Salix ochetophylla Goerz

Salix omeiensis C. K. Schneid.

Salix opsimantha C. K. Schneid.

Salix oreinoma C. K. Schneid.

Salix oreophila Hook. f. ex Andersson

Salix oreophila var. **secta** (Hooker f. ex Andersson) Andersson

Salix oritrepha C. K. Schneid.

Salix paraplesia C. K. Schneid.

Salix pella C. K. Schneid.

Salix pella var. **plocotricha** (C. K. Schneid.) L. He

Salix permollis C. Wang & C. Y. Yu

Salix phaidima C. K. Schneid.

Salix phanera C. K. Schneid.

Salix pilosomicrophylla C. Wang & P. Y. Fu

Salix pilosomicrophylla var. **acuminatomicrophylla** (K. S. Hao ex C. F. Fang & A. K. Skvortsov) L. He

Salix piptotricha Hand.-Mazz.

Salix pseudocalyculata Kimura

Salix pseudospissa Goerz

Salix pseudowolohoensis K. S. Hao ex C. F. Fang & A. K. Skvortsov

Salix psilostigma Andersson

Salix pycnostachya Andersson

Salix radinostachya C. K. Schneid.

Salix radinostachya var. **weixiensis** (C. F. Fang) L. He

Salix rehderiana C. K. Schneid.

Salix resecta Diels

Salix resectoides Hand.-Mazz.

Salix salwinensis Hand.-Mazz. ex Enander

Salix schugnanica Goerz

Salix sclerophylla Andersson

Salix sclerophylla var. **obtusa** (C. Wang & P. Y. Fu) C. F. Fang

Salix sericocarpa Andersson

Salix sikkimensis Andersson

Salix spathulifolia Seemen

Salix sphaeronymphe Goerz

Salix staintoniana A. K. Skvortsov

Salix stomatophora Flod.

Salix taoensis Goerz

Salix tenella C. K. Schneid.

Salix tengchongensis C. F. Fang

Salix tetrasperma Roxb.

Salix thomsoniana Andersson

Salix turanica Nasarow

Salix variegata Franch.

Salix viminalis L.

Salix wilhelmsiana M. Bieb.

Salix wolohoensis C. K. Schneid.

Salix zhegushanica N. Chao

Scolopia crenata (Wight & Arn.) Clos

Xylosma congesta (Lour.) Merr.

Xylosma controversa Clos

Xylosma longifolia Clos

166. VIOLACEAE

Rinorea bengalensis (Wall.) Kuntze

Viola arcuata Blume

Viola belophylla H. Boissieu

Viola betonicifolia Sm.

Viola betonicifolia subsp. **jaunsariensis** (W. Becker) Hara

Viola betonicifolia subsp. **jaunsariensis** var. **cordifolia** Hara

Viola bhutanica Hara

Viola biflora L.

Viola biflora var. **rockiana** (W. Becker) Y. S. Chen

Viola bulbosa Maxim.

Viola cameleo H. Boissieu

Viola canescens Wall.

Viola cinerea var. **stockii** (Boiss.) W. Becker

Viola collina Besser

Viola confertifolia C. C. Chang

Viola davidii Franch.

Viola delavayi Franch.

Viola diffusa Ging.

Viola dimorphophylla Y. S. Chen & Q. E. Yang

Viola dissecta Ledeb.

Viola duclouxii W. Becker

Viola fargesii H. Boissieu

Viola fedtschenkoana W. Becker

Viola fedtschenkoana var. **muzaffarabadensis** W. Becker

Viola forrestiana W. Becker

Viola glaucescens Oudem.

Viola grandisepala W. Becker

Viola hediniana W. Becker

Viola hookeri Thomson ex Hook. f.

Viola inconspicua Blume

Viola indica W. Becker

Viola jangiensis W. Becker

Viola jordani var. **falconeri** (Hook. f. & Thomson) W. Becker

Viola kosanensis Hayata

Viola kunawarensis Royle

Viola macroceras Bunge

Viola moupinensis Franch.

Viola muliensis Y. S. Chen & Q. E. Yang

Viola nuda W. Becker

Viola palmaris Buch.-Ham. ex Ging.

Viola pendulicarpa W. Becker

Viola perpusilla H. Boissieu

Viola philippica Cav.

Viola philippica var. **garhwalensis** W. Becker

Viola pilosa Blume

Viola reichenbachiana Jord. ex Boreau

Viola rupestris F. W. Schmidt

Viola rupestris var. **himalayensis** (W. Becker) S. P. Banerjee & Pramanik

Viola sikkimensis W. Becker

Viola sphaerocarpa W. Becker

Viola striatella H. Boissieu

Viola suavis M. Bieb.

Viola sumatrana Miq.

Viola sylvatica Fries ex Hartm. f.

Viola szetschwanensis W. Becker & H. Boissieu

Viola tenuissima Chang

Viola thomsonii Oudem.

Viola tienschiensis W. Becker

Viola trichopetala Chang

Viola turkestanica Regel & Schmalh.

Viola urophylla Franch.

Viola urophylla var. **densivillosa** Ching J. Wang

Viola wallichiana Ging.

Viola yunnanfuensis W. Becker

167. ACHARIACEAE

Gynocardia odorata Roxb.

Hydnocarpus kurzii (King) Warb.

168. LINACEAE

Anisadenia pubescens Griff.

Anisadenia saxatilis Wall. ex C. F. W. Meissn.

Linum corymbulosum Rchb.

Linum nutans Maxim.

Linum pallescens Bunge

Linum perenne L.

Linum usitatissimum L.

Reinwardtia cicanoba (Buch.-Ham. ex D. Don) Hara

Reinwardtia glandulifera B. Yang

Reinwardtia indica Dumort.

Reinwardtia trigyna (Roxb.) Planch.

169. CALOPHYLLACEAE

Calophyllum polyanthum Wall. ex Planch. & Triana

Mesua assamica (King & Prain) Kosterm.

Mesua ferrea L.

Mesua floribunda (Wall.) Kosterm.

170. CLUSIACEAE

Garcinia acuminata Planch. & Triana

Garcinia anomala Planch. & Triana

Garcinia atroviridis Griff. ex T. Anderson

Garcinia cowa Roxb.

Garcinia esculenta Y. H. Li

Garcinia lanceifolia (G. Don) Roxb.

Garcinia nujiangensis C. Y. Wu & Y. H. Li

Garcinia paniculata Roxb.

Garcinia pedunculata Roxb. ex Buch.-Ham.

Garcinia sopsopia (Buch.-Ham.) Mabb.

Garcinia stipulata T. Anderson

Garcinia xanthochymus Hook. f. ex T. Anderson

Lianthus ellipticifolius (H. L. Li) N. Robson

171. PODOSTEMACEAE

Hydrobryum griffithii (Wall. ex Griff.) Tul.

Zeylanidium olivaceum (Gardner) Engl.

172. HYPERICACEAE

Hypericum acmosepalum N. Robson

Hypericum addingtonii N. Robson

Hypericum ascyron L.

Hypericum assamicum S. N. Biswas

Hypericum bellum H. L. Li

Hypericum choisyanum Wall. ex N. Robson

Hypericum cordifolium Choisy

Hypericum curvisepalum N. Robson

Hypericum daliense N. Robson

Hypericum dyeri Rehder

Hypericum elodeoides Choisy

Hypericum filicaule (Dyer) N. Robson

Hypericum forrestii (Chittenden) N. Robson

Hypericum gracilipes Stapf ex C. E. C. Fisch.

Hypericum gramineum G. Forst.

Hypericum griffithii Hook. f. & Thomson ex Dyer

Hypericum henryi H. Lév & Vaniot

Hypericum himalaicum N. Robson

Hypericum hookerianum Wight & Arn.

Hypericum humifusum L.

Hypericum japonicum Thunb.

Hypericum kingdonii N. Robson

Hypericum lagarocladum N. Robson

Hypericum lancasteri N. Robson

Hypericum latisepalum (N. Robson) N. Robson

Hypericum ludlowii N. Robson

Hypericum maclarenii N. Robson

Hypericum monanthemum Hook. f. & Thomson ex Dyer

Hypericum oblongifolium Choisy

Hypericum perforatum L.

Hypericum petiolulatum Hook. f. & Thomson ex Dyer

Hypericum podocarpoides N. Robson

Hypericum przewalskii Maxim.

Hypericum pseudohenryi N. Robson

Hypericum reptans Hook. f. & Thomson ex Dyer

Hypericum sherriffii N. Robson & D. G. Long

Hypericum subcordatum (R. Keller) N. Robson

Hypericum subsessile N. Robson

Hypericum tenuicaule Hook. f. & Thomson ex Dyer

Hypericum trigonum Hand.-Mazz.
Hypericum uralum Buch.-Ham. ex D. Don
Hypericum wardianum N. Robson
Hypericum wightianum Wall. ex Wight & Arn.
Hypericum williamsii N. Robson

173. GERANIACEAE

Erodium cicutarium (L.) Aiton
Erodium oxyrrhynchum M. Bieb.
Erodium stephanianum Willd.
Erodium tibetanum Edgew.
Geranium camaense C. C. Huang
Geranium canopurpureum Yeo
Geranium carolinianum L.
Geranium charlesii (Aitch. & Hemsl.) Vved.
Geranium christensenianum Hand.-Mazz.
Geranium clarkei Yeo
Geranium collinum Willd.
Geranium dahuricum DC.
Geranium delavayi Franch.
Geranium divaricatum Ehrh.
Geranium donianum Sweet
Geranium farreri Stapf
Geranium himalayense Klotzsch
Geranium hispidissimum (Franch.) R. Knuth
Geranium lambertii Sweet
Geranium lucidum L.
Geranium mascatense Boiss.
Geranium melananthum Andrews
Geranium molle L.
Geranium moupinense Franch.
Geranium nakaoanum Hara
Geranium napuligerum Franch.
Geranium nepalense Sweet
Geranium ocellatum Jacquem. ex Cambess.
Geranium pamiricum Ikonn.
Geranium pogonanthum Franch.
Geranium polyanthes Edgew. & Hook. f.
Geranium pratense L.
Geranium procurrens Yeo
Geranium pusillum L.
Geranium pylzowianum Maxim.
Geranium rectum Trautv.
Geranium rectum var. **glabrata** Trautv.
Geranium refractum Edgew. & Hook. f.
Geranium regelii Nevski
Geranium robertianum L.
Geranium rotundifolium L.
Geranium rubifolium Lindl.
Geranium sibiricum L.
Geranium sinense R. Knuth
Geranium strictipes R. Knuth
Geranium swatense Schönb.-Tem.
Geranium tuberaria Jacquem. ex Cambess.
Geranium umbelliforme Franch.
Geranium wallichianum D. Don ex Sweet
Geranium wardii Yeo
Geranium yunnanense Franch.

174. COMBRETACEAE

Combretum acuminatum Roxb.
Combretum alatum Craib
Combretum flagrocarpum C. B. Clarke
Combretum griffithii Van Heurck & Müll. Arg.
Combretum griffithii var. **yunnanense** (Exell) Turland & C. Chen

Combretum indicum (L.) DeFilipps

Combretum kachinense King & Prain

Combretum nanum Buch.-Ham. ex D. Don

Combretum pilosum Roxb. ex G. Don

Combretum punctatum Blume

Combretum roxburghii Spreng.

Combretum wallichii DC.

Getonia floribunda Roxb.

Terminalia acuminata (Roxb. ex DC.) M. Hassler

Terminalia alata Heyne ex Roth

Terminalia anogeissiana Gere & Boatwr.

Terminalia argyrophylla King & Prain

Terminalia arjuna (Roxb. ex DC.) Wight & Arn.

Terminalia bellirica (Gaertn.) Roxb.

Terminalia catappa L.

Terminalia chebula Retz.

Terminalia citrina Roxb. ex Fleming

Terminalia franchetii Gagnep.

Terminalia myriocarpa Van Heurck & Müll. Arg.

175. LYTHRACEAE

Ammannia auriculata Willd.

Ammannia baccifera L.

Ammannia multiflora Roxb.

Cuphea carthagenensis (Jacq.) J. F. Macbr.

Cuphea procumbens Cav.

Duabanga grandiflora (DC.) Walp.

Lagerstroemia indica L.

Lagerstroemia intermedia Koehne

Lagerstroemia minuticarpa Debberm. ex P. C. Kanjilal

Lagerstroemia parviflora Roxb.

Lagerstroemia reginae Roxb.

Lagerstroemia speciosa (L.) Pers.

Lawsonia inermis L.

Lythrum salicaria L.

Punica granatum L.

Rotala cordata Koehne

Rotala densiflora (Roth) Koehne

Rotala indica (Willd.) Koehne

Rotala mexicana Cham. & Schltdl.

Rotala rosea (Poir.) C. D. K. Cook ex Hara

Rotala rotundifolia (Buch.-Ham. ex Roxb.) Koehne

Rotala rubra (Buch.-Ham. ex D. Don) Hara

Trapa maximowiczii Korsh.

Woodfordia fruticosa (L.) Kurz

176. ONAGRACEAE

Chamerion angustifolium (L.) Holub

Chamerion conspersum (Hausskn.) Holub

Chamerion latifolium (L.) Holub

Chamerion speciosum (Decne.) Holub

Circaea alpina L.

Circaea alpina subsp. **angustifolia** (Hand.-Mazz.) Boufford

Circaea alpina subsp. **imaicola** (Ascherson & Magnus) Kitamura

Circaea alpina subsp. **micrantha** (A. K. Skvortsov) Boufford

Circaea cordata Royle

Circaea micrantha A. Skvortsov

Circaea mollis Siebold & Zucc.

Circaea repens Wall. ex Ascherson & Magnus

Epilobium aitchisonii P. H. Raven

Epilobium amurense Hausskn.

Epilobium augustifolium L.

Epilobium angustifolium subsp. **circumvagum** Mosquin

Epilobium blinii H. Lév

Epilobium brevifolium D. Don

Epilobium brevifolium subsp. **trichoneurum** (Hausskn.) P. H. Raven

Epilobium brevisquamatum P. H. Raven

Epilobium chitralense P. H. Raven

Epilobium clarkeanum Hausskn.

Epilobium conspersum Hausskn.

Epilobium cylindricum D. Don

Epilobium fangii C. J. Chen, Hoch & P. H. Raven

Epilobium glaciale P. H. Raven

Epilobium gouldii P. H. Raven

Epilobium hirsutum L.

Epilobium indicum Hausskn.

Epilobium kingdonii P. H. Raven

Epilobium latifolium L.

Epilobium laxum Royle

Epilobium leiophyllum Hausskn.

Epilobium leiospermum Hausskn.

Epilobium minutiflorum Hausskn.

Epilobium palustre L.

Epilobium pannosum Hausskn.

Epilobium parviflorum Schreber

Epilobium platystigmatosum C. B. Robinson

Epilobium pseudobscurum Hausskn.

Epilobium rhynchospermum Hausskn.

Epilobium royleanum Hausskn.

Epilobium sikkimense Hausskn.

Epilobium staintonii P. H. Raven

Epilobium subcoriaceum Hausskn.

Epilobium tetragonum L.

Epilobium tibetanum Hausskn.

Epilobium trichophyllum Hausskn.

Epilobium wallichianum Hausskn.

Epilobium williamsii P. H. Raven

Ludwigia adscendens (L.) Hara

Ludwigia hyssopifolia (G. Don) Exell

Ludwigia octovalvis (Jacq.) P. H. Raven

Ludwigia perennis L.

Ludwigia prostrata Roxb.

Oenothera biennis L.

Oenothera erythrosepala Borbás

Oenothera rosea L'Her. ex Aiton

Oenothera stricta Ledeb. ex Link

177. MYRTACEAE

Callistemon citrinus (Curtis) Skeels

Corymbia citriodora (Hook.) K. D. Hill & L. A. S. Johnson

Decaspermum fruticosum J. R. Forst. & G. Forst.

Decaspermum parviflorum (Lam.) A. J. Scott

Eucalyptus alba Reinw. ex Blume

Eucalyptus amygdalina Labill.

Eugenia bracteata (Wall.) Roxb.

Eugenia heyneana Duthie

Psidium guajava L.

Syzygium aborense (Dunn) Rathakr. & N. C. Nair

Syzygium anisopetalum (R. Parker) N. P. Balakr.

Syzygium aqueum (Burm. f.) Alston

Syzygium aromaticum (L.) Merr. & L. M. Perry

Syzygium assamicum (Biswas & C. S. Purkay.) Raizada

Syzygium balsameum (Wight) Walp.

Syzygium claviflorum (Roxb.) Wall. ex Steud.

Syzygium cumini (L.) Skeels

Syzygium formosum (Wall.) Masam.

Syzygium forrestii Merr. & L. M. Perry

Syzygium helferi (Duthie) Chantar. & J. Parn.

Syzygium jambos (L.) Alston

Syzygium kurzii (Duthie) N. P. Balakr.

Syzygium malaccense (L.) Merr. & L. M. Perry

Syzygium megacarpum (Craib) Rathakr. & N. C. Nair

Syzygium mishmiense Chatterjee

Syzygium nervosum DC.

Syzygium oblatum (Roxb.) Wall. ex A. M. Cowan & Cowan

Syzygium praecox (Roxb.) Rathakr. & N. C. Nair

Syzygium ramosissimum (Blume) N. P. Balakr.

Syzygium salwinense Merr. & L. M. Perry

Syzygium schmidii Rathakr. & N. C. Nair

Syzygium smalianum (Brandis) D. G. Long

Syzygium stenurum Merr. & L. M. Perry

Syzygium syzygioides (Miq.) Merr. & L. M. Perry

Syzygium tetragonum (Wight) Walp.

Syzygium venosum DC.

Syzygium wallichii (Wight) Walp.

Syzygium xizangense Hung T. Chang & R. H. Miao

178. MELASTOMATACEAE

Blastus cochinchinensis Lour.

Medinilla assamica (C. B. Clarke) C. Chen

Medinilla erythrophylla Wall. ex Lindl.

Medinilla himalayana Hook. f. ex Triana

Medinilla nana S. Y. Hu

Medinilla pauciflora Hook. f. ex Triana

Medinilla rubicunda (Jack) Blume

Medinilla septentrionalis (W. W. Sm.) H. L. Li

Melastoma imbricatum Wall. ex Triana

Melastoma malabathricum L.

Melastoma normale D. Don

Memecylon caeruleum Jack

Memecylon celastrinum Kurz

Osbeckia capitata Benth. ex Walp.

Osbeckia chinensis L.

Osbeckia chinensis var. **angustifolia** (D. Don) C. Y. Wu & C. Chen

Osbeckia crinita Benth. ex C. B. Clarke

Osbeckia glauca Benth.

Osbeckia myrtifolia Blume

Osbeckia nepalensis Hook.

Osbeckia nepalensis var. **albiflora** Lindl.

Osbeckia nutans Wall. ex C. B. Clarke

Osbeckia stellata Buch.-Ham. ex Ker Gawl.

Osbeckia truncata D. Don ex Wight & Arn.

Oxyspora cernua (Roxb.) Hook. f. & Thomson ex Triana

Oxyspora paniculata (D. Don) DC.

Oxyspora vagans (Roxb.) Wall.

Oxyspora yunnanensis H. L. Li

Plagiopetalum esquirolii (H. Lév) Rehder

Sarcopyramis bodinieri H. Lév & Vaniot

Sarcopyramis napalensis Wall.

Sonerila arunachalensis G. S. Giri, A. Pramanik & H. J. Chowdhery

Sonerila erecta Jack

Sonerila khasiana C. B. Clarke

Sonerila kurzii C. B. Clarke

Sonerila maculata Roxb.

Sonerila stricta Hook.

Sporoxeia sciadophila W. W. Sm.

179. CRYPTERONIACEAE

Crypteronia paniculata Blume

180. STAPHYLEACEAE

Euscaphis japonica (Thunb.) Kanitz

Staphylea emodi Wall. ex Brand.

Staphylea holocarpa Hemsl.

Staphylea shweliensis W. W. Sm.

Turpinia affinis Merr. & L. M. Perry

Turpinia cochinchinensis (Lour.) Merr.

Turpinia macrosperma C. C. Huang

Turpinia nepalensis Wall. ex Wight & Arn.

Turpinia pomifera (Roxb.) DC.

181. STACHYURACEAE

Stachyurus chinensis Franch.

Stachyurus cordatulus Merr.

Stachyurus himalaicus Hook. f. & Thomson ex Benth.

Stachyurus himalaicus var. **alatipes** C. Y. Wu

Stachyurus himalaicus var. **microphyllus** C. Y. Wu

Stachyurus yunnanensis Franch.

182. BIEBERSTEINIACEAE

Biebersteinia heterostemon Maxim.

Biebersteinia odora Steph. ex Fisch.

183. NITRARIACEAE

Nitraria schoberi L.

Nitraria schoberi var. **roborowskii** (Kom.) Hadidi

Peganum crithmifolium Retz.

Peganum harmala L.

Peganum harmala var. **grandiflorum** L.

Peganum harmala var. **stenophyllum** Boiss.

184. BURSERACEAE

Canarium bengalense Roxb.

Canarium strictum Roxb.

Garuga floribunda var. **gamblei** (King ex Sm.) Kalkman

Garuga forrestii W. W. Sm.

Garuga pinnata Roxb.

Protium serratum (Wall. ex Colebr.) Engl.

185. ANACARDIACEAE

Anacardium occidentale L.

Buchanania latifolia Roxb.

Choerospondias axillaris (Roxb.) B. L. Burtt & A. W. Hill

Cotinus coggygria var. **glaucophylla** C. Y. Wu

Cotinus nanus W. W. Sm.

Cotinus szechuanensis Pénzes

Dobinea delavayi (Baill.) Baill.

Dobinea vulgaris Buch.-Ham. ex D. Don

Dracontomelon dao (Blanco) Merr. & Rolfe

Drimycarpus racemosus (Roxb.) Hook. f.

Lannea coromandelica (Houtt.) Merr.

Mangifera indica L.

Mangifera khasiana Pierre

Mangifera sylvatica Roxb.

Pegia nitida Colebr.

Pistacia atlantica Desf.

Pistacia chinensis Bunge

Pistacia chinensis subsp. **integerrima** (Stewart ex Brandis) Reichb. f.

Pistacia khinjuk Stocks

Pistacia weinmanniifolia J. Poiss. ex Franch.

Rhus chinensis Mill.

Rhus chinensis var. **roxburghii** (DC.) Rehd.

Rhus dhuna Buch.-Ham. ex Hook. f.

Rhus griffithii Hook. f.

Rhus hookeri Sahni & Bahadur

Rhus parviflora Roxb.

Rhus potaninii Maxim.

Rhus punjabensis J. L. Stewart

Rhus punjabensis var. **pilosa** Engl.

Rhus punjabensis var. **sinica** (Diels) Rehder & E. H. Wilson

Rhus succedanea L.

Rhus teniana Hand.-Mazz.

Rhus wallichii Hook. f.

Rhus wilsonii Hemsl.

Searsia paniculata (Wall. ex G. Don) Moffett

Semecarpus anacardium L. f.

Semecarpus longifolius Bl.

Spondias pinnata (L. f.) Kurz

Toxicodendron acuminatum (DC.) C. Y. Wu & T. L. Ming

Toxicodendron delavayi (Franch.) F. A. Barkley

Toxicodendron delavayi var. **quinquejugum** (Rehder & E. H. Wilson) C. Y. Wu & T. L. Ming

Toxicodendron grandiflorum C. Y. Wu & T. L. Ming

Toxicodendron grandiflorum var. **longipes** (Franch.) C. Y. Wu & T. L. Ming

Toxicodendron griffithii (Hook. f.) Kuntze

Toxicodendron hirtellum C. Y. Wu ex T. L. Ming

Toxicodendron hookeri (K. C. Sahni & Bahadur) C. Y. Wu & T. L. Ming

Toxicodendron hookeri var. **microcarpum** (C. C. Huang ex T. L. Ming) C. Y. Wu & T. L. Ming

Toxicodendron succedaneum (L.) Kuntze

Toxicodendron sylvestre (Siebold & Zucc.) Kuntze

Toxicodendron vernicifluum (Stokes) F. A. Barkley

Toxicodendron wallichii (Hook. f.) Kuntze

Toxicodendron yunnanense var. **longipaniculatum** C. Y. Wu & T. L. Ming

186. SAPINDACEAE

Acer acuminatum Wall. ex D. Don

Acer amplum subsp. **catalpifolium** (Rehder) Y. S. Chen

Acer caesium Wall. ex Brandis

Acer caesium subsp. **giraldii** (Pax) A. E. Murray

Acer campbellii Hook. f. & Thomson ex Hiern

Acer campbellii var. **serratifolium** Banerji

Acer cappadocicum Gled.

Acer cappadocicum var. **brevialatum** W. P. Fang

Acer cappadocicum subsp. **sinicum** (Rehder) Hand.-Mazz.

Acer caudatum Wall.

Acer chienii Hu & W. C. Cheng

Acer davidii Franch.

Acer erianthum Schwer.

Acer flabellatum Rehder

Acer forrestii Diels

Acer franchetii Pax

Acer franchetii var. **megalocarpum** Fang & W. K. Hu

Acer fulvescens Rehder

Acer henryi Pax

Acer heptalobum Diels

Acer hookeri Miq.

Acer kuikiangense Hu & W. C. Cheng

Acer kungshanense W. P. Fang & C. Y. Chang

Acer laevigatum var. **salweenense** (W. W. Sm.) Cowan ex W. P. Fang

Acer laevigatum Wall.

Acer laisuense W. P. Fang & W. K. Hu

Acer laurinum Hassk.

Acer laxiflorum Pax

Acer leipoense W. P. Fang & Soong

Acer maximowiczii Pax

Acer oblogum Wall. ex DC.

Acer oligocarpum Fang & L. C. Hu

Acer oliverianum Pax

Acer osmastonii Gamble

Acer pectinatum Wall. ex Pax

Acer pectinatum subsp. **taronense** (Hand.-Mazz.) A. E. Murray

Acer pentaphyllum Diels

Acer pentapomicum J. L. Stewart ex Brandis

Acer pictum subsp. **macropterum** (W. P. Fang) H. Ohashi

Acer pictum subsp. **mono** (Maxim.) H. Ohashi

Acer pinnatinervium Merr.

Acer platanoides subsp. **turkestanicum** (Pax) P. C. de Jong

Acer pubipetiolatum Hu & W. C. Cheng

Acer shenkanense W. P. Fang ex C. C. Fu

Acer sikkimense Miq.

Acer sikkimense var. **serrulatum** Pax

Acer sinense Pax

Acer stachyophyllum Hiern

Acer stachyophyllum subsp. **betulifolium** (Maxim.) P. C. de Jong

Acer sterculiaceum Wall.

Acer sutchuenense Franch.

Acer thomsonii Miq.

Acer tibetense W. P. Fang

Acer tonkinense Lecomte

Acer wardii W. W. Sm.

Acer wilsonii Rehder

Acer yangbiense Y. S. Chen & Q. E. Yang

Acer yui W. P. Fang

Aesculus assamica Griff.

Aesculus indica (Wall. ex Cambess.) Hook. f.

Allophylus cobbe (L.) Raeusch.

Allophylus chartaceus (Kurz) Radlk.

Allophylus serratus (Hiern) Kurz

Allophylus triphyllus (Burm. f.) Merr.

Allophylus zeylanicus L.

Cardiospermum halicacabum L.

Delavaya trifoliata Franch.

Dimocarpus longan Lour.

Dipteronia sinensis Oliv.

Dodonaea viscosa Jacq.

Dodonaea viscosa subsp. **angustifolia** (L. f.) J. G. West

Eurycorymbus cavaleriei (H. Lév.) Rehder & Hand.-Mazz.

Koelreuteria bipinnata Franch.

Koelreuteria paniculata Laxm.

Lepisanthes erecta (Thwaites) Leenh.

Lepisanthes rubiginosa (Roxb.) Leenh.

Lepisanthes senegalensis (Poir.) Leenh.

Lepisanthes tetraphylla Radlk.

Litchi chinensis Sonn.

Nephelium hypoleucum Kurz

Nephelium mutabile Blume

Nephelium ramboutan-ake (Labill.) Leenh.

Sapindus delavayi (Franch.) Radlk.

Sapindus rarak DC.

Sapindus saponaria L.

Schleichera oleosa (Lour.) Oken

Xerospermum glabratum (Kurz) Radlk.

187. RUTACEAE

Acronychia pedunculata (L.) Miq.

Aegle marmelos (L.) Corrêa

Atalantia caudata Hook. f.

Boenninghausenia albiflora (Hook.) Rchb. ex Meisn.

Citrus aurantiifolia (Christm.) Swingle

Citrus aurantium L.

Citrus cavaleriei H. Lév. ex Cavalier

Citrus hystrix DC.

Citrus jambhiri Lush.

Citrus junos Siebold ex Tanaka

Citrus latipes (Swingle) Tanaka

Citrus limon (L.) Burm. f.

Citrus maxima (Burman) Merr.

Citrus medica L.

Citrus reticulata Blanco

Citrus sinensis (L.) Osbeck

Citrus trifoliata L.

Clausena dentata (Willd.) M. Roem.

Clausena excavata Burm. f.

Clausena heptaphylla (Roxb.) Wight & Arn.

Clausena lansium (Lour.) Skeels

Clausena pentaphylla DC.

Clausena vestita D. D. Tao

Dictamus albus L.

Glycosmis borana V. Naray. ex Tanaka

Glycosmis cyanocarpa (Blume) Spreng.

Glycosmis cymosa (Kurz) V. Naray.

Glycosmis lucida Wall. ex C. C. Huang

Glycosmis pentaphylla (Retz.) DC.

Glycosmis xizangensis (C. Y. Wu & H. Li) D. D. Tao

Haplophyllum dubium Korov.

Haplophyllum gilesii (Hemsl.) C. C. Towns.

Limonia acidissima Groff

Limonia pinnatifolia Houtt.

Melicope lunu-ankenda (Gaertn.) T. G. Hartley

Micromelum integerrimum (Buch.-Ham. ex DC.) Wight & Arn. ex M. Roem.

Micromelum minutum Wight & Arn.

Murraya euchrestifolia Hayata

Murraya koenigii (L.) Spreng.

Murraya paniculata (L.) Jack

Naringi crenulata (Roxb.) Nicolson

Orixa japonica Thunb.

Paramignya griffithii Hook. f.

Paramignya monophylla Wight

Phellodendron chinense C. K. Schneid.

Ruta cordata D. Don

Ruta graveolens L.

Skimmia anquetilia N. P. Taylor & Airy Shaw

Skimmia arborescens T. Anders. ex Gamble

Skimmia arunachalensis Goel & Mehrotra

Skimmia kamengensis Goel & Mehrotra

Skimmia laureola (DC.) Sieb. & Zucc. ex Walp.

Skimmia melanocarpa Rehder & E. H. Wilson

Skimmia multinervia C. C. Huang

Tetradium daniellii (Bennett) T. G. Hartley

Tetradium fraxinifolium (Hook. f.) T. G. Hartley

Tetradium glabrifolium (Champ. ex Benth.) T. G. Hartley

Tetradium ruticarpum (A. Jussieu) T. G. Hartley

Tetradium trichotomum Lour.

Toddalia asiatica (L.) Lam.

Zanthoxylum acanthopodium DC.

Zanthoxylum acanthopodium var. **timbor** Hook. f.

Zanthoxylum armatum DC.

Zanthoxylum bungeanum Maxim.

Zanthoxylum bungeanum var. **pubescens** C. C. Huang

Zanthoxylum bungeanum var. **punctatum** C. C. Huang

Zanthoxylum burkillianum Babu

Zanthoxylum dissitum Hemsl.

Zanthoxylum esquirolii H. Lév.

Zanthoxylum khasianum Hook. f.

Zanthoxylum macranthum (Hand.-Mazz.) C. C. Huang

Zanthoxylum motuoense C. C. Huang

Zanthoxylum multijugum Franch.

Zanthoxylum myriacanthum Wall. ex Hook. f.

Zanthoxylum nepalense Babu

Zanthoxylum nitidum (Roxb.) DC.

Zanthoxylum ovalifolium Wight

Zanthoxylum oxyphyllum Edgew.

Zanthoxylum piasezkii Maxim.

Zanthoxylum pilosulum Rehder & E. H. Wilson

Zanthoxylum pseudoxyphyllum Babu

Zanthoxylum rhetsa (Roxb.) DC.

Zanthoxylum scandens Blume

Zanthoxylum stenophyllum Hemsl.

Zanthoxylum tomentellum Hook. f.

188. SIMAROUBACEAE

Ailanthus altissima (Mill.) Swingle

Ailanthus altissima var. **sutchuenensis** (Dode) Rehder & E. H. Wilson

Ailanthus excelsa Roxb.

Ailanthus integrifolia Lam.

Ailanthus integrifolia subsp. **calycina** (Pierre) Nooteb.

Brucea javanica (L.) Merr.

Brucea mollis Wall. ex Kurz

Picrasma chinensis P. Y. Chen

Picrasma javanica Blume

Picrasma quassioides (D. Don) Benn.

Picrasma quassioides var. **glabrescens** Pamp.

189. MELIACEAE

Aglaia beddomei (Kosterm.) S. S. Jain & R. C. Gaur

Aglaia edulis (Roxb.) Wall.

Aglaia korthalsii Miq.

Aglaia lawii (Wight) C. J. Saldanha

Aglaia perviridis Hiern

Aglaia roxburghiana (Wight & Arn.) Miq.

Aglaia spectabilis (Miq.) S. S. Jain & Bennet

Aphanamixis chittagonga (Miq.) Harid. & R. R. Rao

Aphanamixis polystachya (Wall.) R. Parker

Azadirachta indica A. Juss.

Chisocheton cumingianus subsp. **balansae** (C. DC.) Mabb.

Chukrasia tabularis A. Juss.

Cipadessa baccifera (Roth) Miq.

Dysoxylum excelsum Blume

Dysoxylum gotadhora (Buch.-Ham.) Mabb.

Dysoxylum grande Hiern

Dysoxylum medogense C. Y. Wu & H. Li

Dysoxylum mollissimum Blume

Dysoxylum pallens Hiern

Dysoxylum reticulatum King

Heynea trijuga Roxb.

Melia azedarach L.

Melia dubia Cav.

Melia superba Roxb.

Munronia pinnata (Wall.) W. Theob.

Sphaerosacme decandra (Wall.) Penn.

Toona ciliata M. Roem.

Toona fargesii A. Chev.

Toona sinensis (A. Juss.) M. Roem.

Toona sureni (Blume) Merr.

Walsura robusta Roxb.

Walsura tubulata Hiern

190. DIPENTODONTACEAE

Dipentodon sinicus Dunn

191. MALVACEAE

Abroma angusta (L.) L. f.

Abutilon guineense var. **forrestii** (S. Y. Hu) Y. Tang

Abutilon indicum (L.) Sweet

Abutilon paniculatum Hand.-Mazz.

Abutilon persicum (Burm. f.) Merr.

Abutilon ramosum (Cav.) Guill. & Perr.

Abutilon roseum Hand.-Mazz.

Alcea rosea L.

Bombax ceiba L.

Byttneria crenulata Wall. ex Mast.

Byttneria grandifolia DC.

Byttneria pilosa Roxb.

Colona floribunda (Kurz) Craib

Corchorus aestuans L.

Corchorus capsularis L.

Corchorus depressus (L.) Stocks

Corchorus olitorius L.

Corchorus tridens L.

Corchorus trilocularis L.

Craigia yunnanensis W. W. Sm. & W. E. Evans

Eriolaena spectabilis (DC.) Planch. ex Masters

Eriolaena wallichii DC.

Firmiana colorata (Roxb.) R. Br.

Firmiana fulgens (Wall. ex Mast.) K. Schum.

Firmiana major (W. W. Sm.) Hand.-Mazz.

Firmiana simplex (L.) W. Wight

Gossypium arboreum L.

Gossypium hirsutum L.

Grewia asiatica L.

Grewia brachypoda C. Y. Wu

Grewia denticulata Wall. ex Prain

Grewia eriocarpa Juss.

Grewia hirsuta Vahl

Grewia multiflora A. Juss.

Grewia optiva J. R. Drumm. ex Burret

Grewia sapida Roxb. ex DC.

Grewia sclerophylla Roxb.

Grewia serrulata DC.

Helicteres elongata Wall. ex Mast.

Helicteres hirsuta Lour.

Helicteres isora L.

Helicteres plebeja Kurz

Heritiera macrophylla Wall. ex Kurz

Hibiscus aridicola J. Anthony

Hibiscus caesius Garcke

Hibiscus crinitus (Wall.) G. Don

Hibiscus esculentus L.

Hibiscus fragrans Roxb.

Hibiscus furcatus Willd.

Hibiscus indicus (Burm. f.) Hochr.

Hibiscus lobatus (Murray) Kuntze

Hibiscus macrophyllus Roxb. ex Hornem.

Hibiscus malvaviscus L.

Hibiscus manihot L.

Hibiscus moschatus (Medik.) Salisb.

Hibiscus muliensis (K. M. Feng) Y. S. Chen, comb. et stat. nov. (≡ *Abelmoschus muliensis* K. M. Feng)

Hibiscus mutabilis L.

Hibiscus radiatus Cav.

Hibiscus rosa-sinensis L.

Hibiscus surattensis L.

Hibiscus syriacus L.

Hibiscus vitifolius L.

Kydia calycina Roxb.

Kydia glabrescens Masters

Kydia zizyphifolia Griff.

Lavatera cachemiriana Cambess.

Malva cathayensis M. G. Gilbert, Y. Tang & Dorr

Malva neglecta Wallr.

Malva parviflora L.

Malva pusilla Sm.

Malva sylvestris L.

Malva verticillata L.

Mansonia dipikae C. S. Purkay.

Melochia corchorifolia L.

Microcos paniculata L.

Nayariophyton zizyphifolium (Griff.) D. G. Long & A. G. Mill.

Paradombeya sinensis Dunn

Pentapetes phoenicea L.

Pterospermum acerifolium Willd.

Pterospermum javanicum Jungh.

Pterospermum lanceifolium Roxb.

Pterygota alata (Roxb.) R. Br.

Reevesia pubescens Masters

Reevesia wallichii R. Br.

Sida acuta Burm. f.

Sida cordata (Burm. f.) Borss. Waalk.

Sida cordifolia L.

Sida mysorensis Wight & Arn.

Sida rhombifolia L.

Sida spinosa L.

Sida szechuensis Matsuda

Sida yunnanensis S. Y. Hu

Sterculia balanghas var. **angustifolia** (Roxb. ex G. Don) Mast.

Sterculia euosma W. W. Sm.

Sterculia guttata Roxb. ex G. Don

Sterculia hamiltonii (Kuntze) Adelb.

Sterculia kingii Prain

Sterculia lanceifolia Roxb.

Sterculia maingayi Mast.

Sterculia villosa Roxb.

Thespesia lampas (Cav.) Dalzell & Gibson

Tilia callidonta Hung T. Chang

Tilia chinensis Maxim.

Tilia likiangensis Hung T. Chang

Tilia nobilis Rehder & E. H. Wilson

Tilia paucicostata Maxim.

Tilia paucicostata var. **yunnanensis** Diels

Tilia tuan var. **chenmoui** (W. C. Cheng) Y. Tang

Triumfetta annua L.

Triumfetta cana Blume

Triumfetta obliqua Roth

Triumfetta pilosa Roth

Triumfetta rhomboidea Jacq.

Triumfetta tomentosa Bojer ex Bouton

Urena lobata L.

Urena repanda Roxb. ex Sm.

Wissadula periplocifolia (L.) Thwaites

192. THYMELAEACEAE

Aquilaria agallocha Roxb.

Aquilaria khasiana Hallier f.

Aquilaria malaccensis Lam.

Daphne acutiloba Rehder

Daphne angustiloba Rehder

Daphne aurantiaca Diels

Daphne bholua Buch.-Ham. ex D. Don

Daphne bholua var. **glacialis** (W. W. Sm. & Cave) B. L. Burtt

Daphne brevituba H. F. Zhou ex C. Y. Chang

Daphne composita (L. f.) Gilg

Daphne depauperata H. F. Zhou ex C. Y. Chang

Daphne erosiloba C. Y. Chang

Daphne feddei H. Lév

Daphne gemmata E. Pritz. ex Diels

Daphne holosericea (Diels) Hamaya

Daphne holosericea var. **thibetensis** (Lecomte) Hamaya

Daphne involucrata Wall.

Daphne limprichtii H. Winkl.

Daphne longilobata (Lecomte) Turrill

Daphne ludlowii D. G. Long & S. J. Rae

Daphne macrantha Ludlow

Daphne modesta Rehder

Daphne mucronata Royle

Daphne papyracea Wall. ex Steudel

Daphne penicillata Rehder

Daphne purpurascens S. C. Huang

Daphne retusa Hemsl.

Daphne rosmarinifolia Rehder

Daphne sureil W. W. Sm. & Cave

Daphne tangutica Maxim.

Daphne taylorii Halda

Daphne tenuiflora Bureau & Franch.

Daphne thanguensis Jayanta Ghosh, Mrinmoy Midday, S. K. Dey & D. Maity

Daphne tripartita H. F. Zhou ex C. Y. Chang

Daphne wangiana (Hamaya) Halda

Diarthron vesiculosum (Fisch. & C. A. Mey.) C. A. Mey.

Edgeworthia albiflora Nakai

Edgeworthia gardneri (Wall.) Meisn.

Edgeworthia longipes Lace

Eriosolena composita (L. f.) Tieghem

Stellera chamaejasme L.

Stellera lessertii (Wikstr.) C. A. Mey.

Thymelaea passerina (L.) Coss. & Germ.

Wikstroemia alternifolia Batalin

Wikstroemia baimashanensis S. C. Huang

Wikstroemia canescens Meisn.

Wikstroemia capitatoracemosa S. C. Huang

Wikstroemia delavayi Lecomte

Wikstroemia dolichantha Diels

Wikstroemia haoi Domke

Wikstroemia lamatsoensis Hamaya

Wikstroemia leptophylla W. W. Sm.

Wikstroemia lichiangensis W. W. Sm.

Wikstroemia ligustrina Rehder

Wikstroemia lungtzeensis S. C. Huang

Wikstroemia paxiana H. Winkl.

Wikstroemia scytophylla Diels

Wikstroemia stenophylla E. Pritzel ex Diels

Wikstroemia techinensis S. C. Huang

Wikstroemia zhouana (Halda) C. Shang & S. Liao

193. DIPTEROCARPACEAE

Dipterocarpus gracilis Blume

Dipterocarpus retusus Blume

Hopea shingkeng (Dunn) Bor

Shorea assamica Dyer

Shorea robusta C. F. Gaertn.

Vatica lanceifolia (Roxb.) Blume

194. TROPAEOLACEAE

Tropaeolum majus L.

195. MORINGACEAE

Moringa oleifera Lam.

196. CARICACEAE

Carica papaya L.

197. RESEDACEAE

Oligomeris linifolia (Vahl) J. F. Macbr.

Reseda aucheri Boiss.

198. STIXACEAE

Stixis suaveolens (Roxb.) Pierre

199. CAPPARACEAE

Capparis acutifolia Sweet

Capparis assamica Hook. f. & Thomson

Capparis bodinieri H. Lév.

Capparis cantoniensis Lour.

Capparis decidua (Forsk.) Edgew.

Capparis himalayensis Jafri

Capparis membranifolia Kurz

Capparis multiflora Hook. f. & Thomson

Capparis olacifolia Hook. f. & Thomson

Capparis pachyphylla M. Jacobs

Capparis roxburghii DC.

Capparis sepiaria L.

Capparis sikkimensis Kurz

Capparis spinosa L.

Capparis tenera Dalzell

Capparis yunnanensis Craib & W. W. Sm.

Capparis zeylanica L.

Crateva adansonii subsp. **odora** (Buch.-Ham.) Jacobs

Crateva magna (Lour.) DC.

Crateva religiosa G. Forst.

Crateva unilocularis Buch.-Ham.

Maerua arenaria (DC.) Hook. f. & Thorns.

200. CLEOMACEAE

Arivela viscosa (L.) Raf.

Cleome ariana Hedge & Lamond

Cleome noeana Boiss.

Cleome rutidosperma DC.

Cleome scaposa DC.

Cleoserrata speciosa (Raf.) Iltis

Gynandropsis gynandra (L.) Briq.

Stylidocleome brachycarpa (Vahl ex DC.) Roalson & J. C. Hall

Tarenaya hassleriana (Chodat) Iltis

201. BRASSICACEAE

Acirostrum alaschanicum (Maxim.) Y. Z. Zhao

Alliaria petiolata (M. Bieb.) Cavara & Grande

Alyssum desertorum Stapf

Alyssum turkestanicum Regel & Schmalh.

Aphragmus bouffordii Al-Shehbaz

Aphragmus hinkuensis (Kats. Arai, H. Ohba & Al-Shehbaz) Al-Shehbaz & Warwick

Aphragmus hobsonii (H. Pearson) Al-Shehbaz & Warwick

Aphragmus ladakianus Al-Shehbaz

Aphragmus obscurus (Dunn) O. E. Schulz

Aphragmus ohbana (Al-Shehbaz & Kats. Arai) Al-Shehbaz & Warwick

Aphragmus oxycarpus (Hook. f. & Thomson) Jafri

Aphragmus pygmaeus Al-Shehbaz

Aphragmus serpens (W. W. Sm.) Al-Shehbaz & Warwick

Arabidopsis thaliana (L.) Heynh.

Arabis amplexicaulis Edgew.

Arabis auriculata Lam.

Arabis bijuga Watt

Arabis paniculata Franch.

Arabis pterosperma Edgew.

Arabis pycnocarpa M. Hopkins

Arcyosperma primulifolium (Thomson) O. E. Schulz

Asperuginoides axillaris (Boiss. & Hohen.) Rauschert

Atelanthera perpusilla Hook. f. & Thomson

Baimashania pulvinata Al-Shehbaz

Baimashania wangii Al-Shehbaz

Barbarea intermedia Boreau

Barbarea plantaginea DC.

Barbarea vulgaris W. T. Aiton

Brassica juncea (L.) Czern.

Brassica juncea subsp. **napiformis** (Pailleux & Bois) Gladis

Brassica napus L.

Brassica napus subsp. **napobrassica** (L.) Jafri

Brassica nigra (L.) W. D. J. Koch

Brassica oleracea L.

Brassica rapa L.

Braya fengii (Al-Shehbaz) Al-Shehbaz & D. A. German

Braya forrestii W. W. Sm.

Braya humilis (C. A. Mey.) B. L. Rob.

Braya parvia (C. H. An) Al-Shehbaz & D. A. German

Braya piasezkii (Maxim.) Al-Shehbaz & D. A. German

Braya rosea Bunge

Braya sichuanica Al-Shehbaz

Braya stigmatosa (Franch.) Al-Shehbaz & D. A. German

Braya thomsonii Hook. f.

Braya tibetica Hook. f. & Thomson

Calymmatium draboides (Korsh.) O. E. Schulz

Camelina rumelica Velen.

Camelina sativa (L.) Crantz

Capsella bursa-pastoris (L.) Medik.

Cardamine africana L.

Cardamine calcicola W. W. Sm.

Cardamine calthifolia H. Lév.

Cardamine caroides C. Y. Wu ex W. T. Wang

Cardamine circaeoides Hook. f. & Thomson

Cardamine delavayi Franch.

Cardamine flexuosa With.

Cardamine fragariifolia O. E. Schulz

Cardamine franchetiana Diels

Cardamine gouldii Al-Shehbaz

Cardamine gracilis (O. E. Schulz) T. Y. Cheo & R. C. Fang

Cardamine granulifera (Franch.) Diels

Cardamine griffithii Hook. f. & Thomson

Cardamine hirsuta L.

Cardamine hongdeyuana Al-Shehbaz

Cardamine hydrocotyloides W. T. Wang

Cardamine hygrophila T. Y. Cheo & R. C. Fang

Cardamine impatiens L.

Cardamine ligengiana Al-Shehbaz

Cardamine loxostemonoides O. E. Schulz

Cardamine luxurians (O. E. Schulz) Rashid & H. Ohba

Cardamine macrophylla Willd.

Cardamine microzyga O. E. Schulz

Cardamine multiflora T. Y. Cheo & R. C. Fang

Cardamine multijuga Franch.

Cardamine nepalensis N. Kurosaki & H. Ohba

Cardamine parviflora L.

Cardamine paucifolia Hand.-Mazz.

Cardamine pseudotrifoliolata Al-Shehbaz

Cardamine pulchella (Hook. f. & Thomson) Al-Shehbaz & G. Yang

Cardamine purpurascens (O. E. Schulz) Al-Shehbaz, T. Y. Cheo, L. L. Lou & G. Yang

Cardamine repens (Franch.) Diels

Cardamine rockii O. E. Schulz

Cardamine simplex Hand.-Mazz.

Cardamine stenoloba Hemsl.

Cardamine tangutorum O. E. Schulz

Cardamine tianqingiae Al-Shehbaz & Boufford

Cardamine trifoliolata Hook. f. & Thomson

Cardamine violacea (D. Don) Wall. ex Hook. f. & Thomson

Cardamine yunnanensisb Franch.

Catolobus pendulus (L.) Al-Shehbaz

Chorispora bungeana Fisch. & C. A. Mey.

Chorispora macropoda Trautv.

Chorispora sabulosa Cambess.

Chorispora sibirica (L.) DC.

Chorispora tenella (Pall.) DC.

Christolea crassifolia Cambess.

Conringia planisiliqua Fisch. & C. A. Mey.

Crambe kotschyana Boiss.

Crucihimalaya axillaris (Hook. f. & Thomson) Al-Shehbaz, O'Kane & R. A. Price

Crucihimalaya himalaica (Edgew.) Al-Shehbaz, O'Kane & R. A. Price

Crucihimalaya lasiocarpa (Hook. f. & Thomson) Al-Shehbaz, O'Kane & R. A. Price

Crucihimalaya mollissima (C. A. Mey.) Al-Shehbaz, O'Kane & R. A. Price

Crucihimalaya stricta (Cambess.) Al-Shehbaz, O'Kane & R. A. Price

Crucihimalaya tibetica (Hook. f. & Thomson) Al-Shehbaz, D. A. German & M. Koch

Crucihimalaya wallichii (Hook. f. & Thomson) Al-Shehbaz, O'Kane & R. A. Price

Dendroarabis fruticulosa (C. A. Mey.) D. A. German & Al-Shehbaz

Descurainia sophia (L.) Webb ex Prantl

Didymophysa fedtschenkoana Regel

Dilophia ebracteata Maxim.

Dilophia salsa Thomson

Diplotaxis harra (Forssk.) Boiss.

Dipoma iberideum Franch.

Dontostemon glandulosus (Kar. & Kir.) O. E. Schulz

Dontostemon pinnatifidus (Willd.) Al-Shehbaz & H. Ohba

Draba affghanica Boiss.

Draba alajica Litv.

Draba alshehbazii Klimeš & D. A. German

Draba altaica (C. A. Mey.) Bunge

Draba alticola Kom.

Draba amoena O. E. Schulz

Draba amplexicaulis Franch.

Draba aubrietioides Jafri

Draba bagmatiensis Al-Shehbaz

Draba bartholomewii Al-Shehbaz

Draba bhutanica Hara

Draba cachemirica Gand.

Draba calcicola O. E. Schulz

Draba cholaensis W. W. Sm.

Draba draboides (Maxim.) Al-Shehbaz

Draba elata Hook. f. & Thomson

Draba ellipsoidea Hook. f. & Thomson

Draba eriopoda Turcz. ex Ledeb.

Draba falconeri O. E. Schulz

Draba glomerata Royle

Draba gracillima Hook. f. & Thomson

Draba handelii O. E. Schulz

Draba himachalensis Al-Shehbaz

Draba humillima O. E. Schulz

Draba involucrata (W. W. Sm.) W. W. Sm.

Draba jiulongensis Al-Shehbaz

Draba jucunda W. W. Sm.

Draba koeiei Rech. f.

Draba kongboiana Al-Shehbaz

Draba korschinskyi (O. Fedtsch.) Pohle

Draba ladyginii Pohle

Draba lanceolata Royle

Draba lasiophylla Royle

Draba lichiangensis W. W. Sm.

Draba linearifolia L. L. Chou & T. Y. Cheo

Draba macbeathiana Al-Shehbaz

Draba matangensis O. E. Schulz

Draba melanopus Kom.

Draba mieheorum Al-Shehbaz

Draba mongolica Turcz.

Draba nemorosa L.

Draba nuda (Bél.) Al-Shehbaz & M. Koch

Draba nylamensis Al-Shehbaz

Draba olgae Regel & Schmalh.

Draba oreades Schrenk

Draba oreodoxa W. W. Sm.

Draba pamirica (O. Fedtsch.) Pohle

Draba poluniniana Al-Shehbaz

Draba polyphylla O. E. Schulz

Draba radicans Royle

Draba remotiflora O. E. Schulz

Draba senilis O. E. Schulz

Draba setosa Royle

Draba sherriffii Grierson

Draba sikkimensis (Hook. f. & T. Anderson) Pohle

Draba staintonii Jafri ex Hara

Draba stenobotrys Gilg & O. E. Schulz

Draba stenocarpa Hook. f. & Thomson

Draba sunhangiana Al-Shehbaz

Draba surculosa Franch.

Draba tenerrima O. E. Schulz

Draba tibetica Hook. f. & Thomson

Draba trinervis O. E. Schulz

Draba verna L.

Draba winterbottomii (Hook. & Thomson) Pohle

Draba yuei Al-Shehbaz

Draba yunnanensis Franch.

Draba zangbeiensis L. L. Zhou

Eruca vesicaria subsp. **sativa** (Mill.) Thell.

Erysimum absconditum O. E. Schulz

Erysimum afghanicum Kitam.

Erysimum benthamii Monnet

Erysimum cheiranthoides L.

Erysimum deflexum Hook. f. & Thomson

Erysimum erosum O. E. Schulz

Erysimum forrestii (W. W. Sm.) Polatschek

Erysimum funiculosum Hook. f. & Thomson

Erysimum handel-mazzettii Polatschek

Erysimum limprichtii O. E. Schulz

Erysimum macilentum Bunge

Erysimum melicentae Dunn

Erysimum pachycarpum Hook. f. & Thomson

Erysimum repandum L.

Erysimum robustum D. Don

Erysimum roseum (Maxim.) Polatschek

Erysimum schlagintweitianum O. E. Schulz

Erysimum sisymbrioides C. A. Mey.

Erysimum wardii Polatschek

Euclidium syriacum (L.) W. T. Aiton

Eutrema altaicum (C. A. Mey.) Al-Shehbaz & Warwick

Eutrema boufordii Al-Shehbaz

Eutrema deltoideum (Hook. f. & Thomson) O. E. Schulz

Eutrema edwardsii R. Br.

Eutrema fontanum (Maxim.) Al-Shehbaz & Warwick

Eutrema heterophyllum (W. W. Sm.) Hara

Eutrema himalaicum Hook. f. & Thomson

Eutrema hookeri Al-Shehbaz & Warwick

Eutrema lowndesii (Hara) Al-Shehbaz & Warwick

Eutrema racemosum Al-Shehbaz, G. Q. Hao & J. Q. Liu

Eutrema schulzii Al-Shehbaz & Warwick

Eutrema sherriffii Al-Shehbaz & Warwick

Eutrema tenue (Miq.) Makino

Eutrema verticillatum (Jeffrey & W. W. Sm.) Al-Shehbaz & Warwick

Eutrema wuchengyii (Al-Shehbaz, T. Y. Cheo, L. L. Lu & G. Yang) Al-Shehbaz & Warwick

Eutrema yunnanense Franch.

Goldbachia pendula Botsch.

Goldbachia torulosa DC.

Graellsia chitralensis O. E. Schulz

Hemilophia franchetii Al-Shehbaz

Hemilophia pulchella Franch.

Hemilophia rockii O. E. Schulz

Hemilophia serpens (O. E. Schulz) Al-Shehbaz

Hemilophia sessilifolia Al-Shehbaz

Hornungia procumbens (L.) Hayek

Ianhedgea minutiflora (Hook. f. & Thomson) Al-Shehbaz & O'Kane

Isatis costata C. A. Mey.

Isatis gymnocarpa (Fisch. ex DC.) Al-Shehbaz, Moazzeni & Mumm.

Isatis koelzii Rech. f.

Isatis tinctoria L.

Ladakiella klimesii (Al-Shehbaz) D. A. German & Al-Shehbaz

Leiospora pamirica (Botsch. & Vved.) Botsch. & Pachom.

Lepidium altissimum Rech. f.

Lepidium apetalum Willd.

Lepidium appelianum Al-Shehbaz

Lepidium capitatum Hook. f. & Thomson

Lepidium chalepense L.

Lepidium cordatum Willd. ex Steven

Lepidium cuneiforme C. Y. Wu ex Al-Shehbaz

Lepidium densipuberulum Al-Shehbaz

Lepidium didymum L.

Lepidium latifolium L.

Lepidium obtusum Basiner

Lepidium patrinoides (Regel) Al-Shehbaz & Mumm.

Lepidium pinnatifidum Ledeb.

Lepidium sativum L.

Lepidium silaifolium (Hook. f. & Thomson) Al-Shehbaz & Mumm.

Lepidium virginicum L.

Lepidostemon everestianus Al-Shehbaz

Lepidostemon glaricola Al-Shehbaz

Lepidostemon gouldii Al-Shehbaz

Lepidostemon pedunculosus Hook. f. & Thomson

Lepidostemon rosularis (K. C. Kuan & C. H. An) Al-Shehbaz

Lepidostemon williamsii (Hara) Al-Shehbaz

Litwinowia tenuissima (Pall.) Woronow ex Pavlov

Matthiola chorassanica Bunge ex Boiss.

Matthiola tenera Rech. f.

Megacarpaea bifida Benth.

Megacarpaea delavayi Franch.

Megacarpaea polyandra Benth.

Megadenia pygmaea Maxim.

Nasturtium officinale W. T. Aiton

Neotorularia brevipes (Kar. & Kir.) Hedge & J. Léonard

Neotorularia korolkowii (Regel & Schmalh.) Hedge & J. Léonard

Neslia apiculata Fisch., C. A. Mey. & Avé-Lall.

Noccaea andersonii (Hook. f. & Thomson) Al-Shehbaz

Noccaea cochlearioides (Hook. f. & Thomson) Al-Shehbaz

Noccaea cornuticarpa (Naqshi) Al-Shehbaz

Noccaea flagellifera (O. E. Schulz) Al-Shehbaz

Noccaea griffithiana (Boiss.) F. K. Mey.

Noccaea nepalensis Al-Shehbaz

Noccaea perfoliata (L.) Al-Shehbaz

Noccaea platycarpa (Fisch. & C. A. Mey.) Al-Shehbaz

Noccaea swatensis F. K. Mey.

Noccaea wendelboi (Rech. f.) F. K. Mey.

Noccaea yunnanensis (Franch.) Al-Shehbaz

Notoceras bicorne (Aiton) Amo

Olimarabidopsis pumila (Celak.) Al-Shehbaz, O'Kane & R. A. Price

Parrya axilliflora Jafri

Parrya glabra (Royle) D. A. German & Al-Shehbaz

Parrya minjanensis Rech. f.

Parrya pinnatifida Kar. & Kir.

Parrya podlechii (F. Dvořák) D. A. German & Al-Shehbaz

Parrya turkestanica (Korsh.) N. Busch

Parryodes axilliflora Jafri

Pegaeophyton angustiseptatum Al-Shehbaz, T. Y. Cheo, L. L. Lu & G. Yang

Pegaeophyton minutum Hara

Pegaeophyton nepalense Al-Shehbaz, Arai & H. Ohba

Pegaeophyton purii (D. S. Rawat, L. R. Dangwal & R. D. Gaur) Al-Shehbaz

Pegaeophyton scapiflorum (Hook. f. & Thomson) C. Marquand & Airy Shaw

Pegaeophyton scapiflorum var. **robustum** (O. E. Schulz) Al-Shehbaz, T. Y. Cheo, L. L. Li & G. Yang

Pegaeophyton sulphureum Al-Shehbaz

Pegaeophyton watsonii Al-Shehbaz

Pseudocamelina conwayi (Hemsl.) Al-Shehbaz

Pycnoplinthopsis bhutanica Jafri

Pycnoplinthus uniflorus (Hook. f. & Thomson) O. E. Schulz

Raphanus raphanistrum L.

Raphanus sativus L.

Rapistrum rugosum (L.) All.

Rorippa benghalensis (DC.) Hara

Rorippa cochlearioides (Rath) Al-Shehbaz & Jonsell

Rorippa dubia (Pers.) Hara

Rorippa elata (Hook. f. & Thomson) Hand.-Mazz.

Rorippa globosa (Turcz. ex Fisch. & C. A. Mey.) Hayek

Rorippa indica (L.) Hiern

Rorippa palustris (L.) Besser

Rorippa sylvestris (L.) Besser

Scapiarabis saxicola (Edgew.) M. Koch, R. Karl, D. A. German & Al-Shehbaz

Shangrilaia nana Al-Shehbaz, J. P. Yue & H. Sun

Shehbazia tibetica (Maxim.) D. A. German

Sinapis alba L.

Sinoarabis setosifolia (Al-Shehbaz) R. Karl, D. A. German, M. Koch & Al-Shehbaz

Sisymbriopsis mollipila (Maxim.) Botsch.

Sisymbrium afghanicum Gilli

Sisymbrium altissimum L.

Sisymbrium brassiciforme C. A. Mey.

Sisymbrium heteromallum C. A. Mey.

Sisymbrium irio L.

Sisymbrium loeselii L.

Sisymbrium nepalense Al-Shehbaz

Sisymbrium officinale (L.) Scop.

Sisymbrium orientale L.

Sisymbrium yunnanense W. W. Sm.

Smelowskia annua Rupr.

Smelowskia bartholomewii (Al-Shehbaz) Al-Shehbaz & Warwick

Smelowskia calycina (Stephan) C. A. Mey.

Smelowskia flavissima (Kar. & Kir.) Kar. & Kir.

Smelowskia furcata (Al-Shehbaz) Al-Shehbaz & Warwick

Smelowskia heishuiensis (W. T. Wang) Al-Shehbaz

Smelowskia tibetica (Thomson) Lipsky

Solms-laubachia albiflora (T. Anderson) D. A. German & Al-Shehbaz

Solms-laubachia angustifolia J. P. Yue, Al-Shehbaz & H. Sun

Solms-laubachia baiogoinensis (K. C. Kuan & C. H. An) J. P. Yue, Al-Shehbaz & H. Sun

Solms-laubachia calcicola J. P. Yue, Al-Shehbaz & H. Sun

Solms-laubachia eurycarpa (Maxim.) Botsch.

Solms-laubachia grandiflora J. P. Yue, Al-Shehbaz & H. Sun

Solms-laubachia haranensis (Al-Shehbaz) J. P. Yue, Al-Shehbaz & H. Sun

Solms-laubachia himalayensis (Cambess.) J. P. Yue, Al-Shehbaz & H. Sun

Solms-laubachia jafrii (Al-Shehbaz) J. P. Yue, Al-Shehbaz & H. Sun

Solms-laubachia lanata Botsch.

Solms-laubachia lanuginosa (Hook. f. & Thomson) D. A. German & Al-Shehbaz

Solms-laubachia linearifolia (W. W. Sm.) O. E. Schulz

Solms-laubachia linearis (N. Busch) J. P. Yue, Al-Shehbaz & H. Sun

Solms-laubachia marinellii (Pamp.) D. A. German & Al-Shehbaz

Solms-laubachia mieheorum (Al-Shehbaz) J. P. Yue, Al-Shehbaz & H. Sun

Solms-laubachia minor Hand.-Mazz.

Solms-laubachia mirabilis (Pamp.) J. P. Yue, Al-Shehbaz & H. Sun

Solms-laubachia nepalensis (Hara) J. P. Yue, Al-Shehbaz & H. Sun

Solms-laubachia parryoides (Kurz ex Hook. f. & T. Anderson) D. A. German & Al-Shehbaz

Solms-laubachia platycarpa (Hook. f. & Thomson) Botsch.

Solms-laubachia prolifera (Maxim.) J. P. Yue, Al-Shehbaz & H. Sun

Solms-laubachia pulcherrima Muschl. ex Diels

Solms-laubachia pumila (Kurz) Dvořák

Solms-laubachia retropilosa Botsch.

Solms-laubachia stewartii (T. Anderson) J. P. Yue, Al-Shehbaz & H. Sun

Solms-laubachia sunhangiana J. P. Yue & Al-Shehbaz

Solms-laubachia surculosa (N. Busch) D. A. German & Al-Shehbaz

Solms-laubachia tianbaoshanensis H. L. Chen, Al-Shehbaz, J. P. Yue & H. Sun

Solms-laubachia villosa (Maxim.) D. A. German & Al-Shehbaz

Solms-laubachia xerophyta (W. W. Sm.) H. F. Comber

Solms-laubachia zhongdianensis J. P. Yue, Al-Shehbaz & H. Sun

Stevenia canescens (DC.) D. A. German

Strigosella africana (L.) Botsch.

Strigosella cabulica Boiss.

Strigosella hispida (Litv.) Botsch.

Strigosella intermedia Botsch.

Strigosella strigosa (Boiss.) Botsch.

Strigosella toppinii (O. E. Schulz) Botsch.

Tetracme quadricornis (Stephan ex Willd.) Bunge

Thlaspi arvense L.

Turritis glabra L.

Yinshania acutangula (O. E. Schulz) Y. H. Zhang

Yinshania acutangula var. **wilsonii** (O. E. Schulz) Al-Shehbaz, G. Yang, L. L. Lu & T. Y. Cheo

Yinshania zayuensis Y. H. Zhang

202. BALANOPHORACEAE

Balanophora abbreviata Blume

Balanophora dioica R. Br. ex Royle

Balanophora fargesii (Tieghem) Harms

Balanophora harlandii Hook. f.

Balanophora involucrata Hook. f. & Thomson

Balanophora laxiflora Hemsl.

Balanophora polyandra Griff.

Rhopalocnemis phalloides Junghuhn

203. OLACACEAE

Erythropalum scandens Blume

Olax acuminata Wall. ex Benth.

Olax nana Wall. ex Benth.

Olax scandens Roxb.

Olax wightiana Wall. ex Wight & Arn.

204. OPILIACEAE

Cansjera rheedei J. F. Gmel.

Lepionurus sylvestris Blume

205. SANTALACEAE

Arceuthobium chinense Lecomte

Arceuthobium minutissimum Hook. f.

Arceuthobium oxycedri (DC.) M. Bieb.

Arceuthobium pini Hawksw. & Wiens

Arceuthobium sichuanense (H. S. Kiu) Hawksw. & Wiens

Arceuthobium tibetense H. S. Kiu & W. Ren

Dendrotrophe granulata (Hook. f. & Thomson ex DC.) A. N. Henry & B. Roy

Dendrotrophe platyphylla (Sprengel) N. H. Xia & M. G. Gilbert

Korthalsella japonica (Thunb.) Engl.

Osyris quadripartita Salzm. ex Decne.

Phacellaria compressa Benth.

Pyrularia edulis (Wall.) DC.

Santalum album L.

Thesium bomiense C. Y. Wu ex D. D. Tao

Thesium emodi Hendrych

Thesium himalense Royle ex Edgew.

Thesium hookeri Hendrych

Thesium jarmilae Hendrych

Thesium longiflorum Hand.-Mazz.

Thesium longifolium Turcz.

Thesium orgadophilum P. C. Tam

Thesium pachyrhizum A. DC.

Thesium ramosoides Hendrych

Thesium refractum C. A. Mey.

Thesium remotebracteatum C. Y. Wu & D. D. Tao

Thesium tongolicum Hendrych

Viscum album L.

Viscum album subsp. **meridianum** (Danser) D. G. Long

Viscum articulatum Burm. f.

Viscum diospyrosicola Hayata

Viscum liquidambaricola Hayata

Viscum loranthi Elmer

Viscum monoicum Roxb. ex DC.

Viscum multinerve (Hayata) Hayata

Viscum nudum Danser

Viscum ovalifolium Wall. ex DC.

206. LORANTHACEAE

Dendrophthoe falcata (L. f.) Ettingsh.

Dendrophthoe pentandra (L.) Miq.

Elytranthe albida (Blume) Blume

Elytranthe parasitica (L.) Danser

Elytranthe psilantha (Hook. f.) Danser

Helixanthera coccinea (Jack) Danser

Helixanthera intermedia Danser

Helixanthera ligustrina (Wall.) Danser

Helixanthera parasitica Lour.

Helixanthera scoriarum (W. W. Sm.) Danser

Helixanthera terrestris (Hook. f.) Danser

Loranthus cordifolius Wall.

Loranthus delavayi Tiegh.

Loranthus europaeus Jacq.

Loranthus hypoglaucus Kurz

Loranthus lambertianus Schult. f.

Loranthus longiflorus Desr.

Loranthus odoratus Wall.

Macrosolen cochinchinensis (Lour.) Tiegh.

Macrosolen globosus (Roxb.) Tiegh.

Macrosolen psilanthus (Hook. f.) Danser

Scurrula atropurpurea (Blume) Danser

Scurrula buddleioides (Desr.) G. Don

Scurrula cordifolia (Wall.) G. Don

Scurrula eglandulosa Rajasek.

Scurrula elata (Edgew.) Danser

Scurrula ferruginea (Jack) Danser

Scurrula gongshanensis H. S. Kiu

Scurrula graciliflora (Roxb. ex Schult.) Danser

Scurrula leenhoutsii Rajasek.

Scurrula parasitica L.

Scurrula pulverulenta (Wall.) G. Don

Taxillus caloreas (Diels) Danser

Taxillus delavayi (Tiegh.) Danser

Taxillus kaempferi (DC.) Danser

Taxillus reflexilobus Rajasek.

Taxillus renii H. S. Kiu

Taxillus sericus Danser

Taxillus sutchuenensis (Lecomte) Danser

Taxillus thibetensis (Lecomte) Danser

Taxillus umbellifer (Schult.) Danser

Taxillus vestitus (Wall.) Danser

Tolypanthus involucratus (Roxb.) Blume

207. SCHOEPFIACEAE

Schoepfia fragrans Wall.

Schoepfia griffithii Tiegh. ex Steenis

Schoepfia jasminodora Siebold & Zucc.

208. TAMARICACEAE

Myricaria albiflora Grierson & D. G. Long

Myricaria bracteata Royle

Myricaria davurica (Willd.) Ehrenb.

Myricaria elegans Royle

Myricaria elegans var. **tsetangensis** P. Y. Zhang & Y. J. Zhang

Myricaria germanica subsp. **alopecuroides** (Schrenk) Kitam.

Myricaria germanica subsp. **pakistanica** Qaiser

Myricaria paniculata P. Y. Zhang & Y. J. Zhang

Myricaria prostrata Hook. f. & Thorns. ex Benth. & Hook. f.

Myricaria rosea W. W. Sm.

Myricaria squamosa Desv.

Myricaria wardii C. Marquand

Tamarix aphylla (L.) H. Karst.

Tamarix arceuthoides Bunge

Tamarix dioica Roxb. ex Roth

Tamarix indica Willd.

Tamarix leptostachya Bunge

Tamarix ramosissima Ledeb.

209. PLUMBAGINACEAE

Acantholimon auganum Bunge

Acantholimon chitralicum Rech. f. & Schiman-Czeika

Acantholimon diapensioides Boiss.

Acantholimon erythraeum Bunge

Acantholimon gillii Rech. f. & Köie

Acantholimon lycopodioides (Girard) Boiss.

Acantholimon munroanum Aitch. & Hemsl.

Acantholimon pulchellum Korovin

Acantholimon zaprjagaevii Lincz.

Ceratostigma griffithii C. B. Clarke

Ceratostigma minus Stapf ex Prain

Ceratostigma ulicinum Prain

Ceratostigma willmottianum Stapf

Limonium lacostei (Danguy) Kamelin

Plumbagella micrantha (Ledeb.) Spach

Plumbago capensis Thunb.

Plumbago indica L.

Plumbago zeylanica L.

210. POLYGONACEAE

Atraphaxis pyrifolia Bunge

Atraphaxis spinosa L.

Bistorta affinis (D. Don) Greene

Bistorta amplexicaulis (D. Don) Greene

Bistorta coriacea (Sam.) Yonekura & H. Ohashi

Bistorta diopetes H. Ohba & S. Akiyama

Bistorta emodi (Meisn.) Petrov

Bistorta griffithii (Hook. f.) Grierson

Bistorta jaljalensis H. Ohba & S. Akiyama

Bistorta macrophylla (D. Don) Soják

Bistorta macrophylla var. **stenophylla** (Meisn.) Miyam.

Bistorta milletii H. Lév.

Bistorta milletioides H. Ohba & S. Akiyama

Bistorta officinalis subsp. **pacifica** (Petrov ex Kom.) Yonek.

Bistorta paleacea (Wall.) Yonekura & H. Ohashi

Bistorta perpusilla (Hook. f.) Greene

Bistorta purpureonervosa (A. J. Li) Yonekura & H. Ohashi

Bistorta rubra Yonekura & H. Ohashi

Bistorta sherei H. Ohba & S. Akiyama

Bistorta sinomontana (Sam.) Miyam.

Bistorta subscaposa Petrov

Bistorta suffulta (Maxim.) H. Gross

Bistorta suffulta var. **suffultoides** (A. J. Li) Yonek. & H. Ohashi

Bistorta vacciniifolia (Wall. ex Meisn.) Greene

Bistorta vivipara (L.) Delarbre

Fagopyrum caudatum (Sam.) A. J. Li

Fagopyrum crispatifolium J. L. Liu

Fagopyrum dibotrys (D. Don) Hara

Fagopyrum esculentum Moench

Fagopyrum gilesii (Hemsl.) Hedberg

Fagopyrum kashmirianum Munshi

Fagopyrum leptopodum (Diels) Hedberg

Fagopyrum megacarpum Hara

Fagopyrum tataricum (L.) Gaertn.

Fagopyrum tibeticum (A. J. Li) Adr. Sanchez & Jan. M. Burke

Fallopia aubertii (L. Henry) Holub

Fallopia convolvulus (L.) Á. Löve

Fallopia cynanchoides (Hemsl.) Haraldson

Fallopia cynanchoides var. **glabriuscula** (A. J. Li) A. J. Li

Fallopia dentato-alata (F. Schmidt) Holub

Fallopia dumetorum (L.) Holub

Fallopia filipes (Hara) Holub

Fallopia pterocarpa (Wall. ex Meisn.) Holub

Knorringia sibirica (Laxm.) Tzvelev

Knorringia sibirica subsp. **thomsonii** (Meisn.) S. P. Hong

Koenigia alpina (All.) T. M. Schust. & Reveal

Koenigia campanulata (Hook. f.) T. M. Schust. & Reveal

Koenigia cathayana (A. J. Li) T. M. Schust. & Reveal

Koenigia cyanandra (Diels) Mesicek & Soják

Koenigia delicatula (Meisn.) Hara

Koenigia fertilis Maxim.

Koenigia forrestii (Diels) Mesicek & Soják

Koenigia hookeri (Meisn.) T. M. Schust. & Reveal

Koenigia islandica L.

Koenigia lichiangensis (W. W. Sm.) T. M. Schust. & Reveal

Koenigia limosa (Kom.) T. M. Schust. & Reveal

Koenigia mollis (D. Don) T. M. Schust. & Reveal

Koenigia mollis var. **paniculata** (Blume) T. M. Schust. & Reveal

Koenigia mollis var. **rudis** (Meisn.) M. Hassler

Koenigia nepalensis D. Don

Koenigia pilosa Maxim.

Koenigia pilosa var. **hubertii** (Lingelsh.) T. M. Schust. & Reveal

Koenigia polystachya (Wall. ex Meisn.) T. M. Schust. & Reveal

Koenigia rumicifolia (Royle ex Bab.) T. M. Schust. & Reveal

Koenigia tortuosa (D. Don) T. M. Schust. & Reveal

Koenigia tortuosa var. **tibetana** (Meisn.) T. M. Schust. & Reveal

Oxyria digyna (L.) Hill

Oxyria sinensis Hemsl.

Persicaria amphibia (L.) Delarbre

Persicaria barbata subsp. **stagnina** (Ham. ex Meisn.) Soják

Persicaria breviochreata (Makino) Ohki

Persicaria capitata (D. Don) Gross

Persicaria chinensis (L.) Nakai

Persicaria dichotoma (Blume) S. K. Dixit, Datt & G. P. Roy

Persicaria erectominor (Makino) Nakai

Persicaria filiformis (Thunb.) Nakai

Persicaria glabra (Willd.) M. Gómez

Persicaria glacialis (Meisn.) Hara

Persicaria humilis (Meisn.) Hara

Persicaria hydropiper (L.) Delarbre

Persicaria japonica (Meisn.) Nakai

Persicaria kawagoeana (Makino) Nakai

Persicaria lapathifolia (L.) Delarbre

Persicaria limbata (Meisn.) Hara

Persicaria maculosa S. F. Gray

Persicaria minor (Huds.) Opiz

Persicaria muricata (Meisn.) Nemoto

Persicaria nepalensis (Meisn.) H. Gross

Persicaria orientalis (L.) Spach

Persicaria palmata (Dunn) K. Yonekura & H. Ohashi

Persicaria perfoliata (L.) H. Gross

Persicaria pinetora (Hemsl.) H. Gross

Persicaria posumbu (Buch.-Ham. ex D. Don) H. Gross

Persicaria praetermissa (Hook. f.) Hara

Persicaria pubescens (Bl.) Hara

Persicaria punctata (Ell.) Small

Persicaria runcinata (Buch.-Ham. ex D. Don) H. Gross

Persicaria sagittata (L.) H. Gross

Persicaria sinuata (Royle ex Bab.) Rajbh. & R. Joshi

Persicaria snobarii Munshi & Javeid

Persicaria strigosa (R. Br.) Nakai

Persicaria strindbergii (J. Schust.) Galasso

Persicaria thunbergii (Siebold & Zucc.) H. Gross

Persicaria umbrosa (Sam.) Galasso

Persicaria viscosa (Buch.-Ham. ex D. Don) H. Gross ex Nakai

Polygonum arenastrum Boreau

Polygonum aviculare L.

Polygonum biaristatum Aitch. & Hemsl.

Polygonum cashmiriense H. Gross

Polygonum cognatum subsp. **chitralicum** (Rech. f. & Schiman-Czeika) Qaiser

Polygonum cognatum var. **myrtillifolium** (Kom.) Yurtseva

Polygonum cognatum var. **serpyllaceum** (Jaub. & Spach) Yurtseva

Polygonum corrigioloides Jaub. & Spach

Polygonum forrestii Diels

Polygonum ganderbalensis Munshi & Javeid

Polygonum hastatosagittatum Makino

Polygonum longiseta Bruijn

Polygonum molliaeforme Boiss.

Polygonum nummulariifolium Meisn.

Polygonum paronychioides C. A. Mey. ex Hohen.

Polygonum plebeium R. Br.

Polygonum polycnemoides Jaub. & Spach

Polygonum recumbens Royle ex Bab.

Polygonum rottboellioides Jaub. & Spach

Polygonum rottboellioides var. **tibetica** (Hook. f.) R. R. Stewart

Polygonum roylei Bab.

Polygonum thymifolium Jaub. & Spach

Rheum acuminatum Hook. f. & Thomson

Rheum alexandrae Batalin

Rheum australe D. Don

Rheum delavayi Franch.

Rheum forrestii Diels

Rheum globulosum Gage

Rheum inopinatum Prain

Rheum kialense Franch.

Rheum laciniatum Prain

Rheum lhasaense A. J. Li & P. K. Hsiao

Rheum likiangense Sam.

Rheum moorcroftianum Royle

Rheum nobile Hook. f. & Thomson

Rheum palmatum L.

Rheum przewalskyi Losinsk.

Rheum pumilum Maxim.

Rheum rhomboideum Losinsk.

Rheum spiciforme Royle

Rheum subacaule Sam.

Rheum tanguticum (Maxim. ex Regel) Maxim. ex Balfour

Rheum tibeticum Hook. f.

Rheum webbianum Royle

Rheum yunnanense Sam.

Rumex acetosa L.

Rumex angulatus Rech. f.

Rumex aquaticus L.

Rumex arcuato-ramosus Rech. f.

Rumex chalepensis Mill.

Rumex conglomeratus Murray

Rumex crispellus Rech. f.

Rumex crispus L.

Rumex dentatus L.

Rumex dentatus subsp. **klotzschianus** (Meisn.) Rech. f.

Rumex hastatus D. Don

Rumex nepalensis Spreng.

Rumex palustris Sm.

Rumex panjabensis Vaid & Naithani

Rumex patientia L.

Rumex patientia subsp. **pamiricus** (Rech. f.) Rech. f.

Rumex patientia subsp. **tibeticus** (Rech. f.) Rech. f.

Rumex paulsenianus Rech. f.

Rumex trisetifer Stokes

Rumex vesicarius L.

Rumex yungningensis Sam.

211. DROSERACEAE

Drosera burmanni Vahl

Drosera peltata Sm. ex Willd.

Drosera peltata var. **lunata** Sm.

212. CARYOPHYLLACEAE

Acanthophyllum laxiflorum Boiss.

Acanthophyllum schugnanicum Schischk.

Acanthophyllum stewartii (Thomson ex Edgew. & Hook. f.) Barkoudah

Arenaria acicularis F. N. Williams

Arenaria amdoensis L. H. Zhou

Arenaria auricoma Y. W. Tsui ex L. H. Zhou

Arenaria barbata Franch.

Arenaria barbata var. **hirsutissima** W. W. Sm.

Arenaria baxoiensis L. H. Zhou

Arenaria bomiensis L. H. Zhou

Arenaria brevipetala Y. W. Tsui & L. H. Zhou

Arenaria bryophylla Fernald

Arenaria chamdoensis C. Y. Wu ex L. H. Zhou

Arenaria ciliolata Edgew.

Arenaria ciliolata var. **pendula** Duthie ex F. N. Williams

Arenaria compressa McNeil

Arenaria dawuensis A. J. Li & Q. Ban

Arenaria delavayi Franch.

Arenaria densissima Wall. ex Edgew. & Hook. f.

Arenaria depauperata (Edgew.) Hara

Arenaria dimorphitricha C. Y. Wu ex L. H. Zhou

Arenaria dsharaensis Pax & K. Hoffmann

Arenaria edgeworthiana Majumdar

Arenaria euodonta W. W. Sm.

Arenaria festucoides Benth.

Arenaria festucoides var. **imbricata** Edgew. & Hook. f.

Arenaria filipes C. Y. Wu ex L. H. Zhou

Arenaria fimbriata Mattf.

Arenaria forrestii Diels

Arenaria fridericae Hand.-Mazz.

Arenaria gerzensis L. H. Zhou

Arenaria glanduligera Edgew.

Arenaria glanduligera var. **cernua** F. N. Williams

Arenaria globiflora (Fenzl) Wall. ex Edgew. & Hook. f.

Arenaria griffithii Boiss.

Arenaria griffithii var. **multiflora** (Gilli) Rech. f.

Arenaria haitzeshanensis Y. W. Tsui ex L. H. Zhou

Arenaria inconspicua Hand.-Mazz.

Arenaria inornata W. W. Sm.

Arenaria ionandra Diels

Arenaria ionandra var. **melanotricha** H. F. Comber

Arenaria ischnophylla F. N. Williams

Arenaria kansuensis Maxim.

Arenaria karakorensis Em. Schmid

Arenaria koelzii Rech. f.

Arenaria kumaonensis Maxim.

Arenaria lancangensis L. H. Zhou

Arenaria leucasteria Mattf.

Arenaria littledalei Hemsl.

Arenaria longicaulis C. Y. Wu ex L. H. Zhou

Arenaria longipes C. Y. Wu ex L. H. Zhou

Arenaria longipetiolata C. Y. Wu ex L. H. Zhou

Arenaria longiseta C. Y. Wu ex Z. Xuan

Arenaria longistyla Franch.

Arenaria longistyla var. **eugonophylla** Fernald

Arenaria longistyla var. **pleurogynoides** Diels

Arenaria ludlowii Hara

Arenaria melanandra (Maxim.) Mattf. ex Hand.-Mazz.

Arenaria melandryiformis F. N. Williams

Arenaria melandryoides Edgew.

Arenaria membranisepala C. Y. Wu

Arenaria minima C. Y. Wu ex L. H. Zhou

Arenaria monantha F. N. Williams

Arenaria monilifera Mattf.

Arenaria monosperma F. N. Williams

Arenaria mukerjeeana (Majumdar) Hara

Arenaria napuligera Franch.

Arenaria napuligera var. **monocephala** W. W. Sm.

Arenaria neelgherrensis Wight & Arn.

Arenaria nigricans Hand.-Mazz.

Arenaria nivalomontana C. Y. Wu ex L. H.

Zhou

Arenaria orbiculata Royle ex Edgew. & Hook. f.

Arenaria oreophila Hook. f.

Arenaria paramelanandra Hara

Arenaria pharensis McNeill & Majumdar

Arenaria pogonantha W. W. Sm.

Arenaria polysperma C. Y. Wu ex L. H. Zhou

Arenaria polytrichoides Edgew.

Arenaria przewalskii Maxim.

Arenaria pseudostellaria C. Y. Wu, L. H. Zhou & W. L. Wagner

Arenaria pulvinata Edgew.

Arenaria puranensis L. H. Zhou

Arenaria quadridentata (Maxim.) F. N. Williams

Arenaria ramellata F. N. Williams

Arenaria reducta Hand.-Mazz.

Arenaria rhodantha Pax & K. Hoffmann

Arenaria roborowskii Maxim.

Arenaria rockii Diels

Arenaria roseiflora Sprague

Arenaria rotundifolia M. Bieb.

Arenaria saginoides Maxim.

Arenaria salweenensis W. W. Sm.

Arenaria schneideiana Hand.-Mazz.

Arenaria serpyllifolia L.

Arenaria setifera C. Y. Wu ex L. H. Zhou

Arenaria shannanensis L. H. Zhou

Arenaria sikkimensis Majumdar (syn: *Arenaria debilis* Hook. f., nom. illeg.)

Arenaria smithiana Mattf.

Arenaria spathulifolia C. Y. Wu ex L. H. Zhou

Arenaria stracheyi Edgew.

Arenaria szechuensis F. N. Williams

Arenaria thangoensis W. W. Sm.

Arenaria trichophora Franch.

Arenaria trichophylla C. Y. Wu ex L. H. Zhou

Arenaria tumengelaensis L. H. Zhou

Arenaria weissiana Hand.-Mazz.

Arenaria weissiana var. **bifida** H. Chuang

Arenaria weissiana var. **puberula** C. Y. Wu ex L. H. Zhou

Arenaria xerophila var. **xiangchengensis** (L. H. Zhou) C. Y. Wu

Arenaria xerophila W. W. Sm.

Arenaria yulongshanensis L. H. Zhou ex C. Y. Wu

Arenaria yunnanensis Franch.

Arenaria yunnanensis var. **caespitosa** C. Y. Wu

Arenaria zadoiensis L. H. Zhou

Arenaria zhongdianensis C. Y. Wu

Brachystemma calycinum D. Don

Cerastium alpinum L.

Cerastium arvense subsp. **strictum** Gaudin

Cerastium beeringianum Cham. & Schltdl.

Cerastium cerastoides (L.) Britton

Cerastium dahuricum Fisch.

Cerastium dichotomum L.

Cerastium falcatum Bunge ex Fenzl

Cerastium fontanum Baumg.

Cerastium fontanum subsp. **grandiflorum** (Edgew.) Hara

Cerastium fontanum subsp. **triviale** (Link) Jalas

Cerastium fontanum subsp. **vulgare** (Hartm.) Greuter & Burdet

Cerastium furcatum Cham. & Schltdl.

Cerastium glomeratum Thuill.

Cerastium pusillum Ser.

Cerastium szechuense F. N. Williams

Cerastium thomsonii Hook. f.

Cerastium tianschanicum Schischkin

Cerastium wilsonii Takeda

Cherleria biflora (L.) A. J. Moore & Dillenb.

Cherleria sedoides L.

Dianthus anatolicus Boiss.

Dianthus angulatus Royle

Dianthus barbatus L.

Dianthus cachemericus Edgew.

Dianthus caryophyllus L.

Dianthus chinensis L.

Dianthus crinitus Sm.

Dianthus deltoides L.

Dianthus falconeri Edgew.

Dianthus jacquemontii Edgew. ex Hook. f.

Dianthus kirghizicus Schischk.

Dianthus orientalis Adams

Drymaria cordata (L.) Willd. ex Schult.

Drymaria diandra Blume

Drymaria villosa Schltdl. & Cham.

Gypsophila alsinoides Bunge

Gypsophila cephalotes (Schrenk) Williams

Gypsophila cerastioides D. Don

Gypsophila floribunda (Kar. & Kir.) Turcz. ex Ledeb.

Gypsophila pilosa Huds.

Gypsophila sedifolia Kurz

Herniaria cachemiriana J. Gay

Herniaria glabra L.

Herniaria hirsuta L.

Herniaria incana Lam.

Holosteum kobresietorum Rech. f.

Holosteum umbellatum L.

Lepyrodiclis holosteoides (C. A. Mey.) Fenzl ex Fisch. & C. A. Mey.

Lepyrodiclis stellarioides Fisch. & C. A. Mey.

Lepyrodiclis tenera Boiss.

Lychnis coronaria (L.) Desr.

Lychnis ischnopetala (F. N. Williams) Majumdar

Lychnis senno Siebold & Zucc.

Minuartia ebracteolata Majumdar & G. S. Giri

Myosoton aquaticum (L.) Moench

Petrorhagia alpina (Hablitz) P. W. Ball & Heywood

Polycarpon prostratum (Forssk.) Asch. & Schweinf.

Psammosilene tunicoides W. C. Wu & C. Y. Wu

Pseudostellaria davidii (Franch.) Pax

Pseudostellaria heterantha (Maxim.) Pax

Pseudostellaria heterantha var. **himalaica** Ohwi

Pseudostellaria heterantha var. **nepalensis** (Majumdar) Hara

Pseudostellaria himalaica (Franch.) Pax

Pseudostellaria sylvatica (Maxim.) Pax

Pseudostellaria sylvatica var. **retusa** Ohwi

Pseudostellaria tibetica Ohwi

Sabulina kashmirica (Edgew.) Dillenb. & Kadereit

Sabulina nuristanica (Rech. f.) Dillenb. & Kadereit

Sabulina tenuifolia (L.) Rchb

Sagina apetala Ard.

Sagina japonica (Sw.) Ohwi

Sagina procumbens L.

Sagina purii R. D. Gaur

Sagina saginoides (L.) Karsten

Saponaria griffithiana Boiss.

Silene adenocalyx F. N. Williams

Silene afghanica Rohrb.

Silene amoena L.

Silene apetala Willd.

Silene aprica Turcz. ex Fisch. & C. A. Mey.

Silene arenosa C. Koch

Silene asclepiadea Franch.

Silene atrocastanea Diels

Silene atsaensis (Marquand) Bocquet

Silene baccifera (L.) Roth

Silene batangensis H. Limpr.

Silene bilingua W. W. Sm.

Silene birgittae Bocquet

Silene brahuica Boiss.

Silene breviauriculata Ghaz.

Silene bungei Bocquet

Silene caespitella F. N. Williams

Silene cancellata (Edgew. & Hook. f.) Majumdar

Silene cardiopetala Franch.

Silene cashmeriana (Royle ex Benth.) Majumdar

Silene chodatii Bocquet

Silene chodatii var. **pygmaea** Bocquet

Silene chungtienensis W. W. Sm.

Silene citrina Boiss.

Silene coniflora Nees

Silene conoidea L.

Silene danielii Hadač

Silene davidii (Franch.) Oxelman & Lidén

Silene dawoensis H. Limpr.

Silene delavayi Franch.

Silene dumetosa C. L. Tang

Silene edgeworthii Bocquet

Silene eduardi Bocquet

Silene esquamata W. W. Sm.

Silene falconeri (Rohrb.) Bocquet

Silene falconeriana Benth.

Silene fissicalyx Bocquet & Chater

Silene gallica L.

Silene gonosperma (Rupr.) Bocquet

Silene gonosperma subsp. **himalayensis** (Rohrb.) Bocquet

Silene gracilenta H. Chuang

Silene gracilicaulis C. L. Tang

Silene graminifolia Otth

Silene grandiflora Franch.

Silene grandiflora var. **xerobatica** W. W. Sm.

Silene guntensis B. Fedtsch.

Silene helleboriflora Exell & Bocquet

Silene herbilegorum (Bocquet) Lidén & Oxelman

Silene holosteifolia Bocquet & Chater

Silene huguettiae Bocquet

Silene huguettiae var. **pilosa** C. Y. Wu & H. Chuang

Silene indica Roxb. ex Otth

Silene indica var. **bhutanica** (W. W. Sm.) Bocquet

Silene joerstadii Wendelbo

Silene jugorum (F. N. Williams) Bocquet & Chater

Silene julaensis Grierson

Silene khasiana Rohrb.

Silene kialensis (F. N. Williams) Lidén & Oxelman

Silene kumaonensis F. N. Williams

Silene kunawarensis Benth.

Silene lamarum C. Y. Wu

Silene laxantha Majumdar

Silene lhassana (F. N. Williams) Majumdar

Silene lichiangensis W. W. Sm.

Silene linae Bocquet

Silene lineariloba C. Y. Wu

Silene longicarpophora (Kom.) Bocquet

Silene longicornuta C. Y. Wu & C. L. Tang

Silene longisepala E. Nasir

Silene luciliae Bocquet

Silene madens Majumdar

Silene melanantha Franch.

Silene monbeigii W. W. Sm.

Silene moorcroftiana Wall. ex Benth.

Silene muliensis C. Y. Wu

Silene multifurcata C. L. Tang

Silene namlaensis (Marquand) Bocquet

Silene nangqenensis C. L. Tang

Silene napuligera Franch.

Silene nepalensis Majumdar

Silene nigrescens (Edgew.) Majumdar

Silene nigrescens subsp. **latifolia** Bocquet

Silene oblanceolata W. W. Sm.

Silene otodonta Franch.

Silene phoenicodonta Franch.

Silene platyphylla Franch.

Silene principis Oxelman & Lidén

Silene pseudocashmeriana Bocquet & Chater

Silene pseudo-verticellata E. Nasir

Silene pubicalycina C. Y. Wu

Silene puranensis (L. H. Zhou) C. Y. Wu & H. Chuang

Silene purii Bocquet & N. P. Saxena

Silene rechingeri Bocquet

Silene repens Patrin

Silene rosiflora Kingdon-Ward ex W. W. Sm.

Silene rubricalyx (Marquand) Bocquet

Silene salicifolia C. L. Tang

Silene scopulorum Franch.

Silene setaesperma Majumdar

Silene songarica (Fisch., C. A. Mey. & Avé-Lall.) Bocquet

Silene staintonii Ghaz.

Silene stellariifolia Bocquet & Chater

Silene stewartiana Diels

Silene stewartii (Edgew.) Majumdar

Silene stracheyi Edgew.

Silene subcretacea F. N. Williams

Silene tachtensis Franch.

Silene tenuis Willd.

Silene thomsonii Majumdar

Silene trachyphylla Franch.

Silene tubulosa Oxelman & Lidén

Silene vautierae Bocquet

Silene viridiflora L.

Silene viscidula Franch.

Silene viscosa (L.) Pers.

Silene vulgaris (Moench) Garcke

Silene waltonii F. N. Williams

Silene wardii (Marquand) Bocquet

Silene yetii Bocquet

Silene yunnanensis Franch.

Silene zhongbaensis (L. H. Zhou) C. Y. Wu & C. L. Tang

Silene zhoui C. Y. Wu

Spergula arvensis L.

Spergula fallax (Lowe) E. H. L. Krause

Spergularia diandra (Guss.) Heldr.

Spergularia marina (L.) Bess.

Spergularia media (L.) C. Presl ex Griseb.

Spergularia rubra (L.) J. & C. Presl

Stellaria abaensis H. F. Xu & Z. H. Ma

Stellaria alsine Grimm

Stellaria alsine var. **alpina** (Schur) Hand.-Mazz.

Stellaria alsinoides Boiss. & Buhse

Stellaria aquatica (L.) Scop.

Stellaria arenarioides Shi L. Chen, Rabeler & Turland

Stellaria brachypetala Bunge

Stellaria congestiflora Hara

Stellaria decumbens Edgew.

Stellaria decumbens var. **arenarioides** L. H. Zhou

Stellaria decumbens var. **polyantha** Edgew. & Hook. f.

Stellaria delavayi Franch.

Stellaria depressa Em. Schmid

Stellaria dianthifolia F. N. Williams

Stellaria fontana Popov

Stellaria graminea L.

Stellaria graminea var. **chinensis** Maxim.

Stellaria graminea var. **pilosula** Maxim.

Stellaria graminea var. **viridescens** Maxim.

Stellaria gyangtseensis F. N. Williams

Stellaria gyirongensis L. H. Zhou

Stellaria himalayensis Majumdar

Stellaria infracta Maxim.

Stellaria koelzii Rech. f.

Stellaria lanata Hook. f.

Stellaria lanipes C. Y. Wu & H. Chuang

Stellaria mainlingensis L. H. Zhou

Stellaria media (L.) Vill.

Stellaria media var. **procera** Klatt & Richter

Stellaria monosperma Buch.-Ham. ex D. Don

Stellaria monosperma var. **paniculata** (Edgew.) Majumdar

Stellaria montioides (Edgew. & Hook. f.) Ghaz.

Stellaria neglecta Weihe

Stellaria nepalensis Majumdar & Vartak

Stellaria omeiensis C. Y. Wu & Y. W. Tsui ex P. Ke

Stellaria ovatifolia (M. Mizush.) M. Mizush.

Stellaria palustris Retz.

Stellaria patens D. Don

Stellaria persica Boiss.

Stellaria petiolaris Hand.-Mazz.

Stellaria pilosoides Shi L. Chen, Rabeler & Turland

Stellaria pusilla Em. Schmid

Stellaria reticulivena Hayata

Stellaria sarcophylla Rech. f.

Stellaria semivestita Edgew. ex Edgew. & Hook. f.

Stellaria sikaramensis Rech. f.

Stellaria sikkimensis Hook. f.

Stellaria souliei F. N. Williams

Stellaria subumbellata Edgew.

Stellaria tibetica Kurz

Stellaria turkestanica Schischk.

Stellaria uda F. N. Williams

Stellaria uliginosa Murray

Stellaria uliginosa var. **undulata** (Thunb.) Fenzl

Stellaria umbellata Turcz. ex Kar. & Kir.

Stellaria vestita Kurz

Stellaria vestita var. **amplexicaulis** (Hand.-Mazz.) C. Y. Wu

Stellaria wallichiana Haines

Stellaria webbiana (Benth. ex G. Don) Edgew. & Hook. f.

Stellaria yunnanensis Franch.

Stellaria zangnanensis L. H. Zhou

Thylacospermum caespitosum (Cambess.) Schischk.

Vaccaria hispanica (Mill.) Rauschert

Vaccaria oxyodonta Boiss.

Velezia rigida L.

213. AMARANTHACEAE

Achyranthes aspera L.

Achyranthes aspera var. **argentea** C. B. Clarke

Achyranthes aspera var. **indica** L.

Achyranthes aspera var. **porphyristachya** (Wall. ex Moq.) Hook. f.

Achyranthes bidentata Blume

Acroglochin persicarioides (Poiret) Moq.

Aerva glabrata Hook. f.

Aerva javanica (Burm. f.) Juss. ex Schult.

Aerva javanica var. **bovei** Webb

Aerva lanata (L.) Juss.

Aerva sanguinolenta (L.) Blume

Agriophyllum squarrosum (L.) Moq.

Allmania nodiflora (L.) R. Br. ex Wight

Alternanthera bettzickiana (Regel) G. Nicholson

Alternanthera brasiliana (L.) O. Kuntze

Alternanthera caracasana Kunth

Alternanthera paronychioides A. St.-Hil.

Alternanthera philoxeroides (Mart.) Griseb.

Alternanthera pungens Kunth

Alternanthera sessilis (L.) R. Br. ex DC.

Amaranthus albus L.

Amaranthus blitum L.

Amaranthus blitum subsp. **oleraceus** (L.) Costea

Amaranthus caudatus L.

Amaranthus graecizans L.

Amaranthus hybridus L.

Amaranthus hypochondriacus L.

Amaranthus retroflexus L.

Amaranthus roxburghianus H. W. Kung

Amaranthus spinosus L.

Amaranthus tricolor L.

Amaranthus viridis L.

Archiatriplex nanpinensis G. L. Chu

Atriplex centralasiatica Iljin

Atriplex hortensis L.

Atriplex pamirica Iljin

Atriplex schugnanica Iljin

Axyris hybrida L.

Axyris mira Sukhor.

Axyris prostrata L.

Axyris villosa Moq.

Bassia dasyphylla (Fischer & C. A. Meyer) Kuntze

Bassia hyssopifolia (Pall.) Kuntze

Bassia scoparia (L.) Scott

Beta vulgaris L.

Blitum virgatum L.

Bosea amherstiana (Moq.) Hook. f.

Celosia argentea L.

Celosia cristata L.

Centrostachys aquatica (R. Br.) Moq.

Chenopodiastrum badachschanicum (Tzvelev) S. Fuentes, Uotila & Borsch

Chenopodiastrum murale (L.) S. Fuentes, Uotila & Borsch

Chenopodium album L.

Chenopodium bengalense (Lam.) Spielm. ex Steud.

Chenopodium botrys L.

Chenopodium ficifolium Sm.

Chenopodium foliosum Asch.

Chenopodium giganteum D. Don

Chenopodium glaucum L.

Chenopodium hybridum L.

Chenopodium iljinii Golosk.

Chenopodium karoi (Murr) Aellen

Chenopodium murale L.

Chenopodium nepalense Coll

Chenopodium pallidum Moq.

Chenopodium perttii Sukhor.

Corispermum falcatum Iljin

Corispermum ikramii Aellen

Corispermum lepidocarpum Grubov

Corispermum lhasaense C. P. Tsien & C. G. Ma

Corispermum pamiricum Iljin

Corispermum pamiricum var. **pilocarpum** C. P. Tsien & C. G. Ma

Corispermum pseudofalcatum C. P. Tsien & C. G. Ma

Corispermum tibeticum Iljin

Cyathula capitata Moq.

Cyathula officinalis K. C. Kuan

Cyathula prostrata (L.) Blume

Cyathula tomentosa (Roth) Moq.

Deeringia amaranthoides (Lam.) Merr.

Digera muricata (L.) Mart.

Dysphania ambrosioides (L.) Mosyakin & Clemants

Dysphania aristata (L.) Mosyakin & Clemants

Dysphania bhutanica Sukhor.

Dysphania geoffreyi Sukhor.

Dysphania himalaica Uotila

Dysphania neglecta Sukhor.

Dysphania nepalensis (Colla) Mosyakin & Clemants

Dysphania schraderiana (Roem. & Schult.) Mosyakin & Clemants

Dysphania tibetica (A. J. Li) Uotila

Gomphrena celosioides Mart.

Gomphrena globosa L.

Halogeton glomeratus (M. Bieb.) C. A. Mey.

Halogeton glomeratus var. **tibeticus** (Bunge) Grubov

Halogeton kashmirianus Grey-Wilson & Wadhwa

Haloxylon griffithii subsp. **wakhanicum** (Paulsen) Hedge

Iresine herbstii Hook.

Kali jacquemontii (Moq.) Akhani & Roalson

Kochia prostrata (L.) Schrad.

Kochia scoparia (L.) Schrad.

Kochia stellaris Moq.

Krascheninnikovia ceratoides (L.) Gueldenst.

Krascheninnikovia compacta (Losinsk.) Grubov

Krascheninnikovia ewersmanniana (Losinsk.) Grubov

Microgynoecium tibeticum Hook. f.

Oxybasis glauca (L.) S. Fuentes, Uotila & Borsch

Psilotrichum ferrugineum (Roxb.) Moq.

Pupalia lappacea (L.) Juss.

Pupalia lappacea var. **velutina** (Moq.) Hook. f.

Salsola collina Pall.

Salsola jacquemontii Moq.

Salsola kali subsp. **ruthenica** (Iljin) Soó

Salsola monoptera Bunge

Salsola nepalensis Grubov

Salsola tragus L.

Salsola vermiculata L.

Stilbanthus scandens Hook. f.

Suaeda corniculata (C. A. Mey.) Bunge

Suaeda olufsenii Paulsen

214. GISEKIACEAE

Gisekia pharnaceoides L.

215. AIZOACEAE

Trianthema portulacastrum L.

Zaleya pentandra (L.) Jeffrey

216. PHYTOLACCACEAE

Phytolacca acinosa Roxb.

Phytolacca americana L.

Phytolacca dioica L.

Phytolacca polyandra Batalin

217. NYCTAGINACEAE

Boerhavia coccinea Mill.

Boerhavia diffusa L.

Bougainvillea glabra Choisy

Bougainvillea spectabilis Willd.

Commicarpus chinensis (L.) Heimerl

Commicarpus lantsangensis D. Q. Lu

Mirabilis jalapa L.

Oxybaphus himalaicus Edgew.

Oxybaphus himalaicus var. **chinensis** (Heim.) D. Q. Lu

218. MOLLUGINACEAE

Glinus lotoides L.

Mollugo cerviana (L.) Ser.

Mollugo nudicaulis Lam.

Mollugo pentaphylla L.

Mollugo stricta L.

219. BASELLACEAE

Basella alba L.

220. TALINACEAE

Talinum portulacifolium (Forssk.) Asch. ex Schweinf.

221. PORTULACACEAE

Portulaca oleracea L.

Portulaca pilosa L.

Portulaca quadrifida L.

Talinum cuneifolium Willd.

222. CACTACEAE

Cereus peruvianus (L.) Mill.

Opuntia monacantha (Willd.) Haw.

223. CORNACEAE

Alangium alpinum (C. B. Clarke) W. W. Sm. & Cave

Alangium barbatum (R. Br. ex C. B. Clarke) Baill. ex Kuntze

Alangium chinense (Lour.) Harms

Alangium chinense subsp. **pauciflorum** W. P. Fang

Alangium chinense subsp. **strigosum** W. P. Fang

Alangium chinense subsp. **triangulare** (Wangerin) W. P. Fang

Alangium faberi Oliver

Alangium faberi var. **dolichocarpum** Z. Y. Li

Alangium premnifolium Ohwi

Alangium qingchuanense M. Y. He

Alangium salviifolium (L. f.) Wangerin

Camptotheca acuminata Decne.

Cornus capitata Wall.

Cornus chinensis Wangerin

Cornus controversa Hemsl.

Cornus hemsleyi C. K. Schneid. & Wangerin

Cornus macrophylla Wall.

Cornus macrophylla var. **stracheyi** C. B. Clarke

Cornus oblonga Wall.

Cornus oblonga var. **glabrescens** W. P. Fang & W. K. Hu

Cornus oblonga var. **griffithii** C. B. Clarke

Cornus oligophlebia Merr.

Cornus papillosa W. P. Fang & W. K. Hu

Cornus schindleri Wangerin

Cornus ulotricha C. K. Schneid. & Wangerin

Cornus walteri Wangerin

Davidia involucrata Baill.

Mastixia microcarpa Y. C. Liu & H. Peng

Mastixia pentandra Blume

Mastixia rostrata Blume

Nyssa javanica (Blume) Wangerin

224. HYDRANGEACEAE

Deutzia aspera Rehder

Deutzia bomiensis S. M. Hwang

Deutzia breviloba S. M. Hwang

Deutzia calycosa Rehder

Deutzia calycosa var. **xerophyta** (Hand.-Mazz.) S. M. Hwang

Deutzia compacta Craib

Deutzia corymbosa R. Br. ex G. Don

Deutzia crassidentata S. M. Hwang

Deutzia crassifolia Rehder

Deutzia discolor Hemsl.

Deutzia glaucophylla S. M. Hwang

Deutzia glomeruliflora Franch.

Deutzia hetcrophylla S. M. Hwang

Deutzia hypoglauca Rehder

Deutzia longifolia Franch.

Deutzia macrantha Hook. f. & Thomson

Deutzia monbeigii W. W. Sm.

Deutzia muliensis S. M. Hwang

Deutzia obtusilobata S. M. Hwang

Deutzia pilosa Rehder

Deutzia purpurascens (Franch. ex L. Henry) Rehder

Deutzia rehderiana C. K. Schneid.

Deutzia rubens Rehder

Deutzia setchuenensis Franch.

Deutzia setchuenensis var. **longidentata** Rehder

Deutzia staminea R. Br. ex Wall.

Deutzia subulata Hand.-Mazz.

Deutzia wardiana Zaikonn.

Deutzia yunnanensis S. M. Hwang

Deutzia zhongdianensis S. M. Hwang

Dichroa febrifuga Lour.

Dichroa yunnanensis S. M. Hwang

Hydrangea anomala D. Don

Hydrangea aspera D. Don

Hydrangea bretschneideri Dippel

Hydrangea davidii Franch.

Hydrangea dumicola W. W. Sm.

Hydrangea heteromalla D. Don

Hydrangea hypoglauca Rehder

Hydrangea longipes Franch.

Hydrangea longipes var. **fulvescens** (Rehder) W. T. Wang ex C. F. Wei

Hydrangea macrocarpa Hand.-Mazz.

Hydrangea macrophylla (Thunb.) Ser.

Hydrangea robusta Hook. f. & Thomson

Hydrangea strigosa Rehder

Hydrangea stylosa Hook. f. & Thomson

Hydrangea sungpanensis Hand.-Mazz.

Hydrangea xanthoneura Diels

Philadelphus calvescens (Rehder) S. M. Hwang

Philadelphus delavayi L. Henry

Philadelphus delavayi var. **melanocalyx** Lemoine ex L. Henry

Philadelphus delavayi var. **trichocladus** Hand.-Mazz.

Philadelphus henryi Koehne

Philadelphus incanus Koehne

Philadelphus kansuensis (Rehder) S. Y. Hu

Philadelphus lushuiensis T. C. Ku & S. M. Hwang

Philadelphus purpurascens (Koehne) Rehder

Philadelphus purpurascens var. **venustus** (Koehne) S. Y. Hu

Philadelphus sericanthus Koehne

Philadelphus subcanus Koehne

Philadelphus tetragonus S. M. Hwang

Philadelphus tomentosus Wall. ex G. Don

Pileostegia subansiriana H. B. Naithani & Bennet

Pileostegia viburnoides Hook. f. & Thomson

225. BALSAMINACEAE

Impatiens abbatis Hook. f.

Impatiens acuminata Benth.

Impatiens aganantha Hook. f.

Impatiens amabilis Hook. f.

Impatiens amplexicaulis Edgew.

Impatiens angustiflora Hook. f.

Impatiens annulifer Hook. f.

Impatiens apsotis Hook. f.

Impatiens aquatilis Hook. f.

Impatiens arguta Hook. f. & Thomson

Impatiens arunachalensis Hareesh, A. Joe, M. Sabu & R. Gogoi

Impatiens arunensis Grey-Wilson

Impatiens ashihoi Gogoi & Borah

Impatiens asymmetrica Hook. f.

Impatiens bahanensis Hand.-Mazz.

Impatiens bajurensis S. Akiyama & H. Ohba

Impatiens balfourii Hook. f.

Impatiens balsamina L.

Impatiens barbata H. F. Comber

Impatiens bella Hook. f. & Thomson

Impatiens benthamii Steemis

Impatiens bicolor Royle

Impatiens bicornuta Wall.

Impatiens bivittata Hook. f.

Impatiens brachycentra Kar. & Kir.

Impatiens bracteata Colebr.

Impatiens bracteolata Hook. f.

Impatiens casseabriae Y. H. Tan, S. S. Zhou & B. Yang

Impatiens cathcartii Hook. f.

Impatiens ceratophora H. F. Comber

Impatiens chimiliensis H. F. Comber

Impatiens chinensis L.

Impatiens chiulungensis Y. L. Chen

Impatiens chungtienensis Y. L. Chen

Impatiens cirrhipetala Hook. f.

Impatiens citrina Hook. f.

Impatiens clavicuspis Hook. f. ex W. W. Sm.

Impatiens conaensis Y. L. Chen

Impatiens corchorifolia Franch.

Impatiens cornucopia Franch.

Impatiens cothurnoides Fisch.

Impatiens crassicaudex Hook. f.

Impatiens cristata Wall.

Impatiens cyathiflora Hook. f.

Impatiens cyclosepala Hook. f. ex W. W. Sm.

Impatiens cymbifera Hook. f.

Impatiens dalaiensis Gogoi & Borah

Impatiens decipiens Hook. f.

Impatiens delavayi Franch.

Impatiens desmantha Hook. f.

Impatiens dimorphophylla Franch.

Impatiens discolor DC.

Impatiens divaricata Franch.

Impatiens drepanophora Hook. f.

Impatiens duthiei Hook. f.

Impatiens edgeworthii Hook. f.

Impatiens exilis Hook. f.

Impatiens falcifer Hook. f.

Impatiens flemingii Hook. f.

Impatiens florigera C. B. Clarke ex Hook. f.

Impatiens forrestii Hook. f. ex W. W. Sm.

Impatiens fragicolor C. Marquand & Airy Shaw

Impatiens gamblei Hook. f.

Impatiens gammiei Hook. f.

Impatiens glandulifera Royle

Impatiens glauca Hook. f. & Thomson

Impatiens gongshanensis Y. L. Chen

Impatiens gorepaniensis Grey-Wilson

Impatiens graciliflora Hook. f.

Impatiens gracilipes Hook. f.

Impatiens harae H. Ohba & S. Akiyama

Impatiens harae var. **micrantha** (Hara) H. Ohba & S. Akiyama ex S. Akiyama

Impatiens hengduanensis Y. L. Chen

Impatiens hobsonii Hook. f.

Impatiens holocentra Hand.-Mazz.

Impatiens humilis Hook. f.

Impatiens inayatii Hook. f.

Impatiens infirma Hook. f.

Impatiens infundibularis Hook. f.

Impatiens insignis DC.

Impatiens jaeschkei Hook. f.

Impatiens jurpia Buch.-Ham. ex Hook. f. & Thomson

Impatiens kaliensis Grey-Wilson

Impatiens kathmanduensis Grey-Wilson

Impatiens kharensis S. Akiyama, H. Ohba & Wakab

Impatiens khasiana Hook. f.

Impatiens kingii Hook. f.

Impatiens laevigata Wall. ex Hook. f. & Thomson

Impatiens langeana Hook. f.

Impatiens latiflora Hook. f. & Thomson

Impatiens laxiflora Edgew.

Impatiens lecomtei Hook. f.

Impatiens lemannii Hook. f. & Thornson

Impatiens leptocarpa Hook. f.

Impatiens leptoceras DC.

Impatiens leveillei Hook. f.

Impatiens linghziensis Y. L. Chen

Impatiens lizipingensis Q. Luo

Impatiens longipes Hook. f. & Thomson

Impatiens lutea Hook. f.

Impatiens mallae S. Akiyama & H. Ohba

Impatiens mannii C. B. Clarke ex Hook. f.

Impatiens margaritifera Hook. f.

Impatiens margaritifera var. **humilis** Y. L. Chen

Impatiens margaritifera var. **purpurascens** Y. L. Chen

Impatiens medogensis Y. L. Chen

Impatiens meeboldii Hook. f.

Impatiens microcentra Hand.-Mazz.

Impatiens minimiflora Hook. f.

Impatiens mishmiensis Hook. f.

Impatiens muliensis Y. L. Chen

Impatiens mussotii Hook. f.

Impatiens notolopha Maxim.

Impatiens nubigena W. W. Sm.

Impatiens nummularifolia Hook. f.

Impatiens nyimana C. Marquand & Airy Shaw

Impatiens occultans Hook. f.

Impatiens odontopetala Maxim.

Impatiens oligoneura Hook. f.

Impatiens pahalgamensis Hook. f.

Impatiens paludosa Hook. f.

Impatiens pantlingii Hook. f.

Impatiens paramjitiana Gogoi & Borah

Impatiens parviflora L.

Impatiens pathakiana Gogoi & Borah

Impatiens pianmaensis S. H. Huang

Impatiens platyceras Maxim.

Impatiens podocarpa Hook. f.

Impatiens porrecta Wall. ex Hook. f. & Thomson

Impatiens pradhanii Hara

Impatiens prainii Hook. f.

Impatiens principis Hook. f.

Impatiens procumbens Franch.

Impatiens pseudocitrina Hareesh, M. Sabu & Gogoi

Impatiens pseudokingii Hand.-Mazz.

Impatiens puberula DC.

Impatiens pulchra Hook. f. & Thomson

Impatiens purpurea Hand.-Mazz.

Impatiens putaoensis Y. H. Tan, S. S. Zhou & B. Yang

Impatiens quadriloba K. M. Liu & Y. L. Xiang

Impatiens racemosa DC.

Impatiens racemosa var. **ecalcarata** Hook. f.

Impatiens racemulosa Wall. ex Hook. f. & Thomson

Impatiens radiata Hook. f.

Impatiens rectangula Hand.-Mazz.

Impatiens reidii Hook. f.

Impatiens roingensis Hareesh, A. Joe & M. Sabu

Impatiens rostellata Franch.

Impatiens rubrostriata Hook. f.

Impatiens scabrida DC.

Impatiens scitula Hook. f.

Impatiens scullyi Hook. f.

Impatiens scutisepala Hook. f.

Impatiens serrata Benth. ex Hook. f. & Thomson

Impatiens serratifolia Hook. f.

Impatiens shimianensis G. C. Zhang & Li Bing Zhang

Impatiens siculifer Hook. f.

Impatiens siculifer var. **mitis** Lingelsh. & Borza

Impatiens soulieana Hook. f.

Impatiens spirifer Hook. f. & Thomson

Impatiens stenantha Hook. f.

Impatiens stoliczkai Hook. f.

Impatiens subecalcarata (Hand.-Mazz.) Y. L. Chen

Impatiens sulcata Wall.

Impatiens sunkoshiensis S. Akiyama, H. Ohba & Wakab

Impatiens taronensis Hand.-Mazz.

Impatiens tenerrima Y. L. Chen

Impatiens tenuibracteata Y. L. Chen

Impatiens thiochroa Hand.-Mazz.

Impatiens thomsonii Hook. f.

Impatiens tienchuanensis Y. L. Chen

Impatiens tomentella Hook. f.

Impatiens tortisepala Hook. f.

Impatiens torulosa Hook. f.

Impatiens toxophora Hook. f.

Impatiens trichocladon Hook. f.

Impatiens trilobata Colebr.

Impatiens tripetala Roxb. ex DC.

Impatiens tropaeolifolia Griff. ex Hook. f.

Impatiens tsangshanensis Y. L. Chen

Impatiens tuberculata Hook. f. & Thomson

Impatiens tubifer B. Heyne ex Roxb.

Impatiens uliginosa Franch.

Impatiens uncipetala C. B. Clarke ex Hook. f.

Impatiens unguiculata K. M. Liu & Y. Y. Cong

Impatiens urticifolia Wall.

Impatiens vexillaria Hook. f.

Impatiens violiflora Hook. f.

Impatiens violoides Edgew. ex Hook. f.

Impatiens vittata Franch.

Impatiens wallichii Hook. f.

Impatiens weihsiensis Y. L. Chen

Impatiens williamsii Hara

Impatiens xanthina H. F. Comber

Impatiens xanthina var. **pusilla** Y. L. Chen

Impatiens xanthocephala W. W. Sm.

Impatiens yunnanensis Franch.

226. POLEMONIACEAE

Polemonium caeruleum L.

Polemonium caeruleum var. **himalayanum** Baker

227. LECYTHIDACEAE

Barringtonia acutangula Gaertn.

Careya arborea Roxb.

Careya herbacea Roxb.

228. SLADENIACEAE

Sladenia celastrifolia Kurz

229. PENTAPHYLACACEAE

Adinandra griffithii Dyer

Adinandra kamalae M. K. Pathak, Bhaumik & G. Krishna

Adinandra latifolia L. K. Ling

Anneslea fragrans Wall.

Cleyera japonica var. **wallichiana** (DC.) Sealy

Cleyera ochnacea DC.

Eurya acuminata DC.

Eurya acuminoides P. T. Li

Eurya arunachalensis A. S. Chauhan

Eurya cavinervis Vesque

Eurya cerasifolia (D. Don) Kobuski

Eurya chuekiangensis Hu

Eurya crassilimba Hung T. Chang

Eurya fangii Rehder

Eurya fangii var. **megaphylla** P. S. Hsu

Eurya groffii Merr.

Eurya gungshanensis Hu & L. K. Ling

Eurya handel-mazzettii Hung T. Chang

Eurya impressinervis Kobuski

Eurya inaequalis P. S. Hsu

Eurya jintungensis P. T. Li

Eurya muricata Dunn

Eurya nitida Korth.

Eurya oblonga Y. C. Yang

Eurya oblonga var. **stylosa** Y. C. Yang

Eurya paratetragonoclada Hu

Eurya perserrata Kobuski

Eurya pseudocerasifera Kobuski

Eurya pyracanthifolia P. S. Hsu

Eurya semiserrulata Hung T. Chang

Eurya taronensis P. T. Li

Eurya trichocarpa Korth.

Eurya tsaii Hung T. Chang

Eurya yunnanensis P. S. Hsu

Ternstroemia biangulipes Hung T. Chang

Ternstroemia gymnanthera (Wight & Arn.) Sprague

Ternstroemia gymnanthera var. **wightii** (Choisy) Hand.-Mazz.

230. SAPOTACEAE

Chrysophyllum roxburghii G. Don

Diploknema butyracea (Roxb.) H. J. Lam

Madhuca latifolia (Roxb.) J. F. Macbr.

Madhuca longifolia (Koenig) J. F. Macbr.

Manilkara hexandra (Roxb.) Dubard

Mimusops elengi L.

Palaquium polyanthum (Wall. ex G. Don) Baill.

Sarcosperma arboreum Buch.-Ham. ex C. B. Clarke

Sarcosperma griffithii Hook. f. ex C. B. Clarke

Sarcosperma kachinense King & Pantl.

Xantolis hookeri (Clarke) van Royon

231. EBENACEAE

Diospyros balfouriana Diels

Diospyros cacharensis (Das & P. C. Kanjilal) H. B. Naithani

Diospyros dumetorum W. W. Sm.

Diospyros forrestii J. Anthony

Diospyros kaki Thunb.

Diospyros lanceifolia Roxb.

Diospyros lotus L.

Diospyros malabarica (Desr.) Kostel.

Diospyros montana Roxb.

Diospyros nigricans Wall. ex A. DC.

Diospyros toposia Buch.-Ham.

Diospyros variegata Kurz

232. PRIMULACEAE

Amblyanthopsis membranacea (Wall.) Mez

Amblyanthus glandulosus (Roxb.) A. DC.

Amblyanthus obovatus G. S. Giri, S. K. Das & H. J. Chowdhery

Androsace adenocephala Hand.-Mazz.

Androsace aizoon Duby

Androsace aizoon subsp. **brirensis** Y. Nasir

Androsace akbaitalensis Derganc.

Androsace alashanica var. **zadoensis** Yang & Huang

Androsace alchemilloides Franch.

Androsace axillaris (Franch.) Franch.

Androsace baltistanica Y. Nasir

Androsace bisulca Bureau & Franch.

Androsace bisulca var. **aurata** (Petitm.) Y. C. Yang & R. F. Huang

Androsace bisulca var. **brahmaputrae** (Hand.-Mazz.) G. Sm. & D. Lowe

Androsace brachystegia Hand.-Mazz.

Androsace bulleyana Forrest

Androsace ceruiflora Y. C. Yang & R. F. Huang

Androsace ciliifolia Ludlow

Androsace coronata (Watt) Hand.-Mazz.

Androsace croftii Watt

Androsace croftii var. **scaposa** Santapau & Banerji

Androsace cuscutiformis Franch.

Androsace cuttingii C. E. C. Fisch.

Androsace delavayi Franch.

Androsace dissecta (Franch.) Franch.

Androsace duthieana Knuth

Androsace elatior Pax & K. Hoffm.

Androsace erecta Maxim.

Androsace euryantha Hand.-Mazz.

Androsace flavescens Maxim.

Androsace foliosa Duby

Androsace forrestiana Hand.-Mazz.

Androsace gagnepainiana Hand.-Mazz.

Androsace geraniifolia Watt

Androsace geraniifolia var. **escaposa** Hand.-Mazz.

Androsace geraniifolia var. **hookeriana** Knuth

Androsace geraniifolia var. **pedicellata** Knuth

Androsace geraniifolia var. **setosa** Knuth

Androsace globifera Duby

Androsace gmelini Roem. & Schult.

Androsace gmelinii var. **geophila** Hand.-Mazz.

Androsace graceae Forrest

Androsace gracilis Hand.-Mazz.

Androsace graminifolia C. E. C. Fisch.

Androsace handel-mazzettii Kress (≡ *Androsace lutea* Hand.-Mazz., nom. illeg.)

Androsace harrissii Duthie

Androsace hazarica R. R. Stewart ex Y. Nasir

Androsace hemisphaerica Ludlow

Androsace henryi Oliv.

Androsace henryi subsp. **simulans** C. M. Hu & Y. C. Yang

Androsace himalaica (R. Knuth) Hand.-Mazz.

Androsace hookcriana Klatt

Androsace integra (Maxim.) Hand.-Mazz.

Androsace jacquemontii Duby

Androsace lanuginosa Wall.

Androsace laxa C. M. Hu & Y. C. Yang

Androsace lehmannii Wall. ex Duby

Androsace limprichtii Pax & Hoffm.

Androsace limprichtii var. **laxiflora** (Peteign.) Hand.-Mazz.

Androsace lowariensis Y. J. Nasir

Androsace ludlowiana Hand.-Mazz.

Androsace mairei H. Lév.

Androsace minor (Hand.-Mazz.) C. M. Hu & Y. C. Yang

Androsace mollis Hand.-Mazz.

Androsace mucronifolia Watt

Androsace muscoidea Duby

Androsace nortonii Ludlow ex Stearn

Androsace ojhorensis Y. Nasir

Androsace olgae Ovcz.

Androsace ovalifolia Y. C. Yang

Androsace paxiana R. Knuth

Androsace phaeoblephara Hand.-Mazz.

Androsace podlechii Wendelbo

Androsace pomeiensis C. M. Hu & Y. C. Yang

Androsace rhizomatosa Hand.-Mazz.

Androsace rigida Hand.-Mazz.

Androsace robusta (R. Knuth) Hand.-Mazz.

Androsace robusta var. **gracilis** Y. Nasir

Androsace rockii W. E. Evans

Androsace rotundifolia Hardw.

Androsace rotundifolia subsp. **glandulosa** (Hook. f.) Y. Nasir

Androsace rotundifolia var. **thomsonii** Watt

Androsace russellii Y. Nasir

Androsace sarmentosa Wall.

Androsace selago Hook. f. & Thomson ex Klatt

Androsace sempervivoides Jacq. ex Duby

Androsace septentrionalis L.

Androsace spinulifera (Franch.) Knuth

Androsace squarrosula Maxim.

Androsace staintonii Y. Nasir

Androsace stenophylla (Petitm.) Hand.-Mazz.

Androsace strigillosa Franch.

Androsace studiosorum Kress

Androsace sublanata Hand.-Mazz.

Androsace tanggulashanensis Y. C. Yang & R. F. Huang

Androsace tapete Maxim.

Androsace thomsonii (Watt) Nasir

Androsace umbellata (Lour.) Merr.

Androsace wardii W. W. Sm.

Androsace wilsoniana Hand.-Mazz.

Androsace yargongensis Petitm.
Androsace zambalensis (Petitm.) Hand.-Mazz.
Androsace zayulensis Hand.-Mazz.
Antistrophe oxyantha (A. DC.) A. DC.
Ardisia bhotanica C. B. Clarke
Ardisia colorata Roxb.
Ardisia crenata Sims
Ardisia crispa (Thunb.) A. DC.
Ardisia depressa C. B. Clarke
Ardisia garrettii H. R. Fletcher
Ardisia griffithii C. B. Clarke
Ardisia humilis Vahl
Ardisia hymenandra Wall.
Ardisia icara Wall. ex DC.
Ardisia involucrata Kurz
Ardisia japonica (Horn.) Blume
Ardisia macrocarpa Wall.
Ardisia paniculata Roxb.
Ardisia rhynchophylla C. B. Clarke
Ardisia solanacea Roxb.
Ardisia thomsonii Mez
Ardisia thyrsiflora D. Don
Ardisia virens Kurz
Bryocarpum himalaicum Hook. f. & Thomson
Cortusa brotheri Pax ex Lipsky
Cortusa matthioli subsp. hazarica Y. Nasir
Embelia clarkei Bedd. & Mez ex Mez
Embelia floribunda Wall.
Embelia frondosa (Gamble) D. G. Long
Embelia gamblei Kurz ex C. B. Clarke
Embelia parviflora Wall. ex DC.
Embelia procumbens Hemsl.
Embelia ribes Burm. f.
Embelia robusta Roxb.
Embelia tsjeriam-cottam (Roem. & Schult.) A. DC.
Embelia undulata (Wall.) Mez
Embelia vestita Roxb.
Lysimachia albescens Franch.
Lysimachia alternifolia Wall.
Lysimachia arvensis (L.) U. Manns & Anderb.
Lysimachia assamica C. M. Hu
Lysimachia auriculata Hemsl.
Lysimachia baoxingensis (F. H. Chen & C. M. Hu) C. M. Hu
Lysimachia barystachys Bunge
Lysimachia biflora C. Y. Wu
Lysimachia breviflora C. M. Hu
Lysimachia candida Lindl.
Lysimachia capillipes Hemsl.
Lysimachia chenopodioides Watt ex Hook. f.
Lysimachia christiniae Hance
Lysimachia chungdienensis C. Y. Wu
Lysimachia circaeoides Hemsl.
Lysimachia clethroides Duby
Lysimachia congestiflora Hemsl.
Lysimachia cordifolia Hand.-Mazz.
Lysimachia davurica Ledeb.
Lysimachia debilis Wall.
Lysimachia decurrens G. Forst.
Lysimachia delavayi Franch.
Lysimachia deltoidea var. **cinerascens** Franch.
Lysimachia deltoidea Wight
Lysimachia drymarifolia Franch.
Lysimachia engleri R. Knuth
Lysimachia engleri var. **glabra** (Bonati) F. H. Chen & C. M. Hu
Lysimachia erosipetala F. H. Chen & C. M. Hu
Lysimachia evalvis Wall.
Lysimachia excisa Hand.-Mazz.
Lysimachia ferruginea Edgew.
Lysimachia fistulosa Hand.-Mazz.
Lysimachia glaucina Franch.
Lysimachia hemsleyi Franch.

Lysimachia henryi Hemsl.

Lysimachia japonica Thunb.

Lysimachia laxa Baudo

Lysimachia lichiangensis Forrest

Lysimachia lichiangensis var. **xerophylla** C. Y. Wu

Lysimachia lobelioides Wall.

Lysimachia maritima (L.) Galasso, Banfi & Soldano

Lysimachia medogensis F. H. Chen & C. M. Hu

Lysimachia melampyroides var. **burnelloides** (Pax & Hoffm.) Chen & C. H. Hu

Lysimachia microcarpa C. Y. Wu

Lysimachia miyiensis Y. I. Fang & C. Z. Cheng ex F. H. Chen & C. M. Hu

Lysimachia omeiensis Hemsl.

Lysimachia orbicularis F. H. Chen & C. M. Hu

Lysimachia otophora C. Y. Wu

Lysimachia ovalis (Ruiz & Pav.) U. Manns & Anderb.

Lysimachia paridiformis Franch.

Lysimachia paridiformis var. **stenophylla** Franch.

Lysimachia parvifolia Franch. ex F. B. Forbes & Hemsl.

Lysimachia pentapetala Bunge

Lysimachia platypetala Franch.

Lysimachia prolifera Klatt

Lysimachia pseudotrichopoda Hand.-Mazz.

Lysimachia pteranthoides Bonati

Lysimachia pumila (Baudo) Franch.

Lysimachia pyramidalis Wall.

Lysimachia racemiflora Bonati

Lysimachia reflexiloba Hand.-Mazz.

Lysimachia robusta Hand.-Mazz.

Lysimachia roseola F. H. Chen & C. M. Hu

Lysimachia rubiginosa Hemsl.

Lysimachia shimianensis F. H. Chen & C. M. Hu

Lysimachia silvestrii (Pamp.) Hand.-Mazz.

Lysimachia stellarioides Hand.-Mazz.

Lysimachia stenosepala Hemsl.

Lysimachia stenosepala var. **flavescens** F. H. Chen & C. M. Hu

Lysimachia taliensis Bonati

Lysimachia tengyuehensis Hand.-Mazz.

Lysimachia trichopoda Franch.

Lysimachia tsaii C. M. Hu

Lysimachia tsarongensis Hand.-Mazz.

Lysimachia violascens Franch.

Lysimachia vulgaris L.

Maesa angustifolia A. DC.

Maesa argentea (Wall.) A. DC.

Maesa arunachalensis G. S. Giri, S. K. Das & M. P. Nayar

Maesa castaneifolia Mez

Maesa cavinervis C. Chen

Maesa chisia Buch.-Ham. ex D. Don

Maesa indica (Roxb.) DC.

Maesa longilanceolata C. Chen

Maesa manipurensis Mez

Maesa macrophylla (Wall.) A. DC.

Maesa marioniae Merr.

Maesa maxima Mez

Maesa montana A. DC.

Maesa nayarii G. S. Giri & S. K. Das

Maesa permollis Kurz

Maesa ramentacea (Roxb.) Wall.

Maesa rugosa C. B. Clarke

Maesa truncata Sastry

Maesa ziroensis G. S. Giri & G. D. Pal

Myrsine africana L.

Myrsine capitellata Wall.

Myrsine kwangsiensis (E. Walker) Pipoly & C. Chen

Myrsine seguinii H. Lév.

Myrsine semiserrata Wall.

Omphalogramma brachysiphon W. W. Sm.

Omphalogramma delavayi (Franch.) Franch.

Omphalogramma elegans Forrest

Omphalogramma elwesianum (King ex Watt) Franch.

Omphalogramma forrestii Balf. f.

Omphalogramma minus Hand.-Mazz.

Omphalogramma souliei Franch.

Omphalogramma tibeticum H. R. Fletcher

Omphalogramma vinciflorum (Franch.) Franch.

Pomatosace filicula Maxim.

Primula advena W. W. Sm.

Primula advena var. **argentata** W. W. Sm.

Primula advena var. **concolor** W. W. Sm.

Primula advena var. **euprepes** (W. W. Sm.) A. J. Richards

Primula aemula Balf. f. & Forrest

Primula agleniana Balf. f. & Forrest

Primula aliciae G. Taylor ex W. W. Sm. & H. R. Fletcher

Primula alpicola (W. W. Sm.) Stapf

Primula alsophila Balf. f. & Farrer

Primula ambita Balf. f.

Primula amethystina Franch.

Primula amethystina subsp. **argutidens** (Franch.) W. W. Sm. & Fletcher

Primula amethystina subsp. **brevifolia** (Forrest) W. W. Sm. & Fletcher

Primula anisodora Balf. f. & Forrest

Primula annulata Balf. f. & Kingdon-Ward

Primula apoclita Balf. f. & Forrest

Primula aromatica W. W. Sm. & Forrest

Primula assamica H. R. Fletcher

Primula atrodentata W. W. Sm.

Primula aurantiaca W. W. Sm. & Forrest

Primula aureata H. R. Fletcher

Primula aureata subsp. **fimbriata** A. J. Richards ex Gould

Primula baileyana Kingdon-Ward

Primula barbatula W. W. Sm.

Primula bathangensis Petitm. & Hand.-Mazz.

Primula beesiana Forrest

Primula bella Franch.

Primula bella subsp. **bonatiana** (Petitm.) W. W. Sm. & Forrest

Primula bella subsp. **cyclostegia** (Hand.-Mazz.) W. W. Sm. & Forrest

Primula bella subsp. **sciophila** (Balf. f. & Ward) W. W. Sm. & Forrest

Primula bellidifolia King ex Hook. f.

Primula bhutanica Fletcher

Primula blandula W. W. Sm.

Primula blattariformis Franch.

Primula bomiensis F. H. Chen & C. M. Hu

Primula boothii Craib

Primula boreiocalliantha Balf. f. & Forrest

Primula brachysiphon (W. W. Sm.) Halda

Primula bracteata Franch.

Primula bracteosa Craib

Primula brevicula Balf. f. & Forrest

Primula breviscapa Franch.

Primula bryophila Balf. F. & Farrer

Primula bullata Franch.

Primula bulleyana Forrest

Primula burmanica Balf. f. & Kingdon-Ward

Primula buryana Balf. f.

Primula buryana var. **purpurea** Fletcher

Primula cachemeriana Munro

Primula caldaria W. W. Sm. & Forrest

Primula calderiana Balf. f. & R. E. Cooper

Primula calliantha Franch.

Primula calthifolia W. W. Sm.

Primula candicans W. W. Sm.

Primula capitata Hook.

Primula capitata subsp. **craibeana** (Balf. f. & W. W. Sm.) W. W. Sm. & Forrest

Primula capitata subsp. **crispata** (Balf. f. & W. W. Sm.) W. W. Sm. & Forrest

Primula capitata subsp. **lacteocapitata** (Balf. f. & W. W. Sm.) W. W. Sm. & Forrest

Primula capitata subsp. **mooreana** (Balf. f. & W. W. Sm.) W. W. Sm. & Forrest

Primula capitata subsp. **sphaerocephala** (Balf. f. & Forrest) W. W. Sm. & Forrest

Primula capitellata Boiss.

Primula caveana W. W. Sm.

Primula cawdoriana Kingdon-Ward

Primula cerina H. R. Fletcher

Primula cernua Franch.

Primula chamaedoron W. W. Sm.

Primula chamaethauma W. W. Sm.

Primula chasmophila Balf. f. ex Hutch.

Primula chionantha Balf. f. & Forrest

Primula chionogenes H. R. Fletcher

Primula chrysostoma O. Schwarz

Primula chumbiensis W. W. Sm.

Primula chungensis Balf. f. & Kingdon-Ward

Primula cinerascens Franch.

Primula clarkei Watt

Primula clutterbuckii Kingdon-Ward

Primula cockburniana Hemsl.

Primula coerulea Forrest

Primula concholoba Stapf & Sealy

Primula concinna Watt

Primula cooperi Balf. f.

Primula crassa Hand.-Mazz.

Primula davidii Franch.

Primula dawuensis Y. S. Chen, nom. nov. [≡ *Primula brevis* A. J. Richards (1993), nom. illeg., non Kuntze(1881), ≡ *Primula humilis* Pax & K. Hoffm.(1921), nom. illeg., non Steud.(1840)]

Primula deflexa Duthie

Primula densa Balf. f.

Primula denticulata Sm.

Primula denticulata subsp. **sinodenticulata** (Balf. f. & Forrest) W. W. Sm.

Primula deuteronana Craib

Primula diantha Bureau & Franch.

Primula dickieana Watt

Primula dictyophylla W. W. Sm.

Primula didyma W. W. Sm.

Primula drummondiana Craib

Primula dryadifolia Franch.

Primula dryadifolia subsp. **congestifolia** (Forrest) W. W. Sm. & Forrest

Primula dryadifolia subsp. **cycliophylla** (Balf. f. & Farrer) W. W. Sm. & Forrest

Primula dryadifolia subsp. **jonardunii** (W. W. Sm.) F. H. Chen & C. M. Hu

Primula dryadifolia var. **chlorodryas** (W. W. Sm.) A. J. Richards

Primula dubernardiana Forrest

Primula dueckelmannii Gilli

Primula dumicola W. W. Sm. & Forrest

Primula duthieana Balf. f. & W. W. Sm.

Primula eburnea Balf. f. & R. E. Cooper

Primula edelbergii O. Schwarz

Primula edgeworthii Pax

Primula effusa W. W. Sm. & Forrest

Primula elizabethiae Ludlow ex W. W. Sm.

Primula elliptica Royle

Primula elongata Watt

Primula erosa (Duby) Wall. ex Regel

Primula erratica W. W. Sm.

Primula erythrocarpa Craib

Primula euchaites W. W. Sm.

Primula eucyclia W. W. Sm. & Forrest

Primula euosma Craib

Primula faberi Oliv.

Primula falcifolia Kingdon-Ward

Primula fangii F. H. Chen & C. M. Hu

Primula fasciculata Balf. f. & Kingdon-Ward

Primula fea Kingdon-Ward

Primula fernaldiana W. W. Sm.

Primula filipes Watt

Primula firmipes Balf. f. & Forrest

Primula flabellifera W. W. Sm.

Primula flaccida N. P. Balakr.

Primula flava Maxim.

Primula floribunda Wall.

Primula florida Balf. f. & Forrest

Primula florindae Kingdon-Ward

Primula forbesii Franch.

Primula forrestii Balf. f.

Primula forrestii var. **redolens** (Balf. f. & Kingdon-Ward) A. J. Richards

Primula gambeliana Watt

Primula gemmifera Batalin

Primula gemmifera var. **monantha** W. W. Sm. & Fletcher

Primula genestieriana Hand.-Mazz.

Primula geraniifolia Hook. f.

Primula glabra Klatt

Primula glandulifera Balf. f. & W. W. Sm.

Primula glomerata Pax

Primula gracilenta Dunn

Primula gracilipes Craib

Primula griffithii (Watt) Pax

Primula hazarica Duthie

Primula henricii Bureau & Franch.

Primula heucherifolia Franch.

Primula hilaris W. W. Sm.

Primula hongshanenensis D. W. H. Rankin, Z. D. Fang & H. Sun

Primula hookeri Watt

Primula ianthina Balf. f. & Cave

Primula iljinskii Fed.

Primula inayatii Duthie

Primula inopinata H. R. Fletcher

Primula involucrata Wall. ex Duby

Primula ioessa W. W. Sm.

Primula irregularis Craib

Primula jaffreyana King

Primula jigmediana W. W. Sm.

Primula jucunda W. W. Sm.

Primula kialensis Franch.

Primula kingii Watt

Primula klattii N. P. Balakr.

Primula klaveriana Forrest

Primula kongboensis Kingdon-Ward

Primula lacerata W. W. Sm.

Primula laciniata Pax & K. Hoffm.

Primula lactucoides F. H. Chen & C. M. Hu

Primula laeta W. W. Sm.

Primula latisecta W. W. Sm.

Primula laxiuscula W. W. Sm.

Primula limbata Balf. f. & Forrest

Primula listeri King ex Hook. f.

Primula littledalei Balf. f. & Watt

Primula lungchiensis W. P. Fang

Primula luteoflora X. F. Gao & W. B. Ju

Primula mackliniae A. J. Richards

Primula macrophylla D. Don

Primula macrophylla var. **macrocarpa** (Watt) Pax

Primula macrophylla var. **moorcroftiana** (Wall. ex Klatt) W. W. Sm. & H. R. Fletcher

Primula macrophylla var. **ninguida** (W. W. Sm.) W. W. Sm. & H. R. Fletcher

Primula malacoides Franch.

Primula malvacea Franch.

Primula megalocarpa Hara

Primula melanodonta W. W. Sm.

Primula melanops W. W. Sm. & Kingdon-Ward

Primula membranifolia Franch.

Primula minor Balf. f. & Kingdon-Ward

Primula minutissima Jacquem. ex Duby

Primula mollis Nutt. ex Hook.

Primula morsheadiana Kingdon-Ward
Primula moschophora Balf. f. & Forrest
Primula moupinensisn Franch.
Primula muscarioides Hemsl.
Primula muscoides Hook. f. ex Watt
Primula mystrophylla Balf. f. & Forrest
Primula nana Wall.
Primula nanobella Balf. f. & Forrest
Primula nivalis Pall.
Primula obconica Hance
Primula obliqua W. W. Sm.
Primula obtusifolia Royle
Primula occlusa W. W. Sm.
Primula odontica W. W. Sm.
Primula odontocalyx (Franch.) Pax
Primula optata Farrer ex Balf. f.
Primula orbicularis Hemsl.
Primula oxygraphidifolia W. W. Sm. & Kingdon-Ward
Primula palmata Hand.-Mazz.
Primula pamirica Fed.
Primula pauliana W. W. Sm. & Forrest
Primula petiolaris Wall.
Primula petitmenginii Bonati
Primula pinnatifida Franch.
Primula poissonii Franch.
Primula polonensis Kingdon-Ward
Primula poluninii H. R. Fletcher
Primula polyneura Franch.
Primula praetermissa W. W. Sm.
Primula prattii Hemsl.
Primula prenantha Balf. f. & W. W. Sm.
Primula primulina (Spreng.) Hara
Primula prolifera Wall.
Primula pseudodenticulata Pax
Primula pseudoglabra Hand.-Mazz.
Primula pulchella Franch.
Primula pulchelloides Kingdon-Ward
Primula pulchra Watt
Primula pulverulenta Duthie
Primula pumilio Maxim.
Primula purdomii Craib
Primula pycnoloba Bureau & Franch.
Primula ramzanae W. W. Sm. & H. R. Fletcher
Primula reidii Duthie
Primula reidii var. **williamsii** Ludlow
Primula reptans Hook. f. ex Watt
Primula reticulata Wall.
Primula rhodochroa W. W. Sm.
Primula rhodochroa var. **geraldinae** (W. W. Sm.) F. H. Chen & C. M. Hu
Primula rhodochroa var. **meiotera** W. W. Sm. & H. R. Fletcher
Primula rimicola W. W. Sm.
Primula rockii W. W. Sm.
Primula rosea Royle
Primula rotundifolia Wall.
Primula rubicunda H. R. Fletcher
Primula runcinata C. M. Hu
Primula rupicola Balf. f. & Forrest
Primula russeola Balf. f. & Forrest
Primula sandemaniana W. W. Sm.
Primula sapphirina Hook. f. & Thomson
Primula saturata W. W. Sm. & H. R. Fletcher
Primula scapigera (Hook. f.) Craib
Primula schlagintweitiana Pax
Primula secundiflora Franch.
Primula septemloba Franch.
Primula serratifolia Franch.
Primula sessilis Royle ex Craib
Primula sharmae H. R. Fletcher
Primula sherriffiae var. **ludlowii** (W. W. Sm.) A. J. Richards
Primula sherriffiae W. W. Sm.
Primula sikkimensis Hook.

Primula sikkimensis subsp. **pseudosikkimensis** (Forrest ex Diels) W. W. Sm. & Forrest

Primula sikkimensis subsp. **pudibunda** (W. W. Sm.) W. W. Sm. & Forrest

Primula sikkimensis var. **lorifolia** W. W. Sm.

Primula silaensis Petitm.

Primula sinensis Sabine ex Lindl.

Primula sinolisteri Balf. f.

Primula sinoplantaginea Balf. f.

Primula siphonantha W. W. Sm.

Primula smithiana Craib

Primula soldanelloides Watt

Primula sonchifolia Franch.

Primula soongii F. H. Chen & C. M. Hu

Primula souliei Franch.

Primula spathulifolia Craib

Primula spicata Franch.

Primula stenocalyx Maxim.

Primula stirtoniana Watt

Primula strumosa Balf. f. & R. E. Cooper

Primula stuartii Wall.

Primula subularia W. W. Sm.

Primula szechuanica Pax

Primula taliensis Forrest

Primula tangutica Duthie

Primula tanneri King

Primula tayloriana H. R. Fletcher

Primula tenella King ex Hook. f.

Primula tenuiloba (Watt) Pax

Primula tibetica Watt

Primula tongolensis Franch.

Primula tridentifera F. H. Chen & C. M. Hu

Primula triloba Balf. f. & Forrest

Primula tsariensis W. W. Sm.

Primula tsariensis var. **porrecta** W. W. Sm.

Primula tsongpenii H. R. Fletcher

Primula tzetsouensis Petitm.

Primula umbratilis Balf. f. & R. E. Cooper

Primula umbratilis var. **alba** Fletcher

Primula umbrella Forrest

Primula vaginata Watt

Primula vaginata subsp. **normaniana** (Kingdon-Ward) F. H. Chen & C. M. Hu

Primula valentiniana Hand.-Mazz.

Primula vialii Delavay ex Franch.

Primula vilmoriniana Petitm.

Primula violacea W. W. Sm. & Kingdon-Ward

Primula waddellii Balf. f. & W. W. Sm.

Primula walshii Craib

Primula waltonii Watt ex Balf. f.

Primula warshenewskiana B. Fedtsch.

Primula warshenewskiana subsp. **rhodantha** (Balf. f. & W. W. Sm.) Wendelbo

Primula watsonii Dunn

Primula wattii King

Primula whitei W. W. Sm.

Primula wigramiana W. W. Sm.

Primula wilsonii Dunn

Primula wollastonii Balf. f.

Primula woodwardii Balf. f.

Primula woonyoungiana W. P. Fang

Primula xanthopa Balf. f. & R. E. Cooper

Primula youngeriana W. W. Sm.

Primula yunnanensis Franch.

Primula zambalensis Petitm.

Sadiria boweri Dunn

Sadiria erecta (C. B. Clarke) Mez

Sadiria eugeniifolia (Wall. ex A. DC.) Mez

Sadiria griffithii (C. B. Clarke) Mez

Sadiria solanifolia Mez

Sadiria subsessilifolia M. P. Nayar & G. S. Giri

Samolus valerandi L.

233. THEACEAE

Camellia caudata Wall.

Camellia costei H. Lév.

Camellia fluviatilis Hand.-Mazz.

Camellia forrestii (Diels) Cohen-Stuart

Camellia kissii Wall.

Camellia kissii var. **confusa** (Craib) T. L. Ming

Camellia lawii Sealy

Camellia lutescens Dyer

Camellia mairei var. **lapidea** (Y. C. Wu) Sealy

Camellia oleifera C. Abel

Camellia pachyandra Hu

Camellia pitardii Cohen-Stuart

Camellia punctata (Kochs) Cohen-Stuart

Camellia reticulata Lindl.

Camellia saluenensis Stapf ex Bean

Camellia siangenssi T. K. Paul & Nayar

Camellia sinensis (L.) Kuntze

Camellia sinensis var. **assamica** (J. W. Mast.) Kitam.

Camellia taliensis (W. W. Sm.) Melch.

Camellia tsaii Hu

Camellia villicarpa S. S. Chien

Camellia wardii Kobuski

Camellia wardii var. **muricatula** (Hung T. Chang) T. L. Ming

Camellia yunnanensis (Pitard ex Diels) Cohen-Stuart

Camellia yunnanensis var. **camellioides** (Hu) T. L. Ming

Gordonia excelsa (Blume) Blume

Polyspora chrysandra (Cowan) Hu ex B. M. Barthol. & T. L. Ming

Polyspora longicarpa (Hung T. Chang) C. X. Ye ex B. M. Barthol. & T. L. Ming

Polyspora speciosa (Kochs) B. M. Barthol. & T. L. Ming

Pyrenaria barringtoniifolia Seem.

Pyrenaria diospyricarpa Kurz

Pyrenaria diospyricarpa var. **camelliiflora** (Kurz) S. X. Yang

Pyrenaria khasiana R. N. Paul

Schima argentea E. Pritzel

Schima khasiana Dyer

Schima noronhae Reinw. ex Blume

Schima sericans (Hand.-Mazz.) T. L. Ming

Schima sericans var. **paracrenata** (Hung T. Chang) T. L. Ming

Schima wallichii (DC.) Korthals

Stewartia medogensis J. Li & T. L. Ming

Stewartia pteropetiolata W. C. Cheng

234. SYMPLOCACEAE

Symplocos adenophylla Wall. & G. Don

Symplocos anomala Brand

Symplocos cochinchinensis (Lour.) S. Moore

Symplocos crataegoides Buch.-Ham. ex D. Don

Symplocos dryophila C. B. Clarke

Symplocos ferruginea Roxb.

Symplocos glandulosopunctata Y. F. Wu

Symplocos glauca (Thunb.) Koidz.

Symplocos glomerata King ex C. B. Clarke

Symplocos grandiflora Wall. ex A. DC.

Symplocos hookeri C. B. Clarke

Symplocos lancifolia Siebold & Zucc.

Symplocos leucantha Kurz

Symplocos lucida (Thunb.) Siebold & Zucc.

Symplocos macrophylla Wall. ex A. DC.

Symplocos oxyphylla Wall. ex A. DC.

Symplocos paniculata (Thunb.) Miq.

Symplocos pendula Wight

Symplocos pendula var. **hirtistylis** (C. B. Clarke) Noot.

Symplocos pyrifolia Wall. ex G. Don

Symplocos racemosa Roxb.

Symplocos ramosissima Wall. ex G. Don

Symplocos spicata Roxb.

Symplocos stellaris Brand

Symplocos subsecunda Brand

Symplocos sulcata Kurz

Symplocos sumuntia Buch.-Ham. ex D. Don

Symplocos theifolia D. Don

Symplocos viridissima Brand

Symplocos xylopyrena C. Y. Wu ex Y. F. Wu

235. DIAPENSIACEAE

Berneuxia thibetica Decne.

Diapensia bulleyana G. Forrest ex Diels

Diapensia himalaica Hook. f. & Thomson

Diapensia purpurea Diels

Diapensia wardii W. E. Evans

Diplarche multiflora Hook. f. & Thomson

Diplarche pauciflora Hook. f. & Thomson

236. STYRACACEAE

Alniphyllum fortunei (Hemsl.) Makino

Bruinsmia polysperma (C. B. Clarke) Steenis

Huodendron biaristatum (W. W. Sm.) Rehder

Huodendron tibeticum (J. Anthony) Rehder

Parastyrax lacei (W. W. Sm.) W. W. Sm.

Pterostyrax psilophyllus Diels ex Perkins

Rehderodendron gongshanense Y. C. Tang

Rehderodendron microcarpum K. M. Feng

Sinojackia henryi (Dümmer) Merr.

Styrax grandiflorus Griff.

Styrax hemsleyanus Diels

Styrax hookeri C. B. Clarke

Styrax japonicus Siebold & Zucc.

Styrax limprichtii Lingelsh. & Borza

Styrax perkinsiae Rehder

Styrax roseus Dunn

Styrax serrulatus Roxb.

Styrax suberifolius Hook. & Arn.

Styrax wilsonii Rehder

237. ACTINIDIACEAE

Actinidia arguta (Siebold & Zucc.) Planch. ex Miq.

Actinidia arguta var. **giraldii** (Diels) Vorosch.

Actinidia callosa Lindl.

Actinidia callosa var. **henryi** Maxim.

Actinidia glaucocallosa C. Y. Wu

Actinidia grandiflora C. F. Liang

Actinidia kolomikta (Rupr. & Maxim.) Maxim.

Actinidia melanandra Franch.

Actinidia pilosula (Finet & Gagnep.) Stapf ex Hand. Mazz.

Actinidia polygama (Siebold & Zucc.) Maxim.

Actinidia rubricaulis Dunn

Actinidia strigosa Hook. f. & Thomson

Actinidia tetramera Maxim.

Actinidia umbelloides C. F. Liang

Actinidia venosa Rehder

Actinidia vitifolia C. Y. Wu

Clematoclethra scandens (Franch.) Maxim.

Clematoclethra scandens subsp. **actinidioides** (Maxim.) Y. C. Tang & Q. Y. Xiang

Saurauia armata Kurz

Saurauia cerea Griff. ex Dyer

Saurauia erythrocarpa C. F. Liang & Y. S. Wang

Saurauia fasciculata Wall.

Saurauia griffithii Dyer

Saurauia griffithii var. **annamica** Gagnep.

Saurauia macrotricha Kurz ex Dyer

Saurauia napaulensis DC.

Saurauia polyneura C. F. Liang & Y. S. Wang

Saurauia polyneura var. **paucinervis** J. Q. Li & D. D. Soejarto

Saurauia punduana Wall.

Saurauia roxburghii Wall.

Saurauia rubricalyx C. F. Liang & Y. S. Wang

Saurauia sinohirsuta J. Q. Li & Soejarto

Saurauia tristyla DC.

Saurauia yunnanensis C. F. Liang & Y. S. Wang

238. CLETHRACEAE

Clethra delavayi Franch.

Clethra monostachya Rehder & E. H. Wilson

239. ERICACEAE

Agapetes aborensis Airy Shaw

Agapetes adenobotrys Airy Shaw

Agapetes angulata (Griff.) Hook. f.

Agapetes anonyma Airy Shaw

Agapetes arunachalensis D. Banik & Sanjappa

Agapetes auriculata (Griff.) Benth. & Hook. f.

Agapetes bhutanica N. P. Balakr. & Sud. Chowdhury

Agapetes brachypoda Airy Shaw

Agapetes brandisiana W. E. Evans

Agapetes brevipedicellata Y. H. Tan & S. S. Zhou

Agapetes burmanica W. E. Evans

Agapetes buxifolia Nutt. ex Hook. f.

Agapetes camelliifolia S. H. Huang

Agapetes ciliata S. H. Huang

Agapetes dalaiensis D. Banik & Sanjappa

Agapetes discolor C. B. Clarke

Agapetes epacridea Airy Shaw

Agapetes fasciculiflora Airy Shaw

Agapetes flava (Hook. f.) Sleumer

Agapetes forrestii W. E. Evans

Agapetes graciliflora R. C. Fang

Agapetes griffithii C. B. Clarke

Agapetes haemantha Airy Shaw

Agapetes hillii Brandis

Agapetes hookeri (C. B. Clarke) Sleumer

Agapetes hyalocheilos Airy Shaw

Agapetes incurvata (Griff.) Sleumer

Agapetes incurvata var. **hookeri** (Griff.) Sleumer

Agapetes interdicta (Hand.-Mazz.) Sleumer

Agapetes lacei Craib

Agapetes lacei var. **glaberrima** Airy Shaw

Agapetes lacei var. **tomentella** Airy Shaw

Agapetes leiocarpa S. H. Huang

Agapetes leucocarpa S. H. Huang

Agapetes linearifolia C. B. Clarke

Agapetes listeri (King ex C. B. Clarke) Sleumer

Agapetes lobbii C. B. Clarke

Agapetes macrantha (Hook.) Hook. f.

Agapetes macrophylla C. B. Clarke

Agapetes mannii Hemsl.
Agapetes marginata Dunn
Agapetes medogensis S. H. Huang
Agapetes miniata (Griff.) Hook. f.
Agapetes miranda Airy Shaw
Agapetes mitrarioides Hook. f. ex C. B. Clarke
Agapetes muscorum Airy Shaw
Agapetes neriifolia (King & Prain) Airy Shaw
Agapetes nutans Dunn
Agapetes nuttallii C. B. Clarke
Agapetes oblonga Craib
Agapetes obovata (Wight) Hook. f.
Agapetes odontocera (Wight) Benth. & Hook. f.
Agapetes pensilis Airy Shaw
Agapetes pilifera Hook. f. ex C. B. Clarke
Agapetes pottingeri Prain
Agapetes praeclara C. Marquand
Agapetes praestigiosa Airy Shaw
Agapetes pseudogriffithii Airy Shaw
Agapetes pubiflora Airy Shaw
Agapetes putaoensis Y. H. Tong & N. H. Xia
Agapetes pyrolifolia Airy Shaw
Agapetes refracta Airy Shaw
Agapetes salicifolia C. B. Clarke
Agapetes saligna (Hook. f.) Benth. & Hook. f.
Agapetes serpens (Wight) Sleumer
Agapetes siangensis D. Banik & Sanjappa
Agapetes sikkimensis Airy Shaw
Agapetes similis Airy Shaw
Agapetes smithiana Sleumer
Agapetes spissa Airy Shaw
Agapetes spissiformis Airy Shaw
Agapetes subansirica G. D. Pal & Thoth.
Agapetes subsessilifolia S. H. Huang, H. Sun & Z. K. Zhou
Agapetes subvinacea Airy Shaw
Agapetes toppinii Airy Shaw
Agapetes trianguli Airy Shaw
Agapetes variegata (Roxb.) D. Don ex G. Don
Agapetes vernayana Merr.
Agapetes wardii W. W. Sm.
Agapetes xiana Y. H. Tong
Agapetes xizangensis S. H. Huang
Arctous microphyllus C. Y. Wu
Arctous ruber (Rehder & E. H. Wilson) Nakai
Cassiope abbreviata Hand.-Mazz.
Cassiope argyrotricha T. Z. Hsu
Cassiope dendrotricha Hand.-Mazz.
Cassiope fastigiata (Wall.) D. Don
Cassiope membranifolia R. C. Fang
Cassiope myosuroides W. W. Sm.
Cassiope nana T. Z. Hsu
Cassiope palpebrata W. W. Sm.
Cassiope pectinata Stapf
Cassiope selaginoides Hook. f. & Thomson
Cassiope wardii C. Marquand
Chimaphila japonica Miq.
Chimaphila monticola Andres
Craibiodendron henryi W. W. Sm.
Craibiodendron yunnanense W. W. Sm.
Enkianthus chinensis Franch.
Enkianthus deflexus (Griff.) C. K. Schneid.
Enkianthus deflexus var. **glabrescens** R. C. Fang
Enkianthus pauciflorus E. H. Wilson
Gaultheria brevistipes (C. Y. Wu & T. Z. Hsu) R. C. Fang
Gaultheria cardiosepala Hand.-Mazz.
Gaultheria codonantha Airy Shaw
Gaultheria cuneata (Rehder & E. H. Wilson) Bean
Gaultheria discolor Nutt. ex Hook.
Gaultheria dolichopoda Airy Shaw
Gaultheria dumicola W. W. Sm.

Gaultheria dumicola var. **aspera** Airy Shaw

Gaultheria dumicola var. **petanoneuron** Airy Shaw

Gaultheria fragrantissima Wall.

Gaultheria griffithiana Wight

Gaultheria griffithiana var. **insignis** R. C. Fang

Gaultheria heteromera R. C. Fang

Gaultheria hookeri C. B. Clarke

Gaultheria hookeri var. **angustifolia** C. B. Clarke

Gaultheria hypochlora Airy Shaw

Gaultheria longiracemosa Y. C. Yang

Gaultheria notabilis J. Anthony

Gaultheria nummularioides D. Don

Gaultheria praticola C. Y. Wu ex T. Z. Hsu

Gaultheria prostrata W. W. Sm.

Gaultheria pseudonotabilis H. Li ex R. C. Fang

Gaultheria purpurea R. C. Fang

Gaultheria pyrolifolia Hook. f. ex C. B. Clarke

Gaultheria semi-infera (C. B. Clarke) Airy Shaw

Gaultheria seshagiriana Subba Rao & Kumari

Gaultheria sinensis J. Anthony

Gaultheria sinensis var. **nivea** J. Anthony

Gaultheria straminea R. C. Fang

Gaultheria suborbicularis W. W. Sm.

Gaultheria tetramera W. W. Sm.

Gaultheria thymifolia Stapf ex Airy Shaw

Gaultheria trichophylla Royle

Gaultheria trichophylla var. **eciliata** S. J. Rae & D. G. Long

Gaultheria trichophylla var. **tetracme** Airy Shaw

Gaultheria trigonoclada R. C. Fang

Gaultheria wardii C. Marquand & Airy Shaw

Gaultheria wardii var. **elongata** R. C. Fang

Leucothoe griffithiana C. B. Clarke

Lyonia doyonensis (Hand.-Mazz.) Hand.-Mazz.

Lyonia macrocalyx (J. Anthony) Airy Shaw

Lyonia ovalifolia (Wall.) Drude

Lyonia ovalifolia var. **hebecarpa** (Franch. ex Forbes & Hemsl.) Chun

Lyonia villosa (Wall. ex C. B. Clarke) Hand.-Mazz.

Lyonia villosa var. **sphaerantha** (Hand.-Mazz.) Hand.-Mazz.

Moneses uniflora (L.) A. Gray

Monotropa hypopitys L.

Monotropa uniflora L.

Monotropastrum humile (D. Don) Hara

Orthilia secunda (L.) House

Pieris formosa (Wall.) D. Don

Pyrola atropurpurea Franch.

Pyrola calliantha Andres

Pyrola corbieri H. Lév.

Pyrola decorata Andres

Pyrola forrestiana Andres

Pyrola markonica Y. L. Chou & R. C. Zhou

Pyrola minor L.

Pyrola monophylla Y. L. Chou & R. C. Zhou

Pyrola rotundifolia L.

Pyrola rugosa Andrcs

Pyrola sikkimensis Krisa

Pyrola sororia Andres

Pyrola szechuanica Andres

Rhododendron adenobracteum X. F. Gao & Y. L. Peng

Rhododendron adenogynum Diels

Rhododendron adenosum (Cowan & Davidian) Davidian

Rhododendron aganniphum Balf. f. & Kingdon-Ward

Rhododendron aganniphum var. **flavorufum** (Balf. f. & Forrest) D. F. Chamb.

Rhododendron aganniphum var. **schizopeplum** (Balf. f. & Forrest) T. L. Ming

Rhododendron aganniphum var. **vellereum** (Hutch. ex Tagg) Geng

Rhododendron agastum Balf. f. & W. W. Sm.

Rhododendron agastum var. **pennivenium** (Balf. f. & Forrest) T. L. Ming

Rhododendron albertsenianum Forrest ex Balf. f.

Rhododendron alutaceum Balf. f. & W. W. Sm.

Rhododendron alutaceum var. **iodes** (Balf. f. & Forrest) D. F. Chamb.

Rhododendron alutaceum var. **russotinctum** (Balf. f. & Forrest) D. F. Chamb.

Rhododendron amandum Cowan

Rhododendron amesiae Rehder & E. H. Wilson

Rhododendron amundsenianum Hand.-Mazz.

Rhododendron angustinii subsp. **chasmanthum** (Diels) Cullen

Rhododendron annae subsp. **laxiflorum** (Balf. f. & Forrest) T. L. Ming

Rhododendron anthopogon D. Don

Rhododendron anthopogonoides Maxim.

Rhododendron anthosphaerum Diels

Rhododendron aperantum Balf. f. & Kingdon-Ward

Rhododendron araiophyllum Balf. f. & W. W. Sm.

Rhododendron arboreum Sm.

Rhododendron arboreum subsp. **cinnamomeum** Lindl.

Rhododendron arboreum var. **album** Wall.

Rhododendron arboreum var. **delavayi** (Franch.) D. F. Chamb.

Rhododendron arboreum var. **peramoenum** (Balf. f. & Forrest) D. F. Chamb.

Rhododendron arboreum var. **roseum** Lindl.

Rhododendron argipeplum Balf. f. & R. E. Cooper

Rhododendron argyrophyllum Franch.

Rhododendron arizelum Balf. f. & Forrest

Rhododendron asperulum Hutch. & Kingdon-Ward

Rhododendron asterochnoum Diels

Rhododendron asterochnoum var. **brevipedicellatum** W. K. Hu

Rhododendron atropunicum H. P. Yang

Rhododendron augustinii Hemsl.

Rhododendron augustinii subsp. **chasmanthum** (Diels) Cullen

Rhododendron augustinii subsp. **hardyi** (Davidian) Cullen

Rhododendron auritum Tagg

Rhododendron baileyi Balf. f.

Rhododendron bainbridgeanum Tagg & Forrest

Rhododendron balangense W. P. Fang

Rhododendron balfourianum Diels

Rhododendron balfourianum var. **aganniphoides** Tagg & Forrest

Rhododendron barbatum Wall. ex G. Don

Rhododendron barkamense D. F. Chamb.

Rhododendron basilicum Balf. f. & Forrest

Rhododendron bathyphyllum Balf. f. & Forrest

Rhododendron beanianum Cowan

Rhododendron beesianum Diels

Rhododendron bellissimum D. F. Chamb.

Rhododendron bhutanense D. G. Long & Bowes Lyon

Rhododendron bijiangense T. L. Ming

Rhododendron bonvalotii Franch.

Rhododendron boothii Nuttall

Rhododendron brachyanthum Franch.

Rhododendron brachyanthum subsp. **hypolepidotum** (Franch.) Cullen

Rhododendron bracteatum Rehder & E. H. Wilson

Rhododendron brevipetiolatum M. Y. Fang

Rhododendron bulu Hutch.

Rhododendron bureavii Franch.

Rhododendron caesium Hutch.

Rhododendron callimorphum Balf. f. & W. W. Sm.

Rhododendron callimorphum var. **myiagrum** (Balf. f. & Forrest) D. F. Chamb.

Rhododendron calophytum Franch.

Rhododendron calostrotum Balf. f. & Kingdon-Ward

Rhododendron calostrotum var. **calciphilum** (Hutch. & Kingdon-Ward) Davidian

Rhododendron calvescens Balf. f. & Forrest

Rhododendron calvescens var. **duseimatum** (Balf. f. & Forrest) D. F. Chamb.

Rhododendron camelliiflorum Hook. f.

Rhododendron campanulatum D. Don

Rhododendron campanulatum subsp. **aeruginosum** (Hook. f.) D. F. Chamb.

Rhododendron campanulatum var. **wallichii** D. Don

Rhododendron campylocarpum Hook. f.

Rhododendron campylocarpum subsp. **caloxanthum** (Balf. f. & Farrer) D. F. Chamb.

Rhododendron campylogynum Franch.

Rhododendron capitatum Maxim.

Rhododendron catacosmum Balf. f. ex Tagg

Rhododendron cephalanthum Franch.

Rhododendron cephalanthum subsp. **platyphyllum** (Franch. ex Balf. f. & W. W. Sm.) Cullen

Rhododendron cerasinum Tagg

Rhododendron chamaethomsonii (Tagg) Cowan & Davidian

Rhododendron chamaethomsonii var. **chamaedoron** (Tagg) D. F. Chamb.

Rhododendron chamaethomsonii var. **chamaethauma** (Tagg) Cowan & Davidian

Rhododendron charitopes Balf. f. & Farrer

Rhododendron charitopes subsp. **tsangpoense** (Kingdon-Ward) Cullen

Rhododendron chionanthum Tagg & Forrest

Rhododendron chrysodoron Tagg ex Hutch.

Rhododendron ciliatum Hook. f.

Rhododendron ciliicalyx Franch.

Rhododendron cinnabarinum Hook. f.

Rhododendron cinnabarinum subsp. **tamaense** (Davidian) Cullen

Rhododendron cinnamomeum Wall. ex G. Don

Rhododendron circinnatum Cowan & Kingdon-Ward

Rhododendron citriniflorum Balf. f. & Forrest

Rhododendron citriniflorum var. **horaeum** (Balf. f. & Forrest) D. F. Chamb.

Rhododendron clementinae Forrest ex W. W. Sm.

Rhododendron codonanthum Balf. f. & Forrest

Rhododendron coelicum Balf. f. & Farrer

Rhododendron comisteum Balf. f. & Forrest

Rhododendron concinnoides Hutch. & Kingdon-Ward

Rhododendron concinum Hemsl.

Rhododendron coriaceum Franch.

Rhododendron coryanum Tagg & Forrest

Rhododendron cowanianum Davidian

Rhododendron coxianum Davidian

Rhododendron crinigerum Franch.

Rhododendron crinigerum var. **euadenium** Tagg & Forrest

Rhododendron cuffeanum Craub ex Hutch.

Rhododendron cuneatum W. W. Sm.

Rhododendron cyanocarpum (Franch.) Franch. ex W. W. Sm.

Rhododendron dalhousiae Hook. f.

Rhododendron dalhousiae var. **rhabdotum** (Balf. f. & R. E. Cooper) Cullen

Rhododendron danbaense L. C. Hu

Rhododendron dasycladoides Hand.-Mazz.

Rhododendron dasypetalum Balf. f. & Forrest

Rhododendron davidii Franch.

Rhododendron davidsonianum Rehder & E. H. Wilson

Rhododendron decorum Franch.

Rhododendron decorum subsp. **diaprepes** (Balf. f. & W. W. Sm.) T. L. Ming

Rhododendron decorum subsp. **parvistigmatis** W. K. Hu

Rhododendron dekatanum Cowan

Rhododendron dendricola Hutch.

Rhododendron dendrocharis Franch.

Rhododendron dendrocharis var. **petrocharis** (Diels) Geng

Rhododendron detonsum Balf. f. & Forrest

Rhododendron dichroanthum Diels

Rhododendron dichroanthum subsp. **apodectum** (Balf. f. & W. W. Sm.) Cowan

Rhododendron dichroanthum subsp. **scyphocalyx** (Balf. f. & Forrest) Cowan

Rhododendron dichroanthum subsp. **septentrionale** Cowan

Rhododendron dignabile Cowan

Rhododendron dimitrum Balf. f. & Forrest

Rhododendron diphrocalyx Balf. f.

Rhododendron discolor Franch.

Rhododendron discolor var. **wolongense** (W. K. Hu) Geng

Rhododendron dumicola Tagg & Forrest

Rhododendron eclecteum Balf. f. & Forrest

Rhododendron eclecteum var. **bellatulum** Tagg

Rhododendron edgeworthii Hook. f.

Rhododendron eleganthulum Tagg & Forrest

Rhododendron erastum Balf. f. & Forrest

Rhododendron erosum Cowan

Rhododendron erythrocalyx Balf. f. & Forrest

Rhododendron esetulosum Balf. f. & Forrest

Rhododendron euchroum Balf. f. & Kingdon-Ward

Rhododendron eudoxum Balf. f. & Forrest

Rhododendron eudoxum var. **brunneifolium** (Balf. f. & Forrest) D. F. Chamb.

Rhododendron eudoxum var. **mesopolium** (Balf. f. & Forrest) D. F. Chamb.

Rhododendron eurysiphon Tagg & Forrest

Rhododendron exasperatum Tagg

Rhododendron faberi Hemsl.

Rhododendron facetum Balf. f. & Kingdon-Ward

Rhododendron falconeri Hook. f.

Rhododendron fastigiatum Franch.

Rhododendron fastigiatum var. **impeditum** (Balf. f. & W. W. Sm.) Geng

Rhododendron flavantherum Hutch. & Kingdon-Ward

Rhododendron flavoflorum T. L. Ming

Rhododendron fletcherianum Davidian

Rhododendron floccigerum Franch.

Rhododendron floribundum Franch.

Rhododendron forrestii Balf. f. ex Diels

Rhododendron forrestii subsp. **papillatum** D. F. Chamb.

Rhododendron fragariiflorum Kingdon-Ward

Rhododendron fulgens Hook. f.

Rhododendron fulvum Balf. f. & W. W. Sm.

Rhododendron galactinum Balf. f. ex Tagg

Rhododendron gemmiferum M. N. Philipson & Philipson

Rhododendron genestierianum Forrest

Rhododendron glaucophyllum Rehder

Rhododendron glischrum Balf. f. & W. W. Sm.

Rhododendron glischrum subsp. **rude** (Tagg & Forrest) D. F. Chamb.

Rhododendron gongshanense T. L. Ming

Rhododendron grande Wight

Rhododendron griersonianum Balf. f. & Forrest

Rhododendron griffithianum Wight

Rhododendron habrotrichum Balf. f. & W. W. Sm.

Rhododendron haematodes Franch.

Rhododendron haematodes subsp. **chaetomallum** (Balf. f. & Forrest) D. F. Chamb.

Rhododendron heliolepis Franch.

Rhododendron heliolepis var. **oporinum** (Balf. f. & Kingdon-Ward) A. L. Chang ex R. C. Fang

Rhododendron hemitrichotum Balf. f. & Forrest

Rhododendron heteroclitum H. P. Yang

Rhododendron hillieri Davidian

Rhododendron hippophaeoides Balf. f. & W. W. Sm.

Rhododendron hippophaeoides var. **occidentale** M. N. Philipson & Philipson

Rhododendron hirsutipetiolatum A. L. Chang & R. C. Fang

Rhododendron hirtipes Tagg

Rhododendron hodgsonii Hook. f.

Rhododendron hoi W. P. Fang

Rhododendron hookeri Nuttall

Rhododendron huanum W. P. Fang

Rhododendron hunnewellianum Rehder & E. H. Wilson

Rhododendron hylaeum Balf. f. & Farrer

Rhododendron hypenanthum Balf. f.

Rhododendron insculptum Kingdon-Ward

Rhododendron insigne Hemsl. & E. H. Wilson

Rhododendron intricatum Franch.

Rhododendron intricatum var. **complexum** (Balf. f. & W. W. Sm.) Geng

Rhododendron irroratum Franch.

Rhododendron keleticum Balf. f. & Forrest

Rhododendron kendrickii Nuttall

Rhododendron kesangiae D. G. Long & Rushforth

Rhododendron keysii Nuttall

Rhododendron kongboense L. Rothschild

Rhododendron kyawii Lace & W. W. Sm.

Rhododendron lacteum Franch.

Rhododendron lanatoides D. F. Chamb.

Rhododendron lanatum Hook. f.

Rhododendron lanigerum Tagg

Rhododendron lateriflorum R. C. Fang & A. L. Chang

Rhododendron laudandum Cowan

Rhododendron laudandum var. **nyingchiense** (R. C. Fang & S. H. Huang) Geng

Rhododendron laudandum var. **temoense** Kingdon-Ward ex Cowan & Davidian

Rhododendron lepidostylum Balf. f. & Forrest

Rhododendron lepidotum Wall. ex G. Don

Rhododendron leptocarpum Nuttall

Rhododendron leptopeplum Balf. f. & Forrest

Rhododendron leptothrium Balf. f. & Forrest

Rhododendron leucaspis Tagg

Rhododendron lindleyi T. Moore

Rhododendron lindleyi var. **taggianum** (Hutch.) Geng

Rhododendron longesquamatum C. K. Schneid.

Rhododendron longicalyx M. Y. Fang

Rhododendron longistylum Rehder & E. H. Wilson

Rhododendron lowndesii Davidian

Rhododendron ludlowii Cowan

Rhododendron luhuoense H. P. Yang

Rhododendron lukiangense Franch.

Rhododendron lulangense L. C. Hu & Tateishi

Rhododendron lutescens Franch.

Rhododendron lutescens var. **shimianense** (W. P. Fang & P. S. Liu) Geng

Rhododendron mackenzianum Forrest

Rhododendron maddenii Hook. f.

Rhododendron maddenii subsp. **crassum** (Franch.) Cullen

Rhododendron magnificum Kingdon-Ward

Rhododendron mainlingense S. H. Huang & R. C. Fang

Rhododendron mallotum Balf. f. & Kingdon-Ward

Rhododendron maowenense Ching & H. P. Yang

Rhododendron martinianumn Balf. f. & Forrest

Rhododendron mechukae A. A. Mao & A. Paul

Rhododendron meddianum Forrest

Rhododendron meddianum var. **atrokermesinum** Tagg

Rhododendron medoense W. P. Fang & M. Y. He

Rhododendron megacalyx Balf. f. & Kingdon-Ward

Rhododendron megalanthum M. Y. Fang

Rhododendron megeratum Balf. f. & Forrest

Rhododendron mekongense Franch.

Rhododendron mekongense var. **longipilosum** (Cowan) Cullen

Rhododendron mekongense var. **rubrolineatum** (Balf. f. & Forrest) Cullen

Rhododendron mianningense Z. J. Zhao

Rhododendron microgynum Balf. f. & Forrest

Rhododendron microphyton Franch.

Rhododendron mimetes Tagg & Forrest

Rhododendron miniatum Cowan

Rhododendron minyaense M. N. Philipson & Philipson

Rhododendron miyiense W. K. Hu

Rhododendron mollicomum Balf. f. & W. W. Sm.

Rhododendron monanthum Balf. f. & W. W. Sm.

Rhododendron montroseanum Davidian

Rhododendron moulmainense Hook.

Rhododendron moupinense Franch.

Rhododendron nakotiltum Balf. f. & Forrest

Rhododendron neriiflorum Franch.

Rhododendron neriiflorum var. **agetum** (Balf. f. & Forrest) T. L. Ming

Rhododendron neriiflorum var. **appropinquans** (Tagg & Forrest) W. K. Hu

Rhododendron nigroglandulosum Nitzelius

Rhododendron nitidulum Rehder & E. H. Wilson

Rhododendron nivale Hook. f.

Rhododendron nivale subsp. **australe** M. N. Philipson & Philipson

Rhododendron nivale subsp. **boreale** M. N. Philipson & Philipson

Rhododendron niveum Hook. f.

Rhododendron nuttallii T. J. Booth ex Nutt.

Rhododendron nuttallii var. **sinonuttallii** (Balf. f. & Forrest) Geng

Rhododendron orbiculare Decne.

Rhododendron oreodoxa Franch.

Rhododendron oreodoxa var. **adenostylosum** Fang & W. K. Hu

Rhododendron oreotrephes W. W. Sm.

Rhododendron orthocladum Balf. f. & Forrest

Rhododendron pachypodum Balf. f. & W. W. Sm.

Rhododendron pachytrichum Franch.

Rhododendron papillatum Balf. f. & R. E. Cooper

Rhododendron paradoxum Balf. f.

Rhododendron parmulatum Cowan

Rhododendron pendulum Hook. f.

Rhododendron phaeochrysum Balf. f. & W. W. Sm.

Rhododendron phaeochrysum var. **agglutinatum** (Balf. f. & Forrest) D. F. Chamb.

Rhododendron phaeochrysum var. **levistratum** (Balf. f. & Forrest) D. F. Chamb.

Rhododendron piercei Davidian

Rhododendron pocophorum Balf. f. ex Tagg

Rhododendron pocophorum var. **hemidartum** (Balf. f. ex Tagg) D. F. Chamb.

Rhododendron polycladum Franch.

Rhododendron polylepis Franch.

Rhododendron pomense Cowan & Davidian

Rhododendron populare Cowan

Rhododendron praestans Balf. f. & Forrest

Rhododendron prattii Franch.

Rhododendron preptum Balf. f. & Forrest

Rhododendron primuliflorum Bureau & Franch.

Rhododendron primuliflorum var. **lepidanthum** (Balf. f. & W. W. Sm.) Cowan & Davidian

Rhododendron principis Bureau & Franch.

Rhododendron pronum Tagg & Forrest

Rhododendron proteoides Balf. f. & W. W. Sm.

Rhododendron protistum Balf. f. & Forrest

Rhododendron pruniflorum Hutch. & Kingdon-Ward

Rhododendron przewalskii Maxim.

Rhododendron pseudociliipes Cullen

Rhododendron pseudomaddenii A. A. Mao & M. Bhaumik

Rhododendron pubescens Balf. f. & Forrest

Rhododendron pudorosum Cowan

Rhododendron pudorosum var. **giganteum** (Forrest) D. F. Chamb.

Rhododendron pumilum Hook. f.

Rhododendron puntifolium L. C. Hu

Rhododendron qinghaiense Ching ex W. Y. Wang

Rhododendron racemosum Franch.

Rhododendron ramipilosum T. L. Ming

Rhododendron ramsdenianum Cowan

Rhododendron rex H. Lév.

Rhododendron rex subsp. **fictolacteum** (Balf. f.) D. F. Chamb.

Rhododendron rex subsp. **gratum** (T. L. Ming) M. Y. Fang

Rhododendron rhombifolium R. C. Ching

Rhododendron rigidum Franch.

Rhododendron riparioides (Cullen) Cubey

Rhododendron ririei Hemsl. & E. H. Wilson

Rhododendron rothschildii Davidian

Rhododendron roxieanum Forrest ex W. W. Sm.

Rhododendron roxieanum var. **cucullatum** (Hand.-Mazz.) D. F. Chamb. ex L. C. Hu

Rhododendron roxieanum var. **oreonastes** (D. F. Chamb.) Geng

Rhododendron rubiginosum Franch.

Rhododendron rubiginosum var. **ptilostylum** R. C. Fang

Rhododendron rufescens Franch.

Rhododendron rufum Batalin

Rhododendron rupicola W. W. Sm.

Rhododendron rupicola var. **chryseum** (Balf. f. & Kingdon-Ward) M. N. Philipson & Philipson

Rhododendron rupicola var. **muliense** (Balf. f. & Forrest) M. N. Philipson & Philipson

Rhododendron russatum Balf. f. & Forrest

Rhododendron saluenense Franch.

Rhododendron saluenense var. **prostratum** (W. W. Sm.) R. C. Fang

Rhododendron sanguineum Franch.

Rhododendron sanguineum var. **cloiophorum** (Balf. f. & Forrest) D. F. Chamb.

Rhododendron sanguineum var. **didymoides** Tagg & Forrest

Rhododendron sanguineum var. **didymum** (Balf. f. & Forrest) T. L. Ming

Rhododendron sanguineum var. **haemaleum** (Balf. f. & Forrest) D. F. Chamb.

Rhododendron sanguineum var. **himertum** (Balf. f. & Forrest) D. F. Chamb.

Rhododendron santapaui Sastry et al.

Rhododendron sargentianum Rehder & E. H. Wilson

Rhododendron scabrifolium Franch.

Rhododendron schistocalyx Balf. f. & Forrest

Rhododendron scopulorum Hutch.

Rhododendron seingkuense Kingdon-Ward ex Hutch.

Rhododendron selense Franch.

Rhododendron selense subsp. **dasycladum** (Balf. f. & W. W. Sm.) D. F. Chamb.

Rhododendron selense subsp. **jucundum** (Balf. f. & W. W. Sm.) D. F. Chamb.

Rhododendron semnoides Ragg & Forrest

Rhododendron setiferum Balf. f. & Forrest

Rhododendron setosum D. Don

Rhododendron sherriffii Cowan

Rhododendron shweliense Balf. f. & Forrest

Rhododendron sidereum Balf. f.

Rhododendron siderophyllum Franch.

Rhododendron sikangense W. P. Fang

Rhododendron sinogrande Balf. f. & W. W. Sm.

Rhododendron sinosimulans D. F. Chamb.

Rhododendron smithii Nutt.

Rhododendron souliei Franch.

Rhododendron sperabile Balf. f. & Farrer

Rhododendron sperabile var. **weihsiense** Tagg & Forrest

Rhododendron sperabiloides Tagg & Forrest

Rhododendron sphaeroblastum Balf. f. & Forrest

Rhododendron stewartianum Diels

Rhododendron strigillosum Franch.

Rhododendron strigillosum var. **monosematum** (Hutch.) T. L. Ming

Rhododendron subansiriense D. F. Chamb. & Cox

Rhododendron succothii Davidian

Rhododendron sulfureum Franch.

Rhododendron taggianum Hutch.

Rhododendron taliense Franch.

Rhododendron tanastylum Balf. f. & Kingdon-Ward

Rhododendron tanastylum var. **lingzhiense** M. Y. Fang

Rhododendron tapetiforme Balf. f. & Kingdon-Ward

Rhododendron tatsienense Franch.

Rhododendron tatsienense var. **nudatum** R. C. Fang

Rhododendron telmateium Balf. f. & W. W. Sm.

Rhododendron temenium Balf. f. & Forrest

Rhododendron temenium var. **dealbatum** (Cowan) D. F. Chamb.

Rhododendron temenium var. **gilvum** (Cowan) D. F. Chamb.

Rhododendron tenuifolium R. C. Fang & S. H. Huang

Rhododendron tephropeplum Balf. f. & Farrer

Rhododendron thayerianum Rehder & E. H. Wilson

Rhododendron thomsonii Hook. f.

Rhododendron thomsonii subsp. **lopsangianum** (Cowan) D. F. Chamb.

Rhododendron thymifolium Maxim.

Rhododendron traillianum Forrest & W. W. Sm.

Rhododendron traillianum var. **dictyotum** (Balf. f. ex Tagg) D. F. Chamb.

Rhododendron trichanthum Rehder

Rhododendron trichocladum Franch.

Rhododendron trichocladum subsp. **nepalense** Hara

Rhododendron trichogynum L. C. Hu

Rhododendron trichostomum Franch.

Rhododendron trichostomum var. **ledoides** (Balf. f. & W. W. Sm.) Cowan & Davidian

Rhododendron trichostomum var. **radendum** (W. P. Fang) Geng

Rhododendron trichostomum var. **radinum** (Balf. f. & W. W. Sm.) Cowan & Davidian

Rhododendron triflorum Hook. f.

Rhododendron trilectorum Cowan

Rhododendron tsariense Cowan

Rhododendron tsariense var. **trimoense** Davidian

Rhododendron tubiforme (Cowan & Davidian) Davidian

Rhododendron tubulosum Ching ex W. Y. Wang

Rhododendron uniflorum Hutch. & Kingdon-Ward

Rhododendron uvarifolium Diels

Rhododendron vaccinioides Hook. f.

Rhododendron valentianum Forrest ex Hutch.

Rhododendron venator Tagg ex L. Rostschild

Rhododendron vernicosum Franch.

Rhododendron verruciferum W. K. Hu

Rhododendron vesiculiferum Tagg

Rhododendron virgatum Hook. f.

Rhododendron viridescens Hutch.

Rhododendron viscidifolium Davidian

Rhododendron wallichii Hook. f.

Rhododendron walongense Kingdon-Ward

Rhododendron wardii W. W. Sm.

Rhododendron wardii var. **puralbum** (Balf. f. & W. W. Sm.) D. F. Chamb.

Rhododendron watsonii Hemsl. & E. H. Wilson

Rhododendron wasonii var. **wenchuanense** L. C. Hu

Rhododendron websterianum Rehder & E. H. Wilson

Rhododendron websterianum var. **yulongense** M. N. Philipson & Philipson

Rhododendron wightii Hook. f.

Rhododendron williamsianum Rehder & E. H. Wilson

Rhododendron wiltonii Hemsl. & E. H. Wilson

Rhododendron wongii Hemsl. & E. H. Wilson

Rhododendron xanthocodon Hutch.

Rhododendron xanthostephanum Merr.

Rhododendron xichangense Z. J. Zhao

Rhododendron yungchangense Cullen

Rhododendron yungningense Balf. f. ex Hutch.

Rhododendron yunnanense Franch.

Rhododendron yushuense Z. J. Zhao

Rhododendron zaleucum Balf. f. & W. W. Sm.

Rhododendron zaleucum var. **pubifolium** R. C. Fang

Rhododendron zheguense Ching & H. P. Yang

Vaccinium arbutoides C. B. Clarke

Vaccinium ardisioides Hook. f. ex C. B. Clarke

Vaccinium brachyandrum C. Y. Wu & R. C. Fang

Vaccinium brachybotrys (Franch.) Hand.-Mazz.

Vaccinium bulleyanum (Diels) Sleumer

Vaccinium chaetothrix Sleumer

Vaccinium delavayi Franch.

Vaccinium dendrocharis Hand.-Mazz.

Vaccinium duclouxii (H. Lév.) Hand.-Mazz.

Vaccinium duclouxii var. **pubipes** C. Y. Wu

Vaccinium dunalianum Wight

Vaccinium dunalianum var. **urophyllum** Rehder & E. H. Wilson

Vaccinium fragile Franch.

Vaccinium fragile var. **mekongense** (W. W. Sm.) Sleumer

Vaccinium gaultheriifolium (Griff.) Hook. f. ex C. B. Clarke

Vaccinium glaucoalbum Hook. f. ex C. B. Clarke

Vaccinium iteophyllum Hance

Vaccinium iteophyllum var. **glandulosum** C. Y. Wu & R. C. Fang

Vaccinium kachinense Brandis

Vaccinium kingdon-wardii Sleumer

Vaccinium lanigerum Sleumer

Vaccinium leucobotrys (Nuttall) G. Nicholson

Vaccinium modestum W. W. Sm.

Vaccinium moupinense Franch.

Vaccinium nummularia Hook. f. & Thomson ex C. B. Clarke

Vaccinium papulosum C. Y. Wu & R. C. Fang

Vaccinium pubicalyx Franch.

Vaccinium pubicalyx var. **anomalum** J. Anthony

Vaccinium retusum (Griff.) Hook. f. ex C. B. Clarke

Vaccinium scopulorum W. W. Sm.

Vaccinium sikkimense C. B. Clarke

Vaccinium spiculatum C. Y. Wu & R. C. Fang

Vaccinium subdissitifolium P. F. Stevens

Vaccinium tenuiflorum R. C. Fang

Vaccinium vacciniaceum (Roxb.) Sleumer

Vaccinium vacciniaceum subsp. **glabritubum** P. F. Stevens

Vaccinium venosum Wight

240. ICACINACEAE

Iodes cirrhosa Turcz.

Iodes hookeriana Baillon

Iodes ovalis Blume

Mappianthus hookerianus (Baill.) Sleumer

Mappianthus iodoides Hand.-Mazz.

Miquelia assamica (Griff.) Mast. ex B. D. Jacks.

Miquelia kleinii Meisn.

Natsiatopsis thunbergiifolia Kurz

Natsiatum herpeticum Buch.-Ham. ex Arn.

Nothapodytes nimmoniana (J. Graham) Mabb.

Platea latifolia Blume

241. EUCOMMIACEAE

Eucommia ulmoides Oliver

242. GARRYACEAE

Aucuba himalaica Hook. f. & Thomson

243. RUBIACEAE

Acranthera burmanica Y. H. Tan & B. Yang

Acranthera tomentosa R. Br. ex Hook. f.

Adina cordifolia (Willd. ex Roxb.) Benth. & Hook. f. ex Brandis

Aidia cochinchinensis Lour.

Argostemma sarmentosum Wall.

Argostemma verticillatum Wall.

Asperula oppositifolia Regel & Schmalh.

Asperula setosa Jaub. & Spach

Benkara fasciculata (Roxb.) Ridsdale

Benkara forrestii (J. Anthony) Ridsdale

Benkara griffithii (Hook. f.) Ridsdale

Brachytome hirtellata var. **glabrescens** W. C. Chen

Brachytome wallichii Hook. f.

Canthium glabrum Blume

Catunaregam longispina (Link) Tirveng.

Catunaregam nutans (Roxb.) Tirveng.

Catunaregam spinosa (Thunb.) Tirveng.

Cephalanthus tetrandra (Roxb.) Ridsdale & Bakh. f.

Ceriscoides campanulata (Roxb.) Tirveng.

Chassalia curviflora (Wall.) Thwaites

Chassalia curviflora var. **longifolia** Hook. f.

Chassalia curviflora var. **ophioxyloides** (Wall.) Deb & Mondal

Chassalia staintonii (Hara) Deb & Mondal

Clarkella nana (Edgew.) Hook. f.

Coffea arabica L.

Coffea benghalensis Heyne ex Roem. & Schult.

Damnacanthus indicus Gaertn.

Damnacanthus tsaii Hu

Dentella repens (L.) J. R. Forst. & G. Forst.

Dentella serpyllifolia Wall. ex Craib

Diplospora fruticosa Hemsl.

Fosbergia shweliensis (J. Anthony) Tirveng. & Sastre

Galium acutum Edgew.

Galium acutum var. **himalayense** (Klotzsch & Garcke) R. R. Mill

Galium aparine L.

Galium asperifolium Wall.

Galium asperifolium var. **sikkimense** (Gand.) Cufod.

Galium asperifolium var. **verrucifructum** Cufod.

Galium asperuloides Edgew.

Galium baldensiforme Hand.-Mazz.

Galium blinii H. Lév.

Galium boreale L.

Galium boreale var. **angustifolium** (Freyn) Cufod.

Galium boreale var. **ciliatum** Nakai

Galium boreale var. **kamtschaticum** (Maxim.) Nakai

Galium boreale var. **latifolium** Turcz.

Galium ceratophylloides Hook. f.

Galium chitralensis Nazim.

Galium craticulatum R. R. Mill

Galium dahuricum Turcz. ex Ledeb.

Galium dahuricum var. **densiflorum** (Cufod.) Ehrend.

Galium elegans var. **glabriusculum** Requien ex DC.

Galium elegans Wall.

Galium exile Hook. f.

Galium forrestii Diels

Galium ghilanicum Stapf

Galium glabriusculum Ehrend.

Galium glandulosum Hand.-Mazz.

Galium hirtiflorum Requien ex DC.

Galium hoffmeisteri (Klotzsch) Ehrend. & Schonb.-Tem. ex R. R. Mill

Galium humifusum M. Bieb.

Galium megacyttarion R. R. Mill

Galium nepalense Ehrend. & Schonb.-Tem.

Galium paradoxum Maxim.

Galium paradoxum subsp. **duthiei** Ehrend. & Schonb.-Tem.

Galium pauciflorum Bunge

Galium prattii Cufod.

Galium pusillosetosum Hara

Galium rebae R. R. Mill

Galium saipalense Ehrend. & Schönb.-Tem.

Galium salwinense Hand.-Mazz.

Galium serpylloides Royle ex Hook. f.

Galium sichuanense Ehrend.

Galium spurium L.

Galium subfalcatum Nazim. & Ehrend.

Galium sungpanense Cufod.

Galium tenuissimum M. Bieb.

Galium tricornutum Dandy

Galium vassilczenkoi Pobed.

Galium verum L.

Galium verum var. **trachycarpum** DC.

Gardenia jasminoides J. Ellis

Gardenia turgida Roxb.

Greenea bahadurii R. C. Gaur & Dayal

Haldina cordifolia (Roxb.) Ridsdale

Hedyotis auricularia L.
Hedyotis biflora (L.) Lam.
Hedyotis brachypoda (DC.) Sivar. & Biju
Hedyotis corymbosa (L.) Lam.
Hedyotis diffusa Willd.
Hedyotis griffithii Hook. f.
Hedyotis lineata Roxb.
Hedyotis ovatifolia Cav.
Hedyotis pinifolia Wall. ex G. Don
Hedyotis scandens Roxb.
Hedyotis tetrandra (Roxb.) Craib
Hedyotis uncinella Hook. & Arn.
Hedyotis verticillata (L.) Lam.
Hedyotis vestita R. Br. ex G. Don
Himalrandia lichiangensis (W. W. Sm.) Tirveng.
Himalrandia tetrasperma (Wall. ex Roxb.) T. Yamaz.
Hymenodictyon excelsum (Roxb.) Wall.
Hymenodictyon flaccidum Wall.
Hymenodictyon orixense (Roxb.) Mabb.
Hyptianthera stricta (Roxb.) Wight & Arn.
Ixora acuminata Roxb.
Ixora athroantha Bremek.
Ixora balakrishnii Deb & Rout
Ixora coccinea L.
Ixora cuneifolia Roxb.
Ixora nigricans R. Br. ex Wight & Arn.
Ixora subsessilis Wall. ex G. Don
Ixora tibetana Bremek.
Ixora undulata Roxb.
Kelloggia chinensis Franch.
Khasiaclunea oligocephala (Havil.) Ridsdale
Knoxia corymbosa Willd.
Knoxia roxburghii (Spreng.) M. A. Rau
Knoxia sumatrensis (Retz.) DC.
Kohautia coccinea Royle
Kohautia gracilis (Wall.) DC.
Lasianthus attenuatus Jack
Lasianthus biermannii King ex Hook. f.
Lasianthus henryi Hutch.
Lasianthus hookeri C. B. Clarke ex Hook. f.
Lasianthus japonicus Miq.
Lasianthus japonicus subsp. **longicaudus** (Hook. f.) C. Y. Wu & H. Zhu
Lasianthus lancifolius Hook. f.
Lasianthus linearisepalus C. Y. Wu & H. Zhu
Lasianthus lucidus Blume
Lasianthus micranthus Hook. f.
Lasianthus sikkimensis Hook. f.
Lasianthus wallichii Wight
Lasianthus wardii C. E. C. Fisch. & Kaul
Leptodermis amoena Spring.
Leptodermis brevisepala H. S. Lo
Leptodermis dielsiana H. Winkl.
Leptodermis diffusa Batalin
Leptodermis forrestii Diels
Leptodermis glomerata Hutch.
Leptodermis gracilis C. E. C. Fisch.
Leptodermis gracilis var. **longiflora** H. S. Lo
Leptodermis griffithii Hook. f.
Leptodermis handeliana H. Winkl.
Leptodermis hirsutiflora H. S. Lo
Leptodermis hirsutiflora var. **ciliata** H. S. Lo
Leptodermis kumaonensis R. Parker
Leptodermis lanata H. S. Lo
Leptodermis lanceolata Wall.
Leptodermis limprichtii H. Winkl.
Leptodermis ludlowii Spring.
Leptodermis parvifolia Hutch.
Leptodermis pilosa Diels
Leptodermis pilosa var. **acanthoclada** H. S. Lo ex X. Y. Wen & Q. Lin
Leptodermis pilosa var. **glabrescens** H. Winkl.
Leptodermis potaninii Batalin
Leptodermis pumila H. S. Lo

Leptodermis purdomii Hutch.

Leptodermis rehderiana H. Winkl.

Leptodermis scabrida Hook. f.

Leptodermis schneideri H. Winkl.

Leptodermis scissa H. Winkl.

Leptodermis stapfiana H. Winkl.

Leptodermis umbellata Batalin

Leptodermis velutiniflora H. S. Lo

Leptodermis velutiniflora var. **tenera** H. S. Lo

Leptodermis virgata Edgew.

Leptodermis wilsonii Diels

Leptodermis xizangensis H. S. Lo

Leptodermis yui H. S. Lo

Leptomischus flaviflorus Hareesh, L. Wu & M. Sabu

Luculia grandifolia Ghose

Luculia gratissima (Wall.) Sweet

Luculia pinceana Hook.

Luculia yunnanensis S. Y. Hu

Meyna pubescens (Kurz) Robyns

Meyna spinosa Roxb. ex Link

Mitracarpus hirtus (L.) DC.

Mitragyna parvifolia (Roxb.) Korth.

Mitragyna rotundifolia (Roxb.) Kuntze

Morinda angustifolia Roxb.

Morinda pubescens Sm.

Morinda umbellata L.

Morinda villosa Hook. f.

Mussaenda decipiens H. Li

Mussaenda frondosa L.

Mussaenda glabra Vahl

Mussaenda gongshanensis H. Li

Mussaenda incana Wall.

Mussaenda macrophylla Wall.

Mussaenda roxburghii Hook. f.

Mussaenda taronensis H. Li

Mussaenda treutleri Stapf

Mussaenda wallichii G. Don

Mycetia brevipes F. C. How ex S. Y. Jin & Y. L. Chen

Mycetia listeri Deb

Mycetia longiflora F. C. How ex H. S. Lo

Mycetia longifolia (Wall.) Kuntze

Mycetia mukerjiana Deb & R. M. Dutta

Mycetia nepalensis Hara

Mycetia radiciflora (C. B. Clarke) Airy Shaw

Mycetia stipulata (Hook. f.) Kuntze

Myrioneuron nutans Wall. ex Kurz

Neanotis calycina (Wall. ex Hook. f.) W. H. Lewis

Neanotis gracilis (Hook. f.) W. H. Lewis

Neanotis hirsuta (L. f.) W. H. Lewis

Neanotis ingrata (Wall. ex Hook. f.) W. H. Lewis

Neanotis oxyphylla (Wall. ex G. Don) W. H. Lewis

Neanotis rhombicarpa T. Yamaz.

Neanotis wightiana (Wall. ex Wight & Arn.) W. H. Lewis

Neohymenopogon assamicus (Hook. f.) Bennet

Neohymenopogon oligocarpus (H. L. Li) Bennet

Neohymenopogon parasiticus (Wall.) Bennet

Neolamarckia cadamba (Roxb.) Bosser

Neonauclea gageana (King) Merr.

Neonauclea griffithii (Hook. f.) Merr.

Neonauclea purpurea (Roxb.) Merr.

Neonauclea sessilifolia (Roxb.) Merr.

Nertera sinensis Hemsl. ex F. B. Forbes & Hemsl.

Nostolachma jenkinsii (Hook. f.) Deb & Lahiri

Nostolachma khasiana (Korth.) Deb & Lahiri

Oldenlandia biflora L.

Oldenlandia corymbosa L.

Oldenlandia erecta (Manilal & Sivar.) R. R. Mill

Oldenlandia glabra (R. Br. ex Wall.) Kuntze

Oldenlandia herbacea (L.) Roxb.

Oldenlandia monocephala Kuntze

Oldenlandia pumila (L. f.) DC.

Oldenlandia tenelliflora (Blume) Kuntze

Ophiorrhiza chingii H. S. Lo

Ophiorrhiza debiana Hareesh, A. Joe & M. Sabu

Ophiorrhiza dulongensis H. S. Lo

Ophiorrhiza fasciculata D. Don

Ophiorrhiza heterostyla Dunn

Ophiorrhiza hispida Hook. f.

Ophiorrhiza laevifolia H. S. Lo

Ophiorrhiza longii J. R. I. Wood

Ophiorrhiza lurida Hook. f.

Ophiorrhiza medogensis H. Li

Ophiorrhiza mungos L.

Ophiorrhiza nepalesis Deb & Mondal

Ophiorrhiza nutans C. B. Clarke ex Hook. f.

Ophiorrhiza ochroleuca Hook. f.

Ophiorrhiza pauciflora Hook. f.

Ophiorrhiza repens (Wall. ex G. Don) Bennet

Ophiorrhiza rugosa Wall.

Ophiorrhiza rugosa var. **argentea** (Wall. ex G. Don) Deb & Mondal

Ophiorrhiza rugosa var. **prostrata** (D. Don) Deb & Mondal

Ophiorrhiza staintonii Hara

Ophiorrhiza subcapitata Wall. ex Hook. f.

Ophiorrhiza succirubra King ex Hook. f.

Ophiorrhiza thomsonii Hook. f.

Ophiorrhiza tingens C. B. Clarke ex C. E. C. Fisch.

Ophiorrhiza treutleri Hook. f.

Ophiorrhiza trichocarpa Blume

Ophiorrhiza umbricola W. W. Sm.

Ophiorrhiza villosa Roxb.

Ophiorrhiza wallichii Hook. f.

Ophiorrhiza wattii C. E. C. Fisch.

Paederia cruddasiana Prain

Paederia foetida L.

Pavetta crassicaulis Bremek.

Pavetta indica L.

Pavetta indica var. **glabrescens** (Kurz) Deb & Rout

Pavetta polyantha (Hook. f.) R. Br. ex Bremek.

Pavetta subcapitata Hook. f.

Pavetta tomentosa Roxb. ex Sm.

Pitardella sikkimensis (Hook. f.) Tirveng.

Plocama asperuliformis (Lincz.) M. Backlund & Thulin

Polyura geminata (Wall. ex G. Don) Hook. f.

Psychotria aborensis Dunn

Psychotria adenophylla Wall.

Psychotria burkillii Deb & M. Gangop.

Psychotria calocarpa Kurz

Psychotria denticulata Wall.

Psychotria erratica Hook. f.

Psychotria montana Blume

Psychotria monticola Kurz

Psychotria nicobarica Kurz

Psychotria silhetensis Hook. f.

Psychotria symplocifolia Kurz

Psychotria thomsonii Hook. f.

Psychotria yunnanensis Hutch.

Psydrax dicoccos Gaertn.

Psydrax kingii (Hook. f.) Bridson & Spring.

Randia fasciculata (Roxb.) DC.

Randia longiflora Lam.

Randia sikkimensis Hook. f.

Randia tetrasperma (Roxb.) Benth. & Hook. f. ex Brandis

Randia wallichii Hook. f.

Richardia scabra L.

Rubia alata Wall.

Rubia charifolia Wall. ex G. Don

Rubia chitralensis Ehrend.

Rubia cordifolia L.

Rubia edgeworthii Hook. f.

Rubia gedrosiaca Bornm.

Rubia haematantha Airy Shaw

Rubia himalayensis Klotzsch

Rubia hispidicaulis D. G. Long

Rubia latipetala H. S. Lo

Rubia manjith Roxb. ex Flem.

Rubia membranacea Diels

Rubia oncotricha Hand.-Mazz.

Rubia pallida Diels

Rubia podantha Diels

Rubia pseudogalium Ehrend.

Rubia pterygocaulis H. S. Lo

Rubia salicifolia H. S. Lo

Rubia schumanniana E. Pritz.

Rubia sikkimensis Kurz

Rubia tenuis H. S. Lo

Rubia tibetica Hook. f.

Rubia tinctorum L.

Rubia trichocarpa H. S. Lo

Rubia wallichiana Decne.

Rubia yunnanensis Diels

Saprosma ternatum (Wall.) Hook. f.

Serissa japonica (Thunb.) Thunb.

Spermacoce alata Aubl.

Spermacoce articularis L. f.

Spermacoce exilis (L. O. Williams) C. D. Adams ex W. C. Burger & C. M. Taylor

Spermacoce hispida L.

Spermacoce ocymoides Burm. f.

Spermacoce pusilla Wall.

Spermadictyon suaveolens Roxb.

Spiradiclis bifida Kurz

Spiradiclis cylindrica Wall. ex Hook. f.

Spiradiclis xizangensis H. S. Lo

Tamilnadia uliginosa (Retz.) Tirveng. & Sastre

Tarenna asiatica (L.) Kuntze ex K. Schum.

Tarenna disperma (Hook. f.) Pit.

Tarenna odorata (Roxb.) B. L. Rob.

Tarennoidea wallichii (Hook. f.) Tirveng. & Sastre

Trailliaedoxa gracilis W. W. Sm. & Forrest

Uncaria homomalla Miq.

Uncaria macrophylla Wall.

Uncaria scandens (Sm.) Hutch.

Uncaria sessilifructus Roxb.

Vangueria madagascariensis J. F. Gmel.

Wendlandia appendiculata Wall. ex Cowan

Wendlandia coriacea (Wall.) DC.

Wendlandia exserta (Roxb.) DC.

Wendlandia grandis (Hook. f.) Cowan

Wendlandia heynei (Schult.) Santapau & Merchant

Wendlandia pendula (Wall.) DC.

Wendlandia proxima (D. Don) DC.

Wendlandia puberula DC.

Wendlandia scabra Kurz

Wendlandia sikkimensis Cowan

Wendlandia speciosa Cowan

Wendlandia subalpina W. W. Sm.

Wendlandia tinctoria (Roxb.) DC.

Wendlandia wallichii Wight & Arn.

244. GENTIANACEAE

Canscora andrographioides Griff. ex C. B. Clarke

Canscora decussata (Roxb.) Roem. & Schult.

Canscora diffusa (Vahl) R. Br. ex Roem. & Schult.

Centaurium centaurioides (Roxb.) R. S. Rao & Hemadri

Centaurium pulchellum (Sw.) Druce

Centaurium pulchellum var. **meyeri** (Bunge) Omer

Comastoma borealis (Bunge) T. N. Ho

Comastoma cyananthiflorum (Franch.) Holub

Comastoma cyananthiflorum var. **acutifolium** Ma & H. W. Li

Comastoma disepalum H. W. Li

Comastoma falcata (Turcz. ex Karelin & Kirilov) Toyokuni

Comastoma muliense (Marquand) T. N. Ho

Comastoma pedunculatum (Royle ex D. Don) Holub

Comastoma pseudopulmonarium Omer

Comastoma pulmonarium (Turcz.) Toyokuni

Comastoma stellariifolium (Franch.) Holub

Comastoma tenellum (Rottb.) Toyok.

Comastoma traillianum (Forrest) Holub

Crawfurdia angustata C. B. Clarke

Crawfurdia campanulacea Wall. & Griff. ex C. B. Clarke

Crawfurdia crawfurdioides (Marquand) Harry Sm.

Crawfurdia crawfurdioides var. **iochroa** (Marquand) C. J. Wu

Crawfurdia delavayi Franch.

Crawfurdia dimidiata (Marquand) Harry Sm.

Crawfurdia gracilipes Harry Sm.

Crawfurdia puberula C. B. Clarke

Crawfurdia semialata (Marquand) Harry Sm.

Crawfurdia sessiliflora (Marquand) Harry Sm.

Crawfurdia speciosa Wall.

Crawfurdia thibetica Franch.

Crawfurdia tsangshanensis C. J. Wu

Exacum hamiltonii G. Don

Exacum nanum Klack.

Exacum paucisquamum (C. B. Clarke) Klack.

Exacum teres Wall.

Exacum tetragonum Roxb.

Fagraea ceilanica Thunb.

Gentiana abaensis T. N. Ho

Gentiana albicalyx Burkill

Gentiana albomarginata C. Marquand

Gentiana algida Pallas

Gentiana alii (Omer & Qaiser) T. N. Ho

Gentiana alsinoides Franch.

Gentiana altigena Harry Sm.

Gentiana altorum Harry Sm.

Gentiana amplicrater Burkill

Gentiana anisostemon C. Marquand

Gentiana aprica Decne.

Gentiana aquatica L.

Gentiana aquatica var. **baltistanica** (Omer) Halda

Gentiana arethusae Burkill

Gentiana arethusae var. **delicatula** C. Marquand

Gentiana argentea Royle ex D. Don

Gentiana aristata Maxim.

Gentiana asparagoides T. N. Ho

Gentiana asterocalyx Diels

Gentiana atropurpurea T. N. Ho

Gentiana atuntsiensis W. W. Sm.

Gentiana baoxingensis T. N. Ho

Gentiana bella Franch.

Gentiana bomiensis T. N. Ho

Gentiana bryoides Burkill

Gentiana burkillii Harry Sm.

Gentiana burmensis C. Marquand

Gentiana cachemirica Decne.

Gentiana caelestis (Marquand) Harry Sm.

Gentiana caeruleogrisea T. N. Ho

Gentiana capitata Buch.-Ham. ex D. Don

Gentiana capitata subsp. **harwanensis** (G. Singh) Halda

Gentiana caryophyllea Harry Sm.

Gentiana cephalantha Franch.

Gentiana chinensis Kusnezow

Gentiana choanantha Marquand

Gentiana chungtienensis Marquand

Gentiana clarkei Kusnezow
Gentiana conduplicata T. N. Ho
Gentiana confertifolia C. Marquand
Gentiana crassicaulis Duthie ex Burkill
Gentiana crassula Harry Sm.
Gentiana crassuloides Bureau & Franch.
Gentiana crenulatotruncata (Marquand) T. N. Ho
Gentiana cristata Harry Sm.
Gentiana cuneibarba Harry Sm.
Gentiana curvianthera T. N. Ho
Gentiana curviphylla T. N. Ho
Gentiana dahurica Fischer
Gentiana dahurica var. **campanulata** T. N. Ho
Gentiana damyonensis Marquand
Gentiana daochengensis T. N. Ho
Gentiana decemfida Buch.-Ham. ex D. Don
Gentiana decorata Diels
Gentiana delavayi Franch.
Gentiana deltoidea Harry Sm.
Gentiana dendrologi Marquand
Gentiana depressa D. Don
Gentiana divaricata T. N. Ho
Gentiana dolichocalyx T. N. Ho
Gentiana doxiongshangensis T. N. Ho
Gentiana ccaudata C. Marquand
Gentiana elwesii C. B. Clarke
Gentiana emodii Marquand ex Sealy
Gentiana epichysantha Hand.-Mazz.
Gentiana erectosepala T. N. Ho
Gentiana exigua Harry Sm.
Gentiana exquisita Harry Sm.
Gentiana faucipilosa Harry Sm.
Gentiana filisepala T. N. Ho
Gentiana filistyla Balf. f. & Forrest
Gentiana flexicaulis Harry Sm.
Gentiana formosa Harry Sm.
Gentiana forrestii C. Marquand
Gentiana franchetiana Kusnezow
Gentiana futtereri Diels & Gilg
Gentiana georgei Diels
Gentiana gilvostriata Marquand
Gentiana glabriuscula Harry Sm. ex T. N. Ho
Gentiana globosa T. N. Ho
Gentiana grata Harry Sm.
Gentiana gyirongensis T. N. Ho
Gentiana handeliana Harry Sm.
Gentiana harrowiana Diels
Gentiana harwanensis G. Singh
Gentiana haynaldii Kanitz
Gentiana heleonastes Harry Sm.
Gentiana helophila Balf. f. & Forrest
Gentiana hexaphylla Maxim. ex Kusnezow
Gentiana hicksii T. N. Ho & S. W. Liu
Gentiana himalayaensis T. N. Ho
Gentiana hirsuta Ma & E. W. Ma ex T. N. Ho
Gentiana huxleyi Kusnezow
Gentiana inconspicua Harry Sm.
Gentiana infelix C. B. Clarke
Gentiana intricata Marquand
Gentiana kanchii (D. Maity, S. K. Dey & Adr. Favre) M. Hassler (syn: *Kuepferia kanchii* D. Maity, Sentu K. Dey & Adr. Favre)
Gentiana karelini Griseb.
Gentiana kumaonensis Biswas
Gentiana kurroo Royle
Gentiana lacerulata Harry Sm.
Gentiana lacinulata T. N. Ho
Gentiana lawrencei Burkill
Gentiana lawrencei var. **farreri** (Balf. f.) T. N. Ho
Gentiana laxiflora T. N. Ho
Gentiana leucantha Harry Sm. ex T. N. Ho & S. W. Liu
Gentiana leucomelaena Maxim. ex Kusnezow

Gentiana lhassica Burkill

Gentiana lineolata Franch.

Gentiana linoides Franch.

Gentiana listeri Burkill

Gentiana loureiroi (G. Don) Griseb.

Gentiana ludingensis T. N. Ho

Gentiana macrauchena Marquand

Gentiana maeulchanensis Franch.

Gentiana mailingensis T. N. Ho

Gentiana mairei H. Lév.

Gentiana meiantha (C. B. Clarke) Harry Sm.

Gentiana melandriifolia Franch.

Gentiana membranulifera T. N. Ho

Gentiana membranulifera var. **intermedia** (C. B. Clarke) T. N. Ho

Gentiana micans C. B. Clarke

Gentiana micantiformis Burkill

Gentiana microdonta Franch.

Gentiana microphyta Franch.

Gentiana moniliformis Marquand

Gentiana muscicola Marquand

Gentiana namlaensis Marquand

Gentiana nanobella Marquand

Gentiana napulifera Franch.

Gentiana ninglangensis T. N. Ho

Gentiana nubigena Edgew.

Gentiana nyalamensis T. N. Ho

Gentiana nyalamensis var. **parviflora** T. N. Ho

Gentiana nyingchiensis T. N. Ho

Gentiana obconica T. N. Ho

Gentiana officinalis Harry Sm.

Gentiana olivieri Griseb.

Gentiana oreodoxa Harry Sm.

Gentiana ornata (Wall. ex G. Don) Griseb.

Gentiana otophora Franch.

Gentiana otophoroides Harry Sm.

Gentiana pamirica Grossh.

Gentiana panthaica Prain & Burkill

Gentiana papillosa Franch.

Gentiana parvula Harry Sm.

Gentiana pedicellata (Wall. ex D. Don) Griseb.

Gentiana pentasticta C. Marquand

Gentiana phyllocalyx C. B. Clarke

Gentiana phyllopoda H. Lév.

Gentiana piasezkii Maxim.

Gentiana picta Franch.

Gentiana pluviarum subsp. **subtilis** (Harry Sm.) T. N. Ho

Gentiana praeclara C. Marquand

Gentiana prainii Burkill

Gentiana praticola Franch.

Gentiana prattii Kusnezow

Gentiana pringlei (D. Maity & S. K. Dey) M. Hassler (syn: *Kuepferia pringlei* D. Maity & Sentu K. Dey)

Gentiana producta T. N. Ho

Gentiana prolata Balf. f.

Gentiana prostrata Haenke

Gentiana prostrata var. **ludlowii** (Marquand) T. N. Ho

Gentiana pseudoaquatica Kusnezow

Gentiana pseudosquarrosa Harry Sm.

Gentiana pubiflora T. N. Ho

Gentiana pubigera Marquand

Gentiana purdomii Marquand

Gentiana qiujiangensis T. N. Ho

Gentiana radiata Marquand

Gentiana radicans Harry Sm.

Gentiana recurvata C. B. Clarke

Gentiana rigescens Franch.

Gentiana riparia Kar. & Kir.

Gentiana robusta King ex Hook. f.

Gentiana rubicunda Franch.

Gentiana rubicunda var. **purpurata** (Maxim. ex Kusnezow) T. N. Ho

Gentiana sagarmathae Miyam. & H. Ohba
Gentiana scabrifilamenta T. N. Ho
Gentiana serra Franch.
Gentiana sichitoensis Marquand
Gentiana sikkimensis C. B. Clarke
Gentiana simulatrix Marquand
Gentiana sinoornata Balf. f.
Gentiana sinoornata var. **gloriosa** Marquand
Gentiana siphonantha Maxim. ex Kusnezow
Gentiana souliei Franch.
Gentiana spathulifolia Maxim. ex Kusnezow
Gentiana speciosa (Wall.) Marquand
Gentiana springateana D. Maity
Gentiana squarrosa Ledeb.
Gentiana stellata Turrill
Gentiana stellata var. **acuminata** (C. B. Clarke) T. N. Ho
Gentiana stellulata Harry Sm.
Gentiana stellulata var. **dichotoma** Harry Sm.
Gentiana stipitata Edgew.
Gentiana stipitata subsp. **tizuensis** (Franch.) T. N. Ho
Gentiana stragulata Balf. f. & Forrest
Gentiana straminea Maxim.
Gentiana striata Maxim.
Gentiana striolata T. N. Ho
Gentiana stylosa Biswas
Gentiana subintricata T. N. Ho
Gentiana suborbisepala C. Marquand
Gentiana suborbisepala var. **kialensis** (C. Marquand) T. N. Ho
Gentiana subuliformis S. W. Liu
Gentiana subuniflora C. Marquand
Gentiana syringea T. N. Ho
Gentiana szechenyii Kanitz
Gentiana taliensis Balf. f. & Forrest
Gentiana tatsienensis Franch.
Gentiana ternifolia Franch.
Gentiana tetraphylla Maxim. ex Kusnezow
Gentiana tetrasepala Biswas
Gentiana tetrasticha C. Marquand
Gentiana tianschanica Rupr.
Gentiana tibetica King ex Hook. f.
Gentiana tongolensis Franch.
Gentiana trichotoma Kusnezow
Gentiana trichotoma var. **chingii** (C. Marquand) T. N. Ho
Gentiana tubiflora (Wall. ex G. Don) Griseb.
Gentiana urnula Harry Sm.
Gentiana veitchiorum Hemsl.
Gentiana venusta (G. Don) Griseb.
Gentiana vernayi Marquand
Gentiana viatrix Harry Sm.
Gentiana waltonii Burkill
Gentiana wangchukii E. Aitken & D. G. Long
Gentiana wardii W. W. Sm.
Gentiana wardii var. **emergens** (C. Marquand) T. N. Ho
Gentiana wardii var. **micrantha** Marquand
Gentiana wasenensis Marquand
Gentiana wilsonii Marquand
Gentiana winchuanensis T. N. Ho
Gentiana xanthonannos Harry Sm.
Gentiana yunnanensis Franch.
Gentianella angustiflora Harry Sm.
Gentianella anomala (Marquand) T. N. Ho
Gentianella aurea (L.) Harry Sm.
Gentianella azurea (Bunge) Holub
Gentianella gentianoides (Franch.) Harry Sm.
Gentianella glanduligera Airy Shaw
Gentianella holosteoides Schott & Kotschy ex N. M. Pritch.
Gentianella longicarpa (Gilli) Holub
Gentianella lowndesii Harry Sm.
Gentianella maddeni (C. B. Clarke) Airy Shaw
Gentianella moorcroftiana (Wall. ex G. Don)

Airy Shaw

Gentianella pedunculata (Royle) Harry Sm.

Gentianella pygmaea (Regel & Schmalh.) Harry Sm. ex S. Nilsson

Gentianella staintonii Y. S. Chen, nom. nov. (syn: *Aloitis smithii* S. Omer)

Gentianella stolitzkai (C. B. Clarke) Holub

Gentianella umbellata (M. Bieb.) Holub

Gentianella urnigera E. Aitken & D. G. Long

Gentianella vvedenskyi (Grossh.) Harry Sm.

Gentianopsis barbata (Froel.) Ma

Gentianopsis barbata var. **stenocalyx** H. W. Li

Gentianopsis contorta (Royle) Ma

Gentianopsis detonsa (Rottb.) Ma

Gentianopsis grandis (Harry Sm.) Ma

Gentianopsis paludosa (Munro ex Hook. f.) Ma

Gentianopsis paludosa var. **ovatodeltoidea** (Burkill) Ma

Halenia elliptica D. Don

Jaeschkea canaliculata (Royle ex G. Don) Knobl.

Jaeschkea microsperma C. B. Clarke

Lomatogoniopsis alpina T. N. Ho & S. W. Liu

Lomatogoniopsis galeiformis T. N. Ho & S. W. Liu

Lomatogoniopsis ovatifolia T. N. Ho & S. W. Liu

Lomatogonium bellum (Hemsl.) Harry Sm.

Lomatogonium brachyantherum (C. B. Clarke) Fernald

Lomatogonium caeruleum (Royle) Harry Sm. ex B. L. Burtt

Lomatogonium carinthiacum (Wulfen) A. Br.

Lomatogonium cherukurianum S. K. Dey & D. Maity

Lomatogonium chumbicum (Burkill) Harry Sm.

Lomatogonium forrestii (Balf. f.) Fernald

Lomatogonium forrestii var. **densiflorum** S. W. Liu & T. N. Ho

Lomatogonium gamosepalum (Burkill) Harry Sm.

Lomatogonium graciliflorum Harry Sm.

Lomatogonium himalayense (Klotzsch) E. Aitken

Lomatogonium lloydioides (Burkill) Harry Sm. ex Chater

Lomatogonium longifolium Harry Sm.

Lomatogonium macranthum (Diels & Gilg) Fernald

Lomatogonium micranthum Harry Sm.

Lomatogonium oreocharis (Diels) Marquand

Lomatogonium perenne T. N. Ho & S. W. Liu

Lomatogonium rotatum (L.) Fries ex Nyman

Lomatogonium sikkimense (Burkill) Harry Sm.

Lomatogonium spathulatum (A. Kern.) Fernald

Lomatogonium stapfii (Burkill) Harry Sm.

Lomatogonium zhongdianense S. W. Liu & T. N. Ho

Megacodon stylophorus (C. B. Clarke) Harry Sm.

Metagentiana expansa (Harry Sm.) T. N. Ho & S. W. Liu

Metagentiana leptoclada (Balf. f. & Forrest) T. N. Ho & S. W. Liu

Metagentiana primuliflora (Franch.) T. N. Ho & S. W. Liu

Metagentiana pterocalyx (Franch.) T. N. Ho & S. W. Liu

Metagentiana rhodantha (Franch.) T. N. Ho & S. W. Liu

Metagentiana serra (Franch.) T. N. Ho, S. W. Liu & Shi L. Chen

Pterygocalyx volubilis Maxim.

Sebaea microphylla (Edgew.) Knobl.

Swertia acaulis Harry Sm.

Swertia alata (D. Don) C. B. Clarke

Swertia alba T. N. Ho & S. W. Liu

Swertia alpina U. C. Bhattach. & S. Agrawal

Swertia alternifolia Royle

Swertia angustifolia Buch.-Ham. ex D. Don

Swertia angustifolia var. **pulchella** Burkill

Swertia asarifolia Franch.

Swertia assamensis Harry Sm.

Swertia bifolia Batalin

Swertia bimaculata (Siebold & Zucc.) Hook. f. & Thomson ex C. B. Clarke

Swertia binchuanensis T. N. Ho & S. W. Liu

Swertia calycina Franch.

Swertia candelabrum Harry Sm.

Swertia chirayita (Roxb.) Buch.-Ham. ex C. B. Clarke

Swertia ciliata (D. Don ex G. Don) B. L. Burtt

Swertia cincta Burkill

Swertia conaensis T. N. Ho & S. W. Liu

Swertia cordata (Wall. ex G. Don) C. B. Clarke

Swertia crossoloma Harry Sm.

Swertia cuneata Wall. ex D. Don

Swertia decora Franch.

Swertia delavayi Franch.

Swertia dilatata C. B. Clarke

Swertia divaricata Harry Sm.

Swertia elata Harry Sm.

Swertia endotricha Harry Sm.

Swertia erythrosticta Maxim.

Swertia fasciculata T. N. Ho & S. W. Liu

Swertia forrestii Harry Sm.

Swertia franchetiana Harry Sm.

Swertia grandiflora Harry Sm.

Swertia gyacaensis T. N. Ho & S. W. Liu

Swertia handeliana Harry Sm.

Swertia hispidicalyx Burkill

Swertia hookeri C. B. Clarke

Swertia kingii Hook. f.

Swertia kouitchensis Franch.

Swertia lactea Bunge

Swertia laticalyx J. Shah

Swertia ligengiana T. N. Ho & S. W. Liu

Swertia longifolia Boiss.

Swertia lurida (D. Don) C. B. Clarke

Swertia macrosperma (C. B. Clarke) C. B. Clarke

Swertia marginata Schrenk

Swertia membranifolia Franch.

Swertia multicaulis D. Don

Swertia multicaulis var. **umbellifera** T. N. Ho & S. W. Liu

Swertia mussoti Franch.

Swertia mussoti var. **flavescens** T. N. Ho & S. W. Liu

Swertia nepalensis J. Shah

Swertia nervosa (G. Don) C. B. Clarke

Swertia paniculata Wall.

Swertia paniculata var. **gracilescens** (Harry Sm.) T. N. Ho

Swertia patula Harry Sm.

Swertia petiolata D. Don

Swertia pianmaensis T. N. Ho & S. W. Liu

Swertia pseudohookeri Harry Sm.

Swertia pubescens Franch.

Swertia punicea Hemsl.

Swertia racemosa (Wall. ex Griseb.) C. B. Clarke

Swertia ramosa W. W. Sm.

Swertia rotundiglandula T. N. Ho & S. W. Liu

Swertia scapiformis T. N. Ho & S. W. Liu

Swertia schugnanica Pissjauk.

Swertia souliei Burkill

Swertia speciosa G. Don

Swertia splendens Harry Sm.

Swertia staintonii Harry Sm.

Swertia taylorii J. Shah

Swertia tenuis T. N. Ho & S. W. Liu

Swertia tetragona Edgew.

Swertia tetraptera Maxim.

Swertia thomsonii C. B. Clarke

Swertia tibetica Batalin

Swertia verticillifolia T. N. Ho & S. W. Liu

Swertia virescens Harry Sm.

Swertia wardii Marquand

Swertia wolfgangiana Grüning

Swertia younghusbandii Burkill

Swertia zayuensis T. N. Ho & S. W. Liu

Tripterospermum coeruleum (Hand.-Mazz.) Harry Sm.

Tripterospermum cordatum (Marquand) Harry Sm.

Tripterospermum fasciculatum (Wall.) Chater

Tripterospermum maculatum Adr. Favre, Matuszak & Muellner

Tripterospermum membranaceum (Marquand) Harry Sm.

Tripterospermum nigrobaccatum Hara

Tripterospermum volubile (D. Don) Hara

Veratrilla baillonii Franch.

Veratrilla burkilliana (W. W. Sm.) Harry Sm.

245. LOGANIACEAE

Gardneria angustifolia Wall.

Gardneria ovata Wall.

Mitrasacme pygmaea var. **malaccensis** (Wight) Hara

Mitreola pedicellata Benth.

Mitreola petiolata (J. F. Gmel.) Torr. & A. Gray

Strychnos axillaris Colebr.

Strychnos nux-vomica L.

Strychnos wallichiana Steud. ex A. DC.

246. APOCYNACEAE

Aganosma gracilis Hook. f.

Allamanda cathartica L.

Alstonia mairei H. Lév.

Alstonia neriifolia D. Don

Alstonia scholaris (L.) R. Br.

Alstonia sebusi (Van Heurck & Müll. Arg.) Monach.

Alyxia fascicularis Benth.

Alyxia gracilis (Wall. ex A. DC.) Hook. f.

Amphineurion marginatum (Roxb.) D. J. Middleton

Anodendron paniculatum A. DC.

Apocynum venetum L.

Asclepias curassavica L.

Beaumontia grandiflora Wall.

Biondia crassipes M. G. Gilbert & P. T. Li

Biondia insignis Tsiang

Biondia longipes P. T. Li

Biondia pilosa Tsiang & P. T. Li

Biondia revoluta M. G. Gilbert & P. T. Li

Biondia tsiukowensis M. G. Gilbert & P. T. Li

Biondia yunnanensis (H. Lév.) Tsiang

Brachystelma nepalense (Radcl.-Sm.) Meve

Calotropis acia Buch.-Ham.

Calotropis gigantea (L.) W. T. Aiton

Calotropis procera (Aiton) W. T. Aiton

Calotropis procera subsp. **hamiltonii** (Wight) Ali

Caralluma staintonii Hara

Carissa carandas L.

Carissa opaca Stapf ex Haines

Carissa paucinerva A. DC.

Carissa spinarum L.

Catharanthus roseus (L.) G. Don

Cerbera manghas L.

Ceropegia aridicola W. W. Sm.

Ceropegia bhutanica Hara

Ceropegia dolichophylla Schltr.

Ceropegia dorjei C. E. C. Fisch.

Ceropegia edulis (Edgew.) Bruyns

Ceropegia hookeri C. B. Clarke

Ceropegia hookeri var. **mollis** H. Huber

Ceropegia kachinensis Prain

Ceropegia longifolia Wall.

Ceropegia lucida Wall.

Ceropegia ludlowii H. Huber

Ceropegia macrantha Wight

Ceropegia mairei (H. Lév.) H. Huber

Ceropegia meleagris H. Huber

Ceropegia monticola W. W. Sm.

Ceropegia muliensis W. W. Sm.

Ceropegia paohsingensis Tsiang & P. T. Li

Ceropegia poluniniana Bruyns

Ceropegia pubescens Wall.

Ceropegia sinoerecta M. G. Gilbert & P. T. Li

Ceropegia stenophylla C. K. Schneid.

Ceropegia teniana Hand.-Mazz.

Ceropegia ugeni C. E. C. Fisch.

Ceropegia wallichii Wight

Chonemorpha fragrans (Moon) Alston

Chonemorpha griffithii Hook. f.

Chonemorpha verrucosa (Blume) D. J. Middleton

Cryptolepis buchananii Schult.

Cryptolepis sinensis (Lour.) Merr.

Cynanchum acutum L.

Cynanchum acutum subsp. **sibiricum** (Willd.) Rech. f.

Cynanchum anthonyanum Hand.-Mazz.

Cynanchum auriculatum Royle ex Wight

Cynanchum batangense P. T. Li

Cynanchum bicampanulatum M. G. Gilbert & P. T. Li

Cynanchum canescens (Willd.) K. Schum.

Cynanchum corymbosum Wight

Cynanchum dalhousiae Wight

Cynanchum decipiens C. K. Schneid.

Cynanchum deltoideum Hance

Cynanchum forrestii Schltr.

Cynanchum heydei Hook. f.

Cynanchum inamoenum (Maxim.) Loes.

Cynanchum jacquemontianum Decne.

Cynanchum kingdonwardii M. G. Gilbert & P. T. Li

Cynanchum linearisepalum P. T. Li

Cynanchum longipedunculatum M. G. Gilbert & P. T. Li

Cynanchum lysimachioides Tsiang & P. T. Li

Cynanchum megalanthum M. G. Gilbert & P. T. Li

Cynanchum officinale (Hemsl.) Tsiang & Zhang

Cynanchum otophyllum C. K. Schneid.

Cynanchum rockii M. G. Gilbert & P. T. Li

Cynanchum roylei Wight

Cynanchum szechuanense Tsiang & Zhang

Cynanchum viminale (L.) L.

Cynanchum wallichii Wight

Cynanchum wilfordii (Maxim.) Hook. f.

Decalepis khasiana (Kurz) Ionta ex Kambale

Dischidia bengalensis Colebr.

Dischidia nummularia R. Br.

Dittoceras andersonii Hook. f.

Dregea sinensis Hemsl.

Dregea volubilis (L. f.) Benth. ex Hook. f.

Dregea yunnanensis (Tsiang) Tsiang & P. T. Li

Genianthus crassifolius (Wight & Arn.) Hook. f.

Genianthus laurifolius (Roxb.) Hook. f.

Gongronema nepalense (Wall.) Decne.

Gongronema thomsonii (Hook. f.) K. M. Matthew

Gongronema ventricosum Hook. f.

Gongronema ventricosum var. **bhutanicum** Hara

Gymnema inodorum (Lour.) Decne.

Gymnema macranthum Hook. f.

Gymnema sylvestre (Retz.) R. Br. ex Schult.

Gymnema tingens (Roxb.) Spreng.

Hemidesmus indicus (L) R. Br. ex Schult.

Heterostemma alatum Wight

Heterostemma wallichii Wight

Holarrhena pubescens Wall. ex G. Don

Holostemma ada-kodien Schult.

Holostemma annularium (Roxb.) K. Schum.

Hoya acuminata (Wight) Benth. ex Hook. f.

Hoya arnottiana Wight

Hoya bhutanica Grierson & D. G. Long

Hoya edenii King ex Hook. f.

Hoya fusca Wall.

Hoya globulosa Hook. f.

Hoya lanceolata Wall. ex D. Don

Hoya lii C. M. Burton

Hoya linearis Wall. ex D. Don

Hoya linearis var. **sikkimensis** Hook. f.

Hoya lobbii Hook. f.

Hoya longifolia Wall. ex Wight

Hoya mekongensis M. G. Gilbert & P. T. Li

Hoya obcordata Hook. f.

Hoya parasitica Wall. ex Traill

Hoya pauciflora Wight

Hoya polyneura Hook. f.

Hoya salweenica Tsiang & P. T. Li

Hoya serpens Hook. f.

Hoya siamica Craib

Hoya silvatica Tsiang & P. T. Li

Hoya teretifolia Griff. ex Hook. f.

Hoya thomsonii Hook. f.

Hoya vaccinioides Hook. f.

Ichnocarpus frutescens (L.) W. T. Aiton

Ichnocarpus polyanthus (Blume) P. I. Forster

Kopsia fruticosa (Roxb.) A. DC.

Leptadenia reticulata (Retz.) Wight & Arn.

Marsdenia calesiana Wight

Marsdenia koi Tsiang

Marsdenia lucida Edgew. ex Madden

Marsdenia medogensis P. T. Li

Marsdenia oreophila W. W. Sm.

Marsdenia pulchella Hand.-Mazz.

Marsdenia roylei Wight

Marsdenia stenantha Hand.-Mazz.

Marsdenia tenacissima (Roxb.) Moon

Marsdenia tinctoria R. Br.

Marsdenia tingens (Roxb.) P. I. Forst.

Melodinus khasianus Hook. f.

Melodinus monogynus Roxb.

Myriopteron paniculatum Griff.

Nerium oleander L.

Orthanthera viminea Wight

Oxystelma esculentum (L. f.) Sm.

Parabarium hookeri Pierre

Parameria polyneura Hook. f.

Parsonsia alboflavescens (Dennst.) Mabb.

Pentanura khasiana Kurz

Pentasachme caudatum Wall. ex Wight

Pentasachme pulcherrima Grierson & D. G. Long

Pentasachme wallichii Wight

Pergularia daemia (Forssk.) Chiov.

Pergularia exilis (Colebr.) Spreng.

Periploca aphylla Decne.

Periploca calophylla (Wight) Falc.

Periploca calophylla subsp. **forrestii** Browicz

Periploca forrestii Schltr.

Periploca hydaspidis Falc.

Periploca secamone L.

Plumeria rubra L.

Pottsia laxiflora (Blume) Kuntze

Raphistemma pulchellum (Roxb.) Wall.

Rauvolfia serpentina (L.) Benth. ex Kurz

Rauvolfia tetraphylla L.

Rauvolfia verticillata (Lour.) Baill.

Riocreuxia nepalensis Radcl.-Sm.

Streptocaulon sylvestre Wight

Strophanthus wallichii A. DC.

Tabernaemontana divaricata (L.) R. Br. ex Roem. & Schult.

Telosma cordata (Burm. f.) Merr.

Telosma pallida (Roxb.) Craib

Thevetia linearis Raf.

Toxocarpus aurantiacus C. Y. Wu ex Tsiang & P. T. Li

Toxocarpus himalensis Falc. ex Hook. f.

Trachelospermum asiaticum (Siebold & Zucc.) Nakai

Trachelospermum assamense Woodson

Trachelospermum axillare Hook. f.

Trachelospermum bodinieri (H. Lév.) Woodson

Trachelospermum brevistylum Hand.-Mazz.

Trachelospermum jasminoides (Lindl.) Lem.

Trachelospermum lucidum (D. Don) K. Schum.

Treutlera insignis Hook. f.

Tylophora belostemma (Wall.) Benth. ex Benth. & Hook. f.

Tylophora exilis (Spreng.) Colebr.

Tylophora fasciculata Buch.-Ham. ex Wight

Tylophora forrestii M. G. Gilbert & P. T. Li

Tylophora govanii (Wight) Decne.

Tylophora himalaica Hook. f.

Tylophora hirsuta (Wall.) Wight

Tylophora indica (Burm. f.) Merr.

Tylophora longifolia Wight

Tylophora nana C. K. Schneid.

Tylophora ovata (Lindl.) Hook. ex Steud.

Tylophora rockii M. G. Gilbert & P. T. Li

Tylophora rotundifolia Buch.-Ham. ex Wight

Tylophora tenerrima Wight

Tylophora wallichii (Wight) Hook. f.

Tylophora yunnanensis Schltr.

Urceola linearicarpa (Pierre) D. J. Middleton

Urceola micrantha (Wall. ex G. Don) D. J. Middleton

Urceola rosea (Hook. & Arn.) D. J. Middleton

Urceola tournieri (Pierre) D. J. Middleton

Vallaris indecora (Baillon) Tsiang & P. T. Li

Vallaris solanacea (Roth) Kuntze

Vincetoxicum arnottianum (Wight) Wight

Vincetoxicum canescens (Wind.) Decne.

Vincetoxicum deltoideum (Hook. f.) Jagtap & Singh

Vincetoxicum hirsutum Kuntze

Vincetoxicum hirundinaria Medik.

Vincetoxicum hirundinaria subsp. **glaucum** (Wall. ex Wight) Hara

Vincetoxicum laxum Gren. & Godr.

Wattakaka volubilis (L. f.) Stapf

Wrightia arborea (Dennst.) Mabb.

Wrightia coccinea (Roxb. ex Hornem.) Sims

Wrightia sikkimensis Gamble

Wrightia tomentosa Roem. & Schult.

247. BORAGINACEAE

Actinocarya acaulis (W. W. Sm.) I. M. Johnst.

Actinocarya tibetica Benth.

Anchusa arvensis (L.) M. Bieb.

Anchusa arvensis subsp. **orientalis** (L.) Nordh.

Anchusa azurea Mill.

Antiotrema dunnianum Hand.-Mazz.

Arnebia afghanica (Kitam.) Rech. f. & Riedl

Arnebia benthamii (Wall. ex G. Don) I. M. Johnst.

Arnebia coerulea Schipcz.

Arnebia decumbens (Vent.) Coss. & Kralik

Arnebia euchroma (Royle) I. M. Johnst.

Arnebia euchroma subsp. **andarabica** F. Sadat

Arnebia euchroma subsp. **caespitosa** Rech. f. & H. Riedl

Arnebia guttata Bunge

Arnebia guttata subsp. **griffithii** (Boiss.) Sadat.

Arnebia hispidissima (Lehm.) A. DC.

Arnebia inconspicua Hemsl. & Lace

Arnebia lindbergiana (Rech. f.) I. M. Johnst.

Arnebia nepalensis (Kitam.) Hara

Arnebia simulatrix Riedl

Arnebia speciosa Aitch. & Hemsl.

Asperugo procumbens L.

Bothriospermum hispidissimum Hand.-Mazz.

Bothriospermum tenellum (Hornem.) Fisch. & C. A. Mey.

Buglossoides arvensis (L.) I. M. Johnst.

Buglossoides tenuiflora (L. f.) I. M. Johnst.

Chionocharis hookeri (C. B. Clarke) I. M. Johnst.

Cordia dichotoma G. Forst.

Cordia furcans I. M. Johnst.

Cordia grandis Roxb.

Cordia myxa L.

Cordia vestita Hook. f. & Thomson

Cynoglossum amabile Stapf & J. R. Drumm.

Cynoglossum amabile var. **pauciglochidiatum** Y. L. Liu

Cynoglossum divaricatum Steph. ex Lehm.

Cynoglossum lanceolatum Forssk

Cynoglossum microglochin Benth

Cynoglossum microglochin var. **nervosum** (Benth. ex C. B. Clarke) Y. J. Nasir

Cynoglossum schlagintweitii (Brand) Kazmi

Cynoglossum triste Diels

Cynoglossum wallichii G. Don

Cynoglossum wallichii var. **glochidiatum** (Wall. ex Benth.) Kazmi

Cynoglossum zeylanicum (Vahl) Brand

Decalepidanthus echioides (Royle ex Benth.) Dickoré & Hilger

Decalepidanthus elongatus (Decne.) Dickoré & Hilger

Decalepidanthus moltkioides (Royle) Dickoré & Hilger

Decalepidanthus parviflorus (Decne.) Dickoré & Hilger

Decalepidanthus primuloides (Decne.) Dickoré & Hilger

Decalepidanthus racemosus (Royle ex Benth.) Dickoré & Hilger

Decalepidanthus trollii (Melch.) Dickoré & Hilger

Echium italicum L.

Ehretia acuminata Br.

Ehretia confinis I. M. Johnst.

Ehretia corylifolia C. H. Wright

Ehretia dunniana H. Lév.

Ehretia Laevis Roxb.

Ehretia macrophylla Wall.

Ehretia macrophylla var. **glabrescens** (Nakai) Y. L. Liu

Ehretia obtusifolia Hochst. ex DC.

Ehretia psilosiphon R. R. Mill

Ehretia wallichiana Hook.

Eritrichium afghanicum Rech. f.

Eritrichium axilliflorum Li Bing Zhang & Yi F. Duan

Eritrichium canum (Benth.) Kitam.

Eritrichium canum var. **fruticulosum** (Klotzsch) Y. Nasir

Eritrichium canum var. **patens** (Decne.) Y. Nasir

Eritrichium canum var. **spathulatum** (Benth.) Y. Nasir

Eritrichium deqinense W. T. Wang

Eritrichium echinocaryum (I. M. Johnst.) Y. S. Lian & J. Q. Wang

Eritrichium hemisphaericum W. T. Wang

Eritrichium humillimum W. T. Wang

Eritrichium kangdingense W. T. Wang

Eritrichium laxum I. M. Johnst.

Eritrichium lianyongshanii Li Bing Zhang & Yi F. Duan

Eritrichium longifolium Decne.

Eritrichium longipes Y. S. Lian & J. Q. Wang

Eritrichium medicarpum Y. S. Lian & J. Q. Wang

Eritrichium minimum (Brand) Hara

Eritrichium pamiricum Fedtsch.

Eritrichium pectinatociliatum Y. S. Lian & J. Q. Wang

Eritrichium petiolare W. T. Wang

Eritrichium petiolare var. **subturbinatum** W. T. Wang

Eritrichium qofengense Y. S. Lian & J. Q. Wang

Eritrichium serxuense W. T. Wang

Eritrichium sinomicrocarpum W. T. Wang

Eritrichium tangkulaense W. T. Wang

Eritrichium thomsoni (C. B. Clarke) I. M. Johnst.

Eritrichium thymifolium (DC.) Y. S. Lian & J. Q. Wang

Eritrichium thymifolium subsp. **latialatum** Y. S. Lian & J. Q. Wang

Eritrichium villosum (Ledeb.) Bunge

Eritrichium wangwencaii Li Bing Zhang & Yi F. Duan

Gastrocotyle hispida (Forssk.) Bunge

Hackelia bhutanica R. R. Mill

Hackelia brachytuba (Diels) I. M. Johnst.

Hackelia difformis (Y. S. Lian & J. Q. Wang) Riedl

Hackelia macrophylla (Brand) I. M. Johnst.

Hackelia obtusifolia R. R. Mill

Hackelia uncinatum (Benth.) C. E. C. Fisch.

Heliotropium biannulatum Bunge

Heliotropium bucharicum B. Fedtsch.

Heliotropium cabulicum Bunge

Heliotropium crispum (Lehm.) DC.

Heliotropium dasycarpum Ledeb.

Heliotropium dasycarpum var. **gymnostomum** (Hemsl.) Kazmi

Heliotropium dicricophorum Rech. f. & Riedl

Heliotropium ellipticum Ledeb.

Heliotropium europaeum L.

Heliotropium europaeum var. **lasiocarpum** (Fisch. & C. A. Mey.) Kazmi

Heliotropium griffithii Boiss.

Heliotropium indicum L.

Heliotropium ovalifolium Forssk.

Heliotropium rariflorum Stocks

Heliotropium strigosum Willd.

Heliotropium strigosum subsp. **brevifolium** (Wall.) Kazmi

Heterocaryum szovitsianum (Fisch. & C. A. Mey.) A. DC.

Heterocaryum szovitsianum var. **rigidum** (DC.) Y. J. Nasir

Lappula barbata (M. Bieb.) Gürke

Lappula caespitosa Ching J. Wang

Lappula consanguinea (Fisch. & C. A. Mey.) Gürke

Lappula heteracantha Borbás

Lappula himalayensis C. J. Wang

Lappula microcarpa (Ledeb.) Gürke

Lappula myosotis V. Wolf

Lappula patula (Lehm.) Asch. exGürke

Lappula redowskii (Hornem.) Greene

Lappula semiglabra (Ledeb.) Gurke

Lappula sericata Popov

Lappula shanhsiensis Kitag.

Lappula sinaica (DC.) Asch. & Schweinf.

Lappula stricta (Ledeb.) Gurke

Lappula tadshikorum Popov

Lasiocaryum densiflorum (Duthie) I. M. Johnst.

Lasiocaryum diffusum (A. Brand) I. M. Johnst.

Lasiocaryum munroi (Clarke) I. M. Johnst.

Lasiocaryum trichocarpum (Hand.-Mazz.) I. M. Johnst.

Lepechiniella inconspicua (Brand) Riedl

Lepechiniella microcarpa (Boiss.) Riedl

Lindelofia longiflora (Benth.) Baill.

Lindelofia longiflora var. **falconeri** (C. B. Clarke) Brand

Lindelofia longiflora var. **levingii** (C. B. Clarke) Brand

Lindelofia macrostyla (Bunge) Popov

Lindelofia stylosa (Kar. & Kir.) Brand

Lithospermum erythrorhizon Siebold & Zucc.

Lithospermum officinale L.

Lithospermum zollingeri A. DC.

Mattiastrum asperum (Stocks) Brand

Mattiastrum bungei (Boiss.) Rech. f. & Riedl

Mattiastrum himalayense (Klotzsch) Brand

Mattiastrum himalayense var. **fallax** (Rech. f. & Riedl) Kazmi

Mattiastrum howardii Kazmi

Mattiastrum karakoricum Podlech & Sadat

Mattiastrum thomsonii (C. B. Clarke) Kazmi

Mattiastrum tibeticum (C. B. Clarke) Brand

Metaeritrichium microuloides W. T. Wang

Microcaryum pygmacum (Clarke) I. M. Johnst.

Microula bhutanica (Yamazaki) Hara

Microula blepharolepis (Maxim.) I. M. Johnst.

Microula bothriospermoides W. T. Wang

Microula ciliaris (Bur. & Franch.) I. M. Johnst.

Microula diffusa (Maxim.) I. M. Johnst.

Microula efoveolata W. T. Wang

Microula filicaulis W. T. Wang

Microula floribunda W. T. Wang

Microula forrestii (Diels) I. M. Johnst.

Microula hispidissima W. T. Wang

Microula jilungensis W. T. Wang

Microula leiocarpa W. T. Wang

Microula leucantha W. T. Wang

Microula longituba W. T. Wang

Microula myosotidea (Franch.) I. M. Johnst.

Microula oblongifolia Hand.-Mazz.

Microula oblongifolia var. **glabrescens** W. T. Wang

Microula ovalifolia (Bureau & Franch.) I. M. Johnst.

Microula ovalifolia var. **pubiflora** W. T. Wang

Microula pentagonaula W. T. Yu, S. T. Chen & Z. K. Zhou

Microula polygonoides W. T. Wang

Microula pseudotrichocarpa var. **grandiflora** W. T. Wang

Microula pseudotrichocarpa W. T. Wang

Microula pustulosa (C. B. Clarke) Duthie

Microula rockii I. M. Johnst.

Microula roseiflora W. T. Yu

Microula sikkimensis (C. B. Clarke) Hemsl.

Microula sikkimensis var. **involucriformis** (W. T. Wang) W. T. Yu & Z. K. Zhou

Microula spathulata W. T. Wang

Microula stenophylla W. T. Wang

Microula tangutica Maxim.

Microula tibetica Benth.

Microula tibetica var. **laevis** W. T. Wang

Microula tibetica var. **pratensis** (Maxim.) W. T. Wang

Microula trichocarpa (Maxim.) I. M. Johnst.

Microula trichocarpa var. **lasiantha** W. T. Wang

Microula trichocarpa var. **macrantha** W. T. Wang

Microula turbinata W. T. Wang

Myosotis alpestris F. W. Schmidt

Myosotis alpestris var. **albicans** (Riedl) Y. J. Nasir

Myosotis arvensis (L.) Hill

Myosotis asiatica Schischk. & Serg.

Myosotis laxa subsp. **caespitosa** (Schultz) Hyl. ex Nordh.

Myosotis refracta Boiss.

Myosotis scorpioides L.

Myosotis stricta Link

Myosotis sylvatica Hoffm.

Nonea caspica (Willd.) G. Don

Nonea edgeworthii A. DC

Nonea pulla (L.) DC.

Omphalodes heterophylla Rech. f. & H. Riedl.

Onosma adenopus I. M. Johnst.

Onosma album W. W. Sm. & Jeffrey

Onosma bheriense Hara

Onosma bhutanica (I. M. Johnst.) Grierson & D. G. Long

Onosma bicolor Wall. ex G. Don

Onosma borii C. E. C. Fisch.

Onosma bracteata Wall.

Onosma burmanica Collett & Hemsl.

Onosma chitralicum I. M. Johnst.

Onosma confertum W. W. Sm.

Onosma dichroantha Boiss.

Onosma egregium I. M. Johnst.

Onosma emodi Wall.

Onosma exsertum Hemsl.

Onosma fistulosum I. M. Johnst.

Onosma glomeratum Y. L. Liu

Onosma griersonii R. R. Mill

Onosma griffithii Vatke

Onosma hispida Wall. ex G. Don

Onosma hookeri C. B. Clarke

Onosma hookeri var. **hirsutum** Y. L. Liu

Onosma hookeri var. **longiflorum** (Duthie) A. V. Duthie ex Stapf

Onosma hypoleucum I. M. Johnst.

Onosma khyberiana I. M. Johnst.

Onosma lhokaense Y. He & Q. R. Liu, sp. nov.

Onosma limitaneum I. M. Johnst.

Onosma maaikangense W. T. Wang

Onosma multiramosum Hand.-Mazz.

Onosma nangqenense Y. L. Liu

Onosma paniculatum Bureau & Franch.

Onosma pyramidale Hook. f.

Onosma sinicum Diels

Onosma sinicum var. **farreri** (I. M. Johnst.) W. T. Wang & Y. L. Liu

Onosma thomsonii Clarke

Onosma verruculosum I. M. Johnst.

Onosma waddellii Duthie

Onosma wallichiana Benth. ex Hook. f.

Onosma waltonii Duthie

Onosma wardii (W. W. Sm.) I. M. Johnst.

Onosma xiangchengense W. T. Wang

Onosma yajiangense W. T. Wang

Onosma yamagatae Kitam.

Onosma zayueense Y. L. Liu

Paracaryum intermedium (Fresen.) Lipsky

Paracaryum intermedium var. **papillosum** (DC.) Kazmi

Paracaryum rugulosum (DC.) Boiss.

Phacelia tanacetifolia Benth.

Rochelia bungei (Boiss.) Gurke

Rochelia cardiosepala Bunge

Rochelia laxa I. M. Johnst.

Rochelia leiocarpa Ledeb.

Rochelia macrocalyx Bunge

Rochelia peduncularis Boiss.

Rochelia stylaris Boiss.

Sinojohnstonia moupinensis (Franch.) W. T. Wang

Solenanthus stamineus (Desf.) Wettst.

Stenosolenium saxatile Turcz.

Thyrocarpus sampsonii Hance

Tournefortia hookeri Clarke.

Tournefortia montana Lour.

Trichodesma calycosum Collett & Hemsl.

Trichodesma indicum (L.) R. Br.

Trigonotis amblyosepala Nakai & Kitag.

Trigonotis barkamensis C. J. Wang

Trigonotis bracteata C. J. Wang

Trigonotis caespitosa Banerjee

Trigonotis cavaleriei (H. Lév.) Hand.-Mazz.

Trigonotis cavaleriei var. **angustifolia** Ching J. Wang

Trigonotis cinereifolia Ching J. Wang

Trigonotis corispermoides Ching J. Wang

Trigonotis corispermoides var. **sessilis** W. T. Wang

Trigonotis delicatula Hand.-Mazz.

Trigonotis floribunda I. M. Johnst.

Trigonotis gracilipes I. M. Johnst.

Trigonotis harrysmithii R. R. Mill

Trigonotis heliotropifolia Hand.-Mazz.

Trigonotis hookeri Benth.

Trigonotis leucantha W. T. Wang

Trigonotis logipes W. T. Wang

Trigonotis logiromosa W. T. Wang

Trigonotis microcarpa (A. DC.) Benth. ex C. B. Clarke

Trigonotis muliensis var. **strigosa** W. T. Wang

Trigonotis muliensis W. T. Wang

Trigonotis multicaulis (DC.) Benth. ex C. B. Clarke

Trigonotis ovalifolia (Wall.) Benth. ex C. B. Clarke

Trigonotis peduncularis (Trev.) Benth. ex Baker & Moore

Trigonotis rockii I. M. Johnst.

Trigonotis rotundata I. M. Johnst.

Trigonotis rotundifolia Benth. ex C. B. Clarke

Trigonotis smithii S. P. Banerjee

Trigonotis tibetica (Clarke) I. M. Johnst.

Trigonotis vestita (Hemsl.) I. M. Johnst.

Trigonotis zhuokejiensis W. T. Wang

248. CONVOLVULACEAE

Aniseia martinicensis (Jacq.) Choisy

Argyreia argentea (Roxb.) Sweet

Argyreia atropurpurea (Wall.) Raizada

Argyreia bella (C. B. Clarke) Raizada

Argyreia capitata (Vahl) Choisy

Argyreia capitiformis (Poir.) Ooststr.

Argyreia cymosa (Roxb.) Sweet

Argyreia fulvovillosa C. Y. Wu & S. H. Huang

Argyreia hirsutissima (C. B. Clarke) Raizada

Argyreia hookeri C. B. Clarke

Argyreia nasirii D. Austin

Argyreia nervosa (Burm. f.) Bojer

Argyreia roxburghii Choisy

Argyreia setosa (Roxb.) Choisy

Argyreia sikkimensis (C. B. Clarke) Ooststr.

Argyreia splendens (Hornem.) Sweet

Argyreia venusta Choisy
Argyreia wallichii Choisy
Calystegia hederacea Wall.
Convolvulus ammannii Desr.
Convolvulus arvensis L.
Convolvulus pluricaulis Choisy
Convolvulus pseudocantabricus Schrenk
Cuscuta approximata Bab.
Cuscuta australis R. Br.
Cuscuta campestris Yunck.
Cuscuta capitata Roxb.
Cuscuta chinensis Lam.
Cuscuta europaea L.
Cuscuta europaea var. **indica** Englem.
Cuscuta europaea var. **nepalensis** Yunck.
Cuscuta gigantea Griff.
Cuscuta lehmanniana Bunge
Cuscuta lupuliformis Krock.
Cuscuta macrolepis R. C. Fang & S. H. Huang
Cuscuta monogyna Vahl
Cuscuta reflexa Roxb.
Cuscuta reflexa var. **brachystigma** Englem.
Cuscuta santapaui Banerji & Sitesh Das
Cuscuta violacea Rajput & Syeda
Dichondra micrantha Urban
Dinetus duclouxii (Gagnep. & Courchet) Staples
Dinetus grandiflorus (Wall.) Staples
Dinetus racemosus (Wall.) Sw.
Erycibe laurifolia D. G. Long
Erycibe paniculata Roxb.
Erycibe subspicata Wall. ex G. Don
Evolvulus alsinoides (L.) L.
Evolvulus nummularius (L.) L.
Ipomoea alba L.
Ipomoea aquatica Forssk.
Ipomoea batatas (L.) Lam.
Ipomoea carnea subsp. **fistulosa** (Mart. ex Choisy) D. F. Austin
Ipomoea eriocarpa R. Br.
Ipomoea hirtifolia R. C. Fang & S. H. Huang
Ipomoea indica (Burm.) Merr.
Ipomoea kingii Diels
Ipomoea mauritiana Jacq.
Ipomoea nil (L.) Roth
Ipomoea pes-tigridis L.
Ipomoea plebeia subsp. **indica** Verdc.
Ipomoea purpurea (L.) Roth
Ipomoea quamoclit L.
Ipomoea turbinata Lag.
Merremia cordata C. Y. Wu & R. C. Fang
Merremia emarginata (Burm. f.) Hallier f.
Merremia gemella (Burm. f.) Hallier f.
Merremia hederacea (Burm. f.) Hallier f.
Merremia hirta (L.) Merr.
Merremia hungaiensis (Lingelsh. & Borza) R. C. Fang
Merremia hungaiensis var. **linifolia** (C. C. Huang) R. C. Fang
Merremia kingii Kerr
Merremia mammosa (Lour.) Hallier f.
Merremia poranoides Hallier f.
Merremia sibirica var. **vesiculosa** C. Y. Wu
Merremia umbellata subsp. **orientalis** (Hallier f.) Ooststr.
Merremia vitifolia (Burm. f.) Hallier f.
Merremia yunnanensis (Courchet & Gagnep.) R. C. Fang
Operculina turpethum (L.) Silva Manso
Poranopsis grandiflora Wall.
Poranopsis paniculata (Roxb.) Roberty
Poranopsis racemosa Roxb.
Poranopsis stenoloba Kurz
Rivea hypocrateriformis (Desr.) Choisy
Rivea ornata Choisy

249. SOLANACEAE

Anisodus acutangulus C. Y. Wu & C. Chen

Anisodus luridus Link ex Spreng.

Anisodus tanguticus (Maxim.) Pascher

Atropa acuminata Royle ex Miers

Brunfelsia pauciflora (Cham. & Schltdl.) Benth.

Brunfelsia uniflora (Pohl) D. Don

Capsicum annuum L.

Capsicum microcarpum DC.

Capsicum frutescens L.

Cestrum aurantiacum Lindl.

Cestrum elegans (Brongn. ex Neumann) Schltdl.

Cestrum nocturnum L.

Cestrum parqui L'Her.

Datura innoxia Mill.

Datura metel L.

Datura stramonium L.

Datura suaveolens Humb. & Bonpl. ex Willd.

Hyoscyamus insanus Stocks

Hyoscyamus niger L.

Hyoscyamus pusillus L.

Lycianthes biflora (Lour.) Bitter

Lycianthes biflora subsp. **lysimachioides** (Lour.) Bitter

Lycianthes biflora subsp. **macrodon** (Lour.) Bitter

Lycianthes crassipetala (Wall.) R. R. Mill

Lycianthes laevis subsp. **crassipetala** (Dunal) Bitter

Lycianthes laevis subsp. **subtruncata** (Wall. ex Dunal) Deb

Lycianthes lysimachioides (Wall.) Bitter

Lycianthes macrodon (Wall. ex Nees) Bitter

Lycianthes solitaria C. Y. Wu & A. M. Lu

Lycianthes yunnanensis (Bitter) C. Y. Wu & S. C. Huang

Lycium armatum Griff.

Lycium barbarum L.

Lycium chinense Mill.

Lycium ruthenicum Murray

Lycopersicon esculentum Mill.

Mandragora caulescens C. B. Clarke

Mandragora caulescens subsp. **flavida** Grierson & D. G. Long

Mandragora caulescens subsp. **purpurascens** C. B. Clarke

Nicandra physalodes (L.) Gaertn.

Nicotiana plumbaginifolia Viv.

Nicotiana rustica L.

Nicotiana tabacum L.

Physaliastrum yunnanense Kuang & A. M. Lu

Physaliastrum yunnanense subsp. **bhutanicum** (Grierson & D. G. Long) D'Arcy & Zhi Y. Zhang

Physalis alkekengi L.

Physalis angulata L.

Physalis divaricata D. Don

Physalis peruviana L.

Physalis pubescens L.

Physochlaina macrocalyx Pascher

Physochlaina macrophylla Bonati

Physochlaina praealta (Decne.) Miers

Przewalskia tangutica Maxim.

Solanum aculeatissimum Jacq.

Solanum alatum Moench

Solanum americanum Mill.

Solanum anguivi Lam.

Solanum barbisetum Nees

Solanum dulcamara L.

Solanum elaeagnifolium Cav.

Solanum erianthum D. Don

Solanum ferox L.

Solanum glaucophyllum Desf.

Solanum griffithii (Prain) C. Y. Wu & S. C. Huang

Solanum jasminoides Paxton

Solanum kurzii Prain

Solanum lyratum Thunb.

Solanum mauritianum Scop.

Solanum melongena L.

Solanum myriacanthum Dunal

Solanum nigrum L.

Solanum pittosporifolium Hemsl.

Solanum pseudocapsicum L.

Solanum scabrum Mill.

Solanum seaforthianum Andrews

Solanum septemlobum Bunge

Solanum sisymbrifolium Lam.

Solanum spirale Roxb.

Solanum surattense Burm. f.

Solanum torvum Sw.

Solanum tuberosum L.

Solanum viarum Dunal

Solanum villosum Mill.

Solanum virginianum L.

Solanum wendlandii Hook. f.

Solanum wightii Nees

Withania coagulans (Stocks) Dunal

Withania somnifera (L.) Dunal

250. SPHENOCLEACEAE

Sphenoclea zeylanica Gaertn.

251. HYOROLEACEAE

Hydrolea zeylanica (L.) Vahl

252. CARLEMANNIACEAE

Carlemannia congesta Hook. f.

Carlemannia griffithii Benth.

Carlemannia tetragona Hook. f.

Silvianthus bracteatus Hook. f.

253. OLEACEAE

Chionanthus ramiflorus Roxb.

Chionanthus retusus Lindl. & Paxton

Chionanthus terniflorus Griff.

Forsythia likiangensis Ching & Feng ex P. Y. Bai

Forsythia viridissima Lindl.

Fraxinus ferruginea Lingelsh.

Fraxinus floribunda Wall.

Fraxinus hookeri Wenz.

Fraxinus micrantha Lingelsh.

Fraxinus raibocarpa Regel

Fraxinus sikkimensis (Lingelsh.) Hand.-Mazz.

Fraxinus trifoliolata W. W. Sm.

Fraxinus xanthoxyloides (G. Don) DC.

Jasminum adenophyllum Wall. ex C. B. Clarke

Jasminum amabile Hara

Jasminum amplexicaule Buch.-Ham. ex G. Don

Jasminum arborescens var. **latifolium** Roxb.

Jasminum attenuatum Roxb. ex DC.

Jasminum auriculatum Vahl

Jasminum azoricum L.

Jasminum beesianum Forrest & Diels

Jasminum caudatum Wall. ex Lindl.

Jasminum coarctatum Roxb.

Jasminum dispermum Wall.
Jasminum duclouxii (H. Lév.) Rehder
Jasminum elongatum (Bergius) Willd.
Jasminum grandiflorum L.
Jasminum humile L.
Jasminum humile var. **microphyllum** (L. C. Chia) P. S. Green
Jasminum lanceolarium Roxb.
Jasminum laurifolium Roxb.
Jasminum laurifolium var. **brachylobum** Kurz
Jasminum listeri King ex Gage
Jasminum mesnyi Hance
Jasminum multiflorum (Burm. f.) Andrews
Jasminum nepalense Spreng.
Jasminum nervosum Lour.
Jasminum nudiflorum Lindl.
Jasminum nudiflorum var. **pulvinatum** (W. W. Sm.) Kobuski
Jasminum officinale L.
Jasminum officinale var. **piliferum** P. Y. Bai
Jasminum officinale var. **tibeticum** C. Y. Wu ex P. Y. Bai
Jasminum polyanthum Franch.
Jasminum ritchiei C. B. Clarke
Jasminum sambac (L.) Aiton
Jasminum scandens (Retz.) Vahl
Jasminum seguinii H. Lév.
Jasminum sinense Hemsl.
Jasminum stephanense Lemoine
Jasminum subglandulosum Kurz
Jasminum subhumile W. W. Sm.
Jasminum urophyllum Hemsl.
Ligustrum compactum (Wall. ex G. Don) Hook. f. & Thomson ex Brandis
Ligustrum compactum var. **velutinum** P. S. Green
Ligustrum confusum Decne.
Ligustrum confusum var. **macrocarpum** C. B. Clarke
Ligustrum delavayanum Har.
Ligustrum gamblei Ramamoorthy
Ligustrum gracile Rehder
Ligustrum indicum (Lour.) Merr.
Ligustrum lucidum W. T. Aiton
Ligustrum myrsinites Decne.
Ligustrum nepalense Wall.
Ligustrum quihoui Carrière
Ligustrum robustum subsp. **chinense** P. S. Green
Ligustrum sempervirens (Franch.) Lingelsh.
Ligustrum sinense Lour.
Ligustrum sinense var. **rugosulum** (W. W. Sm.) M. C. Chang
Myxopyrum smilacifolium Blume
Nyctanthes arbor-tristis L.
Olea dioica Roxb.
Olea europaea L.
Olea ferruginea Royle
Olea gamblei C. B. Clarke
Olea laxiflora H. L. Li
Olea paniculata R. Br.
Olea salicifolia Wall. ex G. Don
Olea tsoongii (Merr.) P. S. Green
Osmanthus delavayi Franch.
Osmanthus fragrans Lour.
Osmanthus fragrans var. **longifolius** (DC.) Hara
Osmanthus henryi P. S. Green
Osmanthus suavis King ex C. B. Clarke
Osmanthus yunnanensis (Franch.) P. S. Green
Syringa emodi Wall. ex Royle
Syringa komarowii C. K. Schneid.
Syringa mairei (H. Lév.) Rehder
Syringa pinetorum W. W. Sm.
Syringa pinnatifolia Hemsl.
Syringa sweginzowii Koehne & Lingelsh.
Syringa tibetica P. Y. Pai
Syringa tomentella Bureau & Franch.
Syringa wardii W. W. Sm.
Syringa yunnanensis Franch.

254. GESNERIACEAE

Aeschynanthus acuminatus Wall. ex A. P. DC.

Aeschynanthus angustioblongus W. T. Wang

Aeschynanthus angustissimus (W. T. Wang) W. T. Wang

Aeschynanthus bracteatus Wall. ex A. DC.

Aeschynanthus chiritoides C. B. Clarke

Aeschynanthus deleiensis C. E. C. Fisch.

Aeschynanthus dolichanthus W. T. Wang

Aeschynanthus gracilis Parish ex C. B. Clarke

Aeschynanthus hookeri C. B. Clarke

Aeschynanthus kingii C. B. Clarke

Aeschynanthus lasianthus W. T. Wang

Aeschynanthus lasiocalyx W. T. Wang

Aeschynanthus linearifolius C. E. C. Fisch.

Aeschynanthus lineatus Craib

Aeschynanthus maculatus Lindl.

Aeschynanthus medogensis W. T. Wang

Aeschynanthus micranthus C. B. Clarke

Aeschynanthus mimetes B. L. Burtt

Aeschynanthus monetarius Dunn

Aeschynanthus parasiticus (Roxb.) Wall.

Aeschynanthus parviflorus (D. Don) Spreng.

Aeschynanthus philippinensis C. B. Clarke

Aeschynanthus sikkimensis (C. B. Clarke) Stapf

Aeschynanthus stenosepalus J. Anthony

Aeschynanthus superbus C. B. Clarke

Aeschynanthus tengchungensis W. T. Wang

Aeschynanthus tubulosus J. Anthony

Aeschynanthus wardii Merr.

Ancylostemon aureus (Franch.) B. L. Burtt

Ancylostemon convexus Craib

Ancylostemon gamosepalus K. Y. Pan

Ancylostemon rhombifolius K. Y. Pan

Anna ophiorrhizoides (Hemsl.) B. L. Burtt & R. Davidson

Beccarinda cordifolia (Anthony) B. L. Burtt

Boea clarkeana Hemsl.

Boea herbacea C. B. Clarke

Boea multiflora R. Br.

Boeica clarkei Hareesh, L. Wu, A. Joe & M. Sabu

Boeica filiformis C. B. Clarke

Boeica fulva C. B. Clarke

Briggsia agnesiae (Forrest) Craib

Briggsia aurantiaca B. L. Burtt

Briggsia kurzii (C. B. Clarke) W. E. Evans

Briggsia longicaulis W. T. Wang & K. Y. Pan

Briggsia longifolia Craib

Briggsia muscicola (Diels) Craib

Chirita anachoreta Hance

Chirita bifolia D. Don

Chirita dielsii (Borza) B. L. Burtt

Chirita forrestii J. Anthony

Chirita infundibuliformis W. T. Wang

Chirita lachenensis C. B. Clarke

Chirita macrophylla Wall.

Chirita skogiana Z. Yu Li

Corallodiscus bhutanicus (Craib) B. L. Burtt

Corallodiscus conchifolius Batalin

Corallodiscus cooperi (Craib) B. L. Burtt

Corallodiscus kingianus (Craib) B. L. Burtt

Corallodiscus lanuginosus (Wall. ex R. Br.) B. L. Burtt

Didymocarpus adenocalyx W. T. Wang

Didymocarpus albicalyx C. B. Clarke

Didymocarpus andersonii C. B. Clarke

Didymocarpus aromaticus Wall. ex D. Don

Didymocarpus aurantiacus C. B. Clarke

Didymocarpus bhutanicus W. T. Wang

Didymocarpus cinereus D. Don

Didymocarpus curvicapsa Hilliard

Didymocarpus denticulatus W. T. Wang

Didymocarpus macrophyllus Wall. ex D. Don

Didymocarpus medogensis W. T. Wang

Didymocarpus mortonii C. B. Clarke

Didymocarpus oblongus Wall. ex D. Don

Didymocarpus pedicellatus R. Br.

Didymocarpus podocarpus C. B. Clarke

Didymocarpus praeteritus B. L. Burtt & R. Davidson

Didymocarpus primulifolius D. Don

Didymocarpus pulcher C. B. Clarke

Didymocarpus stenanthos C. B. Clarke

Didymocarpus subpalmatinervis W. T. Wang

Didymocarpus triplotrichus Hilliard

Didymocarpus villosus D. Don

Didymocarpus yunnanensis (Franch.) W. W. Sm.

Didymocarpus zhufengensis W. T. Wang

Epithema carnosum Benth.

Hemiboea omeiensis W. T. Wang

Hemiboea subcapitata C. B. Clarke

Henckelia adenocalyx (Chatterjee) D. J. Middleton & Mich. Möller

Henckelia burttii D. J. Middleton & Mich. Möller

Henckelia calva (C. B. Clarke) D. J. Middleton & Mich. Möller

Henckelia dibangensis (B. L. Burtt, S. K. Srivast. & Mehrotra) D. J. Middleton & Mich. Möller

Henckelia dimidiata (Wall. ex C. B. Clarke) D. J. Middleton & Mich. Möller

Henckelia grandifolia A. Dietr.

Henckelia hookeri (C. B. Clarke) D. J. Middleton & Mich. Möller

Henckelia mishmiensis (Debb. ex Biswas) D. J. Middleton & Mich. Möller

Henckelia oblongifolia (Roxb.) D. J. Middleton & Mich. Möller

Henckelia primulacea (C. B. Clarke) D. J. Middleton & Mich. Möller

Henckelia pumila (D. Don) A. Dietr.

Henckelia punctata (C. B. Clarke) B. L. Burtt

Henckelia speciosa (Kurz) D. J. Middleton & Mich. Möller

Henckelia tibetica (Franch.) D. J. Middleton & Mich. Möller

Henckelia urticifolia (Buch.-Ham. ex D. Don) A. Dietr

Leptobaea multiflora (C. B. Clarke) Benth. ex Gamble

Loxostigma griffithii (Wight) C. B. Clarke

Loxostigma kurzii (C. B. Clarke) Burtt

Loxostigma mekongense (Franch.) B. L. Burtt

Lysionotus atropurpureus Hara

Lysionotus forrestii W. W. Sm.

Lysionotus gamosepalus W. T. Wang

Lysionotus gracilis W. W. Sm.

Lysionotus heterophyllus Franch.

Lysionotus levipes (C. B. Clarke) B. L. Burtt

Lysionotus metuoensis W. T. Wang

Lysionotus palinensis G. D. Pal

Lysionotus pubescens C. B. Clarke

Lysionotus serratus D. Don

Lysionotus sessilifolius Hand.-Mazz.

Lysionotus sulphureus Hand.-Mazz.

Lysionotus wilsonii Rehder

Oreocharis aurantiaca Franch.

Oreocharis cinnamomea J. Anthony

Oreocharis cordatula (Craib) Pellegr.

Oreocharis delavayi Franch.

Oreocharis dentata A. L. Weitzman & L. E. Skog

Oreocharis eximia (Chun ex K. Y. Pan) M. Möller & A. Weber

Oreocharis farreri (W. G. Craib) M. Möller & A. Weber

Oreocharis forrestii (Diels) Skan

Oreocharis georgei J. Anthony

Oreocharis glandulosa (Batalin) M. Möller & A. Weber

Oreocharis henryana Oliv.

Oreocharis lancifolia (Franch.) M. Möller & A. Weber

Oreocharis lancifolia var. **mucronata** (K. Y. Pan) M. Möller & A. Weber

Oreocharis leucantha (Diels) M. Möller & A. Weber

Oreocharis minor (Craib) Pellegr.

Oreocharis obliquifolia (K. Y. Pan) M. Möller & A. Weber

Oreocharis parviflora Lei Cai & Z. K.Wu

Oreocharis primuliflora (Batalin) M. Möller & A. Weber

Oreocharis sichuanica (K. Y. Pan) M. Möller & A. Weber

Oreocharis tubicella Franch.

Oreocharis urceolata (K. Y. Pan) M. Möller & A. Weber

Paraboea multiflora (R. Br.) B. L. Burtt

Paraboea sinensis (Oliv.) B. L. Burtt

Petrocosmea flaccida Craib

Petrocosmea forrestii Craib

Petrocosmea kerrii Craib

Petrocosmea nervosa Craib

Petrocosmea oblata Craib

Petrocosmea sichuanensis W. Y. Chun ex W. T. Wang

Platystemma violoides Wall.

Raphiocarpus sesquifolius (C. B. Clarke) B. L. Burtt

Rhabdothamnopsis sinensis Hemsl.

Rhynchoglossum lazulinum A. S. Rao & J. Joseph

Rhynchoglossum obliquum Blume

Rhynchoglossum omeiense W. T. Wang

Rhynchotechum ellipticum (Wall. ex D. Dietr.) A. DC.

Rhynchotechum vestitum Wall. ex C. B. Clarke

Stauranthera grandifolia Benth.

Stauranthera umbrosa (Griff.) C. B. Clarke

Tetraphyllum bengalense C. B. Clarke

Tremacron forrestii Craib

Trisepalum birmanicum (Craib) B. L. Burtt

Whytockia gongshanensis Yin Z. Wang & H. Li

Whytockia tsiangiana (Hand.-Mazz.) A. Weber

Whytockia wilsonii (A. Weber) Yin Z. Wang

255. PLANTAGINACEAE

Adenosma glutinosum (L.) Druce

Adenosma indianum (Lour.) Merr.

Angelonia salicariifolia Humb. & Bonpl.

Antirrhinum majus L.

Bacopa hamiltoniana (Benth.) Wettst.

Bacopa monnieri (L.) Pennell

Callitriche fehmedianii Majeed Kak & Javeid

Callitriche fuscicarpa Lansdown

Callitriche glareosa Lansdown

Callitriche hermaphroditica L.

Callitriche hermaphroditica subsp. **macrocarpa** (Hegelm.) Lansdown

Callitriche palustris L.

Callitriche stagnalis Scop.

Dopatrium junceum (Roxb.) Buch.-Ham. ex Bentham

Ellisiophyllum pinnatum (Wall. ex Benth.) Makino

Ellisiophyllum pinnatum subsp. **bhutanense** R. R. Mill

Hemiphragma heterophyllum Wall.

Hemiphragma heterophyllum var. **pedicellatum** Hand.-Mazz.

Hippuris vulgaris L.

Kickxia incana (Wall.) Pennell

Kickxia membranacea D. A. Sutton

Kickxia ovata (Boiss.) Kitam.

Kickxia papillosa R. R. Mill

Kickxia ramosissima (Wall.) Janch.

Kickxia subsessilis Pennell

Lagotis alutacea W. W. Sm.

Lagotis alutacea var. **foliosa** W. W. Sm.

Lagotis alutacea var. **rockii** (H. L. Li) P. C. Tsoong ex H. P. Yang

Lagotis brachystachya Maxim.

Lagotis brevituba Maxim.

Lagotis cashmeriana Rupr.

Lagotis chumbica R. R. Mill

Lagotis clarkei Hook. f.

Lagotis crassifolia Prain

Lagotis decumbens Rupr.

Lagotis humilis P. C. Tsoong & H. P. Yang

Lagotis integra W. W. Sm.

Lagotis kongboensis T. Yamaz.

Lagotis kunawurensis (Royle ex Benth.) Rupr.

Lagotis macrosiphon P. C. Tsoong & H. P. Yang

Lagotis nepalensis T. Yamaz.

Lagotis pharica Prain

Lagotis praecox W. W. Sm.

Lagotis ramalana Batalin

Lagotis wardii W. W. Sm.

Lagotis yunnanensis W. W. Sm.

Limnophila aromatica (Lam.) Merr.

Limnophila borealis Y. Z. Zhao & Ma f.

Limnophila chinensis (Osbeck) Merr.

Limnophila connata (Buch.-Ham. ex D. Don) Hand.-Mazz.

Limnophila heterophylla (Roxb.) Benth.

Limnophila indica (L.) Druce

Limnophila micrantha Benth.

Limnophila polyantha Kurz ex Hook. f.

Limnophila repens (Benth.) Benth.

Limnophila rugosa (Roth) Merr.

Limnophila sessiliflora (Vahl) Blume

Linaria bamianica Patzak

Linaria dalmatica (L.) Mill.

Linaria sessilis Kuprian.

Linaria thibetica Franch.

Mecardonia procumbens (Mill.) Small

Misopates orontium (L.) Raf.

Neopicrorhiza minima R. R. Mill

Neopicrorhiza scrophulariiflora (Pennell) D. Y. Hong

Picrorhiza kurroa Royle ex Benth.

Plantago arenaria Waldst. & Kit.

Plantago asiatica L.

Plantago asiatica subsp. **densiflora** (J. Z. Liu) Z. Yu Li

Plantago asiatica subsp. **erosa** (Wall.) Z. Yu Li

Plantago baltistanica Hartmann

Plantago cavaleriei H. Lév.

Plantago depressa Willd.

Plantago exigua Murray

Plantago gentianoides subsp. **griffithii** (Decne.) Rech. f.

Plantago himalaica Pilg.

Plantago lanceolata L.

Plantago major L.

Plantago ovata Forssk.

Scoparia dulcis L.

Scrofella chinensis Maxim.

Veronica alpina subsp. **pumila** (Allioni) Dostá

Veronica anagallis-aquatica L.

Veronica anagallis-aquatica subsp. **michauxii** (Lam.) A. Jelen.

Veronica anagalloides Guss.

Veronica arguteserrata Regel & Schmalh.

Veronica beccabunga subsp. **muscosa** (Korsh.) Jelen.

Veronica biloba Schreb.

Veronica cachemirica Gand.

Veronica campylopoda Boiss.

Veronica cana Wall. ex Benth.

Veronica capitata Royle ex Benth.

Veronica cardiocarpa (Kar. & Kir.) Walp.

Veronica chayuensis D. Y. Hong

Veronica ciliata Fisch.

Veronica ciliata subsp. **cephaloides** (Pennell) D. Y. Hong

Veronica ciliata subsp. **zhongdianensis** D. Y. Hong

Veronica deltigera Wall. ex Benth.

Veronica emodi T. Yamaz.

Veronica eriogyne H. Winkl.

Veronica filipes P. C. Tsoong

Veronica forrestii Diels

Veronica himalensis D. Don

Veronica himalensis subsp. **yunnanensis** (P. C. Tsoong) D. Y. Hong

Veronica hirta Pennell

Veronica javanica Blume

Veronica koelzii Pennell

Veronica lanosa Royle ex Benth.

Veronica lanuginosa Benth. ex Hook. f.

Veronica laxa Benth.

Veronica longipetiolata D. Y. Hong

Veronica oxycarpa Boiss.

Veronica peregrina L.

Veronica persica Poiret

Veronica piroliformis Franch.

Veronica polita Fries

Veronica pusilla Kotschy & Boiss.

Veronica riae H. Winkl.

Veronica robusta (Prain) T. Yamaz.

Veronica rockii H. L. Li

Veronica rockii subsp. **stenocarpa** (H. L. Li) D. Y. Hong

Veronica rubrifolia Boiss.

Veronica salina Schur

Veronica secunda Pennell

Veronica serpyllifolia L.

Veronica stewartii Pennell

Veronica sutchuenensis Franch.

Veronica szechuanica Batalin

Veronica szechuanica subsp. **sikkimensis** (Hook. f.) D. Y. Hong

Veronica tibetica D. Y. Hong

Veronica undulata Wall. ex Jack

Veronica vandellioides Maxim.

Veronica verna L.

Veronica yunnanensis D. Y. Hong

Veronicastrum brunonianum (Benth.) D. Y. Hong

Veronicastrum rhombifolium (Hand.-Mazz.) P. C. Tsoong ex T. L. Chin & D. Y. Hong

Veronicastrum yunnanense (W. W. Sm.) T. Yamaz.

Wulfenia amherstiana Benth.

Wulfenia amherstiana var. **nepalensis** (T. Yamaz.) T. Yamaz.

256. SCROPHULARIACEAE

Buddleja alternifolia Maxim.

Buddleja asiatica Lour.

Buddleja bhutanica T. Yamaz.

Buddleja brachystachya Diels

Buddleja candida Dunn

Buddleja colvilei Hook. f.

Buddleja cooperi W. W. Sm.

Buddleja crispa Benth.

Buddleja davidii Franch.

Buddleja delavayi Gagnep.

Buddleja fallowiana Balf. f. & W. W. Sm.

Buddleja forrestii Diels

Buddleja griffithii C. Marquand

Buddleja hookeri C. Marquand

Buddleja macrostachya Wall. ex Benth.
Buddleja myriantha Diels
Buddleja nivea Duthie
Buddleja officinalis Maxim.
Buddleja paniculata Wall.
Buddleja subcapitata E. D. Liu & H. Peng
Buddleja subherbacea Keenan
Buddleja yunnanensis Gagnep.
Limosella aquatica L.
Oreosolen wattii Hook. f.
Oreosolen williamsii T. Yamaz.
Scrophularia aequilabris P. C. Tsoong
Scrophularia bheriensis T. Yamaz.
Scrophularia calycina Benth.
Scrophularia chasmophila W. W. Sm.
Scrophularia chasmophila subsp. **xizangensis** D. Y. Hong
Scrophularia cooperi R. R. Mill
Scrophularia decomposita Royle ex Benth.
Scrophularia delavayi Franch.
Scrophularia dentata Benth.
Scrophularia diplodonta Franch.
Scrophularia edelbergii Rech. f.
Scrophularia elatior Benth.
Scrophularia himalensis Royle
Scrophularia hypsophila Hand.-Mazz.
Scrophularia koelzii Pennell
Scrophularia laportiifolia T. Yamaz.
Scrophularia lhasaensis D. Y. Hong
Scrophularia lijiangensis T. Yamaz.
Scrophularia mandarinorum Franch.
Scrophularia mapienensis P. C. Tsoong
Scrophularia nudata Pennell
Scrophularia pauciflora Benth.
Scrophularia przewalskii Batalin
Scrophularia scoparia Pennell
Scrophularia souliei Franch.
Scrophularia spicata Franch.
Scrophularia subsessilis R. R. Mill
Scrophularia suffruticosa Pennell
Scrophularia urticifolia Wall. ex Benth.
Scrophularia yunnanensis Franch.
Verbascum chinense (L.) Santapau
Verbascum thapsus L.

257. LINDERNIACEAE

Lindernia anagallis (Burm. f.) Pennell
Lindernia antipoda (L.) Alston
Lindernia ciliata (Colsmann) Pennell
Lindernia crustacea (L.) F. Mueller
Lindernia hookeri (C. B. Clarke) Wettst.
Lindernia micrantha D. Don
Lindernia mollis (Benth.) Wettst.
Lindernia multiflora (Roxb.) Mukerjee
Lindernia nummulariifolia (D. Don) Wettst.
Lindernia oppositifolia (L.) Mukerjee
Lindernia parviflora (Roxb.) Haines
Lindernia procumbens (Krock.) Philcox
Lindernia pusilla (Willd.) Bold.
Lindernia ruellioides (Colsm.) Pennell
Lindernia viscosa (Horn.) Bold.
Picria felterrae Lour.
Torenia asiatica L.
Torenia burttiana R. R. Mill
Torenia cordifolia Roxb.
Torenia diffusa D. Don

Torenia fournieri Linden ex Fourn.

Torenia parviflora Hutch. & Dalzel

Torenia thouarsii (Cham. & Schltdl.) Kuntze

Torenia violacea (Azaola ex Blanco) Pennell

258. PEDALIACEAE

Sesamum indicum L.

259. LAMIACEAE

Achyrospermum densiflorum Blume

Achyrospermum wallichianum Benth. ex Hook. f.

Acinos graveolensis Link.

Agastache rugosa (Fisch. & C. A. Mey.) Kuntze

Ajuga bracteosa Wall. ex Benth.

Ajuga campylantha Diels

Ajuga campylanthoides C. Y. Wu & C. Chen

Ajuga ciliata Bunge

Ajuga forrestii Diels

Ajuga lobata D. Don

Ajuga lupulina Maxim.

Ajuga macrosperma Wall. ex Benth.

Ajuga nipponensis Makino

Ajuga nubigena Diels

Ajuga ovalifolia Bureau & Franch

Ajuga parviflora Benth.

Ajuga sciaphila W. W. Sm.

Alajja rhomboidea (Benth.) Ikonn.

Amethystea caerulea L.

Anisochilus carnosus (L. f.) Benth.

Anisochilus mitis R. A. Clement

Anisochilus pallidus Wall. ex Benth.

Anisochilus polystachyus Benth.

Anisomeles indica (L.) Kuntze

Calamintha debilis (Bunge) Benth.

Callicarpa arborea Roxb.

Callicarpa bodinieri H. Lév.

Callicarpa cathayana H. T. Chang

Callicarpa japonica Thunb.

Callicarpa lobata C. B. Clarke

Callicarpa longifolia Lam.

Callicarpa macrophylla Vahl

Callicarpa pararubella Q. Wang

Callicarpa rubella Lindl.

Callicarpa salicifolia C. Pei & W. Z. Fang

Callicarpa vestita Wall. ex C. B. Clarke

Caryopteris forrestii Diels

Caryopteris glutinosa Rehder

Caryopteris tangutica Maxim.

Caryopteris trichosphaera W. W. Sm.

Chelonopsis albiflora Pax & K. Hoffm. ex Limpr.

Chelonopsis bracteata C. Y. Wu & H. W. Li

Chelonopsis forrestii J. Anthony

Chelonopsis lichiangensis W. W. Sm.

Chelonopsis mollissima C. Y. Wu

Chelonopsis odontochila Diels

Chelonopsis pseudobracteata C. Y. Wu & H. W. Li

Chelonopsis siccanea W. W. Sm.

Chelonopsis souliei (Bonati) Merr.

Clerodendrum bracteatum Wall. ex Walp.

Clerodendrum bungei Steud.

Clerodendrum chinense (Osbeck) Mabb.

Clerodendrum colebrookianum Walp.

Clerodendrum hastatum (Roxb.) Lindl.

Clerodendrum indicum (L.) Kuntze

Clerodendrum japonicum (Thunb.) Sweet

Clerodendrum lindleyi Decne. ex Planch.

Clerodendrum tibetanum C. Y. Wu & S. K. Wu

Clerodendrum trichotomum Thunb.

Clerodendrum viscosum Vent.

Clerodendrum wallichii Merr.

Clerodendrum yunnanense Hu ex Hand.-Mazz.

Clinopodium discolor (Diels) C. Y. Wu & S. J. Hsuan ex H. W. Li

Clinopodium megalanthum (Diels) C. Y. Wu & S. J. Hsuan ex H. W. Li

Clinopodium piperitum (D. Don) Murata

Clinopodium polycephalum (Vaniot) C. Y. Wu & S. J. Hsuan

Clinopodium repens Vell.

Clinopodium umbrosum (M. Bieb.) K. Koch

Clinopodium vulgare L.

Colebrookea oppositifolia Sm.

Coleus forskohlii (Willd.) Briq.

Colquhounia compta W. W. Sm.

Colquhounia seguinii Vaniot

Congea tomentosa Roxb.

Craniotome furcata (Link) Kuntze

Craniotome furcata var. **sikkimensis** D. Maity & Maiti

Discretitheca nepalensis (Moldenke) P. D. Cantino

Dracocephalum bipinnatum Rupr.

Dracocephalum breviflorum Turrill

Dracocephalum bullatum Forrest ex Diels

Dracocephalum calophyllum Hand.-Mazz.

Dracocephalum forrestii W. W. Sm.

Dracocephalum heterophyllum Benth.

Dracocephalum imbricatum C. Y. Wu & W. T. Wang

Dracocephalum isabellae Forrest ex W. W. Sm.

Dracocephalum microflorum C. Y. Wu & W. T. Wang

Dracocephalum moldavica L.

Dracocephalum nutans L.

Dracocephalum paulsenii Briq.

Dracocephalum propinquum W. W. Sm.

Dracocephalum purdomii W. W. Sm.

Dracocephalum stamineum Kar. & Kir.

Dracocephalum taliense Forrest & W. W. Sm.

Dracocephalum tanguticum Maxim.

Dracocephalum tanguticum var. **cinereum** Hand.-Mazz.

Dracocephalum tanguticum var. **nanum** C. Y. Wu & W. T. Wang

Dracocephalum velutinum var. **intermedium** C. Y. Wu & W. T. Wang

Dracocephalum velutinum C. Y. Wu & W. T. Wang

Dracocephalum wallichii Sealy

Dracocephalum wallichii var. **platyanthum** C. Y. Wu & W. T. Wang

Dracocephalum wallichii var. **proliferum** C. Y. Wu & W. T. Wang

Drepanocaryum sewerzowii (Regel) Pojark.

Elsholtzia blanda (Benth.) Benth.

Elsholtzia capituligera C. Y. Wu

Elsholtzia cephalantha Hand.-Mazz.

Elsholtzia ciliata (Thunb.) Hyl.

Elsholtzia concinna Vautier

Elsholtzia cypriani (Pavol.) S. Chow ex P. S. Hsu

Elsholtzia densa Benth.

Elsholtzia eriocalyx C. Y. Wu & S. C. Huang

Elsholtzia eriostachya (Benth.) Benth.

Elsholtzia feddei H. Lév.

Elsholtzia flava (Benth.) Benth.

Elsholtzia fruticosa (D. Don) Rehder

Elsholtzia glabra C. Y. Wu & S. C. Huang

Elsholtzia kachinensis Prain

Elsholtzia luteola Diels

Elsholtzia myosurus Dunn

Elsholtzia pilosa (Benth.) Benth.

Elsholtzia pygmaea W. W. Sm.

Elsholtzia rugulosa Hemsl.

Elsholtzia stachyodes (Link) Raizada & H. O. Saxena

Elsholtzia strobilifera (Benth.) Benth.

Eremostachys acaulis Beck ex Rech. f.

Eremostachys andersii Rech. f.

Eremostachys arctiifolia Popov

Eremostachys ariana Hedge

Eremostachys badakshanica Hedge

Eremostachys bamianica Rech. f.

Eremostachys calophyta Hedge

Eremostachys edelbergii Rech. f.

Eremostachys freitagii Rech. f.

Eremostachys gymnoclada Rech. f. & Köie

Eremostachys hissarica Regel

Eremostachys labiosa Bunge

Eremostachys labiosiformis Knorring

Eremostachys laciniata (L.) Bunge

Eremostachys lindbergii Rech. f.

Eremostachys loasifolia Benth.

Eremostachys rastagalensis Hedge

Eremostachys salangensis Rech. f.

Eremostachys sanglechensis Rech. f.

Eremostachys speciosa Rupr.

Eremostachys stenocalycina Rech. f.

Eremostachys superba Royle ex Benth.

Eremostachys thyrsiflora Benth.

Eremostachys vulnerans Rech. f. & Köie

Eriophyton wallichii Benth.

Eurysolen gracilis Prain

Eusteralis cruciata (Benth.) Panigrahi

Eusteralis stellata (Lour.) Panigrahi

Galeopsis bifida Boenn.

Galeopsis tetrahit L.

Geniosporum coloratum (D. Don) Kuntze

Glechoma complanata (Dunn) Turrill

Glechoma longituba (Nakai) Kuprian.

Glechoma pharica (Prain) Turrill

Glechoma sinograndis C. Y. Wu

Gmelina arborea Roxb.

Gmelina delavayana Dop

Gmelina szechwanensis K. Yao

Gomphostemma melissifolium Wall. ex Benth.

Gomphostemma ovatum Wall. ex Benth.

Gomphostemma parviflorum Wall. ex Benth.

Gomphostemma parviflorum var. **farinosum** Prain

Gomphostemma stellato-hirsutum C. Y. Wu

Gontscharovia popovii (B. Fedtsch. & Gontsch.) Boriss.

Hanceola sinensis (Hemsl.) Kudô

Holmskioldia sanguinea Retz.

Hymenocrater adenothrix Rech. f.

Hymenocrater altimuranus Rech. f.

Hymenocrater sessilifolius Benth.

Hymenopyramis brachiata Wall. ex Griff.

Hypogomphia turkestana Bunge

Hyptis pectinata (L.) Poit.

Hyptis suaveolens (L.) Poit.

Hyssopus seravschanicus Pazij

Isodon adenanthus (Diels) Kudô

Isodon adenolomus (Hand.-Mazz.) Hara

Isodon albopilosus (C. Y. Wu & H. W. Li) Hara

Isodon angustifolius (Dunn) Kudô

Isodon anisochilus (C. Y. Wu) Hara

Isodon assamicus (S. M. Mukerjee) Hara

Isodon atroruber R. A. Clement

Isodon aurantiacus Y. P. Chen & C. L. Xiang

Isodon barbeyanus (H. Lév.) H. W. Li

Isodon brachythyrsus (C. Y. Wu & H. W. Li) Hara

Isodon bulleyanus (Diels) Kudô

Isodon calcicolus (Hand.-Mazz.) Hara

Isodon chionanthus (C. Y. Wu) Hara

Isodon coetsa (Buch.-Ham. ex D. Don) Kudô

Isodon dawoensis (Hand.-Mazz.) Hara

Isodon delavayi C. L. Xiang & Y. P. Chen

Isodon dhankutanus Murata

Isodon eriocalyx (Dunn) Kudô

Isodon flabelliformis (C. Y. Wu) Hara

Isodon flexicaulis (C. Y. Wu & H. W. Li) Hara

Isodon forrestii (Diels) Kudô

Isodon gesneroides (J. Sinclair) Hara

Isodon glutinosus (C. Y. Wu & H. W. Li) Hara

Isodon grandifolius (Hand.-Mazz.) Hara

Isodon grandifolius var. **atuntzeensis** (C. Y. Wu) H. W. Li

Isodon hispidus (Benth.) Murata

Isodon irroratus (Forrest ex Diels) Kudô

Isodon japonicus (Burm. f.) Hara

Isodon kangtingensis (C. Y. Wu & H. W. Li) Hara

Isodon kurzii (Prain) Hara

Isodon leucophyllus (Dunn) Kudô

Isodon lihsienensis (C. Y. Wu & H. W. Li) Hara

Isodon lophanthoides (Buch.-Ham. ex D. Don) Hara

Isodon lophanthoides var. **graciliflorus** (Benth.) Hara

Isodon loxothyrsus (Hand.-Mazz.) Hara

Isodon medilungensis (C. Y. Wu & H. W. Li) Hara

Isodon megathyrsus (Diels) Hara

Isodon melissoides (Benth.) Hara

Isodon mucronatus (C. Y. Wu & H. W. Li) Hara

Isodon muliensis (W. W. Sm.) Kudô

Isodon namikawanus Murata

Isodon oreophilus (Diels) A. J. Paton & Ryding

Isodon oreophilus var. **elongatus** (Hand.-Mazz.) A. J. Paton & Ryding

Isodon oresbius (W. W. Sm.) Kudô

Isodon parvifolius (Batalin) Hara

Isodon pharicus (Prain) Murata

Isodon phyllopodus (Diels) Kudô

Isodon phyllostachys (Diels) Kudô

Isodon pleiophyllus (Diels) Kudô

Isodon polystachys (Y. Z. Sun) Hara

Isodon repens (Wall. ex Benth.) Murata

Isodon rubescens (Hemsl.) Hara

Isodon rugosiformis (Hand.-Mazz.) Hara

Isodon rugosus (Wall. ex Benth.) Codd

Isodon scoparius (C. Y. Wu & H. W. Li) Hara

Isodon scrophularioides (Wall. ex Benth.) Murata

Isodon sculponeatus (Vaniot) Kudô

Isodon secundiflorus (C. Y. Wu) Hara

Isodon setschwanensis (Hand.-Mazz.) Hara

Isodon silvaticus (C. Y. Wu & H. W. Li) H. W. Li

Isodon smithianus (Hand.-Mazz.) Hara

Isodon taliensis (C. Y. Wu) Hara

Isodon tenuifolius (W. W. Sm.) Kudô

Isodon ternifolius (D. Don) Kudô

Isodon wardii (C. Marquand & Airy Shaw) Hara

Isodon weisiensis (C. Y. Wu) Hara

Isodon wikstroemioides (Hand.-Mazz.) Hara

Isodon yuennanensis (Hand.-Mazz.) Hara

Lamium album L.

Lamium amplexicaule L.

Lamium gilongensis H. W. Li

Lamium nepalense Hedge

Lamium staintonii Hedge

Lamium tuberosum Hedge

Leonotis nepetifolia (L.) R. Br.

Leonurus cardiaca L.

Leonurus japonicus Houtt.
Leonurus urticifolius C. Y. Wu & H. W. Li
Leucas aspera (Willd.) Link
Leucas cephalotes (Roth) Spreng.
Leucas ciliata Benth.
Leucas hyssopifolia Benth.
Leucas indica (L.) R. Br. ex Sm.
Leucas lanata Benth.
Leucas mollissima Wall. ex Benth.
Leucas plukenetii (Roth) Spreng.
Leucosceptrum canum W. W. Sm.
Lophanthus elegans Levin
Lophanthus lipskyanus (Lipsky) Ikonn.-Gal. & Nevski
Lophanthus tibeticus C. Y. Wu & Y. C. Huang
Lycopus europaeus L.
Lycopus lucidus Turcz. ex Benth.
Marmoritis pharica (Prain) A. L. Budantzev
Marrubium anisodon C. Koch
Marrubium duabense Murata
Marrubium vulgare L.
Meehania fargesii (H. Lév.) C. Y. Wu
Melissa axillaris (Benth.) Bakh. f.
Melissa flava Benth. ex Wall.
Melissa officinalis L.
Melissa yunnanensis C. Y. Wu & Y. C. Huang
Mentha arvensis L.
Mentha asiatica Boriss.
Mentha longifolia (L.) L.
Mentha royleana Benth.
Mentha spicata L.
Micromeria barosma (W. W. Sm.) Hand.-Mazz.
Micromeria biflora (Buch.-Ham. & D. Don) Benth.
Micromeria biflora var. **hispida** Kitam. & Murata
Micromeria euosma (W. W. Sm.) C. Y. Wu
Micromeria nepalensis Kitam. & Murata
Micromeria wardii Marquand & Airy Shaw
Microtoena bhutanica Stearn
Microtoena delavayi Prain
Microtoena griffithii Prain
Microtoena insuavis (Hance) Prain ex Briq.
Microtoena miyiensis C. Y. Wu & H. W. Li
Microtoena moupinensis (Franch.) Prain
Microtoena muliensis C. Y. Wu & S. J. Hsuan
Microtoena nepalensis Stearn
Microtoena patchoulii (C. B. Clarke ex Hook. f.) C. Y. Wu & S. J. Hsuan
Microtoena prainiana Diels
Microtoena robusta Hemsl.
Microtoena stenocalyx C. Y. Wu & S. J. Hsuan
Microtoena wardii Stearn
Molucella laevis L.
Mosla cavaleriei H. Lév.
Mosla dianthera (Buch.-Ham. ex Roxb.) Maxim.
Nepeta adenophyta Hedge
Nepeta cataria L.
Nepeta ciliaris Benth.
Nepeta clarkei Hook. f.
Nepeta coerulescens Maxim.
Nepeta connata Royle ex Benth.
Nepeta daenensis Boiss.
Nepeta dentata C. Y. Wu & Hsuan
Nepeta discolor Benth.
Nepeta elliptica Royle ex Benth.
Nepeta erecta (Boyle ex Benth.) Berth.
Nepeta floccosa Benth.
Nepeta fordii Hemsl.
Nepeta glutinosa Benth.
Nepeta govaniana Benth.
Nepeta griffithii Hedge
Nepeta hemsleyana Oliver ex Prain

Nepeta hindostana (Roth) Haines

Nepeta jomdaensis H. W. Li

Nepeta kokanica Regel

Nepeta laevigata (D. Don) Hand-Mazz.

Nepeta lamiopsis Benth. ex Hook. f.

Nepeta leucolaena Benth. ex Hook. f.

Nepeta leucophylla Benth.

Nepeta linearis Royle ex Benth.

Nepeta longibracteata Benth.

Nepeta membranifolia C. Y. Wu

Nepeta nepalensis Spreng.

Nepeta nervosa Royle ex Benth.

Nepeta nervosa var. **lutea** Hook. f.

Nepeta pamirensis Franch.

Nepeta paucifolia Mukerjee

Nepeta paulsenii Briq.

Nepeta pinetorum Aitch. & Hemsl.

Nepeta podostachys Benth.

Nepeta prainii Duthie

Nepeta prattii H. Lév.

Nepeta raphanorhiza Benth.

Nepeta royleana R. R. Stewart

Nepeta salviaefolia Royle ex Benth.

Nepeta schmidii Rech. f.

Nepeta sessilis C. Y. Wu & Hsuan

Nepeta souliei H. Lév.

Nepeta spathulifera Benth.

Nepeta staintonii Hedge

Nepeta stewartiana Diels

Nepeta subincisa Benth.

Nepeta sungpanensis C. Y. Wu

Nepeta sungpanensis var. **angustidentata** C. Y. Wu & Y. C. Huang

Nepeta tenuiflora Diels

Nepeta uberrima Rech. f.

Nepeta veitchii Duthie

Nepeta wilsonii Duthie

Nepeta zandaensis H. W. Li

Ocimum americanum L.

Ocimum basilicum L.

Ocimum gratissimum L.

Ocimum tenuiflorum L.

Origanum majorana L.

Origanum vulgare L.

Orthosiphon incurvus Benth.

Orthosiphon incurvus var. **elatus** Benth.

Orthosiphon rubicundus (D. Don) Benth.

Orthosiphon scapiger Benth.

Orthosiphon wattii Prain

Orthosiphon wulfenioides (Diels) Hand.-Mazz.

Otostegia limbata (Benth.) Boiss.

Paraphlomis javanica (Blume) Prain

Perilla frutescens (L.) Britton

Perovskia abrotanoides Kar.

Perovskia atriplicifolia Benth.

Phlomidoschema parviflorum (Benth.) Vved.

Phlomis stewartii (Hook. f.) C. L. Xiang

Phlomis trineura Rech. f.

Phlomoides ambigua (Popov ex Pazij & Vved.) Adylov, Kamelin & Makhm.

Phlomoides atropurpurea (Dunn) Kamelin & Makhm.

Phlomoides betonicoides (Diels) Kamelin & Makhm.

Phlomoides bracteosa (Royle ex Benth.) Kamelin & Makhm.

Phlomoides breviflora (Benth.) Kamelin & Makhm.

Phlomoides canescens (Regel) Adylov, Kamelin & Makhm.

Phlomoides congesta (C. Y. Wu) Kamelin & Makhm.

Phlomoides fimbriata (C. Y. Wu) Kamelin & Makhm.

Phlomoides franchetiana (Diels) Kamelin & Makhm.

Phlomoides hamosa Benth.

Phlomoides hamosa (Benth.) Mathiesen

Phlomoides likiangensis (C. Y. Wu) Kamelin & Makhm.

Phlomoides longiaristata (C. Y. Wu & H. W. Li) Salmaki

Phlomoides longicalyx (C. Y. Wu) Kamelin & Makhm.

Phlomoides macrophylla (Benth.) Kamelin & Makhm.

Phlomoides medicinalis (Diels) Kamelin & Makhm.

Phlomoides megalantha (Diels) Kamelin & Makhm.

Phlomoides melanantha (Diels) Kamelin & Makhm.

Phlomoides milingensis (C. Y. Wu & H. W. Li) Kamelin & Makhm.

Phlomoides muliensis (C. Y. Wu) Kamelin & Makhm.

Phlomoides nana (C. Y. Wu) H. W. Li

Phlomoides ornata (C. Y. Wu) Kamelin & Makhm.

Phlomoides orrestii (Diels) Kamelin & Makhm.

Phlomoides pararotata (Y. Z. Sun) Kamelin & Makhm.

Phlomoides rotata (Benth. ex Hook. f.) Mathiesen

Phlomoides ruptilis (C. Y. Wu) Kamelin & Makhm.

Phlomoides setifera (Bureau & Franch.) Kamelin & Makhm.

Phlomoides setigera (Falc. ex Benth.) Kamelin & Makhm.

Phlomoides spectabilis (Falc. ex Benth.) Kamelin & Makhm.

Phlomoides strigosa (C. Y. Wu) Kamelin & Makhm.

Phlomoides tatsienensis (Bureau & Franch.) Kamelin & Makhm.

Phlomoides tibetica (C. Marquand & Airy Shaw) Kamelin & Makhm.

Phlomoides umbrosa (Turcz.) Kamelin & Makhm.

Phlomoides younghushandii (S. M. Mukerjee) Kamelin & Makhm.

Plectranthus barbatus Andrews

Plectranthus mollis (Aiton) Spreng.

Pogostemon amaranthoides Benth.

Pogostemon andersonii (Prain) Panigrahi

Pogostemon auricularius (L.) Hassk.

Pogostemon benghalensis (Burm. f.) Kuntze

Pogostemon brachystachyus Benth.

Pogostemon brevicorollus Y. Z. Sun

Pogostemon chinensis C. Y. Wu & Y. C. Huang

Pogostemon crassicaulis (Benth.) Press

Pogostemon cruciatus (Benth.) Kuntze

Pogostemon dielsianus Dunn

Pogostemon elsholtzioides Benth.

Pogostemon glaber Benth.

Pogostemon helferi (Hook. f.) Press

Pogostemon hispidocalyx C. Y. Wu & Y. C. Huang

Pogostemon nigrescens Dunn

Pogostemon pumilus (Graham) Press

Pogostemon tuberculosus Benth.

Premna acutata W. W. Sm.

Premna barbata Wall. ex Schauer

Premna bengalensis C. B. Clarke

Premna bracteata Wall. ex C. B. Clarke

Premna crassa Hand.-Mazz.

Premna herbacea Roxb.

Premna interrupta Wall. ex Schauer

Premna latifolia Roxb.

Premna ligustroides Hemsl.

Premna oligantha C. Y. Wu

Premna puberula Pamp.

Premna scandens Roxb.

Premna scoriarum W. W. Sm.

Premna tapintzeana Dop

Premna urticifolia Rehder

Premna wuana Q. Wang

Prunella hispida Benth.

Prunella vulgaris L.

Pseudocaryopteris bicolor (Hardw.) P. D. Cantino

Pseudocaryopteris paniculata (C. B. Clarke) P. D. Cantino

Rostrinucula dependens (Rehder) Kudô

Rotheca nepalense (Moldenke) Q. Wang

Rotheca serrata (L.) Steane & Mabb.

Roylea cinerea (D. Don) Baill.

Rubiteucris palmata (Benth. ex Hook. f.) Kudô

Rubiteucris siccanea (W. W. Sm.) P. D. Cantino

Salvia aegyptiaca L.

Salvia aerea H. Lév.

Salvia alatipetiolata Sun

Salvia amplicalyx E. Peter

Salvia asperata Falc. ex Benth.

Salvia atropurpurea C. Y. Wu

Salvia atrorubra C. Y. Wu

Salvia bifidocalyx C. Y. Wu & Y. C. Huang

Salvia brachyloma E. Peter

Salvia brevilabra Franch.

Salvia bulleyana Diels

Salvia campanulata Wall. ex Benth.

Salvia campanulata var. **codonantha** (E. Peter) E. Peter

Salvia campanulata var. **fissa** E. Peter

Salvia campanulata var. **hirtella** E. Peter

Salvia cana Wall. ex Benth.

Salvia castanea Diels

Salvia cyclostegia E. Peter

Salvia cyclostegia var. **purpurascens** C. Y. Wu

Salvia cynica Dunn

Salvia digitaloides Diels

Salvia digitaloides var. **glabrescens** E. Peter

Salvia dolichantha E. Peter

Salvia evansiana Hand.-Mazz.

Salvia evansiana var. **scaposa** E. Peter

Salvia flava Forrest ex Diels

Salvia flava var. **megalantha** Diels

Salvia grandifolia W. W. Sm.

Salvia handelii E. Peter

Salvia heterochroa E. Peter

Salvia hians Royle ex Benth.

Salvia hylocharis Dies

Salvia japonica var. **multifoliolata** E. Peter

Salvia kiaometiensis H. Lév.

Salvia lankongensis C. Y. Wu

Salvia maximowicziana Hemsl.

Salvia maximowicziana var. **floribunda** E. Peter

Salvia mekongensis E. Peter

Salvia moorcroftiana Wall. ex Benth.

Salvia nubicola Wall. ex Sweet

Salvia omeiana E. Peter

Salvia omeiana var. **grandibracteata** E. Peter

Salvia plebeia R. Br.

Salvia plectranthoides Griff.

Salvia pogonochila Diels

Salvia potaninii Kryl.

Salvia prattii Hemsl.

Salvia przewalskii Maxim.

Salvia przewalskii var. **alba** X. L. Huang & H. W. Li

Salvia przewalskii var. **glabrescens** E. Peter

Salvia przewalskii var. **mandarinorum** (Diels) E. Peter

Salvia przewalskii var. **rubrobrunnea** C. Y. Wu

Salvia rhytidea Benth.

Salvia roborowskii Maxim.

Salvia saxicola Wall. ex Benth.

Salvia schizocalyx E. Peter

Salvia schizochila E. Peter

Salvia sclarea L.

Salvia sikkimensis E. Peter

Salvia sikkimensis var. **chaenocalyx** E. Peter

Salvia smithii E. Peter

Salvia subpalmatinervis E. Peter

Salvia tricuspis Franch.

Salvia trijuga Diels

Salvia virgata Jacq.

Salvia wardii E. Peter

Salvia wuana C. L. Xiang

Salvia yunnanensis C. H. Wright

Schnabelia aureoglandulosa (Vaniot) P. D. Cantino

Schnabelia oligophylla Hand.-Mazz.

Schnabelia terniflora (Maxim.) P. D. Cantino

Scutellaria barbata D. Don

Scutellaria chamaedrifolia Hedge & A. J. Paton

Scutellaria chungtienensis C. Y. Wu

Scutellaria delavayi H. Lév.

Scutellaria discolor Wall. ex Benth.

Scutellaria edelbergii Rech. f.

Scutellaria forrestii Diels

Scutellaria galericulata L.

Scutellaria grossa Wall. ex Benth.

Scutellaria heydei Hook. f.

Scutellaria hypericifolia H. Lév.

Scutellaria indica L.

Scutellaria kingiana Prain

Scutellaria likiangensis Diels

Scutellaria linarioides C. Y. Wu

Scutellaria linearis Benth.

Scutellaria lushuiensis Q. Wang

Scutellaria multicaulis Boiss.

Scutellaria omeiensis C. Y. Wu

Scutellaria orthocalyx Hand.-Mazz.

Scutellaria petiolata Hemsl. & Lace ex Prain

Scutellaria prostrata Jacquem. ex Benth.

Scutellaria pseudotenax C. Y. Wu

Scutellaria quadrilobulata Y. Z. Sun

Scutellaria rehderiana Diels

Scutellaria repens Buch.-Ham. ex D. Don

Scutellaria scandens Buch.-Ham. ex D. Don

Scutellaria scordiifolia Fisch. ex Schrank

Scutellaria shweliensis W. W. Sm.

Scutellaria stocksii Boiss.

Scutellaria tapintzensis C. Y. Wu & H. W. Li

Scutellaria tenax W. W. Sm.

Scutellaria tienchüanensis C. Y. Wu & C. Chen

Scutellaria violacea B. Heyne ex Benth.

Scutellaria weishanensis C. Y. Wu & H. W. Li

Scutellaria wuana C. L. Xiang & F. Zhao

Scutellaria yangbiensis H. W. Li

Scutellaria yingtakensis Y. Z. Sun

Sideritis montana L.

Siphocranion macranthum (Hook. f.) C. Y. Wu

Siphocranion nudipes (Hemsl.) Kudô

Skapanthus oreophilus (Diels) C. Y. Wu & H. W. Li

Stachyopsis maleolens (Rech. f.) Hedge

Stachyopsis oblongata (Schrenk ex Fisch. & C. A. Mey.) Popov & Vved.

Stachys floccosa Benth.

Stachys kouyangensis (Vaniot) Dunn

Stachys melissaefolia Benth.

Stachys melissifolia Benth.

Stachys scaberula Vatke

Stachys sericea Cav.

Stachys setifera C. A. Mey.

Stachys sieboldii Miq.

Stachys tibetica Vatke

Tectona grandis L. f.

Tectona hamiltoniana Wall.

Teucrium grandifolium R. A. Clemen

Teucrium labiosum C. Y. Wu & S. Chow

Teucrium laxum D. Don

Teucrium manghuaense Sun ex S. Chow

Teucrium nanum C. Y. Wu & S. Chow

Teucrium ornatum Hemsl.

Teucrium quadrifarium Buch.-Ham.

Teucrium royleanum Wall. ex Benth.

Teucrium scordium L.

Teucrium simplex Vaniot

Teucrium stocksianum Boiss.

Teucrium viscidum Blume

Teucrium willdenowii Q. Wang

Thuspeinantha persica Briq.

Thymus linearis Benth.

Thymus linearis subsp. **hedgei** Jalas

Thymus linearis subsp. **linearis** Jalas

Tripora divaricata (Maxim.) P. D. Cantino

Vitex burmensis Moldenke

Vitex duclouxii Dop

Vitex negundo L.

Vitex quinata (Loureiro) Williams

Vitex trifolia L.

Zataria multiflora Boiss.

Ziziphora clinopodioides subsp. **afghanica** (Rech. f.) Rech. f.

Ziziphora clinopodioides subsp. **pseudodasyantha** (Rech. f.) Rech. f.

Ziziphora clinopodioides subsp. **rigida** (Boiss.) Rech. f.

Ziziphora tenuior L.

260. MAZACEAE

Dodartia orientalis L.

Lancea hirsuta Bonati

Lancea tibetica Hook. f. & Thomson

Mazus celsioides Hand.-Mazz.

Mazus dentatus Wall. ex Benth.

Mazus humilis Hand.-Mazz.

Mazus lecomtei Bonati

Mazus pumilus (Burm. f.) Steenis

Mazus pumilus var. **delavayi** (Bonati) T. L. Chin ex D. Y. Hong

Mazus rockii H. L. Li

Mazus surculosus D. Don

261. PHRYMACEAE

Erythranthe bhutanica (T. Yamaz.) G. L. Nesom

Erythranthe bodinieri (Vaniot) G. L. Nesom

Erythranthe bracteosa (P. C. Tsoong) G. L. Nesom

Erythranthe nepalensis (Benth.) G. L. Nesom

Erythranthe tenella (Bunge) G. L. Nesom

Erythranthe tibetica (P. C. Tsoong & H. P. Yang) G. L. Nesom

Microcarpaea minima (Koenig) Merr.

Phryma leptostachya var. **oblongifolia** (Koidz.) Honda

262. PAULOWNIACEAE

Paulownia catalpifolia T. Gong ex D. Y. Hong

Paulownia fargesii Franch.

Wightia speciosissima (D. Don) Merr.

263. OROBANCHACEAE

Aeginetia indica L.

Aeginetia pedunculata (Roxb.) Wall.

Alectra arvensis (Benth.) Merr.

Boschniakia himalaica Hook. f. & Thomson

Brandisia rosea W. W. Sm.

Brandisia rosea var. **flava** C. E. C. Fisch.

Buchnera cruciata Buch.-Ham. ex D. Don

Buchnera hispida Buch.-Ham. ex D. Don

Centranthera cochinchinensis var. **nepalensis** (D. Don) Merr.

Centranthera grandiflora Benth.

Centranthera indica (L.) Gamble

Centranthera nepalensis D. Don

Christisonia calcarata Wight

Christisonia hookeri C. B. Clarke

Cistanche tubulosa (Schrenk) Hook. f.

Euphrasia alba Pannell

Euphrasia aristulata Pennell

Euphrasia bhutanica Pugsley

Euphrasia chumbica R. R. Mill

Euphrasia densiflora Pennell

Euphrasia flabellata Pennell

Euphrasia foliosa Pennell

Euphrasia himalayica Wettst.

Euphrasia incisa Pennell

Euphrasia jaeschkei Wettst.

Euphrasia kashmiriana Pugsley

Euphrasia laxa Pennell

Euphrasia melanosticta R. R. Mill

Euphrasia microcarpa Pennell

Euphrasia multiflora Pennell

Euphrasia nepalensis Pugsley

Euphrasia paghmanensis Rech. f.

Euphrasia paucifolia Wettst.

Euphrasia pectinata Tenore

Euphrasia pectinata subsp. **sichuanica** D. Y. Hong

Euphrasia platyphylla Pennell

Euphrasia regelii subsp. **kangtienensis** D. Y. Hong

Euphrasia regelii Wettst.

Euphrasia remota Pennell

Euphrasia schlagintweitii Wettst.

Euphrasia secundiflora Pennell

Euphrasia simplex D. Don

Euphrasia subpetiolaris Pugsley

Gleadovia mupinense Hu

Gleadovia ruborum Gamble & Prain

Lathraea purpurea Cummins ex King

Lathraea squamaria L.

Leptorhabdos parviflora (Benth.) Benth.

Lindenbergia bhutanica T. Yamaz.

Lindenbergia grandiflora (Buch.-Ham. ex D. Don) Benth.

Lindenbergia griffithii Hook. f.

Lindenbergia hookeri C. B. Clarke ex Hook. f.

Lindenbergia indica (L.) Vatke

Lindenbergia muraria (Roxb. ex D. Don) Brühl

Lindenbergia philippensis (Chamisso & Schlecht.) Benth.

Lindenbergia titensis Sikdar & Maiti
Mannagettaea labiata Harry Sm.
Melampyrum klebelsbergianum Soó
Melasma avense (Benth.) Hand.-Mazz.
Orobanche aegyptiaca Pers.
Orobanche alba Stephan
Orobanche amethystea Thuill.
Orobanche amoena C. A. Mey.
Orobanche caesia Rchb.
Orobanche cernua Loefl.
Orobanche cernua var. **nepalensis** Reut.
Orobanche cernua var. **pseudo-clarkei** Jafri
Orobanche clarkei Hook. f.
Orobanche coerulescens Stephan
Orobanche hansii A. Kern.
Orobanche kashmirica C. B. Clarke ex Hook. f.
Orobanche kotschyi Reuter
Orobanche lanuginosa (C. A. Mey.) Beck ex Krylov
Orobanche megalantha Harry Sm.
Orobanche mupinensis Hu
Orobanche orientalis G. Beck
Orobanche psila C. B. Clarke ex Hook. f.
Orobanche purpurea Jacq.
Orobanche sinensis Harry Sm.
Orobanche sinensis var. **cyanescens** (Harry Sm.) Zhi Y. Zhang
Orobanche solmsii C. B. Clarke
Orobanche yunnanensis (Beck) Hand.-Mazz.
Pedicularis alaschanica Maxim.
Pedicularis albertii Regel
Pedicularis albida Pennell
Pedicularis albiflora (Hook. f) Prain
Pedicularis aloensis Hand.-Mazz.
Pedicularis alopecuros Franch. ex Maxim.
Pedicularis amplicollis T. Yamaz.
Pedicularis amplituba H. L. Li
Pedicularis anas Maxim.
Pedicularis angularis P. C. Tsoong
Pedicularis angustiflora H. Limpr.
Pedicularis angustilabris H. L. Li
Pedicularis angustiloba P. C. Tsoong
Pedicularis annapurnensis T. Yamaz.
Pedicularis anomala P. C. Tsoong & H. P. Yang
Pedicularis anserantha T. Yamaz.
Pedicularis aquilina Bonati
Pedicularis armata Maxim.
Pedicularis artiae R. Kr. Singh, Kholia & Sudhakar
Pedicularis aschistorrhyncha C. Marquand & Airy Shaw
Pedicularis atroviridis P. C. Tsoong
Pedicularis atuntsiensis Bonati
Pedicularis aurata (Bonati) H. L. Li
Pedicularis axillaris Franch. ex Maxim.
Pedicularis bambusetorum Hand.-Mazz.
Pedicularis batangensis Bureau & Franch.
Pedicularis bella Hook. f.
Pedicularis bicolor Diels
Pedicularis bicornuta Klotzsch
Pedicularis bidentata Maxim.
Pedicularis bietii Franch.
Pedicularis bifida (Buch.-Ham. ex D. Don) Pennell
Pedicularis binaria Maxim.
Pedicularis bipinnatifida (Pennell) R. R. Mill
Pedicularis bomiensis H. P. Yang
Pedicularis brachycrania H. L. Li
Pedicularis brevifolia D. Don
Pedicularis brevilabris Franch.
Pedicularis brevirostris Pennell
Pedicularis breviscaposa T. Yamaz.
Pedicularis brunoniana Wall.
Pedicularis cabulica Benth.

Pedicularis caeruleo-albescens Wendelbo
Pedicularis canescens P. C. Tsoong
Pedicularis cephalantha Franch. ex Maxim.
Pedicularis cephalanthoides P. C. Tsoong
Pedicularis cernua Bonati
Pedicularis chamissonoides T. Yamaz.
Pedicularis cheilanthifolia Schrenk
Pedicularis chenocephala Diels
Pedicularis chinensis Maxim.
Pedicularis chitralensis Pennell
Pedicularis chorgossica Regel & Winkl.
Pedicularis chumbica Prain
Pedicularis cinerascens Franch.
Pedicularis clarkei Hook. f.
Pedicularis collata Prain
Pedicularis collettii Prain
Pedicularis comptoniaefolia Franch. ex Maxim.
Pedicularis confertiflora Prain
Pedicularis connata H. L. Li
Pedicularis cooperi P. C. Tsoong
Pedicularis coppeyi Bonati
Pedicularis cornigera T. Yamaz.
Pedicularis corydaloides Hand.-Mazz.
Pedicularis corymbifera H. P. Yang
Pedicularis cranolopha Maxim.
Pedicularis craspedotricha Maxim.
Pedicularis crenata Maxim.
Pedicularis crenularis H. L. Li
Pedicularis croizatiana H. L. Li
Pedicularis cryptantha Marquand & Airy Shaw
Pedicularis curvipes Hook. f.
Pedicularis curvituba Maxim.
Pedicularis cyathophylla Franch.
Pedicularis cyathophylloides H. Limpr.
Pedicularis cyclorhyncha H. L. Li
Pedicularis cymbalaria Bonati
Pedicularis cyrtorhyncha Pennell
Pedicularis daltonii Prain
Pedicularis daochengensis H. P. Yang
Pedicularis daucifolia Bonati
Pedicularis davidii Franch.
Pedicularis debilis Franch. ex Maxim.
Pedicularis decora Franch.
Pedicularis decorissima Diels
Pedicularis deltoidea Franch. ex Maxim.
Pedicularis densispica Franch. ex Maxim.
Pedicularis denudata Hook. f.
Pedicularis dhurensis R. R. Mill
Pedicularis dichotoma Bonati
Pedicularis dichrocephala Hand.-Mazz.
Pedicularis diffusa Prain
Pedicularis dissectifolia H. L. Li
Pedicularis dolichantha Bonati
Pedicularis dolichocymba Hand.-Mazz.
Pedicularis dolichoglossa H. L. Li
Pedicularis dolichostachya H. L. Li
Pedicularis duclouxii Bonati
Pedicularis dulongensis H. P. Yang
Pedicularis dunniana Bonati
Pedicularis elegans Franch.
Pedicularis elephantoides Benth.
Pedicularis elevatogaleata T. Yamaz.
Pedicularis elliotii P. C. Tsoong
Pedicularis elwesii Hook. f.
Pedicularis excelsa Hook. f.
Pedicularis exigua H. L. Li
Pedicularis fastigiata Franch.
Pedicularis fengii H. L. Li
Pedicularis filicula Franch. ex Maxim.
Pedicularis filiculiformis P. C. Tsoong
Pedicularis flaccida Prain
Pedicularis fletcheri P. C. Tsoong
Pedicularis flexuosa Hook. f.

Pedicularis floribunda Franch.

Pedicularis forrestiana Bonati

Pedicularis fragarioides P. C. Tsoong

Pedicularis fragilis Prain ex Maxim.

Pedicularis franchetiana Maxim.

Pedicularis furfuracea Wall. ex Benth.

Pedicularis galeata Bonati

Pedicularis gammieana Prain

Pedicularis garckeana Prain ex Maxim.

Pedicularis geosiphon Harry Sm. & P. C. Tsoong

Pedicularis gibbera Prain

Pedicularis giraldiana Diels ex Bonati

Pedicularis glabrescens H. L. Li

Pedicularis globifera Hook. f.

Pedicularis gongshanensis H. P. Yang

Pedicularis gracilicaulis H. L. Li

Pedicularis gracilis Wall. ex Benth.

Pedicularis gracilituba H. L. Li

Pedicularis gruiflora T. Yamaz.

Pedicularis gruina Franch. ex Maxim.

Pedicularis gyirongensis H. P. Yang

Pedicularis gyrorhyncha Franch. ex Maxim.

Pedicularis habachanensis Bonati

Pedicularis hemsleyana Prain

Pedicularis henryi Maxim.

Pedicularis heydei Prain

Pedicularis hicksii P. C. Tsoong

Pedicularis hirtella Franch. ex F. B. Forbes & Hemsl.

Pedicularis hoffmeisteri Klotzsch

Pedicularis hookeriana Wall. ex Benth

Pedicularis humilis Bonati

Pedicularis hypophylla T. Yamaz.

Pedicularis imbricata P. C. Tsoong

Pedicularis imperialis Franch.

Pedicularis inaequilobata P. C. Tsoong

Pedicularis inconspicua P. C. Tsoong

Pedicularis infirma H. L. Li

Pedicularis inflexirostris F. S. Yang, D. Y. Hong & X. Q. Wang

Pedicularis ingens Maxim.

Pedicularis insignis Bonati

Pedicularis instar Prain ex Maxim

Pedicularis integrifolia Hook. f.

Pedicularis jainii Aswal & Mehrotra

Pedicularis kangtingensis P. C. Tsoong

Pedicularis kansuensis Maxim.

Pedicularis kariensis Bonati

Pedicularis kashmiriana Pennell

Pedicularis kawaguchii T. Yamaz.

Pedicularis khasiana (Hook. f.) Pennell

Pedicularis kialensis Franch.

Pedicularis kingii Prain

Pedicularis klotzschii Hurus.

Pedicularis kongboensis P. C. Tsoong

Pedicularis koshiensis T. Yamaz.

Pedicularis labordei Vaniot ex Bonati

Pedicularis lachnoglossa Hook. f.

Pedicularis lamioides Hand.-Mazz.

Pedicularis lamjungensis T. Yamaz.

Pedicularis lanceifolia P. C. Tsoong

Pedicularis lanpingensis H. P. Yang

Pedicularis lasiophrys Maxim.

Pedicularis latibracteata T. Yamaz.

Pedicularis latituba Bonati

Pedicularis laxispica H. L. Li

Pedicularis lecomtei Bonati

Pedicularis leiandra Franch. ex Hemsl.

Pedicularis leptosiphon H. L. Li

Pedicularis likiangensis Franch. ex Maxim.

Pedicularis limprichtiana Hand.-Mazz.

Pedicularis lineata Franch. ex Maxim.

Pedicularis lingelsheimiana H. Limpr.

Pedicularis lobatorostrata T. Yamaz.
Pedicularis longicalyx H. P. Yang
Pedicularis longicaulis Franch. ex Maxim.
Pedicularis longiflora Rudolph
Pedicularis longipedicellata P. C. Tsoong
Pedicularis longipes Maxim.
Pedicularis longipetiolata Franch. ex Maxim.
Pedicularis longistipitata P. C. Tsoong
Pedicularis lophotricha H. L. Li
Pedicularis lopingensis Hand.-Mazz.
Pedicularis ludlowiana P. C. Tsoong
Pedicularis lutescens Franch. ex Maxim.
Pedicularis lyrata Prain ex Maxim.
Pedicularis macilenta Franch.
Pedicularis macrorhyncha H. L. Li
Pedicularis macrosiphon Franch.
Pedicularis magnini Bonati
Pedicularis mahoangensis Bonati
Pedicularis mairei Bonati
Pedicularis maxonii Bonati
Pedicularis mayana Hand.-Mazz.
Pedicularis megalantha D. Don
Pedicularis megalochila H. L. Li
Pedicularis melalimne R. R. Mill
Pedicularis melampyriflora Franch. ex Maxim.
Pedicularis membranacea H. L. Li
Pedicularis merrilliana H. L. Li
Pedicularis metaszetschuanica P. C. Tsoong
Pedicularis meteororhyncha H. L. Li
Pedicularis micrantha H. L. Li
Pedicularis microcalyx Hook. f.
Pedicularis microchilae Franch. ex Maxim.
Pedicularis microloba R. R. Mill
Pedicularis minima P. C. Tsoong & S. H. Cheng
Pedicularis minutilabris P. C. Tsoong
Pedicularis mollis Wall. ex Benth.
Pedicularis monbeigiana Bonati
Pedicularis moupinensis Franch.
Pedicularis mucronulata P. C. Tsoong
Pedicularis muguensis T. Yamaz.
Pedicularis multiflora Pennell
Pedicularis murreeana R. R. Mill
Pedicularis muscicola Maxim.
Pedicularis muscoides H. L. Li
Pedicularis mussotii Franch.
Pedicularis mustanghatana T. Yamaz.
Pedicularis mychophila C. Marquand & Airy Shaw
Pedicularis nagaensis H. L. Li
Pedicularis nana C. E. C. Fisch.
Pedicularis neolatituba P. C. Tsoong
Pedicularis nepalensis Prain
Pedicularis nigra (Bonati) Vaniot ex Bonati
Pedicularis nodosa Pennell
Pedicularis nudicaulis Bonati
Pedicularis nyalamensis H. P. Yang
Pedicularis nyingchiensis H. P. Yang & Tateishi
Pedicularis obliquigaleata W. B. Yu & H. Wang
Pedicularis obscura Bonati
Pedicularis odontocorys T. Yamaz.
Pedicularis odontoloma T. Yamaz.
Pedicularis odontophora Prain
Pedicularis oederi Vahl
Pedicularis oligantha Franch. ex Maxim.
Pedicularis oliveriana Prain
Pedicularis omiiana Bonati
Pedicularis ophiocephala Maxim.
Pedicularis orthocoryne H. L. Li
Pedicularis oxycarpa Franch. ex Maxim.
Pedicularis oxyrhyncha T. Yamaz.
Pedicularis paiana H. L. Li
Pedicularis pantlingii Prain
Pedicularis paradoxa (Prain) T. Yamaz.
Pedicularis pauciflora (Prain) Pennell

Pedicularis paxiana H. Limpr.

Pedicularis pectinata Wall. ex Benth.

Pedicularis pectinatiformis Bonati

Pedicularis pennelliana P. C. Tsoong

Pedicularis pentagona H. L. Li

Pedicularis perpusilla P. C. Tsoong

Pedicularis petitmenginii Bonati

Pedicularis phaceliifolia Franch.

Pedicularis pheulpinii Bonati

Pedicularis pilostachya Maxim.

Pedicularis pinetorum Hand.-Mazz.

Pedicularis platychila P. C. Tsoong

Pedicularis plicata Maxim.

Pedicularis poluninii P. C. Tsoong

Pedicularis polygaloides Hook. f.

Pedicularis polyodonta H. L. Li

Pedicularis porrecta Wall. ex Benth

Pedicularis porriginosa P. C. Tsoong

Pedicularis potaninii Maxim.

Pedicularis praeruptorum Bonati

Pedicularis prainiana Maxim.

Pedicularis princeps Bureau & Franch.

Pedicularis przewalskii Maxim.

Pedicularis pseudocephalantha Bonati

Pedicularis pseudogracilituba P. C. Tsoong

Pedicularis pseudoingens Bonati

Pedicularis pseudomelampyriflora Bonati

Pedicularis pseudomuscicola Bonati

Pedicularis pseudoregeliana P. C. Tsoong

Pedicularis pseudosteiningeri Bonati

Pedicularis pseudoversicolor Hand.-Mazz.

Pedicularis pulchra Paulsen

Pedicularis punctata Decne.

Pedicularis pushpangadanii T. Husain & Arti Garg

Pedicularis pycnantha Boiss.

Pedicularis pygmaea Maxim.

Pedicularis pyramidata Royle ex Benth.

Pedicularis qinghaiensis T. Yamaz.

Pedicularis quadrata Bonati

Pedicularis quxiangensis H. P. Yang

Pedicularis ramosissima Bonati

Pedicularis recurva Maxim.

Pedicularis regeliana Prain

Pedicularis remotiloba Hand.-Mazz.

Pedicularis reptans P. C. Tsoong

Pedicularis retingensis P. C. Tsoong

Pedicularis rex C. B. Clarke ex Maxim.

Pedicularis rhinanthoides Schrenk ex Fisch. & C. A. Mey.

Pedicularis rhizomatosa P. C. Tsoong

Pedicularis rhodotricha Maxim.

Pedicularis rhynchodonta Bureau & Franch.

Pedicularis rhynchotricha P. C. Tsoong

Pedicularis rigida Franch. ex Maxim.

Pedicularis rigidescens T. Yamaz.

Pedicularis rizhaoensis H. P. Yang

Pedicularis roborowskii Maxim.

Pedicularis robusta Hook. f.

Pedicularis rockii Bonati

Pedicularis rotundifolia C. E. C. Fisch.

Pedicularis roylei Maxim.

Pedicularis rudis Maxim.

Pedicularis ruoergaiensis H. P. Yang

Pedicularis rupicola Franch. ex Maxim.

Pedicularis salicifolia Bonati

Pedicularis salviiflora Franch. ex F. B. Forbes & Hemsl.

Pedicularis sanguilimbata R. R. Mill

Pedicularis schizorrhyncha Prain

Pedicularis schugnana B. Fedtsch.

Pedicularis scolopax Maxim.

Pedicularis scullyana Prain ex Maxim

Pedicularis sectifolia T. Yamaz.

Pedicularis semenowii Regel

Pedicularis semitorta Maxim.

Pedicularis sherriffii P. C. Tsoong

Pedicularis sigmoidea Franch. ex Maxim.

Pedicularis sikkimensis Bonati

Pedicularis sima Maxim.

Pedicularis siphonantha D. Don

Pedicularis smithiana Bonati

Pedicularis sorbifolia P. C. Tsoong

Pedicularis souliei Franch.

Pedicularis sparsiflora Bonati

Pedicularis sphaerantha P. C. Tsoong

Pedicularis spicata Pall.

Pedicularis stadlmanniana Bonati

Pedicularis staintonii R. R. Mill

Pedicularis steiningeri Bonati

Pedicularis stenantha Franch.

Pedicularis stenocorys Franch.

Pedicularis stenotheca P. C. Tsoong

Pedicularis stewartii Pennell

Pedicularis streptorhyncha P. C. Tsoong

Pedicularis strobilacea Franch. ex F. B. Forbes & Hemsl.

Pedicularis stylosa H. P. Yang

Pedicularis subulatidens P. C. Tsoong

Pedicularis sunkosiana T. Yamaz.

Pedicularis superba Franch. ex Maxim.

Pedicularis szetschuanica Maxim.

Pedicularis takpoensis P. C. Tsoong

Pedicularis taliensis Bonati

Pedicularis tamurensis T. Yamaz.

Pedicularis tantalorhyncha Franch. ex Bonati

Pedicularis tantalorhynchoides P. C. Tsoong

Pedicularis tapaoensis P. C. Tsoong

Pedicularis tatsienensis Bureau & Franch.

Pedicularis tayloriana P. C. Tsoong

Pedicularis tenacifolia P. C. Tsoong

Pedicularis tenera H. L. Li

Pedicularis tenuicaulis Prain

Pedicularis tenuirostris Benth.

Pedicularis tenuisecta Franch. ex Maxim.

Pedicularis tenuituba Pennell & H. L. Li

Pedicularis terrenoflora T. Yamaz.

Pedicularis thamnophila (Hand.-Mazz.) H. L. Li

Pedicularis tibetica Franch.

Pedicularis tomentosa H. L. Li

Pedicularis tongolensis Franch.

Pedicularis torta Maxim.

Pedicularis triangularidens P. C. Tsoong

Pedicularis trichocymba H. L. Li

Pedicularis trichodonta T. Yamaz.

Pedicularis trichoglossa Hook. f.

Pedicularis trichomata H. L. Li

Pedicularis tricolor Hand.-Mazz.

Pedicularis tristiformis Bonati

Pedicularis tsaii H. L. Li

Pedicularis tsangchanensis Franch. ex Maxim.

Pedicularis tsarungensis H. L. Li

Pedicularis tsekouensis Bonati

Pedicularis tsiangii H. L. Li

Pedicularis tsoongii T. Yamaz.

Pedicularis umbelliformis H. L. Li

Pedicularis urceolata P. C. Tsoong

Pedicularis variegata H. L. Li

Pedicularis verae Vved.

Pedicularis verbenifolia Franch. ex Maxim.

Pedicularis veronicifolia Franch.

Pedicularis verticillata L.

Pedicularis wallichii Bunge ex Walp.

Pedicularis wardii Bonati

Pedicularis weixiensis H. P. Yang

Pedicularis wilsonii Bonati

Pedicularis woodii R. R. Mill

Pedicularis xiangchengensis H. P. Yang

Pedicularis xiqingshanensis H. Y. Feng & J. Z. Sun

Pedicularis xylopoda P. C. Tsoong

Pedicularis yalungensis T. Yamaz.

Pedicularis yamazakiana R. R. Mill

Pedicularis yanyuanensis H. P. Yang

Pedicularis yaoshanensis H. Wang

Pedicularis yarilaica R. R. Mill

Pedicularis yui H. L. Li

Pedicularis yunnanensis Franch. ex Maxim.

Pedicularis zayuensis H. P. Yang

Pedicularis zhongdianensis H. P. Yang

Phtheirospermum glandulosum Benth. ex Hook. f.

Phtheirospermum japonicum (Thunb.) Kanitz

Phtheirospermum tenuisectum Bureau & Franch.

Pterygiella cylindrica P. C. Tsoong

Pterygiella duclouxii Franch.

Pterygiella nigrescens Oliv.

Pterygiella suffruticosa D. Y. Hong

Siphonostegia chinensis Benth.

Sopubia stricta G. Don

Sopubia trifida Buch.-Ham. ex D. Don

Striga asiatica (L.) Kuntze

Striga densiflora (Benth.) Benth.

Striga masuria (Buch.-Ham. ex Benth.) Benth.

Xizangia bartsioides (Hand.-Mazz.) C. Y. Wu & D. D. Tao

264. LENTIBULARIACEAE

Pinguicula alpina L.

Utricularia aurea Lour.

Utricularia australis R. Br.

Utricularia bifida L.

Utricularia brachiata Oliv.

Utricularia caerulea L.

Utricularia christopheri P. Taylor

Utricularia forrestii P. Taylor

Utricularia furcellata Oliv.

Utricularia gibba L.

Utricularia graminifolia Vahl

Utricularia intermedia Hayne

Utricularia kumaonensis Oliv.

Utricularia minor L.

Utricularia multicaulis Oliv.

Utricularia salwinensis Hand.-Mazz.

Utricularia scandens Benj.

Utricularia scandens subsp. **firmula** (Oliv.) Z. Y. Li

Utricularia stellaris L. f.

Utricularia striatula Sm.

Utricularia vulgaris L.

Utricularia vulgaris subsp. **macrorhiza** (Leconte) R. T. Clausen

265. ACANTHACEAE

Acanthus carduaceus Griff.

Acanthus leucostahyus Wall. ex Nees

Andrographis echioides (L.) Nees

Andrographis longipedunculata (Sreem.) L. H. Cramer

Andrographis paniculata (Burm. f.) Nees

Asystasia macrocarpa Nees

Asystasia neesiana (Wall.) Nees

Barleria acanthoides Vahl

Barleria cristata L.

Barleria integrisepala H. P. Tsui

Barleria prionitis L.

Barleria strigosa Willd.

Codonacanthus pauciflorus (Nees) Nees

Dicliptera bupleuroides Nees

Dicliptera chinensis (L.) Juss.

Dicliptera elegans W. W. Sm.

Echinacanthus attenuatus Nees

Eranthemum ciliatum (Craib) Benoist

Eranthemum griffithii (T. Anderson) Bremek. & Nann.-Bremek.

Eranthemum pulchellum Andrews

Eranthemum splendens (T. Anderson) Bremek. & Nann.-Bremek.

Eranthemum tubiflorum (T. Anderson) Radlk. ex Lindau

Eranthemum wardii (W. W. Sm.) N. P. Balakr.

Hygrophila difformis Blume

Hygrophila erecta (Burm. f.) Hochr.

Hygrophila helodes Heine

Hygrophila phlomoides Nees

Hygrophila polysperma (Roxb.) T. Anderson

Hygrophila ringens (L.) R. Br. ex Spreng.

Hygrophila schulli (Buch-Ham.) M. R. Almeida & S. M. Almeida

Hypoestes triflora (Forssk.) Roem. & Schult.

Isoglossa collina (T. Anderson) B. Hansen

Justicia adhatoda L.

Justicia albovelata W. W. Sm.

Justicia atkinsonii T. Anderson

Justicia diffusa Willd.

Justicia gendarussa Burm. f.

Justicia peploides (Nees) T. Anderson

Justicia procumbens L.

Justicia pubigera (Nees) C. B. Clarke

Justicia quadrifaria (Nees) T. Anderson

Justicia quinqueangularis K. D. Koenig ex Roxb.

Justicia siccanea W. W. Sm.

Justicia simplex D. Don

Justicia tukuchensis V. A. W. Graham

Justicia vasculosa (Nees) T. Anderson

Justicia wardii W. W. Sm.

Justicia xerobatica W. W. Sm.

Justicia yunnanensis W. W. Sm.

Lepidagathis cuspidata Nees

Lepidagathis incurva Buch.-Ham. ex D. Don

Lepidagathis prostrata Dalzell

Lepidagathis purpuricaulis Nees

Lepidagathis trinervis Nees

Leptostachya wallichii Nees

Mackaya indica (Nees) Ensermu

Nelsonia canescens (Lam.) Spreng.

Pararuellia delavayana (Baill.) E. Hossain

Peristrophe fera C. B. Clarke

Peristrophe japonica (Thunb.) Bremek.

Peristrophe lanceolaria (Roxb.) Nees

Peristrophe paniculata (Forssk.) Brummitt

Peristrophe speciosa Nees

Peristrophe yunnanensis W. W. Sm.

Petalidium barlerioides (Roth) Nees

Phaulopsis dorsiflora (Retz.) Santapau

Phialacanthus griffithii Benth.

Phlogacanthus guttatus Nees

Phlogacanthus parviflorus T. Anderson

Phlogacanthus pubinervius T. Anderson

Phlogacanthus thyrsiformis (Roxb. ex Hardw.) Mabb.

Phlogacanthus tubiflorus Nees

Phlogacanthus vitellinus (Roxb.) T. Anderson

Phlogacanthus yangtsekiangensis (H. Lév.) C. Xia & Y. F. Deng

Pseuderanthemum latifolium (Vahl) B. Hansen

Pseuderanthemum leptanthus (C. B. Clarke) Lindau

Pseuderanthemum polyanthum (C. B. Clarke ex Oliv.) Merr.

Rhinacanthus beesianus Diels

Rhinacanthus calcaratus (Wall.) Nees

Rhinacanthus grandiflorus Dunn

Ruellia beddomei C. B. Clarke

Ruellia tuberosa L.

Rungia burmanica (C. B. Clarke) Hansen

Rungia himalayensis C. B. Clarke

Rungia khasiana T. Anderson

Rungia pectinata (L.) Nees

Rungia stolonifera C. B. Clarke

Sphinctacanthus grifithii (T. Anderson) Benth.

Staurogyne argentea Wall.

Staurogyne humilis (Nees) Kuntze

Staurogyne spathulata Koord.

Strobilanthes accrescens J. R. I. Wood

Strobilanthes affinis (Griff.) T. Terao ex J. R. I. Wood & J. Benn

Strobilanthes angustifrons C. B. Clarke

Strobilanthes aprica (Hance) T. Anderson

Strobilanthes arenicola W. W. Sm.

Strobilanthes atropurpurea Nees

Strobilanthes atroviridis Y. F. Deng & J. R. I. Wood

Strobilanthes attenuata (Nees) Nees

Strobilanthes auriculata Nees

Strobilanthes bheriensis (Shakya) J. R. I. Wood

Strobilanthes bracteata (Nees) J. R. I. Wood

Strobilanthes candida J. R. I. Wood

Strobilanthes capitata (Nees) T. Anderson

Strobilanthes cruciata (Bremek.) Terao

Strobilanthes cuneata (Shakya) J. R. I. Wood

Strobilanthes cusia (Nees) Kuntze

Strobilanthes cyphantha Diels

Strobilanthes dimorphotricha Hance

Strobilanthes divaricata (Nees) T. Anderson

Strobilanthes echinata Nees

Strobilanthes elongata C. B. Clarke

Strobilanthes euantha J. R. I. Wood

Strobilanthes extensa (Nees) Nees

Strobilanthes fimbriata Nees

Strobilanthes forrestii Diels

Strobilanthes flexa Benoist

Strobilanthes frondosa J. R. I. Wood

Strobilanthes glutinosa Nees

Strobilanthes hamiltoniana (Steud.) Bosser & Heine

Strobilanthes helicta T. Anderson

Strobilanthes henryi Hemsl.

Strobilanthes himalayana J. R. I. Wood

Strobilanthes hirta (Vahl) Blume

Strobilanthes inflata T. Anderson

Strobilanthes inflata var. **aenobarba** (W. W. Sm.) J. R. I. Wood & Y. F. Deng

Strobilanthes inflata var. **gongshanensis** (H. P. Tsui) J. R. I. Wood & Y. F. Deng

Strobilanthes japonica (Thunb.) Miq.

Strobilanthes jennyae J. R. I. Wood

Strobilanthes kachinensis J. R. I. Wood & J. R. Benn.

Strobilanthes khasyana T. Anderson

Strobilanthes kingdonii J. R. I. Wood

Strobilanthes lachenensis C. B. Clarke

Strobilanthes laevigata C. B. Clarke

Strobilanthes lamiifolia (Nees) T. Anderson

Strobilanthes lihengiae Y. F. Deng & J. R. I. Wood

Strobilanthes longispica (H. P. Tsui) J. R. I. Wood & Y. F. Deng

Strobilanthes maculata (Wall.) Nees

Strobilanthes mastersii T. Anderson

Strobilanthes medogensis (H. W. Li) J. R. I. Wood & Y. F. Deng

Strobilanthes multidens C. B. Clarke

Strobilanthes nutans T. Anderson

Strobilanthes oligocephala T. Anderson ex C. B. Clarke

Strobilanthes oresbia W. W. Sm.

Strobilanthes ovata Y. F. Deng & J. R. I. Wood

Strobilanthes oxycalycina J. R. I. Wood

Strobilanthes panichanga (Nees) T Anderson

Strobilanthes parvifolia J. R. I. Wood

Strobilanthes pavala (Roxb.) J. R. I. Wood

Strobilanthes penstemonoides (Nees) T. Anderson

Strobilanthes persicifolia (Lindl.) J. R. I. Wood

Strobilanthes pinetorum W. W. Sm.

Strobilanthes pterygorrhachis C. B. Clarke

Strobilanthes pubiflora J. R. I. Wood

Strobilanthes ramulosa J. R. I. Wood

Strobilanthes rhombifolia C. B. Clarke

Strobilanthes rubescens T. Anderson

Strobilanthes sabiniana Nees

Strobilanthes saccata J. R. I. Wood

Strobilanthes secunda T. Anderson

Strobilanthes setosa J. R. I. Wood

Strobilanthes simonsii T. Anderson

Strobilanthes stolonifera Benoist

Strobilanthes straminea W. W. Sm.

Strobilanthes subnudata C. B. Clarke

Strobilanthes szechuanica (Batalin) J. R. I. Wood & Y. F. Deng

Strobilanthes tamburensis C. B. Clarke

Strobilanthes tenax Dunn

Strobilanthes thomsonii T. Anderson

Strobilanthes tibetica J. R. I. Wood

Strobilanthes tomentosa (Nees) J. R. I. Wood

Strobilanthes tubiflos (C. B. Clarke) J. R. I. Wood

Strobilanthes urophylla Nees

Strobilanthes versicolor Diels

Strobilanthes violifolia T. Anderson

Strobilanthes wardiana J. R. I. Wood

Strobilanthes wilsonii J. R. I. Wood & Y. F. Deng

Strobilanthes yunnanensis Diels

Thunbergia alata Bojer ex Sims

Thunbergia coccinea Wall.

Thunbergia fragrans Roxb.

Thunbergia grandiflora Roxb.

Thunbergia lutea T. Anderson

Thunbergia nepalensis Bh. Adhikari & J. R. I. Wood

Wuacanthus microdonta (W. W. Sm.) Y. F. Deng et al.

266. BIGNONIACEAE

Campsis grandiflora (Thunb.) K. Schum.

Catalpa tibetica Forrest

Fernandoa adenophylla (Wall. ex G. Don) Steenis

Incarvillea altissima Forrest

Incarvillea arguta (Royle) Royle

Incarvillea berezovskii Batalin

Incarvillea compacta Maxim.

Incarvillea delavayi Bureau & Franch.

Incarvillea dissectifoliola Q. S. Zhao

Incarvillea emodi (Wall. ex Royle) Chatterjee

Incarvillea forrestii H. R. Fletcher

Incarvillea himalayensis Grey-Wilson

Incarvillea lutea Bureau & Franch.

Incarvillea mairei (H. Lév.) Grierson

Incarvillea mairei var. **grandiflora** (Wehrh.) Grierson

Incarvillea mairei var. **multifoliolata** (C. Y. Wu & W. C. Yin) C. Y. Wu & W. C. Yin

Incarvillea sinensis Lam.

Incarvillea sinensis var. **przewalskii** (Batalin) C. Y. Wu & W. C. Yin

Incarvillea younghusbandii Sprague

Incarvillea zhongdianensis Grey-Wilson

Oroxylum indicum (L.) Benth. ex Kurz

Pauldopia ghonta (Buch.-Ham. ex G. Don) Steenis

Radermachera gigantea (Blume) Miq.

Radermachera glandulosa (Blume) Miq.

Radermachera sinica (Hance) Hemsl.

Radermachera yunnanensis C. Y. Wu & W. C. Yin

Stereospermum chelonoides (L. f.) DC.

Stereospermum colais (Buch.-Ham. ex Dillwyn) Mabberley

267. VERBENACEAE

Duranta erecta L.

Lantana camara L.

Lantana indica Roxb.

Lippia alba (Mill.) N. E. Br. ex Britton & P. Wilson

Petrea volubilis L.

Phyla nodiflora (L.) Greene

Priva cordifolia (L. f.) Druce

Stachytarpheta jamaicensis (L.) Vahl

Verbena bonariensis L.

Verbena officinalis L.

Verbena tenuisecta Briq.

268. MARTYNIACEAE

Martynia annua L.

269. STEMONURACEAE

Gomphandra tetrandra (Wall.) Sleumer

270. CARDIOPTERIDACEAE

Cardiopteris quinqueloba (Hassk.) Hassk.

271. HELWINGIACEAE

Helwingia chinensis Batal.

Helwingia himalaica Hook. f. & Thomson ex C. B. Clarke

Helwingia japonica (Thunb.) F. Dietr.

272. AQUIFOLIACEAE

Ilex atrata W. W. Sm.

Ilex bioritsensis Hayata

Ilex chinensis Sims

Ilex corallina Franch.

Ilex crenata Thunb.

Ilex cyrtura Merr.

Ilex delavayi Franch.

Ilex dicarpa Y. R. Li

Ilex dipyrena Wall.

Ilex fargesii Franch.

Ilex fargesii subsp. **melanotricha** (Merr.) A. Andrews

Ilex forrestii H. F. Comber

Ilex fragilis Hook. f.

Ilex gansuensis D. Y. Hong

Ilex georgei H. F. Comber

Ilex godajam Colebr. ex Hook. f.

Ilex hookeri King

Ilex intermedia Loes.

Ilex intricata Hook. f.

Ilex kingiana Cockerell

Ilex longecaudata H. F. Comber

Ilex macrocarpa Oliv.

Ilex medogensis Y. R. Li

Ilex micrococca Maxim.

Ilex miguensis S. Y. Hu

Ilex nothofagifolia Kingdon-Ward

Ilex odorata Buch. Ex-Ham. ex D. Don

Ilex pernyi Franch.

Ilex perryana S. Y. Hu

Ilex pseudoodorata Loes.

Ilex rotunda subsp. **excelsa** (Wall.) D. Y. Hong

Ilex shweliensis H. F. Comber

Ilex sikkimensis Kurz

Ilex sublongecaudata C. J. Tseng & S. Liu ex Y. R. Li

Ilex szechwanensis Loes.

Ilex tsiangiana C. J. Tseng

Ilex umbellulata (Wall.) Loes.

Ilex venulosa Hook. f.

Ilex wattii Loes.

Ilex wilsonii Loes.

Ilex wuana T. R. Dudley

Ilex yunnanensis Franch.

273. CAMPANULACEAE

Adenophora capillaris Hemsl.

Adenophora capillaris subsp. **leptosepala** (Diels) D. Y. Hong

Adenophora coelestis Diels

Adenophora dawuensis D. Y. Hong

Adenophora delavayi (Franch.) D. Y. Hong

Adenophora himalayana Feer

Adenophora himalayana subsp. **alpina** (Nannf.) D. Y. Hong

Adenophora jasionifolia Franch.

Adenophora khasiana (Hook. f. & Thomson) Oliv. cx Collet & Hemsl.

Adenophora liliifolioides Pax & K. Hoffm.

Adenophora linearifolia D. Y. Hong

Adenophora lobophylla D. Y. Hong

Adenophora longipedicellata D. Y. Hong

Adenophora potaninii Korsh.

Adenophora stricta subsp. **aurita** (Franch.) D. Y. Hong & S. Ge

Adenophora stricta subsp. **confusa** (Nannf.) D. Y. Hong

Adenophora stricta subsp. **sessilifolia** D. Y. Hong

Adenophora watsonii W. W. Sm.

Adenophora wilsonii Nannf.

Asyneuma chinense D. Y. Hong

Asyneuma fulgens (Wall.) Briq.

Asyneuma fulgens subsp. **forrestii** D. Y. Hong

Asyneuma pakistanicum D. Y. Hong

Asyneuma strictum Wendelbo

Asyneuma thomsonii (Hook. f.) Bormm.

Campanula alsinoides Hook. f. & Thomson

Campanula argyrotricha Wall. ex A. DC.
Campanula aristata Wall.
Campanula calcicola W. W. Sm.
Campanula cana Wall.
Campanula cashmeriana Royle
Campanula chinensis D. Y. Hong
Campanula chrysospleniifolia Franch.
Campanula crenulata Franch.
Campanula dimorphantha Schweinf.
Campanula glomeratoides D. Y. Hong
Campanula immodesta Lammers
Campanula latifolia L.
Campanula leucantha Gilli
Campanula leucoclada Boiss.
Campanula mekongensis Diels ex C. Y. Wu
Campanula microphylloidea D. Y. Hong
Campanula nakaoi Kitam.
Campanula pallida Wall.
Campanula polyclada Rech. f. & Schiman-Czeika
Campanula punctata Lam.
Campanula rotata D. Y. Hong
Campanula staintonii Rech. f. & Schiman-Czeika
Campanula sylvatica Wall.
Campanula tenuissima Dunn
Campanula tristis Kitam.
Campanula wattiana Nayar & Babu
Campanula yunnanensis D. Y. Hong
Codonopsis affinis Hook. f. & Thomson
Codonopsis alpina Nannf.
Codonopsis benthamii Hook. f. & Thomson
Codonopsis bhutanica Ludlow
Codonopsis bomiensis D. Y. Hong
Codonopsis bragaensis Grey-Wilson
Codonopsis bulleyana Forrest ex Diels
Codonopsis campanulataa D. Y. Hong
Codonopsis canescens Nannf.
Codonopsis chimiliensis J. Anthony
Codonopsis chlorocodon C. Y. Wu
Codonopsis clematidea (Shrenk) C. B. Clarke
Codonopsis cordifolioidea P. C. Tsoong
Codonopsis deltoidea Chipp
Codonopsis elliptica D. Y. Hong
Codonopsis farreri J. Anthony
Codonopsis foetens Hook. f. & Thomson
Codonopsis foetens subsp. **nervosa** (Chipp) D. Y. Hong
Codonopsis gongshanica Q. Wang & D. Y. Hong
Codonopsis gracilis Hook. f. & Thomson
Codonopsis hemisphaerica P. C. Tsoong ex D. Y. Hong
Codonopsis henryi Oliv.
Codonopsis hongii Lammers
Codonopsis inflata Hook. f. & Thomson
Codonopsis javanica (Bl.) Hook. f.
Codonopsis javanica subsp. **japonica** (Makino) Lammers
Codonopsis lixianica D. Y. Hong
Codonopsis meleagris Diels
Codonopsis micrantha Chipp
Codonopsis microtubulosa Z. T. Wang & G. J. Xu
Codonopsis obtusa (Chipp) Nannf.
Codonopsis ovata Benth.
Codonopsis pilosula (Franch.) Nannf.
Codonopsis pilosula subsp. **handeliana** (Nannf.) D. Y. Hong & L. M. Ma
Codonopsis reflexa D. Y. Hong
Codonopsis rotundifolia Benth.
Codonopsis subglobosa W. W. Sm.
Codonopsis subscaposa Kom.
Codonopsis subsimplex Hook. f. & Thomson
Codonopsis thalictrifolia Wall.
Codonopsis thalictrifolia subsp. **mollis** (Chipp) D. Y. Hong
Codonopsis tubulosa Kom.

Codonopsis tubulosa subsp. **vadsea** (S. S. Dash & A. A. Mao) D. Y. Hong

Codonopsis viridiflora Maxim.

Codonopsis viridis Wall.

Cyananthus cordifolius Duthie

Cyananthus delavayi Franch.

Cyananthus fasciculatus C. Marquand

Cyananthus flavus C. Marquand

Cyananthus flavus subsp. **montanus** (C. Y. Wu) D. Y. Hong & L. M. Ma

Cyananthus formosus Diels

Cyananthus hookeri C. B. Clarke

Cyananthus incanus Hook. f. & Thomson

Cyananthus inflatus Hook. f. & Thomson

Cyananthus integer Wall. ex Benth.

Cyananthus lichiangensis W. W. Sm.

Cyananthus ligulosus D. Y. Hong

Cyananthus lobatus Wall. ex Benth.

Cyananthus longiflorus Franch.

Cyananthus macrocalyx Franch.

Cyananthus microphyllus Edgew.

Cyananthus pedunculatus C. B. Clarke

Cyananthus sericeus Y. S. Lian

Cyananthus sherriffii Cowan

Cyananthus wardii C. Marquand

Cyclocodon axillaris (Oliv.) W. J. de Wilde & Duyfjes

Cyclocodon lancifolius (Roxb.) Kurz

Cyclocodon parviflorus (Wall. ex A. DC.) Hook. f. & Thomson

Himalacodon dicentrifolius (C. B. Clarke) D. Y. Hong & Q. Wang

Homocodon brevipes (Hemsl.) D. Y. Hong

Homocodon pedicellatus D. Y. Hong & L. M. Ma

Pankycodon purpureus (Wall.) D. Y. Hong & X. T. Ma

Peracarpa carnosa (Wall.) Hook. f. & Thomson

Platycodon grandiflorus (Jacq.) A. DC.

Pseudocodon convolvulaceus (Kurz) D. Y. Hong & H. Sun

Pseudocodon convolvulaceus var. **forrestii** (Diels) D. Y. Hong

Pseudocodon graminifolius (H. Lév.) D. Y. Hong

Pseudocodon grey-wilsonii (J. M. H. Shaw) D. Y. Hong

Pseudocodon hirsutus (Hand.-Mazz.) D. Y. Hong

Pseudocodon petiolatus D. Y. Hong & Q. Wang

Pseudocodon retroserratus (Z. T. Wang & G. J. Xu) D. Y. Hong & Q. Wang

Pseudocodon rosulatus (W. W. Sm.) D. Y. Hong

Pseudocodon vinciflorus (Kom.) D. Y. Hong

Pseudocodon vinciflorus subsp. **dianchuanicus** D. Y. Hong & Q. Wang

Wahlenbergia hirsuta (Edgew.) Tuyn

Wahlenbergia marginata (Thunb.) A. DC.

Wahlenbergia peduncularis (Wall. ex A. DC.) Hook. f. & Thomson

274. LOBELIACEAE

Lobelia alsinoides Lam.

Lobelia angulata G. Forst.

Lobelia doniana Skottsb.

Lobelia drungjiangensis D. Y. Hong

Lobelia erectiuscula Hara

Lobelia foliiiformis T. J. Zhang & D. Y. Hong

Lobelia gaoligongshanica D. Y. Hong

Lobelia heyneana Roth ex Roem. & Schult.

Lobelia montana Reinw. ex Blume

Lobelia nubigena J. Anthony

Lobelia pleotricha Diels

Lobelia pyramidalis Wall.

Lobelia rosea Wall.

Lobelia seguinii H. Lév. & Vaniot

Lobelia sessilifolia Lamb.

Lobelia sichuanensis (Y. S. Lian) D. Y. Hong

Lobelia taliensis Diels

Lobelia terminalis C. B. Clarke

Lobelia zeylanica L.

275. MENYANTHACEAE

Menyanthes trifoliata L.

Nymphoides cristata (Roxb.) Kuntze

Nymphoides indica (L.) Kuntze

Nymphoides peltatum (S. G. Gmelin) Kuntze

276. ASTERACEAE

Acanthocephalus benthamianus Regel & Schmalh.

Acanthospermum hispidum DC.

Achillea alpina L.

Achillea filipendulina Lam.

Achillea millefolium subsp. **chitralensis** Hub.-Mor.

Achillea wilsoniana (Heimerl ex Hand.-Mazz.) Heimerl

Acmella calva (DC.) R. K. Jansen

Acmella oleracea (L.) R. K. Jansen

Acmella paniculata (Wall. ex DC.) R. K. Jansen

Acmella uliginosa (Sw.) Cass.

Adenocaulon himalaicum Edgew.

Adenocaulon nepalense M. Bittmann

Adenostemma lavenia (L.) Kuntze

Adenostemma lavenia var. **angustifolium** (Arn.) Uniyal

Adenostemma lavenia var. **aquaticum** (D. Don) Kuntze

Adenostemma lavenia var. **elata** (D. Don) Hochr.

Adenostemma lavenia var. **latifolium** (D. Don) Panigrahi

Adenostemma lavenia var. **parviflorum** (Blume) Hochr.

Ageratina adenophora (Spreng.) R. M. King & H. Rob.

Ageratina ligustrina (DC.) R. M. King & H. Rob.

Ageratina riparia (Regel) R. M. King & H. Rob.

Ageratum conyzoides L.

Ageratum houstonianum Mill.

Ainsliaea angustifolia Hook. f. & Thomson ex C. B. Clarke

Ainsliaea aptera DC.

Ainsliaea bonatii Beauv.

Ainsliaea crassifolia C. C. Chang

Ainsliaea foliosa Hand.-Mazz.

Ainsliaea fragrans Champ. ex Benth.

Ainsliaea fulvipes Jeffrey & W. W. Sm.

Ainsliaea henryi Diels

Ainsliaea lancifolia Franch.

Ainsliaea latifolia (D. Don) Sch. Bip.

Ainsliaea macrocephala (Mattf.) Y. C. Tseng

Ainsliaea mairei H. Lév.

Ainsliaea pertyoides Franch.

Ainsliaea pertyoides subsp. **albotomentosa** (Beauv.) T. G. Gao

Ainsliaea qianiana S. E. Freire

Ainsliaea reflexa Merr.

Ainsliaea smithii Mattf.

Ainsliaea spicata Vaniot

Ainsliaea yunnanensis Franch.

Ajania adenantha (Diels) Y. Ling & C. Shih

Ajania amphisericea (Hand.-Mazz.) C. Shih

Ajania brachyantha C. Shih

Ajania breviloba (Franch. ex Hand.-Mazz.) Y. Ling & C. Shih

Ajania elegantula (W. W. Sm.) C. Shih

Ajania fruticulosa (Ledeb.) Poljakov

Ajania khartensis (Dunn) C. Shih

Ajania latifolia C. Shih

Ajania myriantha (Franch.) Y. Ling ex C. Shih

Ajania nitida C. Shih

Ajania nubigena (Wall. ex DC.) C. Shih

Ajania penicilliformis (C. Shih) Y. S. Chen, comb. et stat. nov. (≡ *Ajaniopsis penicilliformis* C. Shih)

Ajania potaninii (Krasch.) Poljakov

Ajania przewalskii Poljakov

Ajania purpurea C. Shih

Ajania quercifolia (W. W. Sm.) Y. Ling & C. Shih

Ajania ramosa (C. C. Chang) C. Shih

Ajania remotipinna (Hand.-Mazz.) Y. Ling & C. Shih

Ajania salicifolia (Mattf. ex Rehder & Kobuski) Poljakov

Ajania scharnhorstii (Regel & Schmalh.) Tzvelev

Ajania sericea C. Shih

Ajania tenuifolia (Jacq. ex DC.) Tzvelev

Ajania tibetica (Hook. f. & Thomson ex C. B. Clarke) Tzvelev

Ajania tripinnatisecta Y. Ling & C. Shih

Ajania truncata (Hand.-Mazz.) Y. Ling ex C. Shih

Ajania variifolia (C. C. Chang) Tzvelev

Allardia glabra Decne.

Allardia huegelii Sch. Bip.

Allardia lasiocarpa (G. X. Fu) Bremer & Humphries

Allardia nivea Hook. f. & Thomson ex C. B. Clarke

Allardia stoliczkae C. B. Clarke

Allardia tomentosa Decne.

Allardia tridactylites (Kar. & Kir.) Sch. Bip.

Allardia vestita Hook. f. & Thomson ex C. B. Clarke

Ambrosia polystachya DC.

Anaphalis acutifolia Hand.-Mazz.

Anaphalis aureopunctata Lingelsh. & Borza

Anaphalis aureopunctata var. **atrata** (Hand.-Mazz.) Hand.-Mazz.

Anaphalis aureopunctata var. **tomentosa** Hand.-Mazz.

Anaphalis batangensis Y. L. Chen

Anaphalis bicolor (Franch.) Diels

Anaphalis bicolor var. **longifolia** Chang

Anaphalis bicolor var. **subconcolor** Hand.-Mazz.

Anaphalis bicolor var. **undulata** (Hand.-Mazz.) Ling

Anaphalis bulleyana (Jeffrey) Chang

Anaphalis busua (Buch.-Ham. ex D. Don) DC.

Anaphalis cavei Chatterjee

Anaphalis chitralensis Qaiser & Abid

Anaphalis chlamydophylla Diels

Anaphalis chungtienensis Chen

Anaphalis cinerascens Ling & W. Wang

Anaphalis cinerascens var. **congesta** Ling & W. Wang

Anaphalis contorta (D. Don) Hook. f.

Anaphalis contortiformis Hand.-Mazz.

Anaphalis cooperi Grierson & Spring.

Anaphalis corymbifera C. C. Chang

Anaphalis darvasica Boriss

Anaphalis delavayi (Franch.) Diels

Anaphalis deserti J. R. Drumm.

Anaphalis elegans Y. Ling

Anaphalis flavescens Hand.-Mazz.

Anaphalis flavescens var. **lanata** Y. Ling

Anaphalis gracilis Hand.-Mazz.

Anaphalis gracilis var. **aspera** Hand.-Mazz.

Anaphalis gracilis var. **ulophylla** Hand.-Mazz.

Anaphalis hancockii Maxim.

Anaphalis himachalensis Aswal & Goel

Anaphalis hondae Kitam.

Anaphalis hymenolepis Y. Ling ex W. B. Zhu

Anaphalis lactea Maxim.

Anaphalis larium Hand.-Mazz.

Anaphalis latialata Ling & Y. L. Chen

Anaphalis likiangensis (Franch.) Y. Ling

Anaphalis margaritacea (L.) Benth. & Hook. f.

Anaphalis margaritacea var. **cinnamomea** (DC.) Herder ex Maxim.

Anaphalis monocephala DC.

Anaphalis muliensis (Hand.-Mazz.) Hand.-Mazz.

Anaphalis nepalensis (Spreng.) Hand.-Mazz.

Anaphalis nubigena DC.

Anaphalis oxyphylla Y. Ling & C. Shih

Anaphalis pachylaena Y. L. Chen & Y. Ling

Anaphalis pannosa Hand.-Mazz.

Anaphalis patentifolia Rech. f.

Anaphalis plicata Kitam.

Anaphalis porphyrolepis Y. Ling & Y. L. Chen

Anaphalis rhododactyla W. W. Sm.

Anaphalis royleana DC.

Anaphalis sinica var. **alata** (Maxim.) S. X. Zhu & R. J. Bayer

Anaphalis souliei Diels

Anaphalis spodiophylla Y. Ling & Y. L. Chen

Anaphalis staintonii Georgiadou

Anaphalis stenocephala Y. Ling & C. Shih

Anaphalis subumbellata C. B. Clarke

Anaphalis suffruticosa Hand.-Mazz.

Anaphalis surculosa (Hand.-Mazz.) Hand.-Mazz.

Anaphalis szechuanensis Ling & Y. L. Chen

Anaphalis tenuissima C. C. Chang

Anaphalis tibetica Kitam.

Anaphalis triplinervis (Sims) Sims ex C. B. Clarke

Anaphalis virens C. C. Chang

Anaphalis virgata Thomson ex C. B. Clarke

Anaphalis viridis Cumm.

Anaphalis xylorhiza Sch. Bip. ex Hook. f.

Anaphalis yangii Y. L. Chen & Y. L. Lin

Anaphalis yunnanensis (Franch.) Diels

Anisopappus chinensis (L.) Hook. & Arn.

Anthemis altissima L.

Anthemis cotula L.

Archiserratula forrestii (Iljin) L. Martins

Arctium lappa L.

Artemisia abaensis Y. R. Ling & S. Y. Zhao

Artemisia absinthium L.

Artemisia amygdalina Decne.

Artemisia annua L.

Artemisia austroyunnanensis Y. Ling & Y. R. Ling

Artemisia bhutanica Grierson & Spring.

Artemisia biennis Willd.

Artemisia calophylla Pamp.

Artemisia campbellii Hook. f. & Thomson ex C. B. Clarke

Artemisia capillaris Thunb.

Artemisia caruifolia Buch.-Ham. ex Roxb.

Artemisia cashemirica Kaul & Bakshi

Artemisia chingii Pamp.

Artemisia comaiensis Y. Ling & Y. R. Ling

Artemisia conaensis Y. Ling & Y. R. Ling

Artemisia dalai-lamae Krasch.

Artemisia demissa Krasch.

Artemisia desertorum Spreng.

Artemisia desertorum var. **foetida** (Jacq. ex DC.) Y. Ling & Y. R. Ling

Artemisia desertorum var. **tongolensis** Pamp.

Artemisia divaricata (Pamp.) Pamp.

Artemisia dracunculus L.

Artemisia dracunculus var. **pamirica** (C. Winkl.) Y. R. Ling & Humphries

Artemisia dubia var. **subdigitata** (Mattf.) Y. R. Ling

Artemisia dubia Wall. ex Besser

Artemisia duthreuil-de-rhinsi Krasch.

Artemisia elegantissima Pamp.

Artemisia eriopoda Bunge

Artemisia erlangshanensis Y. Ling & Y. R. Ling

Artemisia flaccida Hand.-Mazz.

Artemisia flaccida var. **meiguensis** Y. R. Ling

Artemisia forrestii W. W. Sm.

Artemisia frigida Willd.

Artemisia fulgens Pamp.

Artemisia giraldii Pamp.

Artemisia gmelinii Weber ex Stechmann

Artemisia gongshanensis Y. R. Ling & Humphries

Artemisia gyangzeensis Y. Ling & Y. R. Ling

Artemisia gyitangensis Y. Ling & Y. R. Ling

Artemisia hedinii Ostenfeld

Artemisia imponens Pampanini

Artemisia incisa Pampanini

Artemisia indica var. **elegantissima** (Pampanini) Y. R. Ling & Humphries

Artemisia indica Willd.

Artemisia japonica Thunb.

Artemisia jilongensis Y. R. Ling & Humphries

Artemisia kangmarensis Y. Ling & Y. R. Ling

Artemisia korovinii Poljakov

Artemisia kuschakewiczii C. Winkl.

Artemisia lactiflora Wall. ex DC.

Artemisia lehmanniana Bunge

Artemisia leucophylla C. B. Clarke

Artemisia leucotricha Ladygin

Artemisia lingyeouruennii L. Shultz & Boufford

Artemisia macrocephala Jacq. ex Bess.

Artemisia mairei H. Lév.

Artemisia mattfeldii Pamp.

Artemisia mattfeldii var. **etomentosa** Hand.-Mazz.

Artemisia minor Jacq. ex Bess.

Artemisia moorcroftiana Wall. ex DC.

Artemisia myriantha Wall. ex Besser

Artemisia myriantha var. **pleiocephala** (Pamp.) Y. R. Ling

Artemisia nanschanica Krasch.

Artemisia neosinensis B. H. Jiao & T. G. Gao

Artemisia nilagirica (C. B. Clarke) Pamp.

Artemisia nortonii Pamp.

Artemisia nujianensis (Y. Ling & Y. R. Ling) Y. R. Ling

Artemisia occidentalisichuanensis Y. R. Ling & S. Y. Zhao

Artemisia occidentalisinensis Y. R. Ling

Artemisia occidentalisinensis var. **denticulata** Y. R. Ling

Artemisia orientalihengduangensis Y. Ling & Y. R. Ling

Artemisia orientalixizangensis Y. R. Ling & Humphries

Artemisia parviflora Buch.-Ham. ex D. Don

Artemisia pengchuoensis Y. R. Ling & S. Y. Zhao

Artemisia persica Boiss.

Artemisia persica var. **subspinescens** (Boiss.) Boiss.

Artemisia pewzowii C. Winkl.

Artemisia phaeolepis Krasch.

Artemisia phyllobotrys (Hand.-Mazz.) Y. Ling & Y. R. Ling

Artemisia prattii (Pamp.) Y. Ling & Y. R. Ling

Artemisia pubescens Ledeb.

Artemisia rhodantha Rupr.

Artemisia robusta (Pamp.) Y. Ling & Y. R. Ling

Artemisia roxburghiana Bess.

Artemisia roxburghiana var. **purpurascens** (Jacq. ex Bess.) Hook. f.

Artemisia rupestris L.

Artemisia rutifolia Steph. ex Spreng.

Artemisia salsoloides Willd.

Artemisia santolinifolia Turcz. ex Krasch.

Artemisia scoparia Waldstein & Kitaibel

Artemisia sericea (Besser) Weber ex Stechmann

Artemisia sichuanensis Y. Ling & Y. R. Ling

Artemisia sichuanensis var. **tomentosa** Y. Ling & Y. R. Ling

Artemisia sieversiana Ehrhart ex Willd.

Artemisia smithii Mattf.

Artemisia speciosa (Pamp.) Y. Ling & Y. R. Ling

Artemisia stechmanniana Bess.

Artemisia stracheyi Hook. f. & Thomson ex C. B. Clarke

Artemisia stricta Edgew.

Artemisia stricta var. **diffusa** (Pamp.) Y. R. Ling & M. G. Gilbert

Artemisia succulentoides Y. Ling & Y. R. Ling

Artemisia tafelii Mattf.

Artemisia tainingensis Hand.-Mazz.

Artemisia tainingensis var. **nitida** (Pamp.) Y. R. Ling

Artemisia tangutica Pamp.

Artemisia tangutica var. **tomentosa** Hand.-Mazz.

Artemisia tecti-mundi Podlech

Artemisia thellungiana Pamp.

Artemisia tournefortiana Rchb.

Artemisia tridactyla Hand.-Mazz.

Artemisia tridactyla var. **minima** Y. R. Ling

Artemisia vachanica Poljakov

Artemisia velutina Pamp.

Artemisia vexans Pamp.

Artemisia viscida (Mattf.) Pamp.

Artemisia viscidissima Y. Ling & Y. R. Ling

Artemisia vulgaris L.

Artemisia vulgaris var. **xizangensis** Y. Ling & Y. R. Ling

Artemisia waltonii J. R. Drumm. ex Pamp.

Artemisia waltonii var. **yushuensis** Y. R. Ling

Artemisia wellbyi Hemsl. & H. Pearson

Artemisia xigazeensis Y. R. Ling & M. G. Gilbert

Artemisia yadongensis Ling & Y. R. Ling

Artemisia younghusbandii J. R. Drumm. ex Pamp.

Artemisia youngii Y. R. Ling

Artemisia yunnanensis Jeffrey ex Diels

Artemisia zayuensis Y. Ling & Y. R. Ling

Artemisia zayuensis var. **pienmaensis** Y. Ling & Y. R. Ling

Artemisia zhongdianensis Y. R. Ling

Askellia corniculata (Regel & Schmalh.) W. A. Weber

Askellia flexuosa (Ledeb.) W. A. Weber

Askellia karelinii (Popov & Schischk. ex Czerep.) W. A. Weber

Askellia lactea (Lipsch.) W. A. Weber

Askellia pygmaea (Ledeb.) Sennikov

Aster albescens (DC.) Wall. ex Hand.-Mazz.

Aster albescens var. **dicolor** Y. Ling

Aster albescens var. **glabratus** (Diels) Boufford & Y. S. Chen

Aster albescens var. **glandulosus** Hand.-Mazz.

Aster albescens var. **gracilior** (Hand.-Mazz.) Hand.-Mazz.

Aster albescens var. **limprichtii** (Diels) Hand.-Mazz.

Aster albescens var. **megaphyllus** Y. Ling

Aster albescens var. **pilosus** Hand.-Mazz.

Aster albescens var. **rugosus** Y. Ling

Aster albescens var. **salignus** (Franch.) Hand.-Mazz.

Aster altaicus Willd.

Aster altaicus var. **canescens** (Nees) Serg.

Aster altaicus var. **hirsutus** Hand.-Mazz.

Aster argyropholis Hand.-Mazz.

Aster argyropholis var. **niveus** Y. Ling
Aster argyropholis var. **paradoxus** Y. Ling
Aster asteroides (DC.) Kuntze
Aster auriculatus Franch.
Aster barbellatus Grierson
Aster batangensis Bureau & Franch.
Aster batangensis var. **staticifolius** (Franch.) Y. Ling
Aster bietii Franch.
Aster bipinnatisectus Ludlow ex Grierson
Aster boweri Hemsl.
Aster brachytrichus Franch.
Aster brevis Hand.-Mazz.
Aster bulleyanus Jeffrey ex Diels
Aster crenatifolius Hand.-Mazz.
Aster diplostephioides (DC.) Benth. ex C. B. Clarke
Aster dolichopodus Y. Ling
Aster eligulatus (Y. Ling ex Y. L. Chen, S. Yun Liang & K. Y. Pan) Brouillet, Semple & Y. L. Chen
Aster falcifolius Hand.-Mazz.
Aster falconeri (C. B. Clarke) Hutch.
Aster falconeri subsp. **nepalensis** Grierson
Aster farreri W. W. Sm. & Jeffrey
Aster flaccidus Bunge
Aster fulgidulus Grierson
Aster fuscescens Bureau & Franch.
Aster fuscescens var. **oblongifolius** Grierson
Aster fuscescens var. **scaberoides** C. C. Chang
Aster gouldii C. E. C. Fisch.
Aster handelii Onno
Aster heliopsis Grierson
Aster hersileoides C. K. Schneid.
Aster heterolepis Hand.-Mazz.
Aster himalaicus C. B. Clarke
Aster hispidus Thunb.
Aster hololachnus Y. Ling ex Y. L. Chen, S. Yun Liang & K. Y. Pan
Aster homochlamydeus Hand.-Mazz.
Aster hypoleucus Hand.-Mazz.
Aster indamellus Grierson
Aster indicus L.
Aster ionoglossus Y. Ling ex Y. L. Chen, S. Yun Liang & K. Y. Pan
Aster jeffreyanus Diels
Aster laka C. B. Clarke
Aster latibracteatus Franch.
Aster lavandulifolius Hand.-Mazz.
Aster likiangensis Franch.
Aster lingulatus Franch.
Aster lixianensis (J. Q. Fu) Brouillet, Semple & Y. L. Chen
Aster longipetiolatus C. C. Chang
Aster megalanthus Y. Ling
Aster molliusculus (Lindl. ex DC.) C. B. Clarke
Aster moupinensis (Franch.) Hand.-Mazz.
Aster neoelegans Grierson
Aster neolanuginosus Brouillet, Semple & Y. L. Chen
Aster oliganthus W. P. Li & Zhi Li
Aster oreophilus Franch.
Aster panduratus Nees ex Walpers
Aster peduncularis subsp. **nepalensis** Grierson
Aster peduncularis Wall. ex Nees
Aster pekinensis (Hance) F. H. Chen
Aster piccolii Hook. f.
Aster platylepis Y. L. Chen
Aster poliothamnus Diels
Aster polius C. K. Schneid.
Aster prainii (J. R. Drumm.) Y. L. Chen
Aster pseudosimplex Brouillet, Semple & Y. L. Chen
Aster pycnophyllus Franch. ex W. W. Sm.
Aster retusus Ludlow
Aster rockianus Hand.-Mazz.
Aster salwinensis Onno

Aster semiprostratus (Grierson) H. Ikeda

Aster senecioides Franch.

Aster setchuenensis Franch.

Aster sikkimensis Hook. f.

Aster sikuensis W. W. Sm. & Farrer

Aster smithianus Hand.-Mazz.

Aster souliei Franch.

Aster souliei var. **limitaneus** (W. W. Sm. & Farrer) Hand.-Mazz.

Aster stracheyi Hook. f.

Aster taliangshanensis Y. Ling

Aster techinensis Y. Ling

Aster thomsonii C. B. Clarke

Aster tientschwanensis Hand.-Mazz.

Aster tongolensis Franch.

Aster tricephalus C. B. Clarke

Aster trichoneurus Y. Ling

Aster trinervius Roxb. ex D. Don

Aster trinervius subsp. **ageratoides** (Turcz.) Grierson

Aster trinervius var. **wattii** (C. B. Clarke) Geierson

Aster tsarungensis (Grierson) Y. Ling

Aster veitchianus Hutch. & J. R. Drumm. ex Y. Ling

Aster verticillatus (Reinwardt) Brouillet, Semple & Y. L. Chen

Aster yunnanensis Franch.

Aster yunnanensis var. **angustior** Hand.-Mazz.

Aster yunnanensis var. **labrangensis** (Hand.-Mazz.) Y. Ling

Aucklandia costus Falc.

Baccharoides anthelmintica (L.) Moench

Bellis perennis L.

Bidens bipinnata L.

Bidens biternata (Lour.) Merr. & Sherff

Bidens cernua L.

Bidens minima Huds.

Bidens parviflora Willd.

Bidens pilosa L.

Bidens tetraspinosa Majeed Kak & Javeid

Bidens tripartita L.

Blainvillea acmella (L.) Philipson

Blumea adenophora Franch.

Blumea aromatica DC.

Blumea axillaris (Lam.) DC.

Blumea balsamifera (L.) DC.

Blumea bifoliata (L.) DC.

Blumea densiflora DC.

Blumea fistulosa (Roxb.) Kurz

Blumea flava DC.

Blumea hamiltonii DC.

Blumea hieraciifolia (D. Don) DC.

Blumea hirsuta (Less.) M. R. Almeida

Blumea lacera (Burm. f.) DC.

Blumea laciniata (Roxb.) DC.

Blumea lanceolaria (Roxb.) Druce

Blumea macrostachya DC.

Blumea martiniana Vaniot

Blumea megacephala (Randeria) C. C. Chang & Y. Q. Tseng

Blumea napifolia DC.

Blumea obovata DC.

Blumea oxyodonta DC.

Blumea paniculata (Willd.) M. R. Almeida

Blumea procera DC.

Blumea pubigera (L.) Merr.

Blumea repanda (Roxb.) Hand.-Mazz.

Blumea riparia DC.

Blumea sessiliflora Decne.

Blumea sikkimensis Hook. f.

Blumea sinuata (Lour.) Merr.

Blumea sonbhadrensis S. Narain, Lata & Juhi Singh

Blumea veronicifolia Franch.

Blumea virens DC.

Bolocephalus saussureoides Hand.-Mazz.

Caesulia axillaris Roxb.

Calendula officinalis L.

Callistephus chinensis (L.) Nees

Carduus crispus L.

Carduus edelbergii Rech. f.

Carduus kumaunensis (Arènes) Kazmi

Carduus pycnocephalus L.

Carduus tenuiflorus W. M. Curtis

Carpesium abrotanoides L.

Carpesium cernuum L.

Carpesium faberi C. Winkl.

Carpesium humile C. Winkl.

Carpesium lipskyi C. Winkl.

Carpesium longifolium F. H. Chen & C. M. Hu

Carpesium macrocephalum Franch. & Savat.

Carpesium minus Hemsl.

Carpesium nepalense Less.

Carpesium nepalense var. **lanatum** (Hook. f. & Thoms. ex C. B. Clarke) Kitam.

Carpesium pedunculosum Wall. ex DC.

Carpesium scapiforme F. H. Chen & C. M. Hu

Carpesium tracheliifolium Less.

Carpesium velutinum C. Winkl.

Carthamus lanatus subsp. **turkestanicus** (Popov) Hanelt

Carthamus tinctorius L.

Catamixis baccharoides Thomson

Cavea tanguensis (J. R. Drumm.) W. W. Sm. & J. Small

Centaurea benedicta (L.) L.

Centaurea calcitrapa L.

Centaurea cyanus L.

Centaurea depressa M. Bieb.

Centaurea iberica Trevir. ex Spreng.

Centaurea lasiopoda Popov & Kult.

Centaurea pulchella Ledeb.

Centaurea ruthenica Lam.

Centipeda minima (L.) A. Braun & Ascherson

Chondrilla maracandica Bunge

Chondrilla phaeocephala Rupr.

Chondrilla setulosa C. B. Clarke ex Hook. f.

Chondrilla yosii Kitam.

Chromolaena odorata (L.) R. M. King & H. Rob.

Chrysanthemum glabriusculum (W. W. Sm.) Hand.-Mazz.

Chrysanthemum hypargyreum Diels

Chrysanthemum inidicum L.

Chrysanthemum lavandulifolium (Fisch. ex Trautv.) Makino

Chrysanthemum lavandulifolium var. **discoideum** Hand.-Mazz.

Cicerbita benthamii (C. B. Cl.) Roohi Bano & Qaiser

Cicerbita crambifolia (Bunge) Beauv.

Cicerbita ladyginii (Tzvelev) N. Kilian

Cicerbita neglecta (Tzvelev) N. Kilian

Cicerbita nepalensis Kitam.

Cicerbita rapunculoides (DC.) Beauverd

Cicerbita roborowskii (Maxim.) Beauverd

Cicerbita zhenduoi (S. W. Liu & T. N. Ho) N. Kilian

Cichorium intybus L.

Cirsium acaule (L.) Scop.

Cirsium argyracanthum DC.

Cirsium arvense (L.) Scop.

Cirsium arvense var. **alpestre** Nägeli

Cirsium botryodes Petrak

Cirsium chlorolepis Petrak ex Hand.-Mazz.

Cirsium chrysolepis C. Shih

Cirsium eriophoroides (Hook. f.) Petr.

Cirsium falconeri (Hook. f.) Petr.

Cirsium fangii Petrak

Cirsium fargesii (Franch.) Diels

Cirsium flavisquamatum Kitam.

Cirsium fuscotrichum C. C. Chang

Cirsium glabrifolium (C. Winkl.) Petrak

Cirsium griffithii Boiss.

Cirsium handelii Petrak

Cirsium henryi (Franch.) Diels

Cirsium japonicum DC.

Cirsium leducei (Franch.) H. Lév.

Cirsium leo Nakai

Cirsium lidjiangense Petrak ex Hand.-Mazz.

Cirsium lineare (Thunb.) Sch. Bip.

Cirsium lipskyi Petrak

Cirsium muliense C. Shih

Cirsium nishiokae Kitam.

Cirsium periacanthaceum C. Shih

Cirsium phulchokiense Kitam.

Cirsium rhizocephalum C. A. Mey.

Cirsium segetum Bunge

Cirsium shansiense Petr.

Cirsium souliei (Franch.) Mattf. ex Rehd. & Kobuski

Cirsium stenobracteatum Gilli

Cirsium subulariforme C. Shih

Cirsium swaticum Petrak

Cirsium verutum (D. Don) Spreng.

Cirsium vulgare (Savi) Ten.

Cirsium vulgare var. **hypoleucum** (DC.) Petrak

Cirsium wallichii DC.

Cissampelopsis buimalia (Buch.-Ham. ex D. Don) C. Jeffrey & Y. L. Chen

Cissampelopsis corifolia C. Jeffrey & Y. L. Chen

Cissampelopsis spelaeicola (Vaniot) C. Jeffrey & Y. L. Chen

Cissampelopsis volubilis (Blume) Miquel

Cosmos bipinnatus Cav.

Cosmos sulphureus Cav.

Cotula anthemoides L.

Cotula hemisphaerica (Roxb.) Wall. ex Benth.

Cousinia auriculata Boiss.

Cousinia brandisii C. B. Clarke

Cousinia buphtalmoides Regel

Cousinia chionophila Rech. f. & Köie

Cousinia chitralensis Rech. f.

Cousinia elegans Aitch. & Hemsl.

Cousinia falconeri Hook. f.

Cousinia kerstanii Bornm.

Cousinia khashensis Rech. f.

Cousinia mattfeldii Bornm.

Cousinia microcarpa Boiss.

Cousinia minuta Boiss.

Cousinia multiloba DC.

Cousinia prolifera Jaub. & Spach

Cousinia racemosa Boiss.

Cousinia scheibeana Bornm.

Cousinia schugnanica Juz.

Cousinia stocksii C. Winkl.

Cousinia takharensis Rech. f.

Cousinia tenella Fisch. & C. A. Mey.

Cousinia thomsonii C. B. Clarke

Cousinia trollii Bornm.

Crassocephalum crepidioides (Benth.) S. Moore

Cremanthodium acernuum R. D. Good

Cremanthodium angustifolium W. W. Sm.

Cremanthodium arnicoides (DC. ex Royle) R. D. Good

Cremanthodium atrocapitatum R. D. Good

Cremanthodium atrocapitatum var. **brachytrichum** (Ling & Liu) L. Wang, C. Ren & Q. E. Yang

Cremanthodium bhutanicum Ludlow

Cremanthodium bomiense L. Wang, C. Ren & Q. E. Yang

Cremanthodium brachychaetum Chang

Cremanthodium brunneopilosum S. W. Liu

Cremanthodium brunneopilosum var. **roseum** (Hand.-Mazz.) L. Wang, C. Ren & Q. E. Yang

Cremanthodium bulbilliferum W. W. Sm.

Cremanthodium bupleurifolium W. W. Sm.

Cremanthodium calcicola W. W. Sm.

Cremanthodium campanulatum Diels

Cremanthodium campanulatum var. **flavidum** S. W. Liu & T. N. Ho

Cremanthodium chungtienense Ling & S. W. Liu

Cremanthodium citriflorum R. D. Good

Cremanthodium conaense S. W. Liu

Cremanthodium cordatum S. W. Liu

Cremanthodium coriaceum S. W. Liu

Cremanthodium cucullatum Ling & S. W. Liu

Cremanthodium cyclaminathum Hand.-Mazz.

Cremanthodium daochengense Ling & S. W. Liu

Cremanthodium decaisnei C. B. Clarke

Cremanthodium delavayi (Franch.) Diels ex H. Lév.

Cremanthodium discoideum Maxim.

Cremanthodium dissectum Griers.

Cremanthodium ellisii (Hook. f.) Kitam.

Cremanthodium farreri W. W. Sm.

Cremanthodium forrestii Jeffrey

Cremanthodium glandulipilosum Y. L. Chen ex S. W. Liu

Cremanthodium glaucum Hand.-Mazz.

Cremanthodium gracillimum W. W. Sm.

Cremanthodium gypsophilium R. D. Good

Cremanthodium helianthus (Franch.) W. W. Sm.

Cremanthodium heterocephalum Y. L. Chen

Cremanthodium hongshanense L. Wang, C. Ren & Q. E. Yang

Cremanthodium humile Maxim.

Cremanthodium laciniatum Y. Ling & Y. L. Chen ex S. W. Liu

Cremanthodium latilobum Y. S. Chen

Cremanthodium liangshanicum L. Wang, C. Ren & Q. E. Yang

Cremanthodium lineare Maxim.

Cremanthodium lineare var. **eligulatum** Ling & S. W. Liu

Cremanthodium lineare var. **roseum** Hand.-Mazz.

Cremanthodium maoxianense L. Wang, C. Ren & Q. E. Yang

Cremanthodium medogense Y. S. Chen

Cremanthodium microphyllum S. W. Liu

Cremanthodium nanum (Decne.) W. W. Sm.

Cremanthodium nepalense Kitam.

Cremanthodium nervosum S. W. Liu

Cremanthodium nobile (Franch.) Diels ex H. Lév.

Cremanthodium nobile var. **yanyuanense** L. Wang, C. Ren & Q. E. Yang

Cremanthodium oblongatum C. B. Clarke

Cremanthodium obovatum Ling & S. W. Liu

Cremanthodium palmatum Benth.

Cremanthodium petiolatum S. W. Liu

Cremanthodium phyllodineum S. W. Liu

Cremanthodium pilosum S. W. Liu

Cremanthodium pinnatifidum Benth.

Cremanthodium pinnatisectum (Ludlow) Y. L. Chen & S. W. Liu

Cremanthodium pirolaefolium (H. Lév.) T. L. Min

Cremanthodium plantagineum Maxim.

Cremanthodium potaninii C. Winkl.

Cremanthodium prattii (Hemsl.) R. D. Good

Cremanthodium principis (Franch.) R. D. Good

Cremanthodium pseudo-oblongatum R. D. Good

Cremanthodium pulchrum R. D. Good

Cremanthodium purpureifolium Kitam.

Cremanthodium reniforme (DC.) Benth.

Cremanthodium rhodocephalum Diels

Cremanthodium sino-oblongatum R. D. Good

Cremanthodium smithianum (Hand.-Mazz.) Hand.-Mazz.

Cremanthodium stenactinium Diels

Cremanthodium stenoglossum Ling & S. W. Liu

Cremanthodium suave W. W. Sm.

Cremanthodium thomsonii C. B. Clarke

Cremanthodium trilobum S. W. Liu

Cremanthodium variifolium R. D. Good

Cremanthodium wardii W. W. Sm.

Cremanthodium yadongense S. W. Liu

Crepidiastrum diversifolium (Ledeb. ex Spreng.) J. W. Zhang & N. Kilian

Crepidiastrum sonchifolium (Maxim.) Pak & Kawano

Crepidiastrum tenuifolium (Willd.) Sennikov

Crepis aitchisonii Boiss.

Crepis bodinieri H. Lév.

Crepis chloroclada Collett & Hemsl.

Crepis dachhigamensis G. Singh

Crepis elongata Babc.

Crepis himalaica Kitam.

Crepis koelzii Babc.

Crepis kotschyana (Boiss.) Boiss.

Crepis lignea (Vaniot) Babc.

Crepis multicaulis Ledeb.

Crepis napifera (Franch.) Babc.

Crepis oreades Schrenk ex Fisch. & C. A. Mey.

Crepis phoenix Dunn

Crepis thomsonii Babc.

Crepis trichocephala (Krasch.) V. V. Nikitin

Crupina vulgaris Pers. ex Cass.

Cyathocline purpurea (Buch.-Ham. ex D. Don) Kuntze

Cymbolaena griffithii (A. Gray) Wagenitz

Dahlia imperialis Roezl ex Ortgies

Dahlia pinnata Cav.

Dichrocephala benthamii C. B. Clarke

Dichrocephala chrysanthemifolia (Blume) DC.

Dichrocephala integrifolia (L. f.) Kuntze

Distephanus forrestii (J. Anthony) H. Robinson & B. Kahn

Dolomiaea baltalensis Dar & Naqshi

Dolomiaea berardioidea (Franch.) C. Shih

Dolomiaea calophylla Y. Ling

Dolomiaea crispoundulata (C. C. Chang) Y. Ling

Dolomiaea edulis (Franch.) C. Shih

Dolomiaea forrestii (Diels) C. Shih

Dolomiaea georgei (Anth.) C. Shih

Dolomiaea platylepis (Hand.-Mazz.) C. Shih

Dolomiaea salwinensis (Hand.-Mazz.) C. Shih

Dolomiaea scabrida (C. Shih & S. Y. Jin) C. Shih

Dolomiaea souliei (Franch.) C. Shih

Dolomiaea taraxacifolia (J. Anthony) Y. S. Chen & Raab-Straube

Dolomiaea wardii (Hand.-Mazz.) Y. Ling

Doronicum briquetii Cavill.

Doronicum calotum (Diels) Q. Yuan

Doronicum falconeri C. B. Clarke ex Hook. f.

Doronicum kamaonense (DC.) Alv. Fern.

Doronicum pardalianches L.

Doronicum stenoglossum Maxim.

Dubyaea amoena (Hand.-Mazz.) Stebbins

Dubyaea atropurpurea (Franch.) Stebbins

Dubyaea chimiliensis (W. W. Sm.) Stebbins

Dubyaea cymiformis C. Shih

Dubyaea emeiensis C. Shih

Dubyaea forrestii Mamgain & R. R. Rao

Dubyaea gombalana (Hand.-Mazz.) Stebbins

Dubyaea hispida DC.

Dubyaea oligocephala (Sch. Bip.) Stebbins

Dubyaea pteropoda Shih

Dubyaea rubra Stebbins

Dubyaea stebbinsii Ludlow

Dubyaea tsarongensis (W. W. Sm.) Stebbins

Duhaldea cappa (Buch.-Ham. ex D. Don) J. Pruski & Anderb.

Duhaldea cuspidata (DC.) Anderb.

Duhaldea eupatorioides (Wall. ex DC.) Steetz

Duhaldea forrestii (J. Anthony) Anderb.

Duhaldea griffithii (C. B. Clarke) Anderb.

Duhaldea latifolia (DC.) R. Dawar & Qaiser

Duhaldea nervosa (Wall. ex DC.) A. Anderberg

Duhaldea pterocaula (Franch.) Anderb.

Duhaldea rubricaulis (Wall. ex DC.) A. Anderberg

Duhaldea simonsii (C. B. Clarke) Anderb.

Echinops chloroleucus Rech. f.

Echinops chlorophyllus Rech. f.

Echinops cornigerus DC.

Echinops echinatus Roxb.

Echinops gedrosiacus Bornm.

Echinops leucographus Bunge

Echinops nanus Bunge

Echinops niveus Wall. ex Royle

Echinops pathanorum Rech. f.

Echinops prionolepis Bornm. & Mattf.

Echinops wakhanicus Rech. f.

Eclipta prostrata (L.) L.

Elephantopus mollis Kunth

Elephantopus scaber L.

Emilia prenanthoides DC.

Enydra fluctuans Lour.

Epilasia hemilasia (Bunge) C. B. Clarke

Erechtites valerianaefolia (Link ex Spreng.) DC.

Erigeron acris L.

Erigeron acris subsp. **phaeocephalus** Rech. f.

Erigeron acris subsp. **politus** (Fries) H. Lindberg

Erigeron angutissimum Lindl. ex DC.

Erigeron annuus (L.) Pers.

Erigeron bellidioides (Buch.-Ham. ex D. Don) Benth. ex C. B. Clarke

Erigeron bonariensis L.

Erigeron breviscapus (Vaniot) Hand.-Mazz.

Erigeron canadensis L.

Erigeron cyanactis Rech. f.

Erigeron himalajensis Vierh.

Erigeron karvinskianus DC.

Erigeron kiukiangensis Y. Ling & Y. L. Chen

Erigeron kumaunensis (Vierh.) Wendelbo

Erigeron kunshanensis Y. Ling & Y. L. Chen

Erigeron lanuginosus Y. L. Chen

Erigeron latifolius Hao Zhang & Zhi F. Zhang

Erigeron leptocladus Rech. f.

Erigeron leucoglossus Y. Ling & Y. L. Chen

Erigeron lonchophyllus Hook.

Erigeron monticolus DC.

Erigeron multifolius Hand.-Mazz.

Erigeron multiradiatus (Lindl. ex DC.) Benth. ex C. B. Clarke

Erigeron patentisquama Jeffrey ex Diels

Erigeron petroiketes Rech. f.

Erigeron porphyrolepis Y. Ling & Y. L. Chen

Erigeron pseuderiocephalus M. Pop.

Erigeron pseudotenuicaulis Brouillet & Y. L. Chen

Erigeron purpurascens Y. Ling & Y. L. Chen

Erigeron roylei DC.

Erigeron semibarbatus DC.

Erigeron sublyratus DC.

Erigeron sumatrensis Retz.

Erigeron uniflorus L.

Eschenbachia aegyptiaca (L.) Brouillet

Eschenbachia blinii (H. Lév.) Brouillet

Eschenbachia japonica (Thunb.) J. Koster

Eschenbachia leucantha (D. Don) Brouillet

Eschenbachia muliensis (Y. L. Chen) Brouillet

Eschenbachia stricta var. **pinnatifida** Kitam.

Eschenbachia stricta Willd.

Ethulia conyzoides L. f.

Eupatorium capillifolium (Lam.) Small ex Porter & Britton

Eupatorium chinense L.

Eupatorium coelestinum L.

Eupatorium heterophyllum DC.

Eupatorium japonicum Thunb.

Eupatorium lindleyanum DC.

Eupatorium longicaule DC.

Eupatorium nodiflorum Wall. ex DC.

Eupatorium repandum Willd.

Faberia faberi (Hemsl.) N. Kilian, Z. H. Wang & J. W. Zhang

Faberia glaucescens (Stebbins) Ying Liu, Y. S. Chen & Q. E. Yang

Faberia lancifolia J. Anthony

Faberia pinnatifida Ying Liu, Yousheng Chen & Boufford

Faberia sinensis Hemsl.

Faberia thibetica (Franch.) Beauv.

Filago arvensis L.

Filago griffithii (A. Gray) Andrés-Sánchez & Galbany

Filago hurdwarica (Wall. ex DC.) Wagenitz

Filago paradoxa Wagenitz

Filago pyramidata L.

Flaveria bidentis (L.) Kunzte

Formania mekongensis W. W. Sm. & J. Small

Frolovia gilesii (Hemsl.) B. A. Sharipova

Gaillardia pulchella Foug.

Galinsoga parviflora Cav.

Galinsoga quadriradiata Ruiz & Pavon

Gamochaeta coarctata (Willd.) Kerguélen

Gamochaeta pensylvanica (Willd.) Cabrera

Gamochaeta purpurea (L.) Cabrera

Garhadiolus minutissimus (Bunge) Kitam.

Gerbera delavayi Franch.

Gerbera gossypina (Royle) Beauv.

Gerbera maxima (D. Don) Beauv.

Gerbera nivea (DC.) Sch. Bip.

Gerbera raphanifolia Franch.

Gerbera rupicola T. G. Gao & D. J. N. Hind

Glebionis coronaria (L.) Cass. ex Spach

Glossocardia bidens (Retz.) Veldkamp

Gnaphalium polycaulon Pers.

Gnaphalium stewartii C. B. Clarke ex Hook. f.

Gnaphalium thomsonii Hook. f.

Gnomophalium pulvinatum (Delile) Greuter

Grangea maderaspatana (L.) Poiret

Guizotia abyssinica (L. f.) Cass.

Gynura bicolor (Roxb. ex Willd.) DC.

Gynura cusimbua (D. Don) S. Moore

Gynura japonica (Thunb.) Juel

Gynura nepalensis DC.

Gynura procumbens (Lour.) Merril

Gynura pseudochina (L.) DC.

Handelia trichophylla (Schrenk ex Fisch. & C. A. Mey.) Heimerl

Helianthus annuus L.

Helianthus tuberosus L.

Hemisteptia lyrata (Bunge) Fisch. & C. A. Mey.

Heteracia szovitzii Fisch. & C. A. Mey.

Hieracium asperrimum D. Don

Hieracium umbellatum L.

Hieracium virosum Pall.

Himalaiella abnormis (Lipsch.) Raab-Straube

Himalaiella afghana (Lipsch.) Raab-Straube

Himalaiella albescens (DC.) Raab-Straube

Himalaiella auriculata (DC.) Raab-Straube

Himalaiella chenopodiifolia (Klatt) Raab-Straube

Himalaiella chitralica (Duthie) Raab-Straube

Himalaiella deltoidea (DC.) Raab-Straube

Himalaiella diffusa (Lipsch.) Y. S. Chen, comb. et stat. nov. (≡ *Saussurea diffusa* Lipsch.)

Himalaiella foliosa (Edgew.) Raab-Straube

Himalaiella heteromalla (D. Don) Raab-Straube

Himalaiella nivea (DC.) Raab-Straube

Himalaiella yakla (C. B. Clarke) Fujikawa & H. Ohba

Hippolytia delavayi (Franch. ex W. W. Sm.) C. Shih

Hippolytia desmantha C. Shih

Hippolytia dolichophylla (Kitam.) K. Bremer & Humphries

Hippolytia glomerata C. Shih

Hippolytia gossypina (C. B. Clarke) C. Shih

Hippolytia kennedyi (Dunn) Y. Ling

Hippolytia senecionis (Jacq. ex Bess.) Poljakov ex Tzvelev

Hippolytia syncalathiformis C. Shih

Hippolytia tomentosa (DC.) Tzvelev

Hippolytia yunnanensis (Jeffrey) C. Shih

Hypochaeris glabra L.

Ifloga spicata (Forssk.) Sch. Bip.

Indocypraea montana (Bl.) Orchard

Inula acuminata DC.

Inula brittanica L.

Inula cutchica C. B. Clarke

Inula griffithii C. B. Clarke

Inula grombczewskii C. Winkl.

Inula helianthus-aquatica C. Y. Wu ex Y. Ling

Inula hookeri C. B. Clarke

Inula japonica Thunb.

Inula kalapani C. B. Clarke

Inula macrosperma Hook. f.

Inula obtusifolia A. Kern.

Inula orientalis Lam.

Inula racemosa Hook. f.

Inula rhizocephala Schrenk

Inula royleana DC.

Inula rubtzovii Gorschk.

Inula salsoloides (Turcz.) Ostenfeld

Inula sericophylla Franch.

Inula serratuloides (Gilli) Grierson

Inula stewartii Abid & Qaiser

Inula verrucosa Klatt

Ixeridium aculeolatum C. Shih

Ixeridium beauverdianum (H. Lév.) L. S. Spring.

Ixeridium gracile (DC.) Pak & S. Kawano

Ixeridium sagittarioides (C. B. Clarke) Pak & S. Kawano

Ixeridium yunnanense C. Shih

Ixeris chinensis (Thunb. ex Thunb.) Nakai

Ixeris polycephala Cass.

Jacobaea nudicaulis (Buch.-Ham. ex Don) B. Nord.

Jurinea baldschuanica C. Winkl.

Jurinea rosulata Klatt

Jurinea tricholepis DC.

Klasea centauroides subsp. **strangulata** (Iljin) L. Martins

Klasea pallida (DC.) Holub

Klasea procumbens (Regel) Holub

Kleinia grandiflora (Wall. ex DC.) N. Rani

Koelpinia linearis Pallas

Koelpinia tenuissima Pavl. & Lipsch.

Lactuca chitralensis (Tuisl) Ghafoor, Qaiser & Roohi Bano

Lactuca dissecta D. Don

Lactuca dolichophylla Kitam.

Lactuca formosana Maxim.

Lactuca indica L.

Lactuca orientalis (Boiss.) Boiss.

Lactuca quercina L.

Lactuca raddeana Maxim.

Lactuca sativa L.

Lactuca serriola L.

Lactuca tatarica (L.) C. A. Mey.

Lactuca undulata Ledeb.

Lagascea mollis Cav.

Lagenophora stipitata (Labill.) Druce

Laggera alata (D. Don) Sch. Bip. ex Oliver

Laggera crispata (Vahl) Hepper & J. R. I. Wood

Lapsana communis L.

Launaea acaulis (Hook. f.) Babc. ex Kerr

Launaea aspleniifolia (Willd.) Hook. f.

Launaea capitata (Spreng.) Dandy

Launaea cassiniana (Jaub. & Spach.) Burkill

Launaea microcephala Hook. f.

Launaea oligocephala (Bornm.) Bornm.

Launaea polyclada Burkill

Launaea procumbens (Roxb.) Ramayya & Rajagopal

Launaea resedifolia (L.) Kuntze

Launaea sarmentosa (Willd.) Kuntze

Launaea secunda Hook. f.

Leibnitzia anandria (L.) Turcz.

Leibnitzia nepalensis (Kunze) Kitam.

Leibnitzia pusilla (DC.) S. Gould. ex Kitam. & S. Gould.

Leibnitzia ruficoma (Franch.) Kitam.

Leontopodium andersonii C. B. Clarke

Leontopodium artemisiifolium (H. Lév.) Beauv.

Leontopodium aurantianum Hand.-Mazz.

Leontopodium brachyactis Gand.

Leontopodium caespitosum Diels

Leontopodium calocephalum (Franch.) Beauv.

Leontopodium dedekensii (Bureau & Franch.) Beauv.

Leontopodium delavayanum Hand.-Mazz.

Leontopodium forrestianum Hand.-Mazz.

Leontopodium franchetii Beauv.

Leontopodium haastioides Hand.-Mazz.

Leontopodium haplophylloides Hand.-Mazz.

Leontopodium himalayanum DC.

Leontopodium jacotianum Beauv.

Leontopodium junpeianum Kitam.

Leontopodium leontopodioides (Willd.) Beauv.

Leontopodium lingianum (Y. L. Chen ex Y. S. Chen & Q. E. Yang) Dickoré

Leontopodium micranthum Ling

Leontopodium monocephalum Edgew.

Leontopodium muscoides Hand.-Mazz.

Leontopodium nanum (Hook. f. & Thoms. ex C. B. Clarke) Hand.-Mazz.

Leontopodium omeiense Y. Ling

Leontopodium pusillum (Beauv.) Hand.-Mazz.

Leontopodium roseum Hand.-Mazz.

Leontopodium sinense Hemsl.

Leontopodium souliei Beauv.

Leontopodium stoechas Hand.-Mazz.

Leontopodium stoloniferum Hand.-Mazz.

Leontopodium stracheyi (Hook. f.) C. B. Clarke ex Hemsl.

Leontopodium suffruticosum Y. L. Chen

Leontopodium villosum Hand.-Mazz.

Leontopodium wilsonii Beauv.

Leucanthemum superbum (Bergmans ex J. W. Ingram) D. H. Kent

Leucanthemum vulgare Lam.

Leucomeris decora Kurz

Leucomeris spectabilis D. Don

Ligularia alatipes Hand.-Mazz.

Ligularia alpigena Pojark.

Ligularia amplexicaulis DC.

Ligularia amplexicaulis var. **nepalensis** S. W. Liu & T. N. Ho

Ligularia anoleuca Hand.-Mazz.

Ligularia atkinsonii (C. B. Clarke) S. W. Liu

Ligularia atroviolacea (Franch.) Hand.-Mazz.

Ligularia botryodes (C. Winkl.) Hand.-Mazz.

Ligularia caloxantha (Diels) Hand.-Mazz.

Ligularia chalybea S. W. Liu

Ligularia changiana S. W. Liu ex Y. L. Chen & Z. Yu Li

Ligularia chimiliensis C. C. Chang

Ligularia confertiflora C. C. Chang

Ligularia cremanthodioides Hand.-Mazz.

Ligularia cuneata S. W. Liu & T. N. Ho

Ligularia curvisquama Hand.-Mazz.

Ligularia cyathiceps Hand.-Mazz.

Ligularia cymbulifera (W. W. Sm.) Hand.-Mazz.

Ligularia cymosa (Hand.-Mazz.) S. W. Liu

Ligularia dictyoneura (Franch.) Hand.-Mazz.

Ligularia discoidea S. W. Liu

Ligularia dux (C. B. Clarke) Y. Ling

Ligularia dux var. **minima** S. W. Liu

Ligularia eriocaulis M. Zhang & L. S. Xu

Ligularia euryphylla (C. Winkl.) Hand.-Mazz.

Ligularia fischeri (Ledeb.) Turcz.

Ligularia franchetiana (H. Lév.) Hand.-Mazz.

Ligularia ghatsukupa Kitam.

Ligularia hodgsonii Hook. f.

Ligularia hookeri (C. B. Clarke) Hand.-Mazz.

Ligularia hookeri var. **polycephala** (R. D. Good) Hand.-Mazz.

Ligularia intermedia Nakai

Ligularia jacquemontiana (Decne.) M. A. Rau

Ligularia jiajinshanensis Y. S. Chen

Ligularia kanaitzensis (Franch.) Hand.-Mazz.

Ligularia kanaitzensis var. **subnudicaulis** (Hand.-Mazz.) S. W. Liu

Ligularia kingiana (W. W. Sm.) R. Mathur

Ligularia lagotis (W. W. Sm.) Hand.-Mazz.

Ligularia lamarum (Diels) C. C. Chang

Ligularia lancifera (J. R. Drumm.) R. C. Srivast. & C. Jeffrey

Ligularia lankongensis (Franch.) Hand.-Mazz.

Ligularia lapathifolia (Franch.) Hand.-Mazz.

Ligularia latihastata (W. W. Sm.) Hand.-Mazz.

Ligularia latiligulata (R. D. Good) Spring.

Ligularia latipes S. W. Liu

Ligularia leesicotal Kitam.

Ligularia lhunzensis Y. S. Chen

Ligularia liatroides (C. Winkl.) Hand.-Mazz.

Ligularia lidjiangensis Hand.-Mazz.

Ligularia lingiana S. W. Liu

Ligularia longifolia Hand.-Mazz.

Ligularia longihastata Hand.-Mazz.

Ligularia × **maoniushanensis** X. Gong & Y. Z. Pan

Ligularia melanocephala (Franch.) Hand.-Mazz.

Ligularia melanothyrsa Hand.-Mazz.

Ligularia microcardia Hand.-Mazz.

Ligularia microcephala (Hand.-Mazz.) Hand.-Mazz.

Ligularia muliensis Hand.-Mazz.

Ligularia myriocephala Ling ex S. W. Liu

Ligularia nelumbifolia (Bureau & Franch.) Hand.-Mazz.

Ligularia nyingchiensis S. W. Liu

Ligularia odontomanes Hand.-Mazz.

Ligularia oligonema Hand.-Mazz.

Ligularia pachycarpa (C. B. Clarke ex Hook. f.) Kitam.

Ligularia paradoxa Hand.-Mazz.

Ligularia petiolaris Hand.-Mazz.

Ligularia phaenicochaeta (Franch.) Hand.-Mazz.

Ligularia phyllocolea Hand.-Mazz.

Ligularia platyglossa (Franch.) Hand.-Mazz.

Ligularia pleurocaulis (Franch.) Hand.-Mazz.

Ligularia potaninii (C. Winkl.) Y. Ling

Ligularia przewalskii (Maxim.) Diels

Ligularia pterodonta C. C. Chang

Ligularia pubifolia S. W. Liu

Ligularia purdomii (Turrill) Chitt.

Ligularia retusa DC.

Ligularia rockiana Hand.-Mazz.

Ligularia ruficoma (Franch.) Hand.-Mazz.

Ligularia rumicifolia S. W. Liu

Ligularia sagitta (Maxim.) Mattf. ex Rehder & Kobuski

Ligularia secunda Y. S. Chen

Ligularia shifangensis G. H. Chen & W. J. Chang

Ligularia stenocephala (Maxim.) Matsum. & Koidz.

Ligularia stenoglossa (Franch.) Hand.-Mazz.

Ligularia subspicata (Bureau & Franch.) Hand.-Mazz.

Ligularia tangutorum Pojarkova

Ligularia tenuicaulis C. C. Chang

Ligularia tenuipes (Franch.) Diels

Ligularia thomsonii (C. B. Clarke) Pojarkova

Ligularia tongkyukensis Hand.-Mazz.

Ligularia tongolensis (Franch.) Hand.-Mazz.

Ligularia transversifolia Hand.-Mazz.

Ligularia tsangchanensis (Franch.) Hand.-Mazz.

Ligularia veitchiana (Hemsl.) Greenm.

Ligularia vellerea (Franch.) Hand.-Mazz.

Ligularia virgaurea (Maxim.) Mattf. ex Rehder & Kobuski

Ligularia virgaurea var. **pilosa** S. W. Liu & T. N. Ho

Ligularia wilsoniana (Hemsl.) Greenm.

Ligularia yunnanensis (Franch.) C. C. Chang

Lipoblepharis urticifolia (Bl.) Orchard

Lipschitziella ceratocarpa (Decne.) Kamelin

Lipschitziella ceratocarpa var. **depressa** (Clarke ex Hook. f.) Karthik. & Moorthy

Matricaria aurea (Loefl.) Sch. Bip.

Matricaria chamomilla L.

Matricaria matricarioides (Less.) Porter ex Britton

Melanoseris atropurpurea (Franch.) N. Kilian & Ze H. Wang

Melanoseris beesiana (Diels) N. Kilian

Melanoseris bracteata (Hook. f. & Thomson ex C. B. Clarke) N. Kilian

Melanoseris brunoniana (Wall. ex DC.) N. Kilian & Ze H. Wang

Melanoseris cyanea (D. Don) Edgew.

Melanoseris decipiens (C. B. Clarke) N. Kilian & Ze H. Wang

Melanoseris graciliflora (DC.) N. Kilian

Melanoseris hirsuta (C. Shih) N. Kilian

Melanoseris jilongensis Ze H. Wang & H. Peng

Melanoseris kashmiriana (Mamgain & R. R. Rao) N. Kilian

Melanoseris lahulensis (Mamgain & R. R. Rao) N. Kilian

Melanoseris leptantha (C. Shih) N. Kilian

Melanoseris lessertiana (DC.) Decne.

Melanoseris likiangensis (Franch.) N. Kilian

Melanoseris macrantha (C. B. Clarke) N. Kilian & J. W. Zhang

Melanoseris macrocephala (C. Shih) N. Kilian & J. W. Zhang

Melanoseris macrorhiza (Royle) N. Kilian

Melanoseris monocephala (C. C. Chang) Ze H. Wang

Melanoseris oligolepis (C. C. Chang ex C. Shih) N. Kilian

Melanoseris pectiniformis (C. Shih) N. Kilian & J. W. Zhang

Melanoseris polyclada (Boiss.) Akhani, N. Kilian & Sennikov

Melanoseris rhombiformis (C. Shih) N. Kilian & Ze H. Wang

Melanoseris sichuanensis (C. Shih) N. Kilian

Melanoseris souliei (Franch.) N. Kilian

Melanoseris taliensis (C. Shih) N. Kilian

Melanoseris tenuis (C. Shih) N. Kilian

Melanoseris triflora (C. C. Chang & C. Shih) N. Kilian

Melanoseris violifolia (Decne.) N. Kilian

Melanoseris yunnanensis (Shih) N. Kilian & Ze H. Wang

Microcephala lamellata (Bunge) Pobed.

Microglossa pyrifolia (Lam.) Kuntze

Mikania micrantha Kunth

Myriactis delavayi Gagnep.

Myriactis nepalensis Less.

Myriactis wallichii Less.

Myriactis wightii DC.

Nabalus tatarinowii (Maxim.) Nakai

Nabalus tatarinowii subsp. **macrantha** (Stebbins) N. Kilian

Nabalus yanyuanensis Y. S. Chen & Rui Ke, sp. nov.

Nannoglottis carpesioides Maxim.

Nannoglottis delavayi (Franch.) Y. Ling & Y. L. Chen

Nannoglottis gynura (C. Winkl.) Ling & Y. L. Chen

Nannoglottis hieraciphylla (Hand.-Mazz.) Y. Ling & Y. L. Chen

Nannoglottis hookeri (Hook. f.) Kitam.

Nannoglottis latisquama Ling & Y. L. Chen

Nannoglottis macrocarpa Y. Ling & Y. L. Chen

Nannoglottis ravida (C. Winkl.) Y. L. Chen

Nannoglottis yunnanensis Hand.-Mazz.

Nemosenecio incisifolius (Jeffrey) B. Nord.

Neobrachyactis anomala (DC.) Brouillet

Neobrachyactis pubescens (DC.) Brouillet

Neobrachyactis roylei (DC.) Brouillet

Neopallasia pectinata (Pallas) Poljakov

Notoseris khasiana (C. B. Clarke) N. Kilian

Notoseris melanantha (Franch.) C. Shih

Notoseris scandens (Hook. f.) N. Kilian

Notoseris trifora (Hemsl.) C. Shih

Notoseris wilsonii (C. C. Chang) C. Shih

Notoseris yakoensis (Jeffrey) N. Kilian

Nouelia insignis Franch.

Olgaea thomsonii (Hook. f.) Iljin

Paraprenanthes dolichophylla (C. Shih) N. Kilian & Z. H. Wang

Paraprenanthes glandulosissima (C. C. Chang) Shih

Paraprenanthes polypodifolia (Franch.) C. C. Chang ex C. Shih

Paraprenanthes sagittiformis C. Shih

Paraprenanthes sororia (Miq.) C. Shih

Paraprenanthes yunnanensis (Franch.) Shih

Parasenecio chola (W. W. Sm.) R. C. Srivast. & C. Jeffrey

Parasenecio leviingii (C. B. Clarke) Y. S. Chen

Parasenecio mortonii R. C. Srivast. & C. Jeffrey

Parasenecio palmatisectus (Jeffrey) Y. L. Chen

Parasenecio quinquelobus (Wall. ex DC.) Y. L. Chen

Parthenium hysterophorus L.

Pentanema glanduligerum (Krasch.) Gorschk.

Pentanema indicum (L.) Ling

Pentanema indicum var. **hypoleucum** (Hand.-Mazz.) Ling

Pentanema vestitum (Wall. ex DC.) Ling

Pertya aitchisonii C. B. Clarke

Pertya markamensis Cai F. Zhang & T. G. Gao

Pertya monocephala W. W. Sm.

Pertya phylicoides Jeffrey

Pertya sinensis subsp. **maximowiczii** (C. Winkl.) Cai F. Zhang & T. G. Gao

Pertya uniflora (Maxim.) Mattf.

Pertya uniflora subsp. **angustifolia** (Y. C. Tseng) Cai F. Zhang & T. G. Gao

Petasites albus (L.) Gaertn.

Petasites kamengicus Deb

Petasites tricholobus Franch.

Petasites versipilus Hand.-Mazz.

Phagnalon acuminatum Boiss.

Phagnalon darvazicum Krasch.

Phagnalon niveum Edgew.

Picris divaricata Vaniot

Picris hieracioides L.

Picris japonica Thunb.

Picris junnanensis V. N. Vassil.

Picris nuristanica Bornm.

Piloselloides hirsuta (Forssk.) C. Jeffrey ex Cufod.

Pluchea lanceolata (DC.) C. B. Clarke

Pluchea linearifolia C. B. Clarke

Pluchea tomentosa DC.

Pluchea wallichiana DC.

Pseudoconyza viscosa (Mill.) D'Arcy

Pseudognaphalium adnatum (DC.) Y. S. Chen

Pseudognaphalium affine (D. Don) Anderb.

Pseudognaphalium chrysocephalum (Franch.) Hilliard & B. L. Burtt

Pseudognaphalium flavescens (Kitam.) Anderb.

Pseudognaphalium hypoleucum (DC.) Hilliard & B. L. Burtt

Pseudognaphalium luteoalbum (L.) Hilliard & B. L. Burtt

Psychrogeton alexeenkoi Krasch.

Psychrogeton amorphoglossus (Boiss.) Novopokr.

Psychrogeton andryaloides (DC.) Novopokr. ex Krasch.

Psychrogeton andryaloides var. **denudatus** (Botsch.) Grierson

Psychrogeton aucheri (DC.) Grierson

Psychrogeton cabulicus Boiss.

Psychrogeton chitralicus Grierson

Psychrogeton olgae Novopokr. ex Nevski

Psychrogeton poncinsii (Franch.) Ling & Y. L. Chen

Pterachaenia stewartii (Hook. f.) R. R. Stewart

Pulicaria angustifolia DC.

Pulicaria chrysantha (Diels) Y. Ling

Pulicaria dysenterica (L.) Bernh.

Pulicaria foliolosa DC.

Pulicaria glaucescens (Boiss.) Jaub. & Spach

Pulicaria gnaphalodes (Vent.) Boiss.

Pulicaria insignis J. R. Drumm. ex Dunn

Pulicaria petiolaris Jaub. & Spach

Pulicaria prostrata Asch.

Pulicaria salviifolia Bunge

Pulicaria vulgaris Gaertn.

Pyrethrum tatsienense (Bureau & Franch.) Ling ex C. Shih

Pyrethrum tatsienense var. **tanacetopsis** (W. W. Sm.) Y. Ling ex C. Shih

Rhaponticum repens (L.) Hidalgo

Richteria djilgense (Franch.) Bremer & Humphries

Richteria pyrethroides Kar. & Kir.

Saussurea acroura Cummins

Saussurea acutisquama Raab-Straube

Saussurea andersonii C. B. Clarke

Saussurea andryaloides (DC.) Sch. Bip.

Saussurea apus Maxim.

Saussurea arenaria Maxim.

Saussurea aster Hemsl.

Saussurea atkinsonii C. B. Clarke

Saussurea austrotibetica Y. S. Chen

Saussurea baoxingensis Y. S. Chen

Saussurea bartholomewii S. W. Liu & T. N. Ho

Saussurea bella Y. Ling

Saussurea bhutanensis Y. S. Chen

Saussurea bhutkesh Fujikawa & H. Ohba

Saussurea bijiangensis Y. L. Chen ex B. Q. Xu, N. H. Xia & G. Hao

Saussurea bodineri H. Lév.

Saussurea bracteata Decne.

Saussurea brunneopilosa Hand.-Mazz.

Saussurea bullata W. W. Sm.

Saussurea caespitosa (DC.) Wall. ex Sch. Bip.

Saussurea candolleana (DC.) Wall. ex Sch. Bip.

Saussurea caudata Franch.

Saussurea centiloba Hand.-Mazz.

Saussurea ceterach Hand.-Mazz.

Saussurea chetchozensis Franch.

Saussurea chinduensis Y. S. Chen

Saussurea chondrilloides C. Winkl.

Saussurea chrysanthemoides F. H. Chen

Saussurea chrysotricha Ludlow

Saussurea ciliaris Franch.

Saussurea cochlearifolia Y. L. Chen & S. Yun Liang

Saussurea colpodes Y. L. Chen & S. Yun Liang

Saussurea columnaris Hand.-Mazz.

Saussurea compta Franch.

Saussurea conaensis (S. W. Liu) Fujikawa & H. Ohba

Saussurea coriacea Y. L. Chen & S. Yun Liang

Saussurea coriolepis Hand.-Mazz.

Saussurea delavayi Franch.

Saussurea depsangensis Pamp.

Saussurea dhwojii Kitam.

Saussurea dielsiana Koidz.

Saussurea dolichopoda Diels

Saussurea donkiah C. B. Clarke ex Spring.

Saussurea dschungdiensis Hand.-Mazz.

Saussurea dulongjiangensis Y. S. Chen

Saussurea durgae C. Jeffrey & R. C. Srivast.

Saussurea dzeurensis Franch.

Saussurea elliptica C. B. Clarke ex Hook. f.

Saussurea epilobioides Maxim.

Saussurea eriocephala Franch.

Saussurea eriostemon Wall. ex C. B. Clarke

Saussurea erubescens Lipsch.

Saussurea euodonta Diels

Saussurea fibrosa King ex W. W. Sm.

Saussurea filicifolia (Hook. f.) Y. S. Chen

Saussurea fistulosa J. Anthony

Saussurea fuscipappa Y. S. Chen

Saussurea glabrescens (Hand.-Mazz.) Y. S. Chen

Saussurea glacialis Herder

Saussurea globosa F. H. Chen

Saussurea gnaphalodes (Royle ex DC.) Sch. Bip.

Saussurea gongriensis Y. S. Chen

Saussurea gossipiphora D. Don

Saussurea graminea Dunn

Saussurea graminea subsp. **ortholepis** (Hand.-Mazz.) Y. S. Chen

Saussurea graminifolia Wall. ex DC.

Saussurea grandiceps S. W. Liu

Saussurea grosseserrata Franch.

Saussurea gyacaensis S. W. Liu

Saussurea gymnocephala (Y. Ling) Raab-Straube

Saussurea hemsleyi Lipsch.

Saussurea hieracioides Hook. f.

Saussurea hookeri C. B. Clarke

Saussurea hylophila Hand.-Mazz.

Saussurea integrifolia Hand.-Mazz.

Saussurea inversa Raab-Straube

Saussurea iodoleuca Hand.-Mazz.

Saussurea ionodasys Hand.-Mazz.

Saussurea irregularis Y. L. Chen & S. Yun Liang

Saussurea japonica (Thunb.) DC.

Saussurea jindongensis Y. S. Chen

Saussurea jiulongensis Y. S. Chen

Saussurea kanaii Fujikawa & H. Ohba

Saussurea kansuensis Hand.-Mazz.

Saussurea katochaete Maxim.

Saussurea kingii C. E. C. Fisch.

Saussurea laminamaensis Kitam.

Saussurea lampsanifolia Franch.

Saussurea lanata Y. L. Chen & S. Yun Liang
Saussurea laneana W. W. Sm.
Saussurea langpoensis Y. S. Chen
Saussurea laniceps Hand.-Mazz.
Saussurea lavrenkoana Lipsch.
Saussurea leiocarpa Hand.-Mazz.
Saussurea leontodontoides (DC.) Sch. Bip.
Saussurea leptolepis Hand.-Mazz.
Saussurea leptophylla Hemsl.
Saussurea leucoma Diels
Saussurea leutoca Hand.-Mazz.
Saussurea lhozhagensis Y. S. Chen
Saussurea lhunzensis Y. S. Chen
Saussurea lhunzhubensis Y. L. Chen & S. Yun Liang
Saussurea liangshanensis Y. S. Chen
Saussurea licentiana Hand.-Mazz.
Saussurea limprichtii Diels
Saussurea linearifolia Ludlow
Saussurea lingulata Franch.
Saussurea longifolia Franch.
Saussurea loriformis W. W. Sm.
Saussurea luae Raab-Straube
Saussurea macrota Franch.
Saussurea medusa Maxim.
Saussurea megacephala C. C. Chang ex Y. S. Chen
Saussurea micradenia Hand.-Mazz.
Saussurea minuta C. Winkl.
Saussurea minutiloba Y. S. Chen
Saussurea montana J. Anthony
Saussurea muliensis Hand.-Mazz.
Saussurea multiloba Y. S. Chen
Saussurea nakikawae Kitam.
Saussurea nana (Pamp.) Pamp.
Saussurea nematolepis Y. Ling
Saussurea nigrescens Maxim.
Saussurea nimborum W. W. Sm.
Saussurea nishiokae Kitam.
Saussurea nyalamensis Y. L. Chen & S. Yun Liang
Saussurea nyingchiensis Y. S. Chen
Saussurea obvallata (DC.) Sch. Bip.
Saussurea ochrochlaena Hand.-Mazz.
Saussurea ovatifolia Y. L. Chen & S. Yun Liang
Saussurea pachyneura Franch.
Saussurea pagriensis Y. S. Chen
Saussurea paleacea Y. L. Chen & S. Yun Liang
Saussurea pantlingiana W. W. Sm.
Saussurea paxiana Diels
Saussurea peduncularis Franch.
Saussurea phaeantha Maxim.
Saussurea picridifolia (Hand.-Mazz.) Y. S. Chen & Qian Yuan
Saussurea pilinophylla Diels
Saussurea pinetorum Hand.-Mazz.
Saussurea pinnatiphylla Grierson & Spring.
Saussurea piptathera Edgew.
Saussurea platyphyllaria Ludlow
Saussurea platypoda Hand.-Mazz.
Saussurea polycephala Hand.-Mazz.
Saussurea polycolea Hand.-Mazz.
Saussurea polypodioides J. Anthony
Saussurea polystichoides Hook. f.
Saussurea poochlamys Hand.-Mazz.
Saussurea populifolia Hemsl.
Saussurea porphyroleuca Hand.-Mazz.
Saussurea przewalskii Maxim.
Saussurea pseudoeriostemon Y. S. Chen
Saussurea pseudojiulongensis Y. S. Chen
Saussurea pseudoleucoma Y. S. Chen
Saussurea pseudolingulata Y. S. Chen
Saussurea pseudomalitiosa Lipsch.
Saussurea pseudoplatyphyllaria Y. S. Chen
Saussurea pseudorockii Y. S. Chen

Saussurea pseudosimpsoniana Y. S. Chen

Saussurea pseudotridactyla Y. S. Chen

Saussurea pteridophylla Hand.-Mazz.

Saussurea pubescens Y. L. Chen & S. Yun Liang

Saussurea pubifolia S. W. Liu

Saussurea pumila C. Winkl.

Saussurea purpurascens Y. L. Chen & S. Yun Liang

Saussurea qamdoensis Y. S. Chen

Saussurea qinghaiensis S. W. Liu & T. N. Ho

Saussurea quercifolia W. W. Sm.

Saussurea ramchaudharyi S. K. Ghimire & H. K. Rana

Saussurea retroserrata Y. L. Chen & S. Yun Liang

Saussurea rockii J. Anthony

Saussurea rolwalingensis Fujikawa & H. Ohba

Saussurea romuleifolia Franch.

Saussurea roylei (DC.) Sch. Bip.

Saussurea roylei subsp. **stenophylla** Lipsch.

Saussurea salsa (Pall.) Spreng.

Saussurea scabrida Franch.

Saussurea schlagintweitii Klatt

Saussurea schlagintweitii subsp. **major** (Hook. f.) Y. S. Chen

Saussurea schultzii Hook. f.

Saussurea semiamplexicarlis Lipsch.

Saussurea semifasciata Hand.-Mazz.

Saussurea semilyrata Franch.

Saussurea sericea Y. L. Chen & S. Yun Liang

Saussurea shangrilaensis Y. S. Chen

Saussurea shuiluoensis Y. S. Chen

Saussurea simpsoniana (Field. & Gardn.) Lipsch.

Saussurea smithiana Hand.-Mazz.

Saussurea sobarocephala Diels

Saussurea sorocephala (Schrenk ex Fisch. & C. A. Mey.) Schrenk

Saussurea souliei Franch.

Saussurea spatulifolia Franch.

Saussurea spicata Ludlow ex Kitam.

Saussurea stella Maxim.

Saussurea stoliczkae C. B. Clarke

Saussurea stricta Franch.

Saussurea sublisquama Hand.-Mazz.

Saussurea subulata C. B. Clarke

Saussurea sughoo C. B. Clarke

Saussurea sugongii S. W. Liu & T. N. Ho

Saussurea superba J. Anthony

Saussurea sylvatica Maxim.

Saussurea taipaiensis Y. Ling

Saussurea tangutica Maxim.

Saussurea taraxacifolia (Lindl. ex Royle) Wall. ex DC.

Saussurea tatsienensis Franch.

Saussurea thomsonii C. B. Clarke

Saussurea thoroldii Hemsl.

Saussurea tibetica C. Winkl.

Saussurea topkegolensis H. Ohba & S. Akiyama

Saussurea tridactyla Sch. Bip. ex Hook. f.

Saussurea tridactyla subsp. **maiduoganla** (S. W. Liu) Y. S. Chen

Saussurea tsoongii Y. S. Chen

Saussurea uliginosa Hand.-Mazz.

Saussurea undulata Hand.-Mazz.

Saussurea uniflora (DC.) Wall. ex Sch. Bip.

Saussurea variiloba Y. Ling

Saussurea velutina W. W. Sm.

Saussurea vestita Franch.

Saussurea vestitiformis Hand.-Mazz.

Saussurea virgata Franch.

Saussurea wardii J. Anthony

Saussurea wellbyi Hemsl.

Saussurea wenchengiae B. Q. Xu, G. Hao & N. H. Xia

Saussurea wernerioides Sch. Bip. ex Hook. f.
Saussurea wettsteiniana Hand.-Mazz.
Saussurea woodiana Hemsl.
Saussurea xianrendongensis Y. S. Chen
Saussurea xiaojinensis Y. S. Chen
Saussurea yangii Y. S. Chen
Saussurea yanyuanensis Y. S. Chen
Saussurea yui Y. S. Chen
Saussurea yunnanensis Franch.
Saussurea yushuensis S. W. Liu & T. N. Ho
Saussurea zayuensis Y. S. Chen
Saussurea zogangensis Y. S. Chen
Scariola orientalis (Boiss.) Soják
Sclerocarpus africanus Jacq.
Scorzonera alba Stewart
Scorzonera codringtonii Rech. f.
Scorzonera ferganica Krasch.
Scorzonera fusca Kitam.
Scorzonera hondae Kitam.
Scorzonera koelpinoides Rech. f.
Scorzonera laciniata L.
Scorzonera pachycephala Podlech & Rech. f.
Scorzonera purpurea L.
Scorzonera pusilla Pall.
Scorzonera raddeana C. Winkl.
Scorzonera tunicata Rech. f. & Koeie
Scorzonera virgata DC.
Senecio acutipinnus Hand.-Mazz.
Senecio albopurpureus Kitam.
Senecio analogus DC.
Senecio arachnanthus Franch.
Senecio asperifolius Franch.
Senecio atrofuscus Grierson
Senecio biligulatus W. W. Sm.
Senecio boluangensis Koyama
Senecio candolleanus Wall. ex DC.
Senecio changii C. Ren & Q. E. Yang
Senecio chungtienensis C. Jeffrey & Y. L. Chen
Senecio cinarifolius H. Lév.
Senecio coriaceisquamus C. C. Chang
Senecio daochengensis Y. L. Chen
Senecio densiserratus C. C. Chang
Senecio desfontainei Druce
Senecio diversipinnus var. **discoideus** C. Jeffrey & Y. L. Chen
Senecio diversipinnus Y. Ling
Senecio dodrans C. Winkl.
Senecio drukensis C. Marquand & Airy Shaw
Senecio dubitabilis C. Jeffrey & Y. L. Chen
Senecio echaetus Y. L. Chen & K. Y. Pan
Senecio faberi Hemsl.
Senecio filifer Franch.
Senecio glaucus L.
Senecio graciliflorus DC.
Senecio kongboensis Ludlow
Senecio korshinskyi Krasch.
Senecio krascheninnikovii Schischk.
Senecio ladakhensis Chowdhery, Uniyal & R. Mathur
Senecio laetus Edgew.
Senecio lhasaensis Y. Ling ex Y. L. Chen, S. Yun Liang & K. Y. Pan
Senecio liangshanensis C. Jeffrey & Y. L. Chen
Senecio lijiangensis C. Jeffrey & Y. L. Chen
Senecio lingianus C. Jeffrey & Y. L. Chen
Senecio megalanthus Y. L. Chen
Senecio mishmi C. B. Clarke
Senecio muliensis C. Jeffrey & Y. L. Chen
Senecio multibracteolatus C. Jeffrey & Y. L. Chen
Senecio multilobus C. C. Chang
Senecio nemorensis L.
Senecio nigrocinctus Franch.
Senecio nodiflorus C. C. Chang
Senecio nudicaulis Buch.-Ham. ex D. Don
Senecio oryzetorum Diels

Senecio panduliformis Kitam.

Senecio paulsenii O. Hoffm.

Senecio pseudodensiserratus T. J. Tong, M. Tang, C. Ren & Q. E. Yang

Senecio pseudomairei H. Lév.

Senecio pteridophyllus Franch.

Senecio ramosus Wall. ex DC.

Senecio raphanifolius Wall. ex DC.

Senecio royleanus DC.

Senecio saussureoides Hand.-Mazz.

Senecio scandens Buch.-Ham. ex D. Don

Senecio spathiphyllus Franch.

Senecio thianschanicus Regel & Schmalh.

Senecio tibeticus Hook. f.

Senecio tricuspis Franch.

Senecio vulgaris L.

Senecio wightianus DC. ex Wight

Senecio wightii (DC.) Benth. ex C. B. Clarke

Senecio yungningensis Hand.-Mazz.

Seriphidium aucheri (Boiss.) Y. Ling & Y. R. Ling

Seriphidium brevifolium (Wall. ex DC.) Y. Ling & Y. R. Ling

Seriphidium chitralense (Podlech) Y. R. Ling

Seriphidium lehmannianum (Bunge) Poljakov

Seriphidium thomsonianum (C. B. Clarke) Y. Ling & Y. R. Ling

Shangwua denticulata (DC.) Raab-Straube & Yu J. Wang

Shangwua denticulata var. **hypoleuca** (DC.) Raab-Straube & Yu J. Wang

Shangwua jacea (Klotzsch) Yu J. Wang & Raab-Straube

Sigesbeckia orientalis L.

Sigesbeckia pubescens (Makino) Makino

Silybum marianum (L.) Gaertn.

Sinacalia caroli (C. Winkl.) C. Jeffrey & Y. L. Chen

Sinacalia davidii (Franch.) H. Koyama

Sinacalia tangutica (Maxim.) B. Nord.

Sinosenecio chienii (Hand.-Mazz.) B. Nord.

Sinosenecio euosmus (Hand.-Mazz.) B. Nord.

Sinosenecio fangianus Y. L. Chen

Sinosenecio homogyniphyllus (Cummins) B. Nord.

Sinosenecio ligularioides (Hand.-Mazz.) B. Nord.

Sinosenecio oldhamianus (Maxim.) B. Nord.

Sinosenecio rotundifolius Y. L. Chen

Sinosenecio subrosulatus (Hand.-Mazz.) B. Nord.

Sinosenecio sungpanensis (Hand.-Mazz.) B. Nord.

Sinosenecio yilingii Y. Liu & Q. E. Yang

Solidago virgaurea L.

Soliva anthemifolia (Juss.) R. Br.

Sonchella dentata (Ledeb.) Sennikov

Sonchella stenoma (Turcz. ex DC.) Sennikov

Sonchus asper (L.) Hill

Sonchus asper subsp. **glaucescens** (L.) Hill

Sonchus brachyotus DC.

Sonchus oleraceus L.

Sonchus wightianus DC.

Sonchus wightianus subsp. **wallianus** (DC.) Boulos

Soroseris depressa (Hook. f. & Thomson) J. W. Zhang, N. Kilian & H. Sun

Soroseris erysimoides (Hand.-Mazz.) C. Shih

Soroseris glomerata (Decne.) Stebbins

Soroseris hookeriana Stebbins

Soroseris pumila Stebbins

Soroseris teres C. Shih

Soroseris umbrella (Franch.) Stebbins

Spathipappus griffithii (C. B. Clarke) Pojarkov

Sphaeranthus senegalensis DC.

Stevia ovata Willd.

Syncalathium chrysocephalum (Shih) Shih

Syncalathium disciforme (Mattf.) Ling

Syncalathium kawaguchii (Kitam.) Y. Ling

Syncalathium porphyreum (C. Marquand & Airy Shaw) Y. Ling

Syncalathium roseum Ling

Synedrella nodiflora (L.) Gaertn.

Synedrella vialis (Less.) A. Gray

Synotis acuminata (Wall. ex DC.) C. Jeffrey & Y. L. Chen

Synotis ainsliaeifolia C. Jeffrey & Y. L. Chen

Synotis alata (Wall. ex DC.) C. Jeffrey & Y. L. Chen

Synotis auriculata C. Jeffrey & Y. L. Chen

Synotis bhot (C. B. Clarke) C. Jeffrey & Y. L. Chen

Synotis birmanica C. Jeffrey & Y. L. Chen

Synotis borii (Raizada) R. Mathur

Synotis brevipappa C. Jeffrey & Y. L. Chen

Synotis brunneo-villosa (Kitam.) C. Jeffrey & Y. L. Chen

Synotis calocephala C. Jeffrey & Y. L. Chen

Synotis cappa (Buch.-Ham. ex D. Don) C. Jeffrey & Y. L. Chen

Synotis cavaleriei (H. Lév.) C. Jeffrey & Y. L. Chen

Synotis chenopodiifolia (DC.) M. Tang, C. Ren & Q. E. Yang

Synotis chingiana C. Jeffrey & Y. L. Chen

Synotis erythropappa (Bureau & Franch.) C. Jeffrey & Y. L. Chen

Synotis glomerata C. Jeffrey & Y. L. Chen

Synotis jowaiensis (N. P. Balakr.) R. Mathur

Synotis kunthiana (Wall. ex DC.) C. Jeffrey & Y. L. Chen

Synotis lucorum (Franch.) C. Jeffrey & Y. L. Chen

Synotis lushaensis (Fischer) C. Jeffrey & Y. L. Chen

Synotis managensis S. Joshi, K. Shrestha & D. Bajracharya

Synotis muliensis Y. L. Chen

Synotis nagensium (C. B. Clarke) C. Jeffrey & Y. L. Chen

Synotis panduriformis C. Jeffrey & Y. L. Chen

Synotis penninervis (H. Koyama) T. J. Tong, M. Tang, C. Ren & Q. E. Yang

Synotis pentantha (Merr.) M. Tang, C. Ren & Q. E. Yang

Synotis pseudo-alata (C. C. Chang) C. Jeffrey & Y. L. Chen

Synotis rhabdos (C. B. Clarke) C. Jeffrey & Y. L. Chen

Synotis rufinervis (DC.) C. Jeffrey & Y. L. Chen

Synotis saluenensis (Diels) C. Jeffrey & Y. L. Chen

Synotis sciatrephes (W. W. Sm.) C. Jeffrey & Y. L. Chen

Synotis setchuenensis (Franch.) C. Jeffrey & Y. L. Chen

Synotis solidaginea (Hand.-Mazz.) C. Jeffrey & Y. L. Chen

Synotis tetrantha (DC.) C. Jeffrey & Y. L. Chen

Synotis triligulata (Buch.-Ham. ex D. Don) C. Jeffrey & Y. L. Chen

Synotis vagans (Wall. ex DC.) C. Jeffrey & Y. L. Chen

Synotis wallichii (DC.) C. Jeffrey & Y. L. Chen

Synotis xantholeuca (Hand.-Mazz.) C. Jeffrey & Y. L. Chen

Synotis yakoensis (Jeffrey ex Diels) C. Jeffrey & Y. L. Chen

Synotis yui C. Jeffrey & Y. L. Chen

Tagetes erecta L.

Tagetes minuta L.

Tagetes tenuifolia Cav.

Tanacetum artemisioides Sch. Bip. ex Hook. f.

Tanacetum atkinsonii (C. B. Clarke) Kitam.

Tanacetum baltistanicum Podlech

Tanacetum chitralense (Podlech) K. Bremer & Humphries

Tanacetum cinerariifolium (Trev.) Sch. Bip.

Tanacetum emodi R. Khan

Tanacetum falconeri Hook. f.

Tanacetum griffithii (C. B. Clarke) Muradyan

Tanacetum pakistanicum Podl.

Tanacetum parthenium (L.) Sch. Bip.

Tanacetum roylei (DC.) Podlech

Tanacetum stoliczkae (C. B. Clarke) R. Khan

Tanacetum tatsienense (Bureau & Franch.) K. Bremer & Humphries

Tanacetum tatsienense var. **tanacetopsis** (W. W. Sm.) Grierson

Taraxacum afghanicum Soest

Taraxacum amabile Soest

Taraxacum apargia Kirschner & Štepánek

Taraxacum apargiiforme Dahlsted

Taraxacum armeriifolium Soest

Taraxacum atrocarpum Kirschner & Štepánek

Taraxacum aurantiacum Dahlstedt

Taraxacum austrotibetanum Kirschner & Štepánek

Taraxacum azzizi Soest

Taraxacum badiocinnamomeum Kirschner & Štepánek

Taraxacum baltistanicum Soest

Taraxacum behzudicum Soest

Taraxacum bessarabicum (Hornem.) Hand.-Mazz.

Taraxacum bhutanicum Soest

Taraxacum bicolor DC.

Taraxacum brevirostre Hand.-Mazz.

Taraxacum calanthodium Dahlstedt

Taraxacum candidatum Kirschner & Štepánek

Taraxacum celsum Kirschner & Štepánek

Taraxacum chionophilum Dahlstedt

Taraxacum chitralense Soest

Taraxacum crepidiforme DC.

Taraxacum dasypodum Soest

Taraxacum dealbatum Hand.-Mazz.

Taraxacum deludens Kirschner & Štepánek

Taraxacum elegans Soest

Taraxacum eriopodum DC.

Taraxacum erioscapum H. Hartmann

Taraxacum forrestii Soest

Taraxacum glaucophylloides Kirschner & Štepánek

Taraxacum glaucophyllum Soest

Taraxacum grypodon Dahlstedt

Taraxacum gulmargense Soest

Taraxacum himalaicum Soest

Taraxacum hooftii Soest

Taraxacum icterinum Kirschner & Štepánek

Taraxacum karakoricum Soest

Taraxacum kashmirense Soest

Taraxacum lanigerum Soest

Taraxacum leucanthum (Ledeb.) Ledeb.

Taraxacum ludlowii Soest

Taraxacum lugubre Dahlstedt

Taraxacum luridum G. E. Haglund

Taraxacum macula Kirschner & Štepánek

Taraxacum mastigophyllum Kirschner & Štepánek

Taraxacum maurocarpum Dahlstedt

Taraxacum minutilobum Popov ex Kovalevsk.

Taraxacum mitalii Soest

Taraxacum mongolicum Hand.-Mazz.

Taraxacum monochlamydeum Hand.-Mazz.

Taraxacum mucronulatum Soest

Taraxacum mutatum Kirschner & Štepánek

Taraxacum nasirii Soest

Taraxacum nepalense Soest

Taraxacum obtusum (Soest) R. Doll

Taraxacum officinale F. H. Wigg.

Taraxacum orientale Kirschner & Štepánek

Taraxacum parvulum DC.

Taraxacum patiens Kirschner & Štepánek

Taraxacum pseudodissimile Soest

Taraxacum pseudoleucanthum Soest

Taraxacum pseudostenoceras Soest

Taraxacum rorippa Soest

Taraxacum scariosum (Tausch) Kirschner & Štepánek

Taraxacum sherriffii Soest

Taraxacum sikkimense Hand.-Mazz.

Taraxacum simulans Kirschner & Štepánek

Taraxacum staintonii Soest

Taraxacum staticifolium Soest

Taraxacum stenoceras Dahlstedt

Taraxacum stenolepium Hand.-Mazz.

Taraxacum suavissimum Kirschner & Štepánek

Taraxacum subcontristans Kirschner & Štepánek

Taraxacum suberiopodum Soest

Taraxacum tibetanum Hand.-Mazz.

Taraxacum turritum Kirschner & Štepánek

Taraxacum vendibile Kirschner & Štepánek

Taraxacum wallichii DC.

Taraxacum wattii Hook. f.

Tephroseris rufa (Hand.-Mazz.) B. Nord.

Tephroseris stolonifera (Cufod.) Holub

Thespis divaricata DC.

Tithonia diversifolia (Hemsl.) A. Gray

Tragopogon badachschanicus Boriss.

Tragopogon dubius Scop.

Tragopogon gracilis D. Don

Tricholepis elongata DC.

Tricholepis furcata DC.

Tricholepis karensium Kurz

Tricholepis nakaoi Kitam.

Tricholepis raghavendrae Saklani & L. B. Chaudhary

Tricholepis roylei Hook. f.

Tricholepis stewartei C. B. Clarke ex Hook. f.

Tricholepis stictophyllum C. B. Clarke

Tricholepis tibetica Hook. f. & Thoms. ex C. B. Clarke

Tridax procumbens L.

Tussilago farfara L.

Uechtritzia lacei (Watt) Jeffrey

Verbesina encelioides (Cav.) Benth. & Hook. f. ex A. Gray

Vernonia albicans DC.

Vernonia anthelmintica (L.) Willd.

Vernonia arborea Buch.-Ham.

Vernonia aspera Buch.-Ham.

Vernonia attenuata DC.

Vernonia blanda DC.

Vernonia cinerea (L.) Less.

Vernonia esculenta Hemsl.

Vernonia extensa DC.

Vernonia parishii Hook. f.

Vernonia revoluta Buch.-Ham.

Vernonia saligna DC.

Vernonia silhetensis (DC.) Hand.-Mazz.

Vernonia squarrosa (D. Don) Less.

Vernonia subsessilis DC.

Vernonia subsessilis var. **macrophylla** Hook. f.

Vernonia talaumifolia Hook. f. & Thomson ex C. B. Clarke

Vernonia vagans DC.

Vernonia volkameriifolia DC.

Wollastonia biflora (L.) DC.

Xanthium spinosum L.

Xanthium strumarium L.

Xanthopappus subacaulis C. Winkl.

Xerochrysum bracteatum (Vent.) Tzvelev

Xylanthemum pamiricum (O. Hoffm.) Tzvelev

Youngia atripappa (Babc.) N. Kilian

Youngia baoxingensis Y. S. Chen

Youngia bifurcata Babc. & Stebbins

Youngia cineripappa (Babc.) Babc. & Stebbins

Youngia conjunctiva Babc. & Stebbins

Youngia cristata C. Shih & C. Q. Cai

Youngia fuscipappa Thwaites

Youngia gongshanensis Y. S. Chen & R. Ke

Youngia gracilipes (Hook. f.) Babc. & Stebbins

Youngia hastiformis C. Shih

Youngia henryi (Diels) Babc. & Stebbins

Youngia heterophylla (Hemsl.) Babc. & Stebbins

Youngia japonica (L.) DC.

Youngia kangdingensis C. Shih

Youngia lanata Babc. & Stebbins

Youngia mairei (H. Lév.) Babc. & Stebbins

Youngia nilgiriensis Babc.

Youngia paleacea (Diels) Babc. & Stebbins

Youngia pilifera C. Shih

Youngia prattii (Babc.) Babc. & Stebbins

Youngia racemifera (Hook. f.) Babc. & Stebbins

Youngia rosthornii (Diels) Babc. & Stebbins

Youngia sericea C. Shih

Youngia silhetensis (DC.) Babc. & Stebbins

Youngia simulatrix (Babc.) Babc. & Stebbins

Youngia terminalis Babc. & Stebbins

Youngia wilsonii (Babc.) Babc. & Stebbins

Youngia yilingii C. Shih

Zinnia elegans Jacq.

Zinnia peruviana L.

Zoegea purpurea Fresen.

277. ADOXACEAE

Adoxa moschatellina L.

Adoxa omeiensis Hara

Adoxa xizangensis G. Yao

Sambucus adnata Wall. ex DC.

Sambucus canadensis L.

Sambucus ebulus L.

Sambucus javanica Blume

Sinadoxa corydalifolia C. Y. Wu, Z. L. Wu & R. F. Huang

Viburnum amplifolium Rehder

Viburnum atrocyaneum C. B. Clarke

Viburnum betulifolium Batalin

Viburnum brachybotryum Hemsl.

Viburnum canum Griff.

Viburnum chingii P. S. Hsu

Viburnum chingii var. **limitaneum** (W. W. Sm.) Q. E. Yang

Viburnum chinshanense Graebn.

Viburnum cinnamomifolium Rehder

Viburnum congestum Rehder

Viburnum colebrookianum Wall. ex DC.

Viburnum corylifolium Hook. f. & Thomson

Viburnum cotinifolium D. Don

Viburnum cotinifolium var. **wallichii** T. R. Dudley

Viburnum cylindricum Buch.-Ham. ex D. Don

Viburnum davidii Franch.

Viburnum dilatatum Thunb.

Viburnum erosum Thunb.

Viburnum erubescens Wall.

Viburnum foetidum Wall.

Viburnum foetidum var. **ceanothoides** (C. H. Wright) Hand.-Mazz.

Viburnum foetidum var. **rectangulatum** (Graebn.) Rehder

Viburnum glomeratum Maxim.

Viburnum glomeratum subsp. **rotundifolium** (P. S. Hsu) P. S. Hsu

Viburnum grandiflorum Wall. ex DC.

Viburnum henryi Hemsl.

Viburnum kansuense Batalin

Viburnum longiradiatum P. S. Hsu & S. W. Fan

Viburnum mullaha Buch.-Ham. ex D. Don

Viburnum nervosum D. Don

Viburnum oliganthum Batalin

Viburnum punctatum Buch.-Ham. ex D. Don

Viburnum schensianum Maxim.

Viburnum sempervirens K. Koch

Viburnum setigerum Hance

Viburnum simonsii Hook. f. & Thomson

Viburnum subalpinum Hand.-Mazz.

Viburnum sympodiale Graebn.

Viburnum tengyuehense (W. W. Sm.) P. S. Hsu

Viburnum ternatum Rehder

Viburnum utile Hemsl.

278. CAPRIFOLIACEAE

Abelia forrestii (Diels) W. W. Sm.

Acanthocalyx alba (Hand.-Mazz.) M. J. Cannon

Acanthocalyx nepalensis (D. Don) M. J. Cannon

Acanthocalyx nepalensis subsp. **delavayi** (Franch.) D. Y. Hong

Dipelta elegans Batalin

Dipelta yunnanensis Franch.

Dipsacus asper Wall. ex DC.

Dipsacus atratus Hook. f. & Thomson ex C. B. Clarke

Dipsacus chinensis Batalin

Dipsacus inermis Wall.

Dipsacus japonicus Miq.

Leycesteria dibangalliensis S. K. Das & G. S. Giri

Leycesteria formosa Wall.

Leycesteria glaucophylla (Hook. f. & Thomson) C. B. Clarke

Leycesteria gracilis (Kurz) Airy Shaw

Leycesteria stipulata (Hook. f. & Thomson) Fritsch

Lonicera acuminata Wall.

Lonicera angustifolia var. **myrtillus** (Hook. f. & Thomson) Q. E. Yang, Landrein, Borosova & J. Osborne

Lonicera angustifolia Wall. ex DC.

Lonicera asperifolia (Decne.) Hook. f. & Thomson

Lonicera caerulea L.

Lonicera calcarata Hemsl.

Lonicera caucasica subsp. **govaniana** (Wall. ex DC.) Hara

Lonicera chrysantha Turcz. ex Ledeb.

Lonicera chrysantha var. **koehneana** (Rehder) Q. E. Yang, Landrein, Borosova & J. Osborne

Lonicera confusa DC.

Lonicera cyanocarpa Franch.

Lonicera ferdinandi Franch.

Lonicera glabrata Wall.

Lonicera heterophylla Decne.

Lonicera hildebrandiana Collett & Hemsl.

Lonicera hispida Pallas ex Schult.

Lonicera hypoleuca Decne.

Lonicera japonica Thunb.

Lonicera korolkovii Stapf.

Lonicera lanceolata Wall.

Lonicera ligustrina Wall.

Lonicera ligustrina var. **yunnanensis** Franch.

Lonicera litangensis Batalin

Lonicera maackii (Rupr.) Maxim.

Lonicera macrantha (D. Don) Spreng.

Lonicera microphylla Willd. ex Schult.

Lonicera minutifolia Kitam.

Lonicera myrtilloides Purpus

Lonicera nervosa Maxim.

Lonicera nigra L.

Lonicera nummularifolia Jaub. & Spach

Lonicera obovata Royle ex Hook. f. & Thomson

Lonicera oreodoxa Harry Sm. ex Rehder

Lonicera porphyrantha (C. Marquand & Airy Shaw) M. P. Nayar & G. S. Giri

Lonicera purpurascens Walp.

Lonicera quinquelocularis Hardw.

Lonicera retusa Franch.

Lonicera rupicola Hook. f. & Thomson

Lonicera rupicola var. **minuta** (Batalin) Q. E. Yang, Landrein, Borosova & J. Osborne

Lonicera rupicola var. **syringantha** (Maxim.) Zabel

Lonicera semenovii Regel

Lonicera setifera Franch.

Lonicera spinosa (Decne.) Jacq. ex Walp.

Lonicera stenantha Pojark.

Lonicera tangutica Maxim.

Lonicera tomentella Hook. f. & Thomson

Lonicera tomentella var. **tsarongensis** W. W. Sm.

Lonicera trichosantha Bureau & Franch.

Lonicera tubuliflora Rehder

Lonicera vaccinioides Rehder

Lonicera webbiana Wall. ex DC.

Lonicera yunnanensis Franch.

Morina bracteata C. Y. Cheng & H. B. Chen

Morina chinensis Y. Y. Pai

Morina chlorantha Diels

Morina coulteriana Royle

Morina kokonorica K. S. Hao

Morina longifolia Wall. ex DC.

Morina ludlowii (M. J. Cannon) D. Y. Hong

Morina polyphylla Wall. ex DC.

Nardostachys jatamansi (D. Don) DC.

Patrinia heterophylla Bunge

Patrinia monandra C. B. Clarke

Patrinia scabiosifolia Link

Patrinia speciosa Hand.-Mazz.

Pterocephalus bretschneideri (Batalin) E. Pritz. ex Diels

Pterocephalus hookeri (C. B. Clarke) Diels

Scabiosa candollei DC.

Scabiosa speciosa Royle

Triosteum himalayanum Wall.

Triosteum pinnatifidum Maxim.

Triplostegia glandulifera Wall. ex DC.

Triplostegia grandiflora Gagnep.

Valeriana barbulata Diels

Valeriana clarkei Briq.

Valeriana daphniflora Hand.-Mazz.

Valeriana fedtschenkoi Coincy

Valeriana ficariifolia Boiss.

Valeriana flagellifera Batalin

Valeriana hardwickii Wall.

Valeriana hengduanensis D. Y. Hong

Valeriana hiemalis Graebn.

Valeriana himalayana Grubov

Valeriana hirticalyx L. C. Chiu

Valeriana jaeschkei C. B. Clarke

Valeriana jaeschkei var. **kaschmiriensis** (Grubov) Y. Nasir

Valeriana jatamansi W. Jones

Valeriana lancifolia Hand.-Mazz.

Valeriana minutiflora Hand.-Mazz.

Valeriana officinalis L.

Valeriana pyrolifolia Decne.

Valeriana sichuanica D. Y. Hong

Valeriana stenoptera Diels

Valeriana stracheyi C. B. Clarke

Valeriana tangutica Batalin

Valeriana trichostoma Hand.-Mazz.

Valerianella dentata (L.) Poll.

Valerianella szovitsiana Fisch. & C. A. Mey.

Zabelia dielsii (Graebn.) Makino

Zabelia triflora (R. Br.) Makino

279. TORRICELLIACEAE

Toricellia angulata Oliv.

Toricellia tiliifolia DC.

280. PITTOSPORACEAE

Pittosporum angustilimbum C. Y. Wu

Pittosporum balansae var. **chatterjeeanum** (Gowda) Z. Y. Zhang & Turland

Pittosporum brevicalyx (Oliver) Gagnep.

Pittosporum glabratum var. **neriifolium** Rehder & E. H. Wilson

Pittosporum glabratum var. **wenxianense** (G. H. Wang & Y. S. Lian) Z. Y. Zhang & Turland

Pittosporum heterophyllum Franch.

Pittosporum johnstonianum Gowda

Pittosporum napaulense (DC.) Rehder & E. H. Wilson

Pittosporum podocarpum Gagnep.

Pittosporum podocarpum var. **angustatum** Gowda

Pittosporum rehderianum Gowda

Pittosporum saxicola Rehder & E. H. Wilson

Pittosporum trigonocarpum H. Lév.

Pittosporum truncatum E. Pritz.

Pittosporum wenxianense G. H. Wang & Y. S. Lian

281. ARALIACEAE

Aralia apioides Hand.-Mazz.

Aralia armata (Wall. ex G. Don) Seem.

Aralia atropurpurea Franch.

Aralia cachemirica Decne.

Aralia caesia Hand.-Mazz.

Aralia delavayi J. Wen

Aralia fargesii Franch.

Aralia foliolosa Seem. ex C. B. Clarke

Aralia gigantea J. Wen

Aralia kingdon-wardii J. Wen, Lowry & Esser

Aralia leschenaultii (DC.) J. Wen

Aralia melanocarpa (H. Lév.) Lauener

Aralia montana Blume

Aralia parasitica (D. Don) J. Wen

Aralia plumosa H. L. Li

Aralia stipulata Franch.

Aralia subcordata (Wall. ex G. Don) J. Wen

Aralia tibetana G. Hoo

Aralia tomentella Franch.

Aralia wilsonii Harms

Aralia yunnanensis Franch.

Brassaiopsis aculeata (Buch.-Ham. ex D. Don) Seem.

Brassaiopsis fatsioides Harms

Brassaiopsis glomerulata (Blume) Regel

Brassaiopsis griffithii C. B. Clarke

Brassaiopsis hainla (Buch.-Ham.) Seem.

Brassaiopsis hispida Seem.

Brassaiopsis magnifica Dunn

Brassaiopsis mitis C. B. Clarke

Brassaiopsis polyacantha (Wall.) R. N. Banerjee

Brassaiopsis shweliensis W. W. Sm.

Brassaiopsis simplicifolia C. B. Clarke

Brassaiopsis tibetana C. B. Shang

Dendropanax burmanicus Merr.

Eleutherococcus baoxinensis (X. P. Fang & C. K. Hsieh) P. S. Hsu & S. L. Pan

Eleutherococcus cissifolius (Griff. ex C. B. Clarke) Nakai

Eleutherococcus lasiogyne (Harms) S. Y. Hu

Eleutherococcus leucorrhizus Oliv.

Eleutherococcus trifoliatus (L.) S. Y. Hu

Eleutherococcus verticillatus (G. Hoo) H. Ohashi

Eleutherococcus wilsonii (Harms) Nakai

Gamblea ciliata C. B. Clarke

Hedera nepalensis K. Koch

Hedera nepalensis var. **sinensis** (Tobler) Rehder

Hydrocotyle handelii H. Wolff ex Hand.-Mazz.

Hydrocotyle himalaica P. K. Mukherjee

Hydrocotyle hookeri (C. B. Clarke) Craib

Hydrocotyle hookeri subsp. **chinensis** (Dunn ex R. H. Shan & S. L. Liou) M. F. Watson & M. L. Sheh

Hydrocotyle javanica Thunb.

Hydrocotyle nepalensis Hook.

Hydrocotyle peltiformis R. Li & H. Li

Hydrocotyle ramiflora Maxim.

Hydrocotyle salwinica R. H. Shan & S. L. Liou

Hydrocotyle sibthorpioides Lam.

Hydrocotyle wilfordii Maxim.

Heteropanax fragrans (Roxb. ex DC.) Seem.

Macropanax dispermus (Blume) Kuntze

Macropanax undulatus (Wall. ex G. Don) Seem.

Merrilliopanax alpinus (C. B. Clarke) C. B. Shang

Merrilliopanax listeri (King) H. L. Li

Merrilliopanax membranifolius (W. W. Sm.) C. B. Shang

Metapanax delavayi (Franch.) J. Wen & Frodin

Panax bipinnatifidus Seem.

Panax japonicus (T. Nees) C. A. Mey.

Panax japonicus var. **angustifolius** (Burkill) C. C. Cheng & Chu

Panax japonicus var. **major** (Burkill) C. Y. Wu & K. M. Feng

Panax japonicus var. **bipinnatifidus** (Seem.) C. Y. Wu & K. M. Feng

Panax pseudoginseng Wall.

Panax stipuleanatus H. T. Tsai & K. M. Feng

Schefflera bengalensis Gamble

Schefflera clarkeana Craib

Schefflera elata (Buch.-Ham.) Harms

Schefflera elliptica (Blume) Harms

Schefflera glabrescens (C. J. Tseng & G. Hoo) Frodin

Schefflera heptaphylla (L.) Frodin

Schefflera hoi (Dunn) R. Viguier

Schefflera hypoleuca (Kurz) Harms

Schefflera impressa (C. B. Clarke) Harms

Schefflera khasiana (C. B. Clarke) R. Viguier

Schefflera leucantha R. Viguier

Schefflera multinervia H. L. Li

Schefflera pueckleri (K. Koch) Frodin

Schefflera rhododendrifolia (Griff.) Frodin

Schefflera venulosa (Wight & Arn.) Harms

Schefflera wallichiana (Wight & Arn.) Harms

Schefflera zhuana Lowry & C. B. Shang

Tetrapanax papyrifer (Hook.) K. Koch

Trevesia palmata (Roxb. ex Lindl.) Vis.

282. APIACEAE

Acronema acronemifolium (C. B. Clarke) H. Wolff

Acronema alpinum S. L. Liou & R. H. Shan

Acronema astrantiifolium H. Wolff

Acronema brevipedicellatum Z. H. Pan & M. F. Watson

Acronema bryophilum Farille & Lachard

Acronema chinense H. Wolff

Acronema chinense var. **humile** S. L. Liou & R. H. Shan

Acronema commutatum H. Wolff

Acronema cryptosciadeum Farille & Lachard

Acronema dyssimetriradiata Farille & S. B. Malla

Acronema edosmioides (H. Boissieu) Pimenov & Kljuykov

Acronema forrestii H. Wolff

Acronema gracile S. L. Liou & R. H. Shan

Acronema graminifolium (W. W. Sm.) S. L. Liou & R. H. Shan

Acronema handelii H. Wolff

Acronema hookeri (C. B. Clarke) H. Wolff

Acronema ioniostyles Farille & Lachard

Acronema johrianum Babu

Acronema minus (M. F. Watson) M. F. Watson & Z. H. Pan

Acronema mukherjeeanum Farille & Lachard

Acronema muscicola (Hand.-Mazz.) Hand.-Mazz.

Acronema nervosum H. Wolff

Acronema paniculatum (Franch.) H. Wolff

Acronema phaeosciadeum Farille & Lachard

Acronema pneumatophobium Farille & Lachard

Acronema pseudotenera P. K. Mukh.

Acronema refugicola Farille & Lachard

Acronema schneideri H. Wolff

Acronema sichuanense S. L. Liou & R. H. Shan

Acronema tenerum (DC.) Edgew.

Acronema xizangense S. L. Liou & R. H. Shan

Acronema yadongense S. L. Liou

Anethum graveolens L.

Angelica apaensis R. H. Shan & C. Q. Yuan

Angelica cyclocarpa (C. Norman) M. Hiroe

Angelica duclouxii Fedde ex H. Wolff

Angelica glauca Edgew.

Angelica indica Pimenov & Kljuykov

Angelica kangdingensis R. H. Shan & F. T. Pu

Angelica laxifoliata Diels

Angelica likiangensis H. Wolff

Angelica longicaudata C. Q. Yuan & R. H. Shan

Angelica longipes H. Wolff

Angelica maowenensis C. Q. Yuan & R. H. Shan

Angelica muliensis C. Y. Liao & X. G. Ma

Angelica nitida H. Wolff

Angelica nubigena (C. B. Clarke) P. K. Mukh.

Angelica paeoniifolia C. Q. Yuan & R. H. Shan

Angelica pinnatiloba R. H. Shan & F. T. Pu

Angelica pseudoselinum H. Boissieu

Angelica sinensis (Oliver) Diels

Angelica sinensis var. **wilsonii** (H. Wolff) Z. H. Pan & M. F. Watson

Anthriscus nemorosus (M. Bieb.) Spreng.

Anthriscus sylvestris (L.) Hoffm.

Apium graveolens L.

Arcuatopterus harae (Pimenov) Pimenov & Ostr.

Arcuatopterus linearifolius M. L. Sheh & R. H. Shan

Arcuatopterus sikkimensis (C. B. Clarke) Pimenov & Ostroumova

Arcuatopterus thalictrioideus M. L. Sheh & R. H. Shan

Athamanta gigantea D. Don

Athamanta teres D. Don

Aulacospermum darwasicum (Lipsky) Schischkin

Aulacospermum stylosum (C. B. Clarke) Rech.

Berula erecta (Huds.) Coville

Bupleurum aitchisonii H. Wolff

Bupleurum alatum R. H. Shan & M. L. Sheh

Bupleurum amplexicaule (Wu) X. J. He & C. B. Wang

Bupleurum baimaense X. G. Ma & X. J. He

Bupleurum candollei Wall. ex DC.

Bupleurum candollei var. **virgatissimum** C. Y. Wu

Bupleurum chaishoui R. H. Shan & M. L. Sheh

Bupleurum commelynoideum H. Boissieu

Bupleurum commelynoideum var. **flaviflorum** R. H. Shan & Yin Li

Bupleurum constancei E. Nasir

Bupleurum dalhousieanum (C. B. Clarke) Koso-Pol.

Bupleurum exaltatum M. Bieb.

Bupleurum falcatum L.

Bupleurum franchetii (H. Boissieu) X. J. He & C. B. Wang

Bupleurum gilesii H. Wolff

Bupleurum gracillimum Klotzsch

Bupleurum hamiltonii N. P. Balakr.

Bupleurum hamiltonii var. **humile** (Franch.) R. H. Shan & M. L. Sheh

Bupleurum himalayense Klotzsch

Bupleurum lanceolatum Wall. ex DC.

Bupleurum longicaule DC.

Bupleurum longicaule var. **amplexicaule** C. Y. Wu ex R. H. Shan & Yin Li

Bupleurum longicaule var. **franchetii** H. Boissieu

Bupleurum malconense R. H. Shan & Yin Li

Bupleurum marginatum Wall. ex DC.

Bupleurum marginatum var. **stenophyllum** (H. Wolff) R. H. Shan & Yin Li

Bupleurum microcephalum Diels

Bupleurum petiolulatum Franch.

Bupleurum petiolulatum var. **tenerum** R. H. Shan & Yin Li

Bupleurum rockii H. Wolff

Bupleurum rupestre Edgew.

Bupleurum shanianum X. G. Ma & X. J. He

Bupleurum sikiangense X. J. He & C. B. Wang

Bupleurum stewartianum Nasir

Bupleurum swatianum Nasir

Bupleurum tenerum (R. H. Shan & Yin Li) X. J. He & C. B. Wang

Bupleurum triradiatum Adams ex Hoffm.

Bupleurum wenchuanense R. H. Shan & Yin Li

Bupleurum yunnanense Franch.

Carum buriaticum Turcz.

Carum carvi L.

Carum diversifolium (DC.) C. B. Clarke

Centella asiatica (L.) Urban

Chaerophyllopsis huai H. Boissieu

Chaerophyllum reflexum Lindl.

Chaerophyllum reflexum var. **acuminatum** (Lindl.) I. C. Hedge & J. M. Lamond

Chaerophyllum villosum DC.

Chamaesium delavayi (Franch.) R. H. Shan & S. L. Liou

Chamaesium jiulongense X. L. Guo & X. J. He

Chamaesium mallaeanum Farille & S. B. Malla

Chamaesium novemjugum (C. B. Clarke) C. Norman

Chamaesium paradoxum H. Wolff

Chamaesium shrestaeanum Farille & S. B. Malla

Chamaesium spatuliferum (W. W. Sm.) C. Norman

Chamaesium thalictrifolium H. Wolff

Chamaesium viridiflorum (Franch.) H. Wolff ex R. H. Shan

Chamaesium wolffianum Fedde ex H. Wolff

Cicuta virosa L.

Cnidium bhutanicum M. F. Watson

Cnidium monnieri (L.) Cusson

Conioselinum acuminatum (Franch.) Lavrova

Conioselinum nematophyllum Pimenov & Kljuykov

Conioselinum nepalense Pimenov & Kljuykov

Conioselinum pseudoangelica (H. Boissieu) Pimenov & Kljuykov

Conioselinum pteridophyllum (Franch.) Lavrova

Conioselinum reflexum Pimenov & Kljuykov

Conioselinum tenuisectum (H. Boissieu) Pimenov & Kljuykov

Conioselinum vaginatum (Spreng.) Thell.

Coriandrum sativum L.

Cortia candollei (DC.) Leute

Cortia depressa (D. Don) C. Norman

Cortia lhasana (H. T. Chang & R. H. Shan) Pimenov & Kljuykov

Cortia papyracea (C. B. Clarke) Leute

Cortia staintoniana Farille & Malla

Cortia vaginata Edgew.

Cortiella caespitosa R. H. Shan & M. L. Sheh

Cortiella cortioides (C. Norman) M. F. Watson

Cortiella hookeri (C. B. Clarke) C. Norman

Cortiella lamondiana Fullarton & M. F. Watson

Cryptotaenia japonica Hassk.

Cuminum cyminum L.

Cyclorhiza peucedanifolia (Franch.) Constance

Cyclorhiza waltonii (H. Wolff) M. L. Sheh & R. H. Shan

Daucus carota L.

Dickinsia hydrocotyloides Franch.

Eriocycla caespitosa (Edgew.) Wolff

Eriocycla thomsonii H. Wolff

Eryngium caucasicum Trautv.

Eryngium foetidum L.

Ferula jaeschkeana Vatke

Ferula kingdon-wardii H. Wolff

Foeniculum vulgare Mill.

Hansenia forbesii (H. Boissieu) Pimenov & Kljuykov

Hansenia forrestii (H. Wolff) Pimenov & Kljuykov

Hansenia oviformis (R. H. Shan) Pimenov & Kljuykov

Hansenia pinnatiinvolucellata (F. T. Pu & Y. P. Wang) Pimenov & Kljuykov

Hansenia weberbaueriana (Fedde ex H. Wolff) Pimenov & Kljuykov

Haplosphaera himalayensis Ludlow

Haplosphaera phaea Hand.-Mazz.

Harrysmithia heterophylla H. Wolff

Heracleum bhutanicum M. F. Watson

Heracleum bivittatum H. Boissieu

Heracleum candicans Wall. ex DC.

Heracleum canescens Lindl.

Heracleum dissectifolium K. T. Fu

Heracleum forrestii H. Wolff

Heracleum franchetii M. Hiroe

Heracleum hemsleyanum Diels

Heracleum henryi H. Wolff

Heracleum kingdonii H. Wolff

Heracleum likiangense H. Wolff

Heracleum nepalense D. Don

Heracleum obtusifolium Wall. ex DC.

Heracleum oreocharis H. Wolff

Heracleum pinnatum C. B. Clarke

Heracleum rapula Franch.

Heracleum scabridum Franch.

Heracleum souliei H. Boissieu

Heracleum stenopteroides Fedde ex H. Wolff

Heracleum stenopterum Diels

Heracleum sublineare C. B. Clarke

Heracleum vicinum H. Boissieu

Heracleum wenchuanense F. T. Pu & X. J. He

Heracleum wolongense F. T. Pu & X. J. He

Heracleum woodii M. F. Watson

Heracleum xiaojinense F. T. Pu & X. J. He

Heracleum yungningense Hand.-Mazz.

Hymenidium album (C. B. Clarke ex H. Wolff) Pimenov & Kljuykov

Hymenidium amabile (Craib & W. W. Sm.) Pimenov & Kljuykov

Hymenidium apiolens (C. B. Clarke) Pimenov & Kljuykov

Hymenidium astrantioideum (H. Boissieu) Pimenov & Kljuykov

Hymenidium bicolor (Franch.) Pimenov & Kljuykov

Hymenidium brunonis (Wall. ex DC.) Lindl.

Hymenidium chloroleucum (Diels) Pimenov & Kljuykov

Hymenidium cristatum (H. Boissieu) Pimenov & Kljuykov

Hymenidium davidii (Franch.) Pimenov & Kljuykov

Hymenidium decurrens (Franch.) Pimenov & Kljuykov

Hymenidium delavayi (Franch.) Pimenov & Kljuykov

Hymenidium dentatum (DC.) Pimenov & Kljuykov

Hymenidium foetens (Franch.) Pimenov & Kljuykov

Hymenidium hedinii (Diels) Pimenov & Kljuykov

Hymenidium heracleifolium (Franch. ex H. Boissieu) Pimenov & Kljuykov

Hymenidium heterosciadium (H. Wolff) Pimenov & Kljuykov

Hymenidium hookeri (C. B. Clarke) Pimenov & Kljuykov

Hymenidium huzhihaoi Pimenov & Kljuykov

Hymenidium lhasanum Pimenov & Kljuykov

Hymenidium linearilobum (W. W. Sm.) Pimenov & Kljuykov

Hymenidium macrochlaenum (Fu Kuntsun & Ho Yehchi) Pimenov & Kljuykov

Hymenidium mieheanum Pimenov & Kljuykov

Hymenidium nanum (Rupr.) Pimenov & Kljuykov

Hymenidium nubigenum (H. Wolff) Pimenov & Kljuykov

Hymenidium pilosum (C. B. Clarke ex H. Wolff) Pimenov & Kljuykov

Hymenidium pulszkyi (Kanitz) Pimenov & Kljuykov

Hymenidium sikkimense (C. B. Clarke) Pimenov & Kljuykov

Hymenidium stellatum (D. Don) Pimenov & Kljuykov

Hymenidium szechenyii (Kanitz) Pimenov & Kljuykov

Hymenidium virgatum Pimenov & Kljuykov

Hymenidium wilsonii (H. Boissieu) Pimenov & Kljuykov

Hymenidium wrightianum (H. Boissieu) Pimenov & Kljuykov

Hymenidium yunnanense (Franch.) Pimenov & Kljuykov

Hymenolaena candollei DC.

Indoschulzia garhwalica (H. Wolff) Pimenov & Kljuykov

Indoschulzia hameliana (Farille & Malla) Pimenov & Kljuykov

Kailashia xizangensis (H. T. Chang & R. H. Shan) Pimenov & Kljuykov

Keraymonia cortiformis Cauwet & S. B. Malla

Keraymonia nipaulensis Cauwet & Farille

Keraymonia pinnatifolia M. F. Watson

Keraymonia triradiata Cauwet & Farille

Kitagawia macilenta (Franch.) Pimenov

Kitagawia pilifera (Hand.-Mazz.) Pimenov

Kitagawia praeruptora (Dunn) Pimenov

Lalldhwojia acronemifolia (H. Wolff) M. F. Watson ex D. G. Long

Lalldhwojia cooperi M. A. Farille

Lalldhwojia pastinacifolia Pimenov & Kljuykov

Lalldhwojia staintonii Farille

Ligusticopsis acaulis (R. H. Shan & M. L. Sheh) Pimenov

Ligusticopsis acuminata (Franch.) Leute

Ligusticopsis angelicifolia (Franch.) Leute

Ligusticopsis brachyloba (Franch.) Leute

Ligusticopsis capillacea (H. Wolff) Leute

Ligusticopsis coniifolia (Wall. ex DC.) Pimenov & Kljuykov

Ligusticopsis cryptotaenia (H. Boissieu) Lavrova

Ligusticopsis daucoides (Franch.) Lavrova ex Pimenov & Kljuykov

Ligusticopsis deqenensis (R. H. Shan & Z. H. Pan) Pimenov

Ligusticopsis dielsiana (H. Wolff) Pimenov & Kljuykov

Ligusticopsis franchetii (H. Boissieu) Leute

Ligusticopsis hispida (Franch.) Lavrova & Kljuykov

Ligusticopsis integrifolia (H. Wolff) Leute

Ligusticopsis involucrata (Franch.) Lavrova

Ligusticopsis likiangensis (H. Wolff) Lavrova & Kljuykov

Ligusticopsis longicalycia (M. L. Sheh) Pimenov & Kljuykov

Ligusticopsis modesta (Diels) Leute

Ligusticopsis multivittata (Franch.) Leute

Ligusticopsis oliveriana (H. Boissieu) Lavrova

Ligusticopsis pteridophyllum (Franch.) Leute

Ligusticopsis purpurascens Lavrova & Kljuykov

Ligusticopsis scapiformis (H. Wolff) Leute

Ligusticopsis wallichiana (DC.) Pimenov & Kljuykov

Ligusticopsis xizangensis (Z. H. Pan & M. L. Sheh) Pimenov & Kljuykov

Ligusticum afghanicum Rech. f.

Ligusticum angelicifolium Franch.

Ligusticum gongshanense F. T. Pu, R. Li & H. Li

Ligusticum levisticifolium H. Wolff

Ligusticum littledalei Fedde ex H. Wolff

Ligusticum mucronatum (Schrenk) Leute

Ligusticum nepalense D. Don

Ligusticum rechingerianum (Leute) R. H. Shan & F. T. Pu

Ligusticum yanyuanense F. T. Pu

Meeboldia achilleifolia (DC.) P. K. Mukh. & Constance

Melanosciadium bipinnatum (R. H. Shan & F. T. Pu) Pimenov & Kljuykov

Nothosmyrnium japonicum var. **sutchuenense** H. Boissieu

Nothosmyrnium xizangense R. H. Shan & T. S. Wang

Nothosmyrnium xizangense var. **simpliciorum** R. H. Shan & T. S. Wang

Notopterygium pinnatiinvolucellum F. T. Pu & Y. P. Wang

Notopterygium tenuifolium M. L. Sheh & F. T. Pu

Oenanthe benghalensis (Roxb.) Kurz

Oenanthe hookeri C. B. Clarke

Oenanthe javanica (Blume) DC.

Oenanthe linearis Wall. ex DC.

Oenanthe thomsonii C. B. Clarke

Oreocome arguta (Lindl.) Pimenov & Kljuykov

Oreocome limprichtii (H. Wolff) Pimenov & Kljuykov

Oreocome striata (DC.) Pimenov & Kljuykov

Oreocomopsis dochenensis (W. W. Sm.) Pimenov & Kljuykov

Oreocomopsis stelliphora (Cauwet & Farille) Pimenov & Kljuykov

Oreocomopsis xizangensis Pimenov & Kljuykov

Osmorhiza aristata (Thunb.) Rydb.

Ostericum grosseserratum (Maxim.) Kitag.

Ostericum muliense (R. H. Shan & F. T. Pu) Pimenov & Kljuykov

Ostericum scaberulum (Franch.) R. H. Shan & C. Q. Yuan

Peucedanum dhana Wall. ex C. B. Clarke

Peucedanum diversifolium H. Wolff

Peucedanum nagpurense (C. B. Clarke) Prain

Peucedanum nanum R. H. Shan & M. L. Sheh

Peucedanum nepalense P. K. Mukh.

Peucedanum pubescens Hand.-Mazz.

Peucedanum rubricaule R. H. Shan & M. L. Sheh

Peucedanum songpanense R. H. Shan & F. T. Pu

Peucedanum violaceum R. H. Shan & M. L. Sheh

Physospermopsis alepidioides (H. Wolff & Hand.-Mazz.) R. H. Shan

Physospermopsis bhutanensis Farille & S. B. Malla

Physospermopsis delavayi (Franch.) H. Wolff

Physospermopsis handelii (H. Wolff ex Hand.-Mazz.) Pimenov & Kljuykov

Physospermopsis muliensis R. H. Shan & S. L. Liou

Physospermopsis nana (Franch.) Pimenov & Kljuykov

Physospermopsis obtusiuscula (DC.) C. Norman

Physospermopsis rubrinervis (Franch.) C. Norman

Physospermopsis shaniana C. Y. Wu & F. T. Pu

Pimpinella acuminata (Edgew.) C. B. Clarke

Pimpinella anisum L.

Pimpinella atropurpurea C. Y. Wu ex R. H. Shan & F. T. Pu

Pimpinella bisinuata H. Wolff

Pimpinella candolleana Wight & Arn.

Pimpinella caudata (Franch.) H. Wolff

Pimpinella chungdienensis C. Y. Wu ex R. H. Shan et al.

Pimpinella coriacea (Franch.) H. Boissieu

Pimpinella diversifolia DC.

Pimpinella flaccida C. B. Clarke

Pimpinella grisea H. Wolff

Pimpinella heliosciadea H. Boissieu

Pimpinella henryi Diels

Pimpinella inundata (Farille & S. B. Malla) P. K. Mukh. & Constance

Pimpinella kingdon-wardii H. Wolff

Pimpinella nyingchiensis Z. H. Pan & K. Yao

Pimpinella pimpinellisimulacrum (Farille & Malla) Farille

Pimpinella purpurea (Franch.) H. Boissieu

Pimpinella rhomboidea Diels

Pimpinella rhomboidea var. **tenuiloba** R. H. Shan & F. T. Pu

Pimpinella rubescens (Franch.) H. Wolff

Pimpinella scaberula (Franch.) H. Boissieu

Pimpinella scaberula var. **ambrosiifolia** (Franch.) H. Wolff

Pimpinella sikkimensis C. B. Clarke

Pimpinella silvatica Hand.-Mazz.

Pimpinella tibetanica H. Wolff

Pimpinella tongloensis P. K. Mukh.

Pimpinella urceolata Watt ex Banerji

Pimpinella wallichii C. B. Clarke

Pimpinella xizangensis R. H. Shan & F. T. Pu

Pimpinella yunnanensis (Franch.) H. Wolff

Pleurospermum albimarginatum H. Wolff

Pleurospermum microphyllum H. Wolff

Pleurospermum microsciadium H. Wolff

Pleurospermum souliei H. Wolff

Pternopetalum botrychioides (Dunn) Hand.-Mazz.

Pternopetalum caespitosum R. H. Shan

Pternopetalum cuneifolium (H. Wolff) Hand.-Mazz.

Pternopetalum davidii Franch.

Pternopetalum delavayi (Franch.) Hand.-Mazz.

Pternopetalum gracillimum (H. Wolff) Hand.-Mazz.

Pternopetalum molle (Franch.) Hand.-Mazz.

Pternopetalum monophyllum J. B. Tan & X. J. He

Pternopetalum porphyronotum J. B. Tan

Pternopetalum senii Deb & R. M. Dutta

Pternopetalum subalpinum Hand.-Mazz.

Pternopetalum tanakae (Franch. & Sav.) Hand.-Mazz.

Pternopetalum trichomanifolium (Franch.) Hand.-Mazz.

Pternopetalum vulgare (Dunn) Hand.-Mazz.

Pterocyclus angelicoides (DC.) Klotzsch

Pterocyclus forrestii (Diels) Pimenov & Kljuykov

Pterocyclus rivulorum (Diels) H. Wolff

Pterocyclus rotundatus (DC.) Pimenov & Kljuykov

Sanicula astrantiifolia H. Wolff

Sanicula chinensis Bunge

Sanicula giraldii H. Wolff

Sanicula hacquetioides Franch.

Sanicula lamelligera Hance

Sanicula rugulosa Diels

Sanicula serrata H. Wolff

Sanicula tienmuensis var. **pauciflora** R. H. Shan & F. T. Pu

Scaligeria hirtula (Regel & Schmalh.) Lipsky ex Korovin

Schulzia albiflora (Kar. & Kir.) Popov

Schulzia bhutanica M. F. Watson

Semenovia gyirongensis Q. Y. Xiao & X. J. He

Semenovia malcolmii (Hemsl. & H. Pears.) Pimenov

Semenovia thomsonii (C. B. Clarke) Manden.

Semenovia torilifolia (H. Boissieu) Pimenov

Seseli buchtormense (Spreng.) W. D. J. Koch

Seseli diffusum (Roxb. ex Sm.) Santapau & Wagh

Seseli nudum (Lindl.) Pimenov & Kljuykov

Seseli nudum var. **purpurascens** (R. H. Shan & C. C. Yuan) Pimenov & Kljuykov

Seseli purpureovaginatum R. H. Shan & M. L. Sheh

Seseli roylei (Lindl.) Pimenov & Kljuykov

Seseli stewartii (M. Hiroe) Pimenov & Kljuykov

Seseli veitchii (H. Boissieu) Pimenov

Seseli yunnanense Franch.

Sinocarum cruciatum (Franch.) H. Wolff ex R. H. Shan & F. T. Pu

Sinocarum cruciatum var. **linearilobum** (Franch.) R. H. Shan & F. T. Pu

Sinocarum dolichopodum (Diels) H. Wolff ex R. H. Shan & F. T. Pu

Sinocarum filicinum H. Wolff

Sinocarum minus M. F. Watson

Sinocarum normanianum (Cauwet & Farille) Farille

Sinocarum pauciradiatum R. H. Shan & F. T. Pu

Sinocarum pityophilum (Diels) H. Wolff

Sinocarum pulchellum C. Norman ex M. F. Watson

Sinocarum schizopetalum (Franch.) H. Wolff ex R. H. Shan & F. T. Pu

Sinocarum schizopetalum var. **bijiangense** (S. L. Liou) X. T. Liu

Sinocarum shanii Pimenov & Kljuykov

Sinocarum sikkimense (P. K. Mukh.) P. K. Mukh. & Constance

Sinocarum vaginatum H. Wolff

Sinocarum wolffianum (Fedde ex H. Wolff) A. K. Mukh. & Constance

Sinodielsia cuneata (H. Wolff) Pimenov & Kljuykov

Sinodielsia delavayi (Franch.) Pimenov & Kljuykov

Sinodielsia thibetica (H. Boissieu) Kljuykov & P. K. Mukh.

Sinolimprichtia alpina H. Wolff

Sium frigidum Hand.-Mazz.

Sphallerocarpus gracilis (Besser ex Trevir.) Koso-Pol.

Spuriopimpinella arguta (Diels) X. J. He & Z. X. Wang

Stenocoelium depressum (R. H. Shan & M. L. Sheh) Pimenov & Kljuykov

Tetrataenium lallii (C. Norman) Cauwet, Carb. & M. Farille

Tetrataenium olgae (Regel & Schmalh.) Manden.

Tetrataenium wallichii (DC.) Manden.

Tetrataenium yunnanense (Franch.) Manden. ex Q. Y. Xiao & X. J. He

Tongoloa elata H. Wolff

Tongoloa fortunatii (H. Boissieu) Pimenov & Kljuykov

Tongoloa gracilis H. Wolff

Tongoloa loloensis (Franch.) H. Wolff

Tongoloa pauciradiata H. Wolff

Tongoloa rockii H. Wolff

Tongoloa rubronervis S. L. Liou

Tongoloa silaifolia (H. Boissieu) H. Wolff

Tongoloa smithii H. Wolff

Tongoloa taeniophylla (H. Boissieu) H. Wolff

Tongoloa tenuifolia H. Wolff

Tongoloa zhongdianensis S. L. Liou

Tordyliopsis brunonis DC.

Torilis japonica (Houtt.) DC.

Trachydium involucellatum R. H. Shan & F. T. Pu

Trachydium napiferum H. Wolff

Trachydium roylei Lindl.

Trachydium simplicifolium W. W. Sm.

Trachydium souliei H. Boissieu

Trachydium subnudum C. B. Clarke ex H. Wolff

Trachydium szechuanense H. Wolff

Trachyspermum ammi (L.) Sprague

Trachyspermum anethifolium (D. Don) H. Wolff

Tricholaser cachemiricum (C. B. Clarke) Alava

Turgenia latifolia (L.) Hoffm.

Vicatia atrosanguinea (Kar. & Kir.) P. K. Mukh. & Pimenov

Vicatia coniifolia Wall. ex DC.

Vicatia nepalensis Kljuykov

List of nomenclatural novelties published in this volume

Ajania penicilliformis (C. Shih) Y. S. Chen, **comb. et stat. nov.** (≡ *Ajaniopsis penicilliformis* C. Shih)

Hibiscus muliensis (K. M. Feng) Y. S. Chen, **comb. et stat. nov.** (≡ *Abelmoschus muliensis* K. M. Feng)

Himalaiella diffusa (Lipsch.) Y. S. Chen, **comb. et stat. nov.** (≡ *Saussurea diffusa* Lipsch.)

Justicia cyclostegia (Hand.-Mazz.) Y. Tong & Y. F. Deng, comb. nov.

Phlomoides brevidentata (H. W. Li) C. L. Xiang, comb. nov.

Phlomoides cashmeriana (Royle ex Benth.) C. L. Xiang, comb. nov.

Phlomoides nyalamensis (H. W. Li) C. L. Xiang, comb. nov.

Primula dawuensis Y. S. Chen, **nom. nov.** [≡ *Primula brevis* A. J. Richards (1993), nom. illeg., non Kuntze (1881); ≡ *Primula humilis* Pax & K. Hoffm. (1921), nom. illeg., non Steud. (1840)]

Index to scientific names

D

U

V

General References

Chowdhery H J, Giri G S, Pramanik A. 2009. Materials for the Flora of Arunachal Pradesh. Vol. 3. Botanical Survey of India, Ministry of Environment & Forests.

Christenhusz M J M, Zhang X C, Schneider J. 2011a. A linear sequence of extent families and genera of lycophytes and ferns. Phytotaxa, 19: 7-54.

Christenhusz M J M, Reveal J L, Farjon A, et al. 2011b. A new classification and linear sequence of extant gymnosperms. Phytotaxa, 19: 55-70.

Favre A, Matuszak S, Sun H, et al. 2014. Two new genera of Gentianinae (Gentianaceae): Sinogentiana and Kuepferia supported by molecular phylogenetic evidence. Taxon, 63: 343-354.

Fraser-Jenkins C R. 2008. Taxonomic Revision of Three Hundred Indian Subcontinental Pteridophytes with a Revised Census-list. Bishen Singh Mahendra Pal Singh, Dehra Dun.

Gao Y D, Gao X F. 2016. Accommodating Nomocharis in Lilium (Liliaceae). Phytotaxa, 277: 205-210.

Ghosh S R. 2004. The Pteridophytic Flora of Eastern India. Vol. 1. Botanical Survery of India, Kolkata.

Giri G S, Pramanik A, Chowdhery H J. 2008. Materials for the Flora of Arunachal Pradesh. Vol. 2. Botanical Survey of India, Ministry of Environment & Forests.

Global Carex Group. 2015. Making *Carex* monophyletic (Cyperaceae, tribe Cariceae): a new broader circumscription. Botanical Journal of the Linnean Society, 179: 1-42.

Grierson A J C, Long D G. 1983-2002.. Flora of Bhutan. Vol. 1-6. Royal Botanic Garden Edinburgh, Edinburgh.

Haidar A, Qaiser M. 2018. Contribution to the flora of Pakistan: Hedysarum alii and H. shahjinalense sp. nov. (Fabaceae) and three new records. Nordic Journal of Botany, 36(7): e01742.

Hajra P K, Verma D M, Giri G S. 1996. Materials for the Flora of Arunachal Pradesh. Vol. 1. Botanical Survey of India, Ministry of Environment & Forests.

Haston E, Richardson J E, Stevens P, et al. 2009. The Linear Angiosperm Phylogeny Group (LAPG) III: a linear sequence of the families in APG III. Botanical Journal of the Linnean Society, 161: 128-131.

Hong D Y. 2016-2017. Flora of Pan-Himalaya. Vols. 30, 47, 48(2), 48(3), and drafts of some families. Science Press, Beijing.

Hong D Y, Blackmore S. 2013. Plants of China. Science Press, Beijing.

Jagtap A P, Singh N P. 1999. Fascicles of Flora of India. Fascicle 24. Botanical Survey of India, Ministry of Environment & Forests (Also Fascicles 19, 20, 21, 22, 23).

Jin W T, Xiang X G, Jin X H. 2015. Generic delimitation of Orchidaceae from China: current situation and perspective. Biodiversity Science, 23: 237-242.

Kress W J, DeFilipps R A, Farr E R, et al. 2003. A Checklist of the Trees, Shrubs, Herbs, and Climbers of Myanmar. Contributions from the United States National Herbarium, 45: 1-590.

Li D Z. 2018. A Dictionary of the Families and Genera of Chinese Vascular Plants. Science Press, Beijing.

Liu B, Ye J F, Liu S, et al. 2015. Families and genera of Chinese angiosperms: a synoptic classification based on APG III. Biodiversity Science, 23: 225-231.

Mabberley D J. 2008. Mabberley's Plant-Book. 3rd ed. Cambridge University Press, Cambridge.

Mittermeier R A, Gil P R, Hoffman M. 2004. Hotspots Revisited: Earth's Biologically Richest and Most Endangered Terrestrial Ecoregions. CEMEX/Agrupaciaon Sierra Madre, Mexico City.

Naithani H B. 1990. Flowering Plants of India, Nepal & Bhutan. Surya Publications, Dehra Dun.

Podlech D. 2012. Checklist of the Flowering Plants of Afghanistan. München.

Press R, Shrestha K K, Sutton DA. 2000. Annotated Checklist of Flowering Plants of Nepal. Natural History Museum, London (updated version available online at efloras.org).

Singh N P, Singh D K, Uniyal B P. 2002.. Flora of Jammu & Kashmir. Vol. 1. Botanical Survey of India, Ministry of Environment & Forests.

Singh N P, Sanjappa M. 2006. Fascicles of Flora of India. Fascicle 23. Botanical Survey of India, Ministry of Environment & Forests.

Singh S, Panigrahi G. 2005. Ferns and Fern-allies of Arunachal Pradesh. Vol. 1-2. Dehra Dun.

Stewart R R. 1972. An Annotated Catalogue of the Vascular Plants of West Pakistan and Kashmir. Fakhri Printing Presss, Karachi.

The Angiosperm Phylogeny Group. 2016. An update of the Angiosperm Phylogeny Group classification for the orders and families of flowering plants: APG IV. Botanical Journal of the Linnean Society, 181: 1-20.

Wang W T. 1993-1994. Vascular Plants of the Hengduan Mountains. Vol. 1-2. Science Press, Beijing.

Wu Z Y, Raven P, Hong D Y. 1994-2013. Flora of China. Science Press, Missouri Botanical Garden Press, Beijing and St Louis (available online at efloras.org).

Zhu X Y. 2015. Nomenclatural novelties and new synonyms of Leguminosae in China. Biodiversity Science, 23: 247-251.

Website resources used for editing the catalogue

Angiosperm Phylogeny Website

http://www.mobot.org/MOBOT/Research/APweb/welcome.html

Catalogue of Life (or Species 2000 China Node)

http://www.sp2000.org.cn

China Species Library

http://species.sciencereading.cn

Chinese Virtual Herbaium

http://www.cvh.ac.cn

eFlora of India

http://efloraindia.nic.in/efloraindia/eFloraHomePage.action

eFloras.org

http://www.efloras.org

Flora Himalaya Database

http://www.leca.univ-savoie.fr/db/florhy/infos.html

Flora of Nepal

http://www.floraofnepal.org/home

iFlora

http://iflora.cn

International Plant Names Index

http://www.ipni.org

JSTOR

http://plants.jstor.org/stable

The Plant List

http://www.theplantlist.org

www.sciencep.com
ISBN 978-7-03-061953-2
9 787030 619532 >
Price: 360.00 Yuan

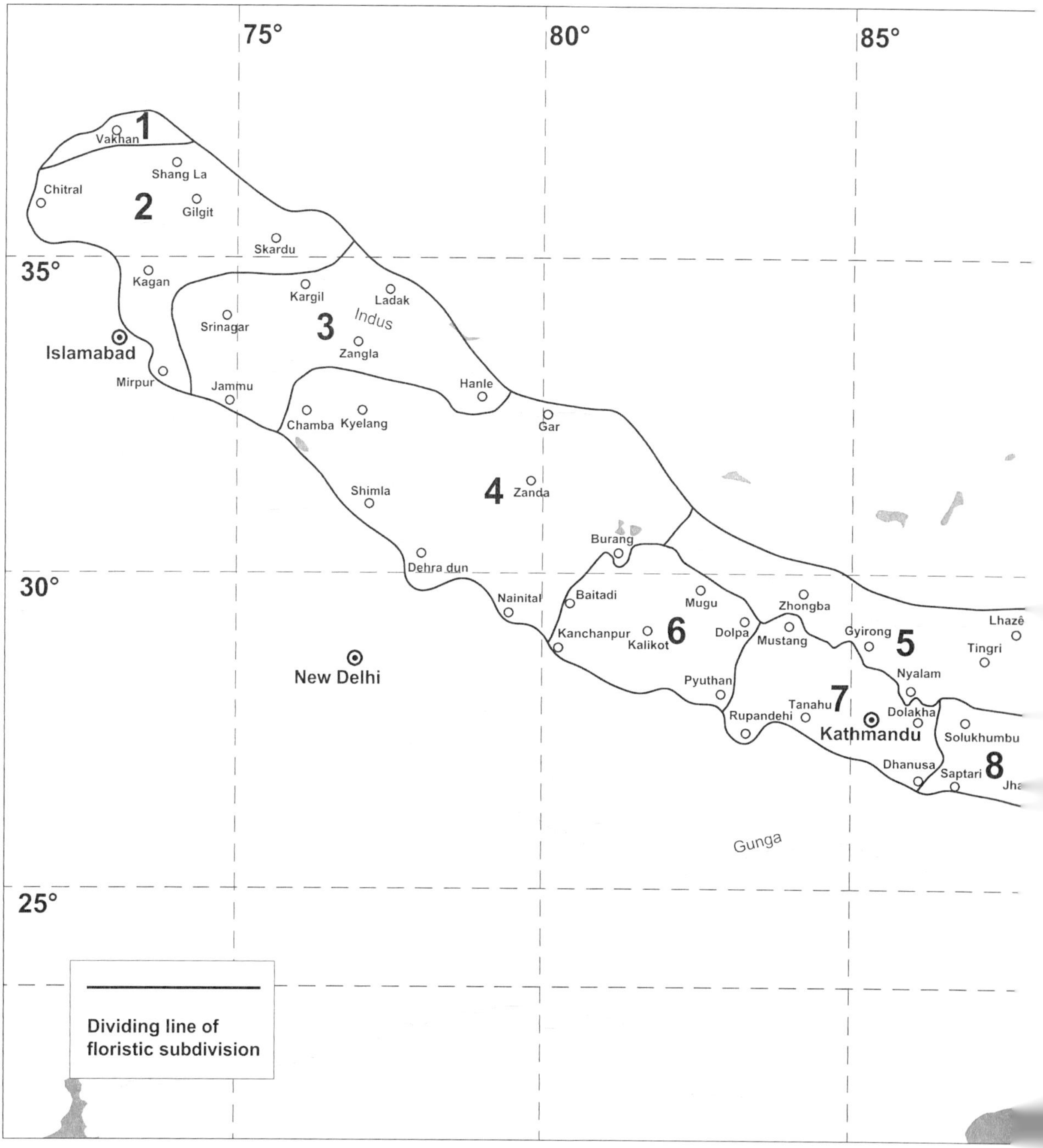

Subdivisions o

1.Vakhan; 2.N Pakistan; 3. Jammu & Kashmir; 4. U Ganga & Indus; 5
10. Bhutan; 11. M Yarlung Zangbo; 12. L Yarlung Zangbo; 13. Yarlung Za